ADAPTATION IN PLANT BREEDING

Developments in Plant Breeding

VOLUME 4

Adaptation in Plant Breeding

Selected Papers from the XIV EUCARPIA Congress on Adaptation in Plant Breeding held at Jyväskylä, Sweden from July 31 to August 4, 1995

Edited by

PETER M.A. TIGERSTEDT

Department of Plant Biology, University of Helsinki, Finland

Chapters indicated with an asterisk in the table of contents were first published in *Euphytica*, Volume 92: 1–2, 1996

Springer-Science+Business Media, B.V.

Library of Congress Cataloging-in-Publication Data

Adaptation in plant breeding / edited by Peter M.A. Tigerstedt.
p. cm. -- (Developments in plant breeding ; v. 4)
ISBN 978-90-481-4708-3 ISBN 978-94-015-8806-5 (eBook)
DOI 10.1007/978-94-015-8806-5
1. Plant breeding. 2. Plants--Adaptation. 3. Crops--Adaptation.
I. Tigerstedt, Peter M. A. II. Series.
SB123.A335 1996
631.5'23--dc20 96-15903

ISBN 978-90-481-4708-3

Printed on acid-free paper

Contents

Preface

When I first suggested the theme 'Adaptation in Plant Breeding' to the EUCARPIA board in 1977 at the General Assembly in Madrid, it was put on a long waitinglist. Looking at the list of published EUCARPIA proceedings today, I think they form a formidable update of the scientific development in plant breeding.

EUCARPIA is celebrating its 40-year anniversary in 1996. I am pleased to see the theme 'adaptation' in honour of this anniversary. The theme is covered in the proceedings by means of 10 chapters, each introduced by a keynote and followed by 2–3 presentations taken from research in agriculture, horticulture or forestry. The importance of adaptation becomes evident when looking at the chapters. Perhaps one could point at some particular trends in plant breeding of today that have caused special interest in adaptation. Firstly, the global concern of protecting the environment and its biodiversity has put special emphasis on genetic resources. Adaptation to marginal conditions and to highly sophisticated man made ecosystems are of high priority in discussions on sustainability. Secondly, the world has become aware of the dangers of using high amounts of chemicals in the production chains. Thus one is looking for better adaptation to environmental stresses and to genetic alternatives to resistance to pest and disease. Thirdly, global climatic change has caused increasing interest in how quickly plants can respond adaptively to such changes in the environment. Also there is a growing interest in the width of adaptation and plant breeders have used the terms 'wide' and 'narrow' adaptation to indicate differences.

I hope the proceedings of XIV EUCARPIA 1995 'Adaptation in Plant Breeding' may be of value for many specialists around the world in their endeavour to breed new plants that meet the demands of a growing world human population.

Helsinki, Finland
Peter M.A. Tigerstedt

P.M.A. Tigerstedt (Ed.), Adaptation in Plant Breeding, 1–11, 1997.

Genetic basis of the evolution of adaptedness in plants

R.W. Allard
University of California at Davis, Davis, CA 95616, U.S.A.

Key words: adaptedness, epistasis, marker assisted dissection, overdominance, purifying selection

Summary

Adaptedness is both complexly inherited and much affected by environment: consequently the genetic mechanisms that have led to improvements in adaptedness have been difficult to identify and to quantify. Recently it has been shown that 'marker assisted dissection' of adaptedness based on changes in the frequencies of discretely inherited alleles of loci of various kinds (e.g. allozyme, restriction fragment, microsatellite loci) is practicable. I will illustrate marker assisted analysis of the genetic basis of adaptedness with a sample of allozyme data from three species groups, two heavily selfing groups (two wild *Avena* species and barley) and one outcrossing species (corn, maize). The results lead to three main conclusions: (1) that the single most important genetic mechanism in all three species groups was the assembly of favorable epistatic combinations of alleles of different loci by means of recurring cycles of selection, intercrossing superior selects, and inbreeding to near homozygosity leading to stable superior multilocus genotypes adapted to specific habitats; (2) that exploitation of favorable interactions among alleles of the same locus played a significant role in tetraploid *A. barbata* and probably also in single-cross maize hybrids; (3) that purifying selection (elimination of deleterious alleles) played a small role in all three species groups. These results indicate that marker alleles provide applied breeders with effective ways to identify, track, and incorporate regions of chromosomes with favorable effects of adaptedness into improved cultivars.

Introduction

Modern cultivars are much better adapted and much more productive in agricultural environments than their wild ancestors. Adaptedness and productivity are, however, both complexly inherited traits and much affected by environment so that we still know little about what happened genetically during the domestication process. In recent years it has been found that examination of changes in the frequencies of discretely distinguishable alleles and genotypes of morphological, disease resistance, allozyme, restriction fragment, and microsatellite loci that occurred during the domestication process allow deductions concerning the genetic mechanisms that have led to improved adaptedness (e.g. Clegg & Allard, 1972; Allard, 1988, 1990; Perez de la Vega, Garcia & Allard, 1991; Garcia et al., 1991; Allard et al., 1993; Saghai Maroof et al., 1994; Cluster & Allard, 1995). In all cases that have been studied in detail it has been found that population behavior can be explained on the basis of selection acting on the chromosomal segments in which the marker loci reside; thus marker assisted dissection of the genetic basis of adaptedness is feasible and a substantial literature has built up, too large to review in the time allotted. Consequently I will limit discussion to only one type of marker (allozymes) and to three representative species. First I will examine changes in allelic and genotypic frequencies that occurred in the diploid Slender Wild Oat (*Avena hirtula*) and its tetraploid descendant (*A. barbata*) as these wild species evolved over thousands of generations in the Mediterranean Basin. Then I will turn to allelic and genotypic frequency changes that occurred during the domestication and spread of barley (*Hordeum vulgare*) and (*Zea mays*) over hundreds of generations in cultivation. The two *Avena* species and barley are self fertilized (99%) whereas in maize populations the mating system is one of about 90% outcrossing and 10% selfing. The results lead to the conclusion that the single most important genetic mechanism involved in all three species groups was the development, and the stabilization by inbreeding, of favorable epistatic combinations of alleles of different loci to produce stable multilocus genotypes that provide superior adaptedness in specific environments. A second but less important genetic mechanism was purifying selection (elimination of deleterious alleles). A third genetic mechanism, namely exploitation of favorable interactions among alleles of the same locus (parallel to overdominance) was important in *A. barbata*, a diploidized tetraploid, and to a lesser extent in corn, but overdominance was of little or no importance in populations of diploid *A. hirtula* or barley (no cost effective method of producing F1 hybrid seed had been found in barley).

Avena hirtula (Ah) and *A. barbata (Ab)*

Diploid *Ah* (n = 7) chromosomes) is indigenous to the Mediterranean Basin where it occurs in sparse and more or less disjunct stands. *Ah* is often less robust than *Ab* but the two taxa are usually so similar that they cannot be distinguished on the basis of visible phenotypic characteristics alone. Tetraploid *Ab* (n = 14 chromosomes) is much more widely distributed. It occurs in massive stands throughout the Mediterranean Basin and across the Middle East to Nepal. It has also been a successful colonizer in areas with Mediterranean-like climates throughout the world. Tetraploid *Ab* is clearly much more widely adapted and vigorous than diploid *Ah*.

Ab was introduced to California accidentally from Southwestern Spain about 200 generations ago but *Ah* has not been found in California. In the past half century or so most plants of *Ab* have been found in two major climatic zones in California: (1) a semi-arid zone (250–350 mm annual rainfall) including nearly all of Southern and Central California, and (2) a generally mesic zone (350–1,500 mm annual rainfall) extending > 1,000 km northward in the Pacific coastal ranges from 36° N to Northern Oregon (45° N). Nearly all populations in the xeric zone are monomorphic for two simply inherited variants (glabrous leaf sheaths and dark lemmas) whereas nearly all populations in the mesic zone are monomorphic for hairy leaf sheaths and light lemmas. However mesic patches located within the xeric zone (usually in higher elevation, higher rainfall sites), as well as xeric patches located within the mesic zone (often on steep southwest facing slopes) are nearly always polymorphic for the two pairs of variants. These contrasting ecogeographical patterns suggested that the two morphic pairs might 'mark' and perhaps also contribute to the adaptedness of two 'ecotypes' (Turesson, 1922) that had developed in response to differing selective pressures operating in the xeric and mesic habitats.

In 1930, with guidance from the late W.W. Mackie (University of California, Berkeley) I started experiments to test the above hypothesis. The initial experiments were conducted in several small internally heterogeneous sites (each ca. one hectare in area) located ca. 100 km southeast of San Francisco in an area where the xeric and mesic zones form interfaces with each other. Several small subsites within each site were classified into one of five categories, extending from xeric through three intermediate categories to mesic: assignment to these five categories was based strictly on various physical features of the subsite (e.g. slope, exposure to the sun, edaphic characteristics). Each year during the following decade (1931–1940) > 100 plants from each subsite were classified for the simply inherited leaf-sheath and lemma-color variants and also measured for several quantitative traits such as height, dry weight and number of seeds produced on the primary tiller. The main finding was that the xeric and mesic subsites were nearly always monomorphic for the glabrous-dark or the hairy-light morphs, respectively, but that intermediate subsites were polymorphic in frequencies correlated with degree of xerism. Another main finding was that all of the quantitative traits measured were highly variable within all subsites and none were significantly correlated with degree of xerism. These results suggested that additional simply-inherited 'Mendelizing' markers were likely to be more informative in future studies of ecotypic differentiation than continuously-varying complexly-inherited quantitative characters. However, no additional simply-inherited markers had been found by the time World-War II made it necessary to suspend studies of *Ab*. On return from W.W.II I shifted the ecogenetic studies based on simply inherited 'markers' to crop plants such as barley, beans, sorghum and maize because many simply-inherited markers and highly-isogenic recombinant inbred lines were available in these species.

The critical breakthrough pertaining to simply-inherited markers came in 1960 with the introduction of enzyme electrophoresis to plant genetics (Schwartz, 1960). In short order many technologically advantageous electrophoretically-distinguishable *codominant* allozyme markers had been identified and pressed into

Figure 1. Geographical locations of 10 populations of *A. hirtula* (A-I) and 50 populations of *A. barbata* (1–50).

service in ecogenetic studies in several crop and wild species, including *Ab* and *Ah*. However, the studies carried out in the 1960s, 1970s and 1980s usually focused on only one or two of the several genetic mechanisms that were ultimately found to be jointly responsible for shaping the internal genetic structure, as well as the overall adaptive landscapes of various species. Consequently, I will cite data from only a few recent studies that took into account simultaneously the most important genetic mechanisms that had been identified earlier in more narrowly-focused investigations.

Table 1 gives a sample of allelic and genotypic frequencies for 14 representative codominant allozyme loci in 10 Spanish diploid and 50 Spanish tetraploid populations distributed as shown in Figure 1 (Allard et al., 1993). Note that allelic diversity appears to be greater in the tetraploid than in the diploid e.g. *Ab* has three alleles of *Pgd2* but *Ah* has only one. Counts made on all 14 loci show this is indeed the case: 52 alleles (3.7/locus) were observed in *Ab* but only 38 (2.7/locus) in *Ah*. However only 29 of the 38 alleles of *Ah* were also present in *Ab* indicating that nine alleles of *Ah* were lost during or following the polyploidization process. This raised a question: which alleles were lost and which survived? The answer was clear cut: the nine least-frequent alleles of *Ah*, which ranged in overall frequency from 0.001 to 0.07, were all lost whereas the 29 most-frequent alleles of *Ah*, which ranged in frequency from 0.11 to 1.0, all survived. This raised another question, where did the 23 alleles (52 - 29 = 23) of the tetraploid not present in the diploid come from? Two facts, first that *Ab* is a fully diploidized tetraploid (Hutchinson et al., 1983) and second, that all of the 'new' alleles of *Ab* were found in non-segregating heteroallelic quadriplexes suggest the answer. Locus *Pgd2* (Table 1) serves as a model for the sequence of events that may account for the greater allelic diversity of the tetraploid. Diploid *Ah* is monomorphic for allele *1* of *Pgd2* in Spain (also worldwide); consequently the original quadriplex formed by polyploidization of duplex *11* was almost certainly *11,11*. Theory (Wright, 1937) indicates that the probability is very small that any mutant, even a highly heterotic mutant such as allele *2*, will become established in a small selfing diploid population. In the tetraploid, however, mutational events are equally likely in either of the two genomes and population sizes are typically much larger than in the diploid; hence more than *twice* as many mutations of allele *1* to allele *2* would likely occur in *11,11* quadriplexes than in the *11* duplexes of the diploid. Furthermore, in a diploidized tetraploid such as *Ab*, each such mutation will lead to a *11,21* (or *12,11*) quadriplex heterozygous in one genome, but homozygous in the other genome, and on selfing for four or five generations, at least 1/2 of the progeny are expected to be heterotic true-breeding *11,22* quadriplexes. Consequently it is much more likely that novel favorable mutants would have been incorporated into diploidized tetraploid *Ab* than into diploid *Ah*, in which selfing will rapidly reduce heterozygosity to zero. Subsequent mutations of allele *1* to allele *3* in *11,22* quadriplexes may have led to quadriplex *22,33* of *Pgd2*, which, although less successful than the original *11,11*, or the derivative *11,22* quadriplex, found some sites in which it was able to survive.

Several other patterns of quadriplex formation are illustrated in Table 1. Four other loci of Table 1 (*Mdh3 Acp1, Acp2, Pgd2*) followed a pattern similar to *Pgd2*; each formed a single highly-successful heteroallelic quadriplex (in some cases from two alleles not present in the diploid), as well as one or more less successful heteroallelic or homoallelic quadriplexes. Locus *Mdh1* was unique. It formed a single heteroallelic quadriplex that is now completely monomorphic worldwide. The *intralocus* interaction between pairs of homozygous alleles (*11,22*) of this locus (parallel to overdominance) clearly led to a remarkable level of adaptedness that has thus far withstood all challenges from novel mutant alleles in all environments worldwide. Another pat-

Table 1. Single-locus duplex and quadriplex frequencies in 10 Spanish diploid (*Ah*) populations and 50 Spanish tetraploid (*Ab*) distributed as shown in Figure 1. (Adapted from Allard et al., 1993)

Locus	Diploid			Tetraploid		
	Duplex	f	No. sites in which fixed	Quadriplex	f	No. sites in which fixed
Pgd2[a]	*11*	1.00	10/10	*11,11*	0.12	1/50
				11,22	0.84	24/50
				22,33	*0.04*	0/50
					1.00	
Mdh1	*11*	1.00	10/10	*11,22*	1.00	50/50
Est1[b]	*11*	0.07	0/10	*33,33*	0.15	1/50
	22	0.24	0/10	*55,55*	0.14	0/50
	55	0.44	2/10	*11,33*	0.04	0/50
	66	0.03	0/10	*22,33*	0.19	2/50
	77	*0.19*	0/10	*22,55*	0.15	1/50
		0.97		*33,55*	0.04	0/50
				33,66	0.03	0/50
				55,77	*0.20*	4/50
					0.94	
Pgm1[c]	*11*	1.00	10/10	*11,11*	1.00	50/50

[a] Loci *Pgd2, Mdh3, Acp1, Acp2* and *Pgd1* have similar patterns.
[b] Loci *Est1, Lap1, Prx1* and *Pgi* have similar patterns.
[c] Loci *Pgm1, Got2, Mdh2* and *Got1* have identical patterns.
[f] Estimates based on 754 diploid plants (1,508 alleles) and 4,751 tetraploid plants (19,004 alleles).

tern of quadriplex formation is that of four loci (*Est1, Lap1, Prx1, Pgi1*) that are highly polymorphic in the diploid. Each of these loci formed several homoallelic and several heteroallelic quadriplexes. Still another pattern of quadriplex formation is that of loci *Pgm1, Got2, Mdh2*, and *Got1*. All of these loci are monomorphic, or very nearly so, in Spain (also worldwide) for a single allele in the diploid and also for the same allele in the tetraploid. Evidently these alleles code for some essential function such that they confer adaptedness in homoallelic state superior to that conferred by any other alleles, whether homoallelic or heteroallelic, that have arisen in either *Ah* or *Ab*. Overall the net effect of polyploidization, followed by diploidization, was a great increase in allelic diversity in the tetraploid. For the majority of loci the increases in allelic diversity provided not only opportunities for favorable intralocus interactions (parallel to overdominance) but also greatly increased opportunities for exploitation of favorable epistatic interactions among alleles of different loci. As an example, consider a diploid plant genotypically *11* for one locus (say *Lap1*) and *55* for another locus (say *Est1*). Heterozygotes are so rare in the diploid that exploitation of possible overdominant intraallelic interactions in *15* heterozygotes seems quite unlikely: also only a single epistatic interaction (*11* × *55*) is possible. However, in a tetraploid population that includes individuals genotypically *22,33* for *Lap1* and *55,77* for *Est1*, six interactions are possible, two intralocus (overdominant) interactions, *22* × *33* for *Lap1* and *55* × *77* for *Est1*, as well as four pairwise interlocus epistatic interactions, *22* × *55*, *22* × *77*, *33* × *55 and 33* × *77*. All of these combinations are either fixed or nearly fixed in all seven populations of the cold Central Northern Plateau of Spain, which suggests that these particular interactions contribute to adaptedness in the habitat of that region. Many other non-random associations of particular duplexes and quadriplexes with particular regions and also with specific sites within regions were also observed, which suggests that specific duplexes and quadriplexes confer superior adaptedness in particular regions as well as in specific sites within regions.

Many different two-locus, three-locus, up to 14-locus interactions are also possible in both the diploid and the tetraploid. Patterns in the diploid are, however, much less complex than in the tetraploid. Consequently I now turn to the 14-locus genotypes observed in

Table 2. Most frequent 14-locus genotypes in 10 diploid sites. (Adapted from Allard et al., 1993)

Site	Locus											N[b]	f[c]
	Pgm1[a]	*Got1*	*Mdh2*	*Mdh3*	*Acp2*	*Pgd1*	*Pgi1*	*Acp1*	*Prx1*	*Lap1*	*Est1*		
1	11	11	11	22	11	11	11	55	55	11	55	1	1.00
2	11	11	11	66	11	11	44	11	22	22	55	1	1.00
3	11	11	11	22	22	11	22	11	11	11	77^P	2	0.93
4	11	22^P	11	22	11	11	11^P	22^P	22	44^P	55^P	8	0.71
5	11	11	11	22	11	11^P	44^P	11	22	22^P	55^P	8	0.51
6	11	11	11	22	11	55^P	11	55^P	44^P	44^P	11^P	13	0.45
7	11	11	11	22	11	22^P	44^P	11^P	22^P	22^P	22^P	20	0.26
8	11	11	11	22^P	11^P	11^P	11^P	22^P	11^P	77^P	77^P	17	0.21
9	11	11	11	22	55^P	11^P	22^P	22^P	22^P	34	22^P	34	0.17
10	11	11	11^P	22	22^P	22^P	22^P	11^P	11^P	11^P	55^P	36	0.12

[a] Also *Got2, Mdh1, Pgd2.*
[b] Number of 14-locus genotypes observed at each site.
[c] Frequency within site of most frequent 14-locus genotypes.
^{P} Site polymorphic for 14-locus genotypes due to polymorphism at locus indicated.

the diploid to give a feel for some of the two-locus and higher-order interactions that have been successful. In total 107 different 14-locus genotypes were observed in the 10 Spanish populations of *Ah*. The main features of both within-population and between-population differentiation can be deduced from Table 2, which gives the frequencies of the single most-frequent 14-locus genotype, and the total number of 14-locus genotypes, at each site. The array of 14-locus genotypes present at each site differed from that at each other site. Only one 14-locus genotype was present in Sites 1 and 2; the 14-locus genotype at these two sites was identical at nine loci but differed at five loci (*Mdh3, Pgi1, Acp1, Prx1, Lap1*). It is possible that differences in adaptedness between these two genotypes may have been due entirely to the main effects of these five loci but the very large number of different patterns in which the alleles of the 14-locus genotypes occur in the diploid suggest that two-locus, three-locus and higher-order interactions were also involved. In fact discrete log-linear analyses showed that many multilocus interactions are highly significant statistically but others are nonsignificant (Perez de la Vega et al., 1991; Garcia et al., 1991). Within-population diversity varied from none for Sites 1 and 2, to small for Site 3 (only two 14-locus genotypes present), to substantial for Sites 4 and 5 (eight different 14-locus genotypes present), to high for the five remaining sites (13 to 36 different 14-locus genotypes present). Hence opportunities for epistatic interactions were much greater in Sites 3–10 than in Sites 1 and 2. Population sizes were large at each site so that the effects of genetic drift are expected to be small. Also frequent seed exchange occur between sites (due largely to agricultural activities). Such exchanges are expected to limit genetic differentiation among sites: hence it seems likely that the substantial genetic differences that occurred among sites were due, not to genetic drift, but rather to selection sorting out the single best-adapted 14-locus genotype in the two monomorphic sites, as well as the particular mixes of 14-locus genotypes observed in the polymorphic sites. Rapid changes in the frequencies of 14-locus genotypes are known to occur from generation to generation within populations, particularly in highly polymorphic sites, but the production of new 14-locus genotypes is almost certainly a slow process. The reasons are as follows. The 99% of selfing that occurs in *Ah* (also *Ab*) forces all loci to near homozygosity with the result that *no* recombination takes place generation after generation between the great majority of loci, even loci located on different chromosomes. At the same time the ca. 1% of outcrossing that occurs leads to short bursts of segregation and recombination that produce novel genotypes, some of which may be superior to existing genotypes. If an outcross plant is heterozygous for n loci, half of its selfed progeny are expected to be heterozygous for these loci in the next generation. Thereafter selfing will reduce heterozygosity by 1/2 in each succeeding generation; thus the series for heterozygosity within each lineage descended from any single outcross individual is 1/2, 1/4, 1/8, 1/16 so that > 90% of the recombination occurs in the first four generations following an outcross. Assuming no fitness differences and large numbers of individuals within lineages, each lineage is soon expected to approach an equilibrium featuring 2^n equally frequent homozygous lines. However it is likely that selection will favor the more fit genotypes and eliminate the less fit, and also that many genotypes will drift out of each lineage. Regardless, the survivors will quickly be driven to homozygosity by the selfing and all loci within each lineage, whether located on the same or different chromosomes, will thereafter behave as if they are tightly linked with recombination value ca. 0.01 (Allard, 1975). Thus any two loci, (say a and b), no matter where they are located within the genome, will behave as a single functional locus c, and they will manifest some degree of pseudo-dominance up to pseudo-*over*dominance whenever $c_1 = a_1b_2$, $c_2 = a_2b_1$, and $c_1c_1 < c_1c_2 > c_2c_2$. Selfing is thus a simple and highly effective way of exploiting all favorable epistatic interactions among alleles of different loci, whether the loci are physically linked on the same chromosome or not. The 1% of outcrossing provides a small but steady supply of novel multilocus genotypes, some of which may be superior, whereas the 99% of selfing restricts recombination sufficiently to protect previously existing, as well as newly arisen favorable epistatic combinations of alleles, from breakup due to segregation. This raises another question: at what rate of outcrossing does the resulting segregation destroy existing population structure? Experiments with barley show that increasing outcrossing rates artificially from the normal 1% to as little as 5%, even for a single generation, leads to dramatic alterations in population structure accompanied by large decreases in seed yields that persist for many generations after selfing is allowed to resume. Evidently even low outcrossing rates can lead to 'outbreeding depression' in populations of heavy selfers.

Turning to *Ab*, the number of 14-locus genotypes observed in the 50 Spanish populations exceeded 440, more than four-fold as many as in *Ah*. Fourteen-locus genetic structure was much more complex than in *Ah* and also appeared to be more closely attuned to specific environments. Consequently I will limit discussion to a single point that emerged more clearly from the *Ab* data than the *Ah* data, namely that very close associations developed between distinctive habitats and certain single 14-locus genotypes, as well as between other specific habitats and particular polymorphic mixtures of 14-locus genotypes. One set of such associations were those with the habitats of the cold, high elevation North Central Spanish Plateau (CNP, Figure 1). A specific 14-locus genotype (Table 3) was most frequent in all seven sites of that region (Sites 43–48, 33); this particular 14-locus genotype was monomorphic (f = 1.00) in Site 43 and nearly monomorphic (f = 0.99) in Site 45, the two highest and coldest sites of the region. Two things happened in the five lower elevation sites of the CNP: (1) the frequency of the 'cold-tolerant' 14-locus genotype fell off (as in Site 46) and/or (2) several additional loci became polymorphic (as in Site 33). The 'cold tolerant' genotype was entirely absent in still lower and less cold areas peripheral to the CNP. In Site 1 (PNP, Figure 1) it was replaced by a 14-locus genotype that differed at only one locus (quadriplex *11,44* of *Acp2* was replaced by quadriplex *33,44*) and in Site 3 in which quadriplex *11,11* of *Prx1* was replaced by quadriplex *11,22*. At the same time these two sites also became polymorphic for several other loci leading to substantial changes in the patterns of multilocus epistatic interactions. In brief, progressively larger environmental changes were accompanied by progressively greater restructuring of the 14-locus genetic makeup of the populations. Evidently natural selection sorted out the single 14-locus genotype that is best adapted in the coldest areas and integrated many different 14-locus genotypes into complex entities that provide superior adaptedness in the less cold environments of lower elevation sites. However among the 50 Spanish populations of *Ab* only four were monomorphic for a single 14-locus genotype whereas more than 20 different 14-locus genotypes were present in a great majority of the populations (Allard et al., 1993). This indicates that superior adaptedness in *Ab*, as in all other heavily selfing species that have been studied in detail, is nearly always associated with substantial genetic diversity featuring complex mixtures of highly-fit non-segregating multilocus genotypes. Heavily selfing populations are thus usually highly variable genetically: they clearly are not stuck in evolutionary 'blind alleys' as has frequently been stated. In fact their genetic system is magnificently equipped to develop and to incorporate novel superior genotypes while at the same time preserving the mix of homozygous nonsegregating genotypes that has evolved over time in response to fluctuating selective pressures exerted by the specific environment they occupy.

Table 3 also gives the 14-locus genotypes of *Ab* in four of seven habitats for which genotype-habitat associations have been established in California. Note that the present day ancestral and colonial populations are closely similar in allelic composition when compared on a locus-by-locus basis: the predominant alleles of Spain are also predominant in the colonial populations. However, the ancestral and colonial populations differ

Table 3. Single most-frequent 14-locus genotypes of *Ab* in the Central Northern Plateau (CNP) and the Peripheral Northern Plateau (PNP) of Spain and in four habitats in California. (Adapted from Allard et al., 1993)

Region and site	Locus *Pgm1*[a]	*Mdh1*	*Pgd2*	*Mdh3*	*Acp2*	*Pgd1*	*Acp1*	*Prx1*	*Lap1*	*Est1*	f[b]
CNP-43	*11,11*	*11,22*	*11,22*	*11,22*	*11,44*	*11,33*	*11,22*	*11,11*	*22,33*	*55,77*	1.00
CNP-45	*11,11*	*11,22*	*11,22*	*11,22*	*11,44*	*11,33*[P]	*11,22*	*11,11*[P]	*22,33*	*55,77*	0.99
CNP-46	*11,11*	*11,22*	*11,22*	*11,22*	*11,44*[P]	*11,33*[P]	*11,22*	*11,11*	*22,33*	*55,77*[P]	0.68
CNP-33	*11,11*	*11,22*	*11,22*[P]	*11,22*[P]	*11,44*[P]	*11,33*[P]	*11,22*	*11,11*[P]	*22,33*[P]	*55,77*[P]	0.91
PNP-1	*11,11*	*11,22*	*11,22*[P]	*11,22*[P]	*33,44*[P]	*11,33*[P]	*11,22*[P]	*11,11*[P]	*22,33*[P]	*55,77*[P]	0.48
PNP-3	*11,11*	*11,22*	11,22[P]	*11,22*[P]	*11,44*[P]	*11,33*[P]	*11,11*[P]	*11,22*[P]	*22,33*	*55,77*[P]	0.33
											rDNA
CA-M[c]	*11,11*	*11,22*	*11,22*	*11,22*	*33,44*	*22,22*	*11,33*	*11,22*	*11,33*	*55,55*	*15,9, 8, 7*
CA-X[c]	*11,11*	*11,22*	*11,22*	*11,22*	*33,44*	*11,11*	*11,22*	*11,33*	*11,22*	*33,55*	*13, 10, 8, 7*
CA-H[c]	*11,11*	*11,22*	*11,22*	*11,22*	*33,44*	*11,22*	*11,22*	*11,33*	*11,22*	*33,35*	*10, 8,7*
CA-JR[c]	*11,11*	*11,22*	*11,22*	*11,22*	*33,44*	*11,22*	*11,22*	*11,22*	*11,22*	*33,55*	*12, 8, 7*

[a] Also *Got2, Mdh3, Got1, Pgi1*.
[b] Frequency within site of single most-frequent 14-locus genotype.
[P] Locus polymorphic within site.
[c] Polymorphic in areas where M, X, H and/or JR habitats form interfaces with each other.

in genotypic configurations in two main ways. First, the predominant alleles of Spain often occur in different combinations within quadriplexes in California and second and more striking, the 14-locus genotypes of California are entirely different from those of Spain. Another difference, not documented in Table 3, is that the *rare* alleles of Spain were replaced in California by a nearly entirely different set of *rare* alleles that evidently arose recently as a result of mutations that occurred within the colonial Californian populations. Rare alleles have little effect on adaptedness; consequently it appears that changes in allelic frequency, which are often considered to be the elemental process of evolution, played only a minor role in the adaptive changes that occurred in the colonial populations. Instead, a different process, namely reorganization of the *predominant* alleles of the ancestral populations into novel combinations adapted to specific habitats in California, played the major role. This dramatic reorganization obviously occurred quickly because *Ab* has been widespread in California for no more than 150 generations. The process of adaptation clearly continues up to the present; many populations are highly polymorphic in California, especially in areas where different habitats adjoin one another and frequent hybridizations occur among genetically different ecotypes.

Table 4. Numbers of alleles of allozyme and rDNA loci in 26 collections of wild barley and from cultivated barleys in three stages of domestication

	No. of loci	No. of alleles	Mean no. alleles/locus	Relative no. alleles/locus
[a] Wild barley	20	103	5.15	100
[b] Landraces	20	55	2.75	53
[c] CCII F_7–F_9	26	41	1.58	31
CCII F_{53}	26	37	1.42	28
[c] CCXXI F_4	25	42	1.68	33
CCXXI F_{22}	25	38	1.52	30
[d] Californian cultivars	25	36	1.44	28

[a] From 26 Middle Eastern collections.
[b] Eighteen landraces from the Middle East.
[c] Composite Crosses II and XXI, respectively, were synthesized from a worldwide sample of 28 early 20th Century cultivars and from 6,200 accessions of the USDA world barley collection. Both populations were grown annually in California in plots sufficiently large to avoid genetic drift, harvested in bulk without conscious selection and the next generation was sown from a random sample of seeds from the previous harvest.
[d] Nine modern Californian cultivars adapted in different ecogeographical regions.

Wild and cultivated barley

Table 4 gives the numbers of alleles observed in wild barley (*Hordeum vulgare* ssp. *spontaneum*), and in cultivated barley (*H. vulgare* ssp. *vulgare*) in four distinct

Table 5. Frequency of alleles of 23 allozyme loci in 94 Mexican landrace collections of maize, in the 30 most popular inbreds, and in the six inbreds most widely used in elite single crosses. (Adapted from Doebley et al., 1985)

Allelic frequency in Mexican races	% of Mexican collections in which observed	No. of alleles in frequency classes	No. surviving alleles into:	
			Set of 30 inbreds	Set of six inbreds
0.99–0.78	100%	18	18 (100%)	18 (100%)
0.68–0.21	85%	10	10 (100%)	6 (60%)
0.19–0.05	36%	13	10 (77%)	5 (38%)
0.05–0.01	12%	25	9 (36%)	4 (16%)
< 0.01	2%	97	2 (1%)	1 (1%)

[a] Reid heterotic group: B73, B37, A632; Lancaster heterotic group C103, Oh43, Mo17.

stages of domestication. In the transition from wild barley to so-called primitive Middle Eastern landraces numbers of alleles/locus *decreased* by about one half. However progressively smaller decreases in numbers of alleles/locus occurred in the transitions from Middle Eastern landraces, to early 20th Century cultivars, to the elite Californian barley cultivars, all of which are closely similar in allelic composition. It became apparent early in studies of the Mendelian inheritance of alleles of the loci of Table 4 that patterns of allelic frequency change in segregating families are good predictors of the probable adaptive values of different alleles of most loci. In such Mendelian studies, observed ratios in F_2 families nearly always deviated significantly from expected 3:1 or 1:2:1 ratios; frequent (predominant) alleles were nearly always in excess and infrequent alleles were in deficiency. Thus, even under the conditions of little or no plant-to-plant competition under which the Mendelian studies were conducted, rare and infrequent alleles nearly always behaved as subvitals. This was also the case in large populations (such as *CCII*) in which interplant competition was intense and genetic drift was negligible. However alleles that were present in intermediate frequencies in the wild, as well as in landraces and early 20th Century varieties, usually followed a quite different pattern of allelic frequency change in large experimental populations. Increases (or decreases) in the frequencies of such alleles were usually slow (0.01 per generation). Also the superior (inferior) adaptive properties of allozyme alleles that ultimately became predominant (or infrequent) were rarely if ever detectable in visual comparisons of pairs of highly isogenic recombinant inbred lines, one of which carried one allele and the other an alternative allele of the locus under investigation. Furthermore, seed yields of such isogenic lines measured in head-to-head small-plot comparisons were usually not significantly different. Although differences between such isogenic lines were clearly subtle, natural selection clearly recognized superior vs. inferior adaptive properties of individual alleles or the very short chromosome segments in which these alleles reside.

Studies of dynamic change have also provided clues regarding another feature of the evolution of cultivated barleys, namely, how populations that are so similar in allelic composition as the landraces, the early 20th Century varieties and more particularly the modern elite cultivars can differ so widely in adaptedness and performance. Comparisons of multilocus genetic structure have shown that populations in different evolutionary stages differ relatively little, either in allelic composition or allelic frequencies, but that they differ widely in the *combinations* in which surviving alleles occur in the multilocus genotype of each cultivar. I will now illustrate this point more precisely with data from corn.

Maize

It is widely accepted that Southwestern Mexico was the cradle of domestication of maize and that North American maize was derived from Mexican races. Table 5 gives the frequencies of 163 alleles of 23 allozyme loci in 94 collections representing 34 different Mexican races. Among these 163 alleles 18 (11%) were present in all 94 Mexican collections at overall high frequencies from 0.78 to 0.99. *All* of these ubiquitous and frequent alleles survived into the 30 most-popular U.S. public inbred lines and *all* of them also survived into the six inbred lines that were used as parents of virtually all U.S. public single-cross hybrids. The situation was the opposite for the 97 alleles (60%) that were present in frequencies < 0.01. These alleles were rarely present in more than one or two Mexican races; also only two among these 97 rare alleles survived into the 30 most popular inbreds and only one survived into the six elite inbreds. The remaining 48 (29%) among the total of 163 alleles were present in Mexican races (also in Bolivian races) in intermediate frequencies. All of the 10 alleles present in relatively high frequencies (0.21 to 0.68) and 10 of the 13 alleles present in moderate frequencies (0.05 to 0.19) survived into the 30 most-popular inbreds; however only about half of these alleles survived into the set of six most-widely-used inbred lines and none were fixed in these six elite lines. Thus, in maize, as in *Avena* and in barley, frequent (predominant) alleles contribute to adaptedness in many habitats and survive many cycles of selection in cultivation whereas alleles that are present in intermediate frequencies overall appear to be useful in some environments but not in others; rare alleles of allozyme loci appear to be of little value anywhere. The data for precisely identifiable alleles of maize as well as barley, therefore support the notion that plant breeding has led to a reduction in allelic diversity; however the data also indicate that the reduction in allelic diversity that occurred is due largely to purifying selection rather than to 'erosion' of useful genetic variation.

The data of Table 6 show that substantial reorganization occurred at the single-locus level during the breeding of the elite public single-cross hybrids and that the reorganization that occurred at single loci ran nearly entirely counter to expectations based on the widely-accepted hypothesis that heterosis results from advantage of heterozygotes over homozygotes. Under the overdominance hypothesis it is expected that the two parental inbreds of high performing single-cross hybrids will have diverged maximally in allelic frequency so that many loci will be heterozygous in single crosses. However, among the 23 loci of Table 6 maximum divergence occurred only for locus *Glu1*; all three inbreds of the Reid heterotic group carry allele *1* and all three inbreds of the Lancaster heterotic group carry allele *2* so that the Lancaster × Reid single crosses are all *12* heterozygotes for *Glu1*. In sharp contrast, both the Lancaster and Reid inbreds carry the same allele for 14 of the 23 loci (61%) so that all Lancaster × Reid single crosses are *11* homozygotes for these 14 loci. The six most-widely-used inbreds were sometimes fixed for different alleles of the eight remaining loci; consequently some Lancaster × Reid single crosses were homozygous and some were heterozygous for these loci. Overall the data of Table 6 show that the great majority of loci (79%) are homozygous and only 21% were heterozygous in the most-widely-grown public single-cross hybrids. The high proportion of homozygotes and the low proportion of heterozygotes at the single-locus level cast doubts about the importance of overdominance in promoting high performance.

It is well known that F_2 seed produced on F_1 single-cross hybrids produce about 10% to 20% lower yields than the F_1 single-cross hybrids themselves. All plants of an F_1 single-cross population are genetically identical so that saving seed produced on open-pollinated single-cross plants is equivalent to a single generation of selfing. Hence producing seed from single-cross populations reduces heterozygosity by half at each heterozygous locus. Is this reduction in heterozygosity the cause of the large decrease in seed yield? Probably not, because it seems unlikely that reducing the frequency of heterozygotes from 1.0 to 0.5 at only 21% of loci would cause such dramatic decreases in performance.

The data of Table 6 show further that extraordinary reorganization took place at the 23-locus level during the breeding of the most-widely-grown public single crosses. Millions of different 23-locus allozyme genotypes are present in the Mexican races, in U.S. open-pollinated varieties of maize, and also in the double-cross hybrids that can be produced from the 30 most popular inbreds, but the number of 23-locus allozyme genotypes had been reduced dramatically to only four in the most-widely-grown public single crosses. Apparently repeated testing for favorable interactions among alleles from different gene pools, such as the Lancaster and Reid gene pools, combined with periodic stabilizing of favorable combinations by inbreeding to homozygosity, reduced the milions of 23-locus genotypes in the Mexican races, open-pollinated varieties and double-cross hybrids to only four 23-locus genotypes observed in the most-widely-grown public single-cross hybrids. Note, however, that very large numbers of pairwise, three-locus, four-locus and higher-order epistatic interactions are possible among the 23 loci. Do favorable interlocus epistatic interactions in fact play a major role in the superior performance of the elite public single crosses? About one in five of the allozyme loci (21%) of the public single crosses is heterozygous and hence it seems likely that segregation and recombination among such loci might quickly dismantle the single 23-locus monomorphic genotype of each of the elite public single-cross populations. Examination of the allozyme genotypes of pop-

Table 6. Twenty-three locus allozyme genotypes in F_1 single-crosses among the six most-widely used-public-inbred lines of maize in North America[a]. (Adapted from Goodman & Stuber, 1980)

Fourteen[b]	Nine polymorphic loci[c]									Twenty-three
monomorphic loci	1	2	3	4	5	6	7	8	9	locus genotypes
11	*12*	*11*	*11*	*12*	*12*	*11*	*11*	*11*	*11*	I
11	*12*	*11*	*11*	*11*	*11*	*12*	*12*	*11*	*33*	II
11	*12*	*12*	*12*	*11*	*11*	*11*	*12*	*12*	*13*	III
11	*12*	*12*	*11*	*11*	*11*	*12*	*12*	*12*	*12*	IV

[a] Reid heterotic group: B73, B37, A632; Lancaster heterotic group C103, Oh43, Mo17.
[b] *Adh1, Cat3, Ep, Got1, Got2, Got3, Idh1, Mdh1, Mdh2, Mdh4, Mdh5, Me, Mmm, Pgm1.*
[c] 1 *Glu1*, 2 *Idh2*, 3 *Pgd1*, 4 *Pgd1*, 5 *Pgd2*, 6 *Pgm2*, 7 *Phi1*, 8 *Est8*, 9 *Acp1*.

ulations derived from seed produced on single crosses showed that many different 23-locus genotypes were present in such F_2 populations. In fact when a sample of 50 or more F_2 progeny of the elite F_1 single crosses was examined no single 23-locus genotype was observed more than once in any F_2 progeny; also in no case did the unique 23-locus genotype of the F_1 single cross appear in its F_2 progeny. Thus in a single generation and recombination apparently completely dismantled the single 23-locus genotype of the most-elite public single-cross hybrids and replaced this genotype with a complex mix of less fit genotypes.

Implications for plant breeders

The main messages for plant breeders that emerge from the above results can be summarized in terms of three generalizations. Generalization 1. The most useful genetic resources are modern elite cultivars and their close relatives, especially materials that are adapted in the local environment or closely similar environments. Natural selection in combination with breeder-directed selection in farmers' fields and breeders' nurseries practiced over large numbers of generations clearly increased the frequencies of favorable alleles and of favorable lower-order combinations of alleles, and at the same time decreased the frequencies of less favorable alleles and less favorable lower-order combinations of alleles. But having many favorable alleles and favorable lower-order combinations present is not enough. Higher-order combinations are also very important. This leads to Generalization 2. Once favorable multiallelic combinations have been developed for a given habitat it is important that such combinations be preserved and enhanced. The most effective way to preserve and enhance favorable combinations is to cross elite materials with elite relatives. Also selfing such crosses is preferable to sib crossing because the wider the cross the greater the chance that segregation will dismantle favorable multilocus combinations. However need may arise ultimately to introgress exotic alleles into the elite materials. The difficulties of doing this depend on the number and the heritability of the alleles to be introgressed. This leads to Generalization 3. Discretely inherited markers often provide breeders with effective ways of identifying, tracking and incorporating regions of chromosomes with favorable effects on adaptedness into elite materials. I will illustrate this with two examples, one relatively simple and one more complex.

The relatively simple example involves a photoperiod-sensitive male-sterile (*pms*) rice plant that was found as a spontaneous mutant in a Chinese cultivar in 1973. Early studies indicated that this mutant could be used to propagate itself under short-day conditions and also to produce F_1 hybrid seeds by interplanting it with normal fertile lines under long-day conditions. Thus *pms* rice appeared to offer great opportunities for replacing the widely used 'three-line' (male-sterile, maintainer, restorer) system with a more efficient 'two-line' system. However, it turned out to be less straightforward than first thought because it was found that, in addition to photoperiod, temperature plays an important role in the fertility of many rice genotypes. The solution turned out to be marker-assisted-dissection of fertility, which revealed that nearly all of the *pms* effect, including the temperature effect, was governed by two genetic loci, one very tightly linked to an RFLP marker on chromosome 7 and another very tightly linked to another RFLP marker on Chromosome 3 (Zhang et al., 1994). This made it possible, in developing *pms* lines, to identify and track the *pms* effect precisely so that laborious, costly and time-consuming measurements of fertility were unnecessary. Chinese rice breeders anticipate that many different elite two-line hybrids adapted in different habitats will soon be available for commercial production.

The second example involves marker dissection of adaptedness in barley Composite Cross II in California. CCII was synthesized in 1928 by compositing hybrid seeds from all possible pairwise combinations (378) among carefully chosen barley varieties representing all major barley growing regions of the world. All of the 30 or so markers ultimately studied in CCII turned out to have substantial effects on adaptedness. Interactions between genotypes and environments were, however, very complex. Early studies at Davis were especially confusing – some markers steadily increased and others steadily decreased in frequency while others varied in behavior, sometimes increasing, sometimes decreasing but sometimes not changing in frequency. The reasons emerged when CCII was grown in areas of California where rainfall was either consistently limiting or rarely limiting. In such areas nearly all markers behaved quite consistently; the inconsistency at Davis stemmed from the fact that about one year out of three featured either severe moisture stress, limited moisture stress, or no moisture stress. Thus, at Davis, CCII chased a different multilocus adaptive peak in about one year out of three, whereas in several other locations CCII chased the same adaptive peak consistently so that it was soon possible to identify the multilocus marker genotype appropriate to each habitat. The interactions among marker loci were complicated but far less difficult to identify and quantify than the interactions of quantitatively inherited characters with each other and with environmental variables. I anticipate that marker assisted breeding will provide helpful guidance in developing varieties with superior adaptedness in the several diverse habitats in which barley is grown in California.

Acknowledgement

Dedicated to Göte Turesson on the 100th anniversary of his birth.

References

Allard, R.W., 1975. The mating system and microevolution. Genetics 79: 115–126.

Allard, R.W., 1988. Genetic changes associated with the evolution of adaptedness in cultivated plants and their wild progenitors. Jour Hered 79: 225–238.

Allard, R.W., 1990. The genetics of host-pathogen coevolution: Implications for genetic resource conservation. Jour Hered 81: 1–6.

Allard, R.W., P. García, L.E. Sáenz de Miera & M. Pérez de la Vega, 1993. Evolution of multilocus genetic structure in *Avena hirtula* and *Avena barbata*. Genetics 135: 1125–1139.

Clegg, M.T. & R.W. Allard, 1972. Patterns of genetic differentiation in the slender wild oat species *Avena barbata*. Proc Natl Acad Sci USA 69: 1820–1824.

Cluster, P.D. & R.W. Allard, 1995. Evolution of ribosomal (DNA) genetic structure in colonial Californian populations of *Avena barbata*. Genetics 139: 941–954.

Doebley, J.F., M.M. Goodman & C.W. Stuber, 1985. Isozyme variation in the races of corn from Mexico. Amer J Bot 72: 629–639.

García, P., M.I. Morris, L.E. Sáenz de Miera, R.W. Allard, M. Pérez de la Vega & G. Ladizinsky, 1991. Genetic diversity and adaptedness in tetraploid *Avena barbata* and its diploid ancestors *A. hirtula* and *A. wiestii*. Proc Natl Acad Sci USA 88: 1207–1211.

Goodman, M.M. & C.W. Stuber, 1980. Genetic identification of lines and crosses using isoenzyme electrophoresis. Annual Corn Sorghum Res Conf Proc 35: 10–31.

Hutchinson, E.S., S.C. Price, A.L. Kahler, M.I. Morris & R.W. Allard, 1983. An experimental verification of segregation theory in a diploidized tetraploid: esterase loci in *Avena barbata*. J Heredity 74: 381–383.

Pérez de la Vega, M., P. García & R.W. Allard, 1991. Multilocus genetic structure of ancestral Spanish and colonial Californian populations of *Avena barbata*. Proc Natl Acad Sci USA 88: 1201–1206.

Saghai Maroof, M.A., R.M. Biyashev, G.P. Yang, Q. Zhang & R.W. Allard, 1994. Extraordinarily polymorphic microsatellite DNA in barley: Species diversity, chromosomal locations and population dynamics. Proc Natl Acad Sci USA 91: 5466–5470.

Schwartz, D., 1960. Genetic studies of mutant isozymes in maize. Synthesis of hybrid enzymes by heterozygotes. Proc Natl Acad Sci USA 88: 1202–1206.

Turesson, G., 1922. The genotypical response of plant species to the habitat. Hereditas 3: 211–211; 331–348.

Wright, S., 1937. The distribution of gene frequencies in populations. Proc Natl Acad Sci USA 23: 307–320.

Zhang, Q., B.Z. Shen, X.K. Dai, M.H. Mei, M.A. Saghai Maroof & Z.B. Li, 1994. Using bulked extremes and recessive class to map genes for photoperiod-sensitive genic male sterility in rice. Proc Natl Acad Sci USA 91: 8675–8679.

P.M.A. Tigerstedt (Ed.), *Adaptation in Plant Breeding*, 13–19, 1997.

Evolution and adaptedness in a facultatively apomictic grass, *Poa pratensis* L.

Andrea Mazzucato[1], Mario Falcinelli[2] & Fabio Veronesi[3]
[1] *Dipartimento di Agrobiologia e Agrochimica, Facoltà di Agraria, Via S. Camillo de Lellis, 01100 Viterbo, Italy;*
[2] *Istituto di Miglioramento Genetico Vegetale, Facoltà di Agraria, Borgo XX Giugno 74, 06100 Perugia, Italy;*
[3] *Dipartimento di Biotecnologie Agrarie ed Ambientali, Facoltà di Agraria, Via Brecce Bianche – Monte d'Ago, 60100 Ancona, Italy*

Key words: apomixis, Kentucky bluegrass, plant adaptedness, *Poa pratensis*

Summary

Facultative apomixis confers great adaptedness and evolutive potential on biotypes belonging to the *Poa pratensis* L. agamic complex. Pioneering and recent research have dealt with and sometimes elucidated different aspects of apomictic expression, such as cytological mechanisms, distribution patterns, genetic control and environmental effects, in this species. We carried out an investigation on the mode of reproduction expressed in Italian natural populations, cultivars and selected genotypes together with an extensive time (56 years) and space (six countries) review in order to obtain a comprehensive picture of apomixis expression in *P. pratensis*. Histograms of the estimations were prepared. They showed that, due to breeder and natural selection, both cultivars and natural populations very frequently expressed a high degree of apomixis. Variations in the degree of apomixis across generations, different pollen donors, environments and estimation levels were generally low. It would, therefore, seem that the tenet which holds that apomixis in *P. pratensis* is strongly influenced by external factors should be reconsidered. Despite the fact that little is known of the genetic, physiological and molecular control underlying apomixis, the overall picture that emerges from our study helps to explain how the balance between sexual and asexual reproduction confers extreme versatility, adaptedness and evolutive capacity on this remarkable grass.

Introduction

Poa pratensis L. deserves the attention of plant scientists not only for its turf and forage aptitudes, but also for its remarkable evolutionary and adaptative capacities. Thousands of years of introgression and natural selection have evolved *P. pratensis* into one of the most variable, buffered and tolerant species (Clausen, 1961). Two connected characteristics are considered the secret of its success; genomic plasticity and reproductive versatility. *P. pratensis* reproduces mostly through aposporous apomixis, which fixes and propagates successful genotypes across generations (Akerberg, 1939; Grazi et al., 1961), but progeny can also originate sexually, in a ratio that is under the genetic control of the mother plant (Brown, 1941). Whether individuals are sexually or apomictically produced they are usually referred to as aberrants or maternals in facultative apomicts (Bashaw & Funk, 1987). From the adaptative point of view, apomixis is a means of multiplying the presently adapted gene combinations, while sexuality ensuring a reservoir of variation to be selected for in the evolutionary process (Clausen et al., 1947).

Over the last decades, much literature has dealt with and sometimes elucidated different aspects of apomixis expression in *P. pratensis*, including cytological mechanisms, distribution patterns, genetic control and the influence of the environment. Nevertheless, the regulation of reproduction in this rather complex species is far from being fully understood. The asexual aptitudes of a given genotype can be estimated on the mother plant, by quantifying the production of 2n egg cells (apospory) or the failure of fertilization (parthenogenesis) or both. More frequently the mode of reproduction is estimated by counting the number of maternal and aberrant individuals in the progeny on the basis of morphological, karyological or biochemical descriptors. However, definition of the mode of reproduction is a difficult task. While apospory and parthenogenesis are genetically controlled steps of the process, many other factors may shift the ratio between maternals and aberrants in the progeny. Competitive events that take place in the ovule between the aposporic initial(s) and the legitimate meiotic cells depend on the timing in development (Nygren, 1951; Grazi et al., 1961), which can be influenced by the environment. Pollen availability, especially at the onset of anthesis, is likely to favour fertilization of both n and 2n egg cells (Bashaw & Funk, 1987). In progeny testing, the relative fertility and vigour of both the maternal and the aberrant fractions eventually alter their numeric ratio during embryogenesis, seed germination, transplant and plant growth (Åkerberg, 1942; Nygren, 1951; Grazi et al., 1961). Last but not least, accuracy and precision of operators and techniques are important in classifying progeny plants. However, although conservatism is usually exercised in classification of aberrants (Brittingham, 1943), the borderline of aberrancy is obviously a subjective measure. Moreover, the efficiency of genetic markers (morphological traits, isozymes and molecular markers) in discriminating aberrant plants on the same progeny has proved variable (Mazzucato et al., 1995).

This survey aims to summarize past and recent results of apomixis research in *P. pratensis* to give an overall view of its expression patterns and the extent to which external factors may influence the reproductive process.

Materials and methods

Forty-six studies that estimated the degree of apomixis in *P. pratensis* from different aspects with varying techniques, including the traditional morphological progeny test, isozyme analysis and cytoembryological investigation, and covered a period of 56 years (1939–1995) worldwide were reviewed and tabulated together with the results of a recent research carried out at the Istituto di Miglioramento Genetico Vegetale, University of Perugia, Italy. Only data useful for making comparison are included in the figures. The mode of reproduction is expressed as the estimated percentage of apomixis and reported as a single value, mean value or range, as appropriate. The weaknesses inherent in comparing data obtained in different decades and locations with different aims and techniques were allowed for.

Because many more data on this topic lie hidden in old volumes, regional bulletins and difficult to access seed company files, such a survey cannot be exhaustive. However, the results highlight certain general trends of scientific interest that should increase our understanding of the adaptative strategies adopted by *P. pratensis*.

Results

Patterns of apomixis expression in cultivars, genotypes from natural populations and selected genotypes

Due to breeder selection, the degree of apomixis expressed in cultivars is usually high and stable (Figure 1a). Early reports based on morphological progeny tests indicate values above 95% in the cultivars Fylking and Merion (Julén, 1954; Duich & Musser, 1959). Thirteen out of 14 estimations cited for cultivar registration (Murray & Powell, 1976; Jacklin et al., 1977, 1988; Bailey et al., 1978; Rewinski et al., 1978; Bohnert & Yungen, 1982; Meyer et al., 1982, 1987; Meier et al., 1985; Brilman et al., 1987, 1989; Ensign, 1987; Kaerwer et al., 1989) reported values above 90%. Lower values (average 83%) were reported by investigators who attempted to increase cross pollination in the mother plant for breeding purposes (Hintzen & van Wijk, 1985). Complete apomixis was reported in only one instance, for the cultivar Späths by Pommer (1972), who also reported the lowest value for the cultivar Steinacher (44%). The introduction of more discriminative techniques, such as isozyme progeny testing, did not produce consistent decrements in the estimated values of apomixis in cultivars (Wu & Jampates, 1986; den Nijs, 1990; Mazzucato, 1995a).

Estimations carried out on genotypes belonging to natural populations from established turfs from the United States, Canada and Europe demonstrated that apomixis tends to be highly expressed in wild genotypes of *P. pratensis* (Figure 1b). The degree of apomixis was above 95% in 51% of the 297 estimations reviewed. A considerable percentage of estimations (20%) in this material, unlike that from cultivars, led to a conclusion of absolute apomixis. The lowest apomixis values (29%) were reported in material of German (Pommer, 1972) and Italian (Mazzucato, 1995b) origin. Genotype frequency followed an increasing trend from the 36–40% to the 91–95% apomixis class, while

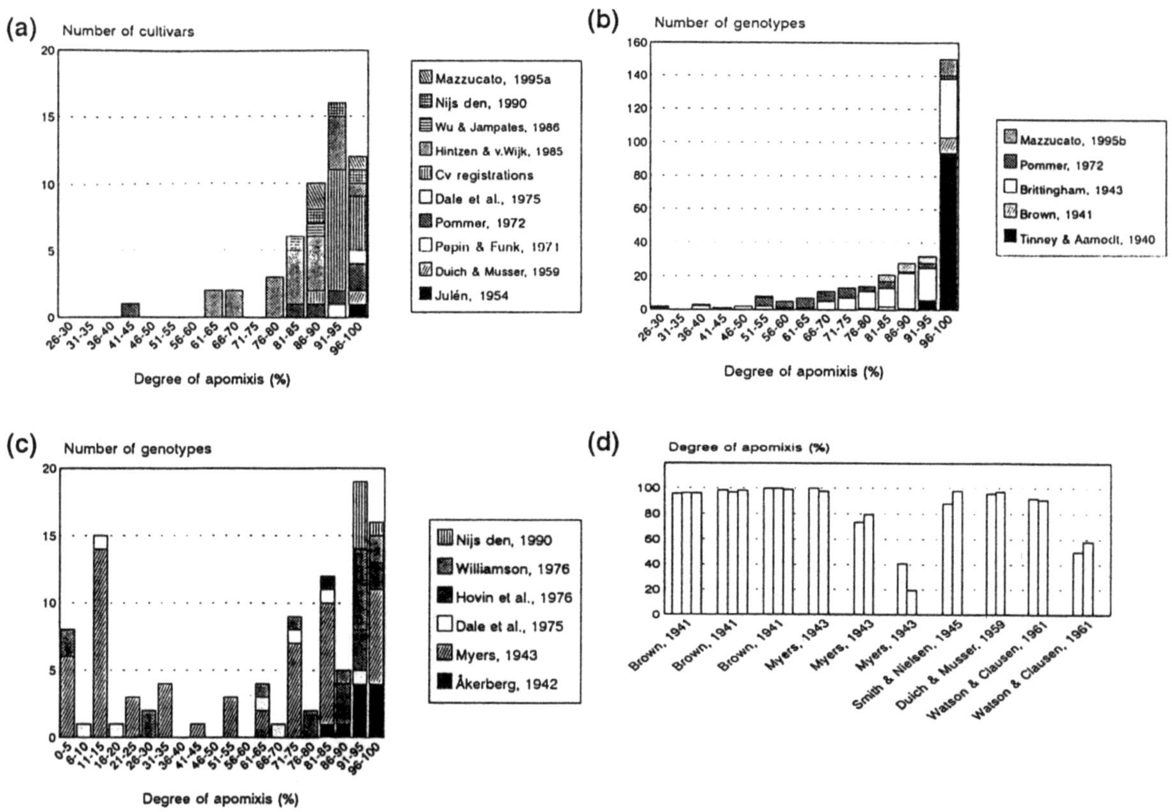

Figure 1. Degree of apomixis expressed in *Poa pratensis* in **(a)** commercial cultivars (for cultivar registration references see text), **(b)** genotypes belonging to natural populations or established turfs, **(c)** genotypes selected for contrasting modes of reproduction, and **(d)** variation in the degree of apomixis in *P. pratensis* genotypes estimated under uniform conditions across generations.

the 96–100% class was by far the most represented (Figure 1b).

Data collected from researchers who selected for contrasting modes of reproduction showed that, in this species, reproductive behaviours range from complete sexuality to nearly absolute apomixis (Figure 1c). Highly or completely sexual clones were selected and described by several investigators (Myers, 1943; Dale et al., 1975; Williamson, 1976).

Stability of the degree of apomixis across generations, different pollen donors, environments and levels of estimation

Variations in estimations of the mode of reproduction under uniform conditions across generations, which were available for different selected clones and cultivars (Brown, 1941; Myers, 1943; Smith & Nielsen, 1945; Duich & Musser, 1959; Watson & Clausen, 1961), were mostly below 10% (Figure 1d). The greatest difference found in the same genotype (21.2%) was reported by Myers (1943), whose classification of reproduction modes was based on rather broad frequency classes which were averaged to construct the histogram.

A few data on estimations based on separate reproductive parts of the same plant were also available. Nielsen & Smith (1951) reported a maximum apomixis difference of 23% in progeny tests of seed from terminal and lower-level panicles, whereas Han (1971) found no variations when he compared the reproductive potential of differently positioned florets.

The range of variations in the degree of apomixis obtained comparing estimations made after open pollination and/or selfing and/or crossing or after crossing with different male parents was always low (Figure 2a). Whereas Åkerberg (1939) and Brittingham (1943) averaged their data over a number of observations, Nygren (1951) and Pepin & Funk (1971, 1974) based their comparisons on single genotypes. Oddly, Brown (1941) reported substantially greater sexuality in selfed than in open pollinated plants, but unfortunately did

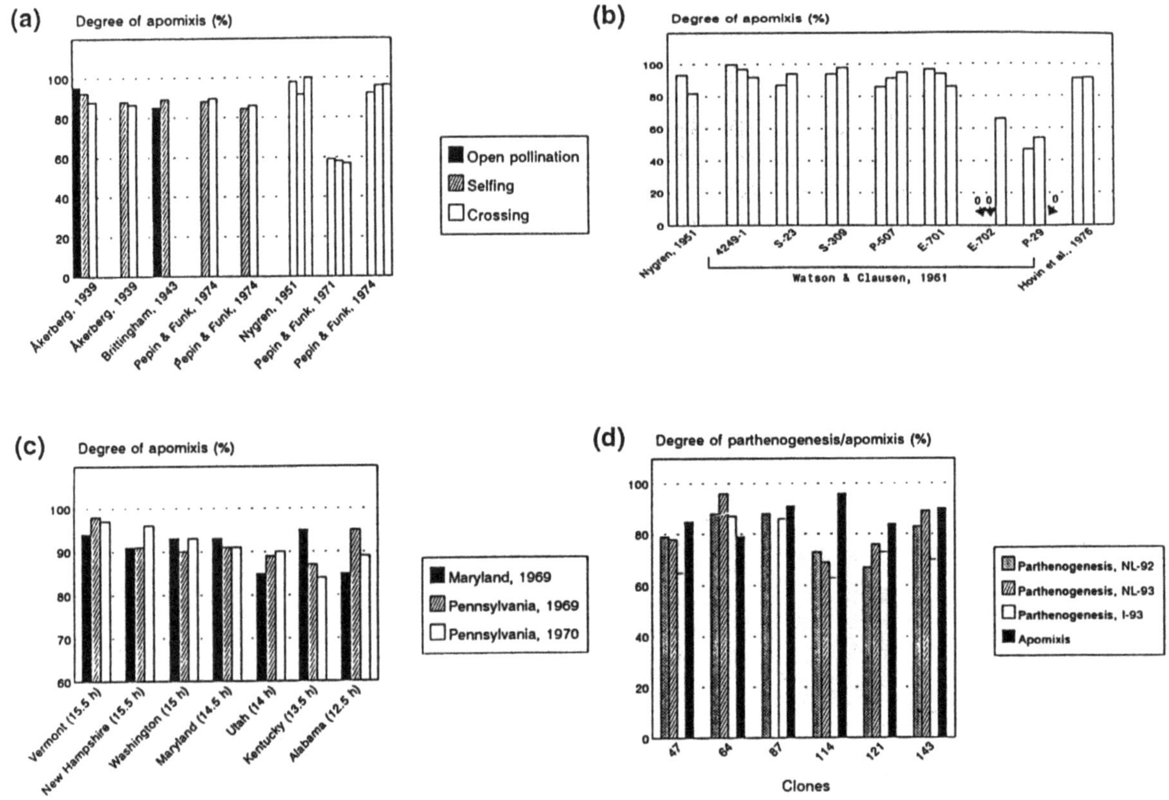

Figure 2. Variation of the degree of apomixis in *Poa pratensis* genotypes due to **(a)** different pollen donors, **(b)** progeny testing in different locations and **(c)** different environment of seed production (average of nine clones, x axis sorted by decreasing photoperiod at the onset of anthesis) and three locations of evaluation (from Hovin et al., 1976), and **(d)** variation in the degree of parthenogenesis estimated in different locations in The Netherlands and Italy (NL, I) and years (1992, 1993) and degree of apomixis estimated after progeny test in six selected clones (from Mazzucato et al., 1996).

not quantify this behaviour. The highest variation of the degree of apomixis (8.5%) in the reports reviewed was estimated by Nygren (1951), after crossing *P. pratensis* with two different *P. scabrella* (Thurb.) Benth. parents.

From the point of view of the stability of apomixis across environments, it is important to consider whether the experimental differences were due to progeny testing environment, to seed production environment, or both. The shift in the degree of apomixis resulting from seed produced in uniform conditions being progeny tested in different locations (and years) was negligible according to several studies (Figure 2b). Consistent shifts were only reported for two partially apomictic clones that were classified as completely sexual when seed was evaluated far away from the region of origin (Watson & Clausen, 1961): E-702, a Scottish selection whose seed was produced at Edinburgh, gave no progeny by apomixis when tested in two United States locations and conversely an American clone, P-29, was completely sexual when evaluated at Edinburgh (Figure 2b).

In other experiments, the mother plants flowered and produced seed under different conditions. Plants that flowered in greenhouses were generally more sexual than those grown in the field, but the differences, when quantified, were small (Grazi et al., 1961; Han, 1971; Murray & Powell, 1976). Variations in gibberellic acid treatment, photoperiod, light intensity and nitrogen application had no effect on the degree of apomixis, but temperature shock increased sexuality (Han, 1971). Differently, Williamson (1981a, 1981b) reported that varying light regimes during flowering have an influence on the mode of reproduction. Low shifts in the mode of reproduction were also reported by Hovin et al. (1976) when seed was produced at different locations (Figure 2c). Increments in mean sexuality were associated with a short photoperiod, low latitudes and wide overall anthesis range. The seed produced in locations with these characteristics, such as Kentucky and Alabama, also gave the highest variations when progeny was tested in different environments (Figure 2c).

Estimation of parthenogenesis in two different locations and seasons by the auxin test (Technique proposed by Matzk, 1991) also showed how the environment can influence part of the apomictic process. In the results reported by Mazzucato et al. (1996) the highest differences in estimated parthenogenesis (clones 47 and 143, Figure 2d) were found across locations situated in The Netherlands and Italy, while differences across years in The Netherlands were below 10%. The parthenogenesis values were lower for four of six clones in the Italian environment.

Comparison of evaluations of single steps of the apomictic process (i.e. apospory or parthenogenesis) with apomixis as assessed at the progeny level also furnished interesting information. Estimated apomixis was 20–30% higher than apospory in two clones studied by Nygren (1951). Similarly, Mazzucato et al. (1996) have found that apomixis tends to be higher than estimated parthenogenesis (Figure 2d).

Discussion

Comparison of the frequency histograms constructed from estimations obtained in cultivars (Figure 1a) with those obtained in natural populations (Figure 1b) leads to the conclusion that natural selection acts as effectively, or more effectively, than breeders in favouring genotypes with a very high degree of apomixis. Although genotypes with widely varying degrees of apomixis are present in the species (Figure 1c), it is clear that natural selection forces sexually reproducing plants to recover high asexual behaviour. While apomixis ensures the multiplication of adapted genotypes, the retained sexual process permits occasional hybridization and has an evident role in evolution (Clausen, 1961). In this way, facultative apomicts not only follow the strategy adopted by strict selfers, but also ensure the maintenance of high levels of heterozygosity and polyploidy.

In contrast to the commonly hold belief that, in *P. pratensis*, apomixis is strongly influenced by external factors, our survey of the experimental data generally revealed only small variations in the degree of apomixis. The consequences of external disturbancies, such as different pollen donors and environments, were seldom greater than those detected when the same genotype was evaluated under uniform conditions. Experimental error due to human judgement or technical constraints may substantially overlap with the ranges due to external factors (Williamson, 1976; Mazzucato et al., 1995).

Nevertheless the mode of reproduction is clearly sensitive to the environment, though to a low extent. The overall trend seems to indicate an increased sexuality when mother plants flower under unusual conditions (greenhouse, short photoperiod, low latitudes). Temperature and photoperiod are crucial factors at this point, because they can influence the timing of processes in the ovule. In this sense an environment different from the one where the plants were adapted automatically produces a higher ratio of recombinant individuals. This behaviour must be taken into consideration during cultivar selection and multiplication, if the working environment is likely to differ from that of germplasm collection. When apomixis is quantified at the progeny level, the fact that the maternal offspring of adapted plants is supported by high fitness, while the aberrant fraction and the whole progeny of non-adapted plants can be strongly selected by natural and artificial (unintentional) pressures must be allowed for. In offtype embryos or seedlings, grossly unbalanced chromosome numbers or gene combinations may influence survival due to extreme weakness leading to non-vitality. The relative fitness of the two fractions is somehow related to the environment of evaluation, as the balance of post-embryogenesis viability acts strongly at this point (Watson & Clausen, 1961). A further explanation would be that a higher 'combining ability', i.e. a genetic arrangement more suitable for gene exchange, rather than the presence of alleles for sexuality, is responsible for the survival of a higher ratio of sexually produced individuals in certain genotypes.

Our review highlights certain aspects of the adaptative reproduction strategies of *P. pratensis* and leads to the conclusion that the often reported complexity of apomixis expression in this species should be reconsidered, since the genetic control appears to be very strict and stable. While external factors do exert an influence on the mode of reproduction, their effect is small and does not greatly alter the reproductive aptitudes of individuals. The picture that emerges from our survey therefore supports the recent proposal that apomixis is a simple process. If, indeed, this is the case in *P. pratensis*, optimism is warranted that apomixis can be fully understood and the knowledge gained applied to manipulating its breeding.

Acknowledgements

The research was supported by the Ministero dell'Università e della Ricerca Scientifica e Tecnologica, funds 40%, 'Il controllo dei sistemi riproduttivi nel miglioramento genetico delle piante agrarie', local coordinator Prof. M. Falcinelli.

References

Åkerberg, E., 1939. Apomictic and sexual seed formation in *Poa pratensis*. Hereditas 25: 359–370.

Åkerberg, E., 1942. Cytogenetic studies in *Poa pratensis* and its hybrid with *Poa alpina*. Hereditas 28: 1–126.

Bailey, R.H., B.L. Rose, W.A. Meyer, K.J. McVeigh & R.C. Funk, 1978. Registration of 'Brunswick' Kentucky bluegrass. Crop Sci 18: 912.

Bashaw, E.C. & R.C. Funk, 1987. Apomictic grasses. In: W.R. Fehr (Ed.), Principles of cultivar development. Vol. 2, Crop Species, pp. 40–82. Macmillan, London.

Bohnert, O. & J.A. Yungen, 1982. Registration of 'Pacific' Kentucky bluegrass. Crop Sci 22: 1084–1085.

Brilman, L.A., A.W. Jacklin, R.H. Hurley, B.B. Clarke & C.R. Funk, 1987. Registration of 'Nassau' Kentucky bluegrass. Crop Sci 27: 1084–1085.

Brilman, L.A., A.W. Jacklin, A.D. Brede, J.J. Zajac & R.C. Funk, 1989. Registration of 'Liberty' Kentucky bluegrass. Crop Sci 29: 825–826.

Brittingham, W.H., 1943. Type of seed formation as indicated by the nature and extent of variation in Kentucky bluegrass, and its practical implications. J Agric Res 67: 225–264.

Brown, W.L., 1941. The cytogenetics of *Poa pratensis*. Ann Missouri Bot Gard 28: 493–516.

Clausen, J., 1961. Introgression facilitated by apomixis in polyploid poas. Euphytica 10: 87–94.

Clausen, J., D.D. Keck & W.M. Hiesey, 1947. Partial apomixis: an evolutionary labyrinth. Carnegie Institute of Washington Year Book, 46 (1946/47): 101–103.

Dale, M.R., M.K. Ahmed, G. Jelenkovic & R.C. Funk, 1975. Characteristics and performance of interspecific hybrids between Kentucky bluegrass and Canada bluegrass. Crop Sci 15: 797–799.

Duich, J.M. & H.B. Musser, 1959. The extent of aberrants produced by 'Merion' Kentucky bluegrass, *Poa pratensis* L., as determined by first and second generation progeny test. Agronomy J 51: 421–424.

Ensign, R.D., 1987. Registration of 'Alene' Kentucky bluegrass. Crop Sci 27: 149.

Grazi, F., M. Umaerus & E. Åkerberg, 1961. Observations on the mode of reproduction and the embryology of *Poa pratensis*. Hereditas 47: 489–541.

Han, S.J., 1971. Effects of genetic and environmental factors on apomixis and the characteristics of non maternal plants in Kentucky bluegrass (*Poa pratensis* L.). Herbage Abstracts 41: 297.

Hintzen, J.J. & A.J.P. van Wijk, 1985. Ecotype breeding and hybridization in Kentucky bluegrass (*Poa pratensis* L.). In: F. Lemaire (Ed.) Proceedings of the Fifth International Turfgrass Research Conference, pp. 213–219.

Hovin, H.W., C.C. Berg, E.C. Bashaw, R.C. Buckner, D.R. Dewey, G.M. Dunn, C.S. Hoveland, C.M. Rincker & G.M. Wood, 1976. Effects of geographic origin and seed production environments on apomixis in Kentucky bluegrass. Crop Sci 16: 635–638.

Jacklin, A.W., J.H. Thorne, R.E. Engel, W.K. Dickson & R.C. Funk, 1977. Registration of 'Glade' Kentucky bluegrass. Crop Sci 17: 187.

Jacklin, A.W., L.A. Brilman, S.J. Patterson, B.B. Clarke & R.C. Funk, 1988. Registration of 'Classic' Kentucky bluegrass. Crop Sci 28: 188.

Julén, G., 1954. Observations on X-rayed *Poa pratensis*. Acta Agr Scand 4: 585–593.

Kaerwer, H.E., D.M. Casnoff, J. Burr & R.C. Funk, 1989. Registration of 'Aspen' Kentucky bluegrass. Crop Sci 29: 1086–1087.

Matzk, F., 1991. A novel approach to differentiate embryos in the absence of endosperm. Sex Plant Reprod 4: 88–94.

Mazzucato, A., 1995a. Italian germplasm of *Poa pratensis* L. I. Variability and mode of reproduction. J Genet Breed 49: 111–118.

Mazzucato, A., 1995b. Italian germplasm of *Poa pratensis* L. II. Isozyme progeny test to characterize genotypes for their mode of reproduction. J Genet Breed 49: 119–126.

Mazzucato, A., G. Barcaccia, M. Pezzotti & M. Falcinelli, 1995. Biochemical and molecular markers for investigating the mode of reproduction in the facultative apomict *Poa pratensis* L. Sex Plant Reprod 8: 133–138.

Mazzucato, A., A.P.M. den Nijs & M. Falcinelli, 1996. Estimation of parthenogenesis frequency in Kentucky bluegrass using auxin-induced parthenocarpic seeds. Crop Sci 36: 9–16.

Meier, V.D., J.A. Long, P.E. Dade & C.R. Funk, 1985. Registration of 'Bristol' Kentucky bluegrass. Crop Sci 25: 197.

Meyer, W.A., C.A. Rose, B.L. Rose & C.R. Funk, 1982. Registration of 'Columbia' Kentucky bluegrass. Crop Sci 22: 1084.

Meyer, W.A., C.A. Rose-Fricker, B.L. Rose, C.J. Petersen & C.R. Funk, 1987. Registration of 'Challenger' Kentucky bluegrass. Crop Sci 27: 1085–1086.

Murray, J.J. & J.B. Powell, 1976. Registration of Belturf Kentucky bluegrass germplasm. Crop Sci 16: 886.

Myers, W.M., 1943. Second generation progeny test of the method of reproduction in Kentucky bluegrass (*Poa pratensis* L.). J Am Soc Agron 35: 413–419.

Nielsen, E.L. & D.C. Smith, 1951. Dimorphic panicle formation in *Poa pratensis*. Bot Gaz 112: 534–535.

Nijs, A.P.M. den, 1990. Experimenting with apomixis and sexuality in *Poa pratensis*. Apomixis Newsletter 2: 52–54.

Nygren, A., 1951. Embryology of *Poa*. Carnegie Institute of Washington Year Book, 50 (1950/51): 113–115.

Pepin, G.W. & C.R. Funk, 1971. Intraspecific hybridization as a method of breeding Kentucky bluegrass (*Poa pratensis* L.) for turf. Crop Sci 11: 445–448.

Pepin, G.W. & C.R. Funk, 1974. Evaluation of turf, reproductive and disease-response characteristics in crossed and selfed progenies of Kentucky bluegrass. Crop Sci 14: 356–359.

Pommer, G., 1972. Untersuchungen zur Züchtung der Wiesenrispe, *Poa pratensis* L., unter besonderer Berücksichtigung der Art der Samenbildung. Z Pflanzenzüchtg 67: 279–304.

Rewinski, T.F., A.M. Radko, W.K. Wiley, M.C. Pick & R.C. Funk, 1978. Registration of 'Touchdown' Kentucky bluegrass. Crop Sci 18: 163.

Smith, D.C. & E.L. Nielsen, 1945. Morphological variation in *Poa pratensis* L. as related to subsequent breeding behavior. J Am Soc Agron 37: 1033–1040.

Tinney, F.W. & G.S. Aamodt, 1940. The progeny test as a measure of the type of seed development in *Poa pratensis*. J Hered 31: 457–464.

Watson, P. & J. Clausen, 1961. Phenotypic responses to contrasting environments in the genus *Poa*. Report of the Scottish Plant Breeding Station, pp. 64–78.
Williamson, C.J., 1976. Problems in the identification and utilisation of interspecific hybrids of *Poa* in a plant breeding programme. PhD thesis, University of Edinburgh.
Williamson, C.J., 1981a. The influence of light regimes during floral development on apomictic seed production and on variability in resulting seed progenies of *Poa ampla* and *Poa pratensis*. New Phytol 87: 769–783.
Williamson, C.J., 1981b. Variability in seedling progenies and the effect of light regimes during seed production on interspecific hybrids of *Poa*. New Phytol 87: 785–797.
Wu, L. & R. Jampates, 1986. Chromosome number and isozyme variation in Kentucky bluegrass cultivars and plants regenerated from tissue culture. Cytologia 51: 125–132.

P.M.A. Tigerstedt (Ed.), Adaptation in Plant Breeding, 21–26, 1997.

Unfecund, gigantic mutant of oats (*Avena sativa*) shows fecundity overdominance and difference in DNA methylation properties

Hannu Ahokas[1]
Division of Genetics, Department of Biosciences, P.O. Box 56, FIN-00014 University of Helsinki, Finland;
[1] *present address: Plant Breeding Section, ARC, FIN-31600 Jokioinen, Finland*

Key words: adaptation, *Avena sativa*, cytosine methylation, fecundity, gigantism, heterozygosity, oats

Summary

The fecundity heterosis of the *gig*/+ oats (*Avena sativa*) was examined in a population showing segregation of the gene under the long photoperiods in S Finland. Segregating isolines of cv Hannes and of cv Kyrö were raised at two plant densities and the grain yield of heterozygotes and nonmutant homozygotes were evaluated. The heterozygotes in mean outyielded the homozygotes. The overdominance of grain number explains the persistance of this mutant at a relatively high frequency in oats in Finland, though the fecundity of the extremely late homozygous mutant is zero. Previous and current data indicate that the heterozygotes show heterosis of fecundity depending on genetic background, season or other environmental conditions. Grafting of cut stems of nonmutants onto wounded nodes of mutants did not transfer any observable florescence inductor. Hypomethylation of DNA, as revealed with HpaII digestion, occurred in topmost leaves of non-mutants, being absent or occurring to a lesser extent in the comparative leaves of mutants.

Introduction

The oats (*Avena sativa* L.) heterozygotes of the gigantic gene (+/*gig*) produced 2.5 times the kernel numbers of nonmutant cv Hannes under wide space conditions (Ahokas, 1976a). The heterozygotes had a three-day longer vegetative development than the nonmutant, whereas they reached maturity with a delay of about one day. It could not be concluded, if the material used for crosses was isogenic for other quantitatively important genes. The isogeny was raised by permitting self-pollination until the F_6 generation in two mid-early cultivars, Hannes and Kyrö (Rekunen, 1970), each segregating for an independently-appeared gene for gigantism. The data below support the existence of overdominance or monogenic heterosis of grain yield. The homozygous gigantic plants are poorly heading and extremely late-maturing and being eliminated from the seed crop (Ahokas, 1976a).

Several different gigantic mutants in oats have been described (Schegalow, 1922; Schwanitz, 1954; Schwanitz & Pirson, 1955; Zillinsky, 1959; Kivi, 1968; Gielo, 1969; Ahokas, 1976a; Kibite & Taylor, 1994; Taylor et al., 1995). The *gig* locus studied by Kibite & Taylor (1994) shows linkage with the multiflorous spikelet (*Mf-1*) and also with the hullessness (*N-1*) locus. The gigantism in the cross of Black Mesdag CI 1877 × Marvellous CI 7023 was attributed by Rothman & Bowman (1963) to hybrid vigour appearing at a southern latitude for oats (Mississippi, USA) and under a 13-h photoperiod. The F_1, I produced between Black Mesdag and Marvellous, failed to show gigantism under a northern (61° N) long-day season, with up to 24-h photoperiods (Ahokas, unpublished).

The methylation of DNA is known to control the gene activity of many plant genes showing usually tissue-specific expression (Waterhouse et al., 1986; Watson et al., 1987; Ngernprasirtsiri et al., 1989; Sano et al., 1989; Riggs & Chrispeels, 1990; Palmgren et al., 1991; Blom Sørensen, 1992; Burn et al., 1993; Sardana et al., 1993; Lund et al., 1995). A difference observed between the methylation level of the DNA in the top leaves in this study may be a consequence of the gigantism.

The transfer of a hypothetical, diffusable florescence inductor through freshly made wounds from nonmutants to mutants was tested.

Materials and methods

Genotypes and heterosis evaluation

The Hannes oats used to evaluate heterosis in the season of 1986 were F_6 segregants of a single F_2 plant, which was a derivative of the cross HA68-2 × Hannes, where HA68-2 is a *gig/gig* mutant line of Hannes (Ahokas, 1976a). The gigantic Kyrö line was obtained from Mr. O. Ulvinen, M.Sc., State Seed Testing Station, Helsinki. One single plant of the Kyrö seed sample was a nonmutant, while the rest were gigantic. The offtype, a putative back-mutation, displayed an F_2 segregation of gigantism in the next generation, otherwise being uniform. Derivatives corresponding to the F_6 generation of this plant were used for evaluation in the season of 1986.

The soil was gytja clay of pH 6.2 in SE Finland. A 16-7-13 (NPK) fertilizer was drilled (500 kg/ha) before planting in a direction crosswise to the plant rows. Below, the generation of seed is marked with parentheses (Ahokas, 1976b). The $F_{(6)}$ seeds were hand planted at two densities, viz. 2.5 cm between seeds and 15 cm between rows or 10 cm between seeds and 15 cm between rows bordered by other rows of oats. The $F_{(6)}$ seeds came from individual plants of segregating families. Weeding was done by hand. No herbicides or pesticides were used. The F_6 plants were harvested individually from segregating offspring rows, leaving the ends of the rows unharvested. The number of plants per row was recorded to estimate the actual mean area per plant. The individual plants were pulled out from the soil, allowed to dry in binds at room temperature and were carefully threshed by hands; their yields were weighed after a four-month storage period in paper bags at about 40% relative humidity and 20–24° C. The F_7 progenies of the individual plants were grown to test the genotypes of the *gig* locus at F_6. These seeds were hand planted on a field occupied by barley during the previous season.

Nodal grafting

Root exudate from non-mutants were transferred into stems of mutants via wounded nodes. Twelve nonmutants and five mutants per a 10 litre soil pot were planted in the greenhouse. At the five-leaf stage of some of the mutants, a slice was removed from the side of the node 3–7 cm above the soil level with a sterile scalpel. A nearby nonmutant stem was cut at one of the basal nodes, and its stump-end was attached onto the freshly sliced node with a stripe of Sealon Film (Fuji Film). A new donor stem was attached to the renewed wound-area three days later. In controls, the nodes were wounded and a Sealon Film stripe was tied around.

DNA extraction and fractionation

DNA was extracted from greenhouse grown plants. The comparative leaf laminas of a few plants were pooled to make about 10 g samples, frozen with liquid nitrogen and stored at - 70° C. Tissues were ground in liquid nitrogen, and then continued principally following the method of Wagner et al. (1987). After ethanolic ammonium acetate treatment, the pellet was shortly dried in vacuum, suspended in Tris-EDTA (10 and 1 mM, respectively) of pH 7.4. The batches were usually extracted with Tris-neutralized phenol (BRL) four times. The DNA containing phase was then extracted once with chloroform, and once with diethylether and its concentration determined with Hoechst 33258 in a TKO 100 fluorometer (Hoefer). The restriction endonucleases used were BstNI and EcoRII (Stratagene), and HpaII and MspI (Pharmacia or Promega). Unmethylated lambda DNA (*dac*$^-$, *dcm*$^-$) was purchased from Promega, and molecular weight standard, λ DNA-HindIII/ϕX174 DNA-HincII, were from Pharmacia. The incubation of 0.8 to 1.5 μg DNA with 20 U of an enzyme usually lasted for 4 or 4.5 hours at the appropriate temperature. The samples were heated for 5 min at 65° C and fractionated by electrophoresis on 0.6 or 0.7% SeaPlaque agarose (FMC) in 1 × TBE buffer and ethidium bromide (0.5 mg/l).

Results

The results presented in Table 1 confirm the heterotic effect of the *gig* gene on grain yield. Although only the yield differences by weight in cv Kyrö were statistically significant, all four experimental items consistently displayed a higher yield when the *gig* gene was heterozygous. If the 13.6 percent point lower grain mass measured in F_1 heterozygotes with Hannes (Ahokas, 1976a) is assumed, there is significant overdominance of fecundity also in Hannes material.

Table 1. Comparisons of yields of heterozygotes and nonmutant homozygotes in two genetic backgrounds, cv Kyrö and cv Hannes

Cultivar tested	Mean space per plant cm^2	Genotype of *gig* gene	Number of plants	Mean yield per plant g ± SEM	Pairwise significance inside spacings P	Relative yield in pairs (+/+ item = 100)	Estimated grain no per *gig*/+ plants	
							+/+ item = 100	P
Kyrö	43	+/+	76	1.25 ± 0.06	< 0.005			
	43	*gig,,c*/+	135	1.52 ± 0.07		122	Unknown	
	127	+/+	46	3.20 ± 0.20	< 0.001			
	127	*gig,,c*/+	66	4.29 ± 0.26		134	Unknown	
Hannes	43	+/+	57	1.52 ± 0.11	> 0.25			< 0.01
	43	*gig,,a*/+	145	1.64 ± 0.06		107	124	
	143	+/+	44	4.40 ± 0.36	> 0.5			< 0.05
	143	*gig,,a*/+	97	4.63 ± 0.29		105	122	

Table 2. Effect of nodal wound grafting of gigantic stems by nonmutants (mean ± SEM)

	Kyrö			Hannes		
	Non-mutant control	Gigantic grafted	Gigantic control	Non-mutant control	Gigantic grafted	Gigantic control
Number of observations	16	22	6	25	18	11
Days to flag-leaf emergence	51 ± 0.6	85 ± 0.7	87 ± 2.0	53 ± 0.4	90 ± 0.5	90 ± 0.9
Number of nodes on terminated main stem	5.3 ± 0.1	8.9 ± 0.1	9.2 ± 0.4	5.2 ± 0.1	9.6 ± 0.1	9.9 ± 0.2

The space-planted F_6 population contained two thirds of heterozygous or segregating rows of plants and one third of unsegregating rows. In the entire population, the frequency of *gig*/*gig* genotypes was 0.167 and the frequency of the *gig* allele was 0.333. In relative terms, the heterosis of yield was not much affected by the spacing in the 1986 season (Table 1).

The nodal grafting did not transfer any florescence inductor from the dominant nonmutants to the recessive mutants, or the factor was not transferred in an effective amount (Table 2).

When studying the digestibility of different DNA samples isolated from gigantic and corresponding nonmutant leaf tissues, the results with the restriction endonuclease isoschizomers MspI and HpaII showed that CCGG sites are usually methylated in oats leaf tissue in a way undigestible with HpaII. HpaII enzyme is sensitive to all tested methylations of cytosines at the cleavage site (Nelson & McClelland, 1991; Nelson et al., 1993). At the booting stage of the nonmutants, the eighth, topmost leaf laminas, which had fully expanded appeared to carry DNA which was partially digested also with HpaII resulting richly in fragments of 400 to 1500 base pairs in nonmutant samples but was practically undigestible in DNA of gigantic Hannes, and slightly digestible in DNA of gigantic Kyrö (Figure 1). The plants were grown side by side in the greenhouse. Though environmental interactions can not be ruled out, the difference is rather determined by genotype and tissue or organ, since the second leaf laminas from the top did not show the difference with HpaII digestions, but the sites were similarly undigestible with HpaII (Figure 1), though still digestible with MspI. Digestion with the neoisoschizomer pair, BstNI and EcoRII, showed that their common recognition site CC(T/A)GG, was generally digested with BstNI also in the DNA of the top-leaf laminas, EcoRII being unable to notable digestion in any tried oat DNA samples. The decrease in HpaII site methylation seems to be an adult plant phenomenon, since it could not be distinguished at seedling stage.

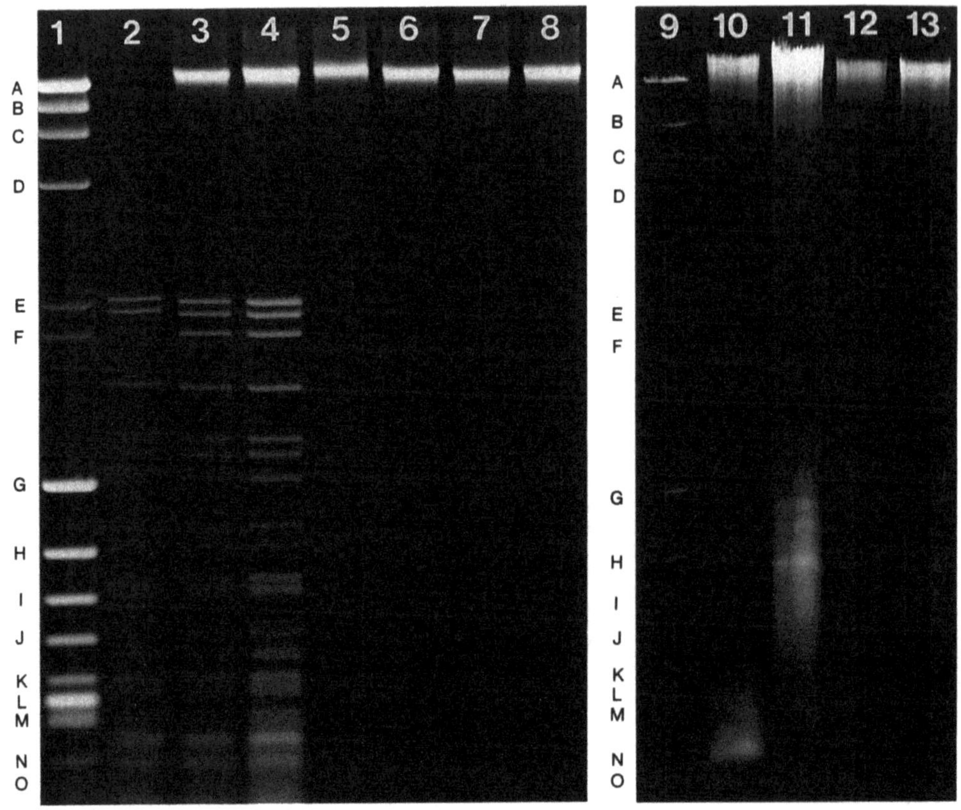

Figure 1. Digested DNAs fractionated with agarose gel electrophoresis. Tracks 2 through 8 and 10 through 13 contain DNAs digested with HpaII. Tracks 2, 3, and 4 have unmethylated lambda DNA (2 μg each) showing the absence of inhibition on HpaII by oat DNAs (tracks 3 and 4). Oat DNAs are as follows: 3, 5 and 10, gigantic Hannes; 4, 6 and 11, nonmutant Hannes; 7 and 12 gigantic Kyrö; 8 and 13, nonmutant Kyrö. Oat DNAs (about 0.4 μg) in samples 3 through 8 are from the seventh leaf laminas from the top, and about 0.7 μg DNA in samples in 10 through 13, from the topmost, eighth leaf laminas from the respective plants collected simultaneously. Tracks 1 and 9 are molecular size markers, 23100, 9420, 6560, 4360, 2320, 2030, 1060, 770, 610, 495, 392, 345, 341, 335 and 297 base pairs, indicated with A through O, respectively.

Discussion

Homozygous gigantic mutants may be useful as forage oats, if its seed could be produced economically. Evidently, lower latitudes will not provide an environment permitting such seed production, since it appeared that the gigantic line of cv Hannes made a weak, headless development when tried during two seasons at a lower latitude (45° N) in St. Paul, Minnesota, USA (P.G. Rothman, pers commun). Likewise, a 12-h photoperiod led nonmutant Hannes oats to phenocopy partly the *gig* mutant but did not induce an earlier flowering in the *gig* mutant (Ahokas, 1976a). Some other oat cultivars grown in Finland were strongly delayed by 9-, 10- or 12-h photoperiods (Doroshenko & Rasumov, 1929; Pohjakallio, 1951). A (9 + 1)-, 12- or 15-h photoperiod delayed heading efficiently in all spring-seeded *A. sativa* cultivars studied (Wiggans & Frey, 1955; Sampson & Burrows, 1972).

Florescence was not induced in the gigantic mutant by either cold vernalization, gibberellins or infection with *Ustilago avenae* (Pers.) Rostr. (Ahokas, 1976a). It is probable that there is no lack of a small diffusible or transported molecule concerned in the florescence induction of the gigantic mutant as judged from grafting and other mentioned experiments.

Extraneous factors as some contaminating polysaccharides may inhibit the digestion by restriction endonucleases (Do & Adams, 1991). Such factors don't simply explain the present observed difference dependent on tissue and genotype. The oat DNAs did not either contain any inhibitor preventing the digestion of unmethylated lambda DNA (Figure 1). The relationship between development rythm and DNA methylation extent is not yet understood. In the present material, there is at least leaf-position specificity and evident genotypic effect in the DNA methylation of many HpaII (MspI) sites in the lamina DNA of oats. The lack of digestion with EcoRII may mean general methylation in the inner C of the concerned recognition site.

The highest mean value of fecundity measured in *gig*/+ plants, 2.51 times that of the nonmutant (Ahokas, 1976a), would fix the *gig* allele at about 0.26 level in the family in an environment constantly of that kind and without selection. The values obtained in 1986 (Table 1) would fix the allele at a lower level, about 0.005. Temporal overdominance may in reality cause fluctuation between these estimates.

There may be some frequency-dependent or allelophatic phenomena in such genetic mixtures, which could be realized e.g. by the volatile compound(s) produced by the mutant (see Ahokas, 1976a), or by shadowing and by competition.

As the heterozygotes *gig*/+ are evidently heterotic, the question arises to whether the stable genetic constitution of formula *gig+*/*gig+* or *gig gig+*/*gig gig+* will also be heterotic. Such gene duplication or triplication might result from an induced rearrangement or, in the future, such constitutions may be produced through gene transfers of a cloned or synthetized sequence.

Acknowledgements

The work was partly done under the auspices of the Academy of Finland.

References

Ahokas, H., 1976a. Genetic and morphologic characteristics of the gigantic mutant of oats, *Avena sativa* L. J Sci Agric Soc Finland 48: 90–105.

Ahokas, H., 1976b. A way to mark the generation of the seed. Barley Genet Newsl 6: 96.

Blom Sørensen, M., 1992. Methylation of B-hordein genes in barley endosperm is inversely correlated with gene activity and affected by the regulatory gene Lys3. Proc Nat Acad Sci USA 89: 4119–4123.

Burn, J.E., D.J. Bagnall, J.D. Metzger, E.S. Dennis & W.J. Peacock, 1993. DNA methylation, vernalization, and the initiation of flowering. Proc Nat Acad Sci USA 90: 287–291.

Do, N. & R.P. Adams, 1991. A simple technique for removing plant polysaccharide contaminants from DNA. BioTechniques 10: 162, 164, 166.

Doroshenko, A.V. & V.I. Rasumov, 1929. Photoperiodism of some cultivated forms in connection with their geographical origin. Trud Prikl Bot Genet Sel 22: 219–276.

Gielo, S., 1969. Giant-type mutations in oats. Genet Polon 10: 62–68.

Kibite, S. & J.S. Taylor, 1994. Inheritance and linkage relationships of genes conditioning hullessness, multiflorous spikelet, and gigantism in oat (*Avena sativa* L.). Canad J Plant Sci 74: 497–500.

Kivi, E., 1968. Ähtärin 'ihmekaura' mielenkiintoinen mutaatio. Ominaisuuksia sensaatiomaisesti liioiteltu. Maaseudun Tulevaisuus 52 (98): 1, 10.

Lund, G., J. Messing & A. Viotti, 1995. Endosperm-specific demethylation and activation of specific alleles of α-tubulin genes of Zea mays L. Mol Gen Genet 246: 716–722.

Nelson, M. & M. McClelland, 1991. Site-specific methylation: effect on DNA modification methyltransferases and restriction endonucleases. Nucleic Acids Res 19: 2045–2071.

Nelson, M., E. Raschke & M. McClelland, 1993. Effect of site-specific methylation on restriction endonucleases and DNA modification methyltransferase. Nucleic Acids Res 21: 3139–3154.

Ngernprasirtsiri, J., R. Chollet, H. Kobayashi, T. Sugiyama & T. Akazawa, 1989. DNA methylation and the differential expression of C_4 photosynthesis genes in mesophyll and bundle sheath cells of greening maize leaves. J Biol Chem 264: 8241–8248.

Palmgren, G., O. Mattsson & F.T. Okkels, 1991. Specific levels of DNA methylation in various tissues, cell lines, and cell types of *Daucus carota*. Plant Physiol 95: 174–178.

Pohjakallio, O., 1951. Über den Einfluss der Umweltfaktoren auf die Dauer der Zeit von der Aussaat bis zum Ähren-(Rispen-)Schieben bei Sommergetreide. Soc Sci Fenn Comm Biol 11 (6): 1–18.

Rekunen, M., 1970. Kaura – Oats. Five-Year Rep Hankkija Plant Breed Inst Siemenjulkaisu 1970: 51–58.

Riggs, C.D. & M.J. Chrispeels, 1990. The expression of phytohemagglutinin genes in *Phaseolus vulgaris* is associated with organ-specific DNA methylation patterns. Plant Mol Biol 14: 629–632.

Rothman, P.G. & D.H. Bowman, 1963. An observation on hybrid vigor in oats. Crop Sci 3: 184.

Sampson, D.R. & V.D. Burrows, 1972. Influence of photoperiod, short-day vernalization, and cold vernalization on days to heading in *Avena* species and cultivars. Canad J Plant Sci 52: 471–482.

Sano, H., I. Kamada, S. Youssefian & H. Wabiko, 1989. Correlation between DNA undermethylation and dwarfism in maize. Biochim Biophys Acta 1009: 35–38.

Sardana, R., M. O'Dell & R. Flavell, 1993. Correlation between the size of the intergenic regulatory region, the status of cytosine methylation of rRNA genes and nucleolar expression in wheat. Mol Gen Genet 236: 155–162.

Schegalow, S., 1922. Das Erscheinen des Gigantismus beim Hafer. Z Indukt Abstamm Vererb 29: 207–208.

Schwanitz, F., 1954. Einige Beobachtungen zur Blütenbiologie und zur Sexualität diploider und polyploider Gigaspflanzen. Gartenbauwissenschaft 19: 73–90.

Schwanitz, F. & H. Pirson, 1955. Chromosomengrösse, Zellgrösse und Zellzahl bei einigen diploiden Gigaspflanzen. Züchter 25: 221–229.

Taylor, J.S., E.C. Yeung, S. Kibite & R.P. Pharis, 1995. Growth and development in an oat mutant expressing gigantism. Crop Sci 35: 688–691.

Wagner, D.B., G.R. Furnier, M.A. Saghai-Maroof, S.M. Williams, B.P. Dancik & R.W. Allard, 1987. Chloroplast DNA polymorphisms in lodgepole and jack pines and their hybrids. Proc Natl Acad Sci USA 84: 2097–2100.

Waterhouse, R.N., D. Boulter & J.A. Gatehouse, 1986. An organ-specific hypomethylation of cotyledon genomic rDNA in *Pisum sativum* L. FEBS Lett 209: 223–226.

Watson, J.C., L.S. Kaufman & W.F. Thompson, 1987. Developmental regulation of cytosine methylation in the nuclear ribosomal RNA genes of *Pisum sativum*. J Mol Biol 193: 15–26.

Wiggans, S.C. & K.J. Frey, 1955. Photoperiodism in oats. Proc Iowa Acad Sci 62: 125–130.

Zillinsky, F.J., 1959. Monster mutant in oats. Cereal News 4 (2): 7–11.

P.M.A. Tigerstedt (Ed.), Adaptation in Plant Breeding, 27–38, 1997.

Plant genetic adaptedness to climatic and edaphic environment

M. Pérez de la Vega
Universidad de León, Area de Genética, 24071 León, Spain

Key words: adaptedness, climate, soil, isozymes

Summary

Genetic adaptation implies the shaping of population and species gene pools in response to environmental challenges. The two components of the abiotic land environment are climate and soil, both of which determine much of the evolutionary adaptedness of plants as, besides representing a set of surrounding physical, chemical and sometimes limiting traits, they determine the availability of nutrients and energy, of which they are the immediate source. Ecogeographical distribution of species and ecotypes and different physiological mechanisms and developmental patterns are good evidence of plant adaptedness to soil and climate. However, it is not always easy to determine the underlying genetics of adaptive processes, because 1) environmental factors to which the plants are responding are not always evident and are sometimes too complex, 2) several genes may be involved in the response to a given environmental factor, and 3) the same gene/s may be involved in different adaptive responses. In particular, data on *Avena* species and temperature as a key environmental factor will be used to illustrate some examples of climatic and edaphic adaptedness. Temperature affects the genetic evolution and geographical distribution of all organisms, and a great deal of evidence indicates that species and populations are genetically adapted to different temperature regimes. Isozymes and other molecular markers have helped in the understanding of the genetic basis of adaptedness. There are many examples of correlation between isozyme and DNA-marker variation and environmental differences. For many population geneticists, isozyme markers are just genetic markers with little or no direct involvement in adaptation. However, metabolic processes are controlled by enzymes, influenced by the environment and used to react in response to it. Evidence that isozymes, and perhaps other molecular polymorphisms, are directly involved in adaptedness will be also presented. Molecular genetic analyses at gene and population levels are opening the ways to a better understanding of plant genetic adaptation.

Introduction

There is no doubt that plant species are adapted to climatic and edaphic conditions. The biogeographic distribution of plant species and, for instance, the colonization of serpentine soils are evidence of the evolution to adapt to different environments. I am also sure that there is general agreement that, within the natural distribution boundaries of a given species, different genotypes adapt differently to different environments. But there is no general agreement on the importance of different genetic components and mechanisms, physiological responses and environmental factors on adaptedness, perhaps because it depends on the genetic constitution of species and populations (itself a result of evolutionary processes) and the particular habitats.

Plants follow environmental changes in two ways: 1) by individual processes of phenotypic plasticity expression or, 2) by populational processes over generations (evolutionary changes). However, for many biologists, evolutionists and plant breeders it is debatable how much of the response to environmental conditions is due to genotypic differences and how much is due to phenotypic plasticity. Quantitative characters present a particular difficulty since they often show high plasticity. Over the last decades allozyme and DNA markers have become more and more frequently used in populational and evolutionary studies and as an aid in breeding programs. Unlike quantitative characters, isozyme and DNA markers are generally not affected by the environment and are under simple genetic controls. They therefore represent an advantage in the study of adaptive populational processes although, as we shall see later, they also have drawbacks.

According to Allard (1988), adaptedness is the degree to which an organism is able to live and reproduce in a given set of environments, the state of being adapted, and adaptation is the process of becoming adapted or more adapted. For crop species adaptation can sometimes by interpreted in a different way because crop adaptation is oriented to optimize productivity, and productivity is always related to the ability to live but not always to the ability to reproduce (e.g., many horticultural crops). In any event, for both crop and wild species, adaptedness means matching the successive developmental stages to climatic and edaphic resources and, where unfavourable extremes are unavoidable, minimizing their coincidence with more vulnerable stages (Roberts et al., 1993).

Phenotypic plasticity was defined by Bradshaw (1965) as 'the amount by which the expressions of individual characters of a genotype are changed by different environments'. But much confusion exists as to what phenotypic plasticity is. Two major types of phenotypic responses have been considered in the literature: 1) the norm of reaction, that is, the array of phenotypes that are produced by a single genotype across a range of environments and, 2) 'noisy plasticity', the variability in phenotype within the current environment rather than the production of a particular phenotype. The first type supposes that the environments involved are repeatable and predictable, while the second is the result of the response to largely unpredictable variability within environments. The benefits obtained through 'noisy plasticity' are likely to be very different from the advantages afforded a genotype by having a particular set of phenotypes across a range of environments (Via, 1994). There are additional causes of confusion: 1) It is not always easy to know what kind of phenotypic plasticity is referred to by authors. 2) Both kinds of plasticities, norm of reaction and noisy plasticity, cover a number of processes in response to environmental conditions, often including reversible physiological changes and changes during development that are often irreversible. 3) Imprecise use of terms, for instance, the term acclimation is used to describe the increase of resistance to harsh environmental conditions after exposure to periods of sub-lethal stress, but it is often used to describe any phenotypic adjustment to seasonal changes (Hällgren & Oquist, 1990; Hoffman & Parsons, 1991).

Adaptation implies changes in population genetic pools in response to environmental changes. Plasticity is sometimes, and in particular 'noisy plasticity', a buffering mechanism against environmental changes, whereby plastic genotypes can respond to environmental changes with few or no changes in the population genetic pool. But adaptedness and plasticity are not opposite concepts. Plasticity can itself be an adaptive mechanism under genetic control since the ability to respond to environmental signals is genetically based and the extent of that response can evolve by natural selection (Via 1994; Emery et al., 1994). For some authors plasticity is useful in particular in the adaptation to environments with very unpredictable or irregular changes shorter than the population generation time (Bradshaw, 1965; Platenkamp, 1990).

Climate and soil are the two sets of abiotic factors plants have to cope with. Each of them influences the distribution of plant species and the adaptive genetic pool of populations in different ways. They are not totally independent either of each other or in their effects on plants, e.g., the ability of plants to withstand high temperature stress is regulated in part by the soil water availability (Burke, 1990). Edaphic characteristics change slowly in the course of time but sometimes very fast in space. They fit the coarse-grained spatial variation in which individuals complete development in only one of the patch types. On the other hand, climatic variables can change very fast in short periods of time but usually much more slowly in space. Climatic components are cyclic and predictable (i.e., photoperiod), or cyclic but unpredictable or irregular (i.e., frost, rainfall, etc.), at least in temperate climates. Adaptation to environmental conditions with predictable spatial and/or temporal variation (or no variation) is more straightforward since both direction and intensity of selection are constant or nearly so, while adaptation to unpredictable and irregular conditions is more difficult owing to the variability in selection pressure.

The evaluation of adaptedness and/or plasticity has been approached in different ways, depending chiefly upon the scientific background of researchers. While ecologically oriented researchers prefer reciprocal transplant experiments, geneticists search for consistent associations between particular genotypes and environments, and quantitative geneticists and plant breeders rely on the analysis of variance and genotype × environment interactions. Since I include myself in the second group of researchers above mentioned, I shall mainly approach the topics of adaptation to climatic and edaphic conditions from the search for genotype-environment associations.

Climatic adaptations

The full understanding of adaptation mechanisms is difficult for many different reasons. 1) In many cases, the environmental factor to which the organism is responding is either not evident or is too complex (that is, there are several factors acting on the populations) to be easily measured. 2) Not all the developmental stages are equally sensitive to environmental factors. 3) Several genetic and physiological systems seem to be involved in the control of response to some environmental factors. 4) The same genes and response mechanisms can be adaptive to two or more different environmental variables. Surely at any moment, each population is being affected by several environmental factors, among which the most intense or stressful one determines the predominant selective trend, probably masking the effect of many other factors. An example of this was provided by an experimental barley population in which in addition to total rainfall, several environmental factors (temperature during flowering and/or maturation among them) affected selective values at different loci (Allard et al., 1992). The response to temperature provides a good example of the complexity of genetic mechanisms implied in adaptation. Thus, Shonnard & Gepts (1994) pointed out that heat tolerance during flower bud formation and pod filling in beans (*Phaseolus vulgaris*) displayed continuous variation, indicating quantitative inheritance, and also cytoplasmic effects including interactions of cyto plasmic with nuclear genes. But they pointed out (see references in Shonnard & Gepts, 1994) that previous studies on beans had reported that heat tolerance at pod set was due to one or two epistatic genes depending upon accessions, or to polygenic quantitative inheritance. Likewise, in cowpea, tolerance of flower bud development to high temperature was due to one recessive gene, but tolerance during pollen formation was controlled by a single dominant gene. Palta & Simon (1993) pointed out that while in the past there has been general agreement that freeze resistance involved multiple genes, recent results suggest that as few as three or four genes or linkage groups could be responsible for freeze tolerance and capacity to cold-acclimate, at least in wild *Solanum* species and *Pisum sativum*. Likewise they indicated that in several crop species it is possible to demonstrate that freeze tolerance and the capacity to cold-acclimate segregated independently. This means that, at least in part, the two mechanisms are controlled by different gene sets. Most probably, the response to temperature, like other environment components, involves several major and many quantitative genes plus epistatic interactions. What we see in each segregation is a portion of the whole puzzle, the effect of the genes in which the parental lines differ. Each piece of research gives us information on some genes and their effects, not on all the genes involved in a given adaptive response. But the converse is also true, a given gene may be involved in the response to several environmental variables. Thus, for instance, in *Arabidopsis* genes *rd29A* and *rd29B* are induced, though differentially, by dehydration, low temperature and high salt treatments (Yamaguchi-Shinozaki & Shinozaki, 1994). Work has begun to bear fruit on the solution of some adaptive puzzles but none is close to completion.

Life cycle components are esential to plant adaptation. One of the most important adaptive mechanisms in plants, and perhaps the most important avoidance mechanism, is seasonality, the matching of favourable periods to life cycle stages. Annual species may avoid periods of potentially damaging environmental conditions by completing their life cycle entirely within the period of favourable conditions. Perennial species, on the other hand, often have the capability of going into a dormant state. A clear example of seasonality, adaptedness and its influence in yield is flowering time. The flowering phase is especially vulnerable to environmental stress, and has a direct effect on harvest time and final yield in many important crops. The time when a crop flowers is determined, almost exclusively, by the genetically controlled responses to day length and temperature (Roberts et al., 1993), but it is possible to select the response. Lentils and wheat are two of many possible examples. After domestication, the dissemination of the lentil crop from West Asia to lower latitudes depended on selection for intrinsic earliness and reduced sensitivity to photoperiod; and to higher latitudes on a modest reduction in photoperiod sensitivity and an increase in temperature sensitivity (Erskine et al., 1994). In fact, earliness is one of the most important adaptations in some harsh environments. Likewise, Kato & Yokoyama (1992) indicated that early heading in Asian wheat landraces was a selective response to avoid reduction of grain yield caused by water deficit in areas with less than 500 mm of rainfall, while in other areas this response was to avoid pre-harvest sprouting due to rain in early summer, constituting an example of how the same response mechanism can be of adaptive value to different environmental conditions.

Since fitness differences among genotypes tend to be maximized under extreme environmental conditions (Parsons, 1987), adaptive variability is expected to fulfil at least two premises: 1) favourable environments are correlated with higher genetic variability within and between populations, especially if they are heterogeneous, and 2) adaptedness is more easily observable in less favourable environments, in which the greater selection pressure eliminates less adapted genotypes and reduces genetic variability.

I would like to illustrate these points with our data on slender wild oats. Log-linear analyses of the associations among 14 allozyme loci in Spanish populations of *Avena barbata* have indicated that both temperature and rainfall are significantly associated with specific allelic combinations at three or four loci, and that temperature and rainfall affect many of these multilocus allelic combinations in the same direction in Spain and in California; e.g., in both countries higher or lower rainfall and temperature are significantly associated with the respective allele combinations at significant two- and three-locus terms (*Pgd1-Lap1-Est1, Pgd2-Prx1-Mdh1* and *Acp1-Mdh1* with temperature, and *Pgd1-Prx1-Acp1, Pgd1-Acp1-Est1* and *Pgd1-Prx1-Mdh1* with rainfall) of the multilocus structure model obtained by a multivariate log-linear analysis (Pérez de la Vega et al., 1991). But both the more diverse and more patchy environments of Spain produce a much more complicated pattern of structuring genetic variability than exists in California (García et al., 1989; Allard et al., 1993; Allard, this issue). This is particularly the case in southern Spain where there is a favourable climate for many diploid, tetraploid, and hexaploid *Avena* species. In this area, with more favourable climatic conditions and higher habitat diversity, genetic variability within and between populations is higher than in the less favourable (in fact diploid species except *A. hirtula* are not present, and even this species is very infrequent) and more uniform northern area. Thus, the average number of *A. barbata* fourteen-locus associations per population in the southern and northern areas are 13.2 and 6.8, respectively.

Allard has already mentioned the 'cold tolerant' 14-locus genotype of *A. barbata* which is highly frequent in the northern part of the Spanish Central Plateau. The adaptedness to this harsh cold area is clearer when an association of four loci is taken into consideration. The four-locus genotype, here designated as 'northern genotype', is *Pgd1 1133, Lap1 2233, Prx1 1111* and *Est1 5577*. These four loci are polymorphic in Spain and in California, and they are part of the allozyme five-locus association involved in environmental adaptations in California. The 'northern genotype' is exclusively found inside or very close to the boundary of the area in which there are more than 80 days of frost per year and the average spring temperature is rather low (average of lower temperatures during May below 6° C), and the genotype frequency drops to zero less than 30 km outside this boundary without apparent changes in any other main environmental variable (Figure 1). Additional evidence supporting the adaptedness of this particular four-loci genotype to low temperature is provided by a set of 10-year data from a population from León, a locality near the above-mentioned boundary. This population shows a relatively high polymorphism, as expected because of its favourable (mesic fields with deep soil) and micro-heterogeneous habitat (open and shadow zones, hay fields and path-sides). The fluctuation of the four-locus genotype frequency in this population closely follows the fluctuation of average lower temperatures during the previous season (from October to June). That is to say, the genotype frequency increases if the season previous to the generation in which adult plants are collected is cold, and conversely decreases if it is mild. These results agree with expectations if in the cooler seasons the fitness of the 'northern genotype' increases leaving more offspring with this genotype. *A. barbata* is an annual self-fertilizing species whose populations are almost exclusively formed by homozygous individuals thus, except for a very few if any seeds originated by outcrossing, all the offspring of a given plant carry its genotype. Therefore, the next season the northern genotype will represent a higher proportion of the plant census on the population. Similar changes, in which selection appeared to move the population in the direction of a multilocus structure appropriate to the set of environmental factors acting each year, was described for an experimental barley population, although in this case the main factor seemed to be rainfall (Allard et al., 1992).

A correlation between the distribution of multilocus associations and temperature has also been observed in Spanish winter landraces of cultivated oats (*A. sativa*) (Pérez de la Vega et al., 1994). In this case two alternative allelic multilocus associations involving up to six loci (*Adh1, Lap1, Pgm1, Mdh1, Mdh2*, and *Prx1*) were correlated respectively with the warmer areas of Spain (named Badajoz association) and the cooler

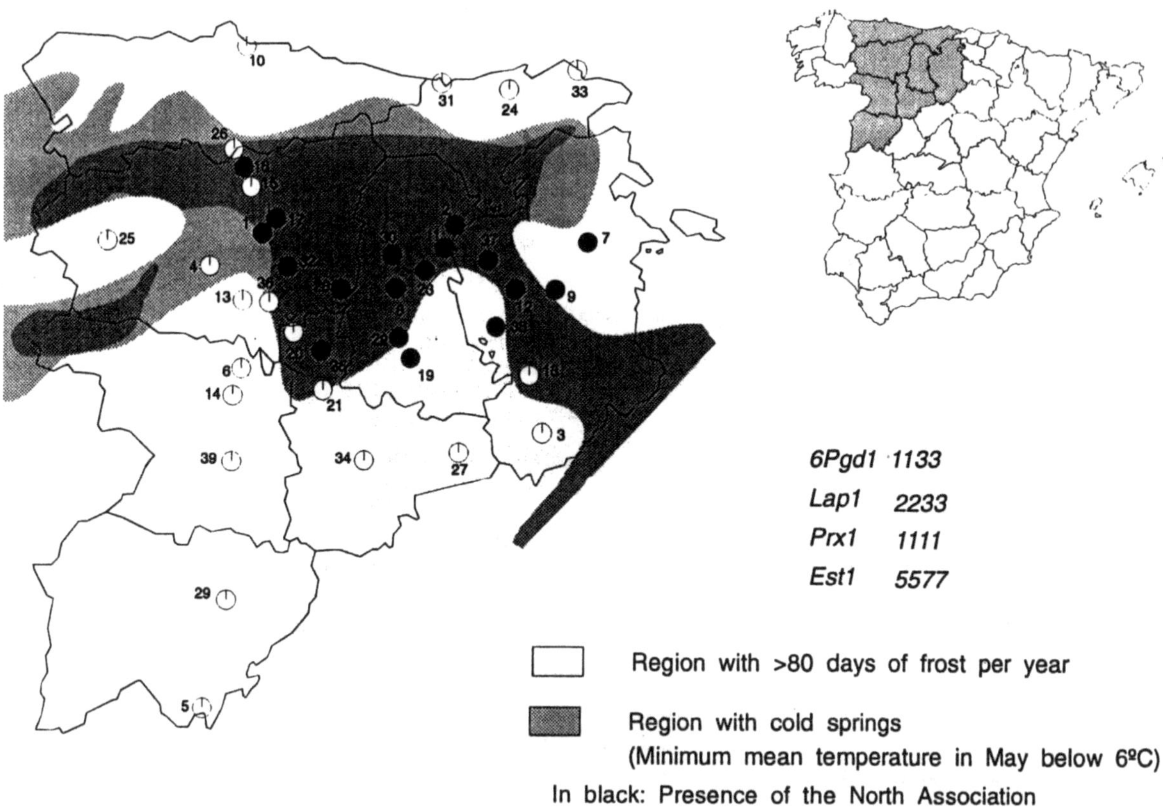

Figure 1. Distribution of *A. barbata* populations in which the northern genotype was found (black circles).

ones (named Soria association). Multivariate log-linear analyses indicated that allelic combinations at four loci (*Lap1, Pgm1, Mdh1* and *Mdh2*) were tightly associated among themselves and in relation to environmental factors. A summary of the data of the five provinces represented by 20 or more populations in the analysed collection is compiled in Table 1. These and previously published (Pérez de la Vega et al., 1994) data suggest that temperature, and in particular a threshold in the number of days of frost, determines the preferential distribution of the alternative multilocus genotypes. The four-locus Badajoz association could easily be extended to a six-locus one by adding loci *Prx1* and *Adh1*, since 93.30% of the four-locus plants of these five provinces have also the six-locus allelic association. On the contrary, only 39.21% of the plants with the four-locus Soria association were included in the extended six-locus one. This fact, and data on the geographical distribution of four- and six-locus genotypes in Spain (Pérez de la Vega et al., 1994) suggest the hypothesis that while allelic combinations at the four loci are responsible for the adaptation to temperature regimes, adaptation to other environmental factors in Southern Spain involves alleles at two additional loci.

Our data on Spanish *Avena* species point to temperature as the environmental variable of major underlying significance in Spain. The most decisive factor in the multilocus genotype adaptive distribution of *A. barbata* in California is rainfall, but perhaps rainfall and its temporal distribution is not such a limiting factor in Spain, so temperature then emerges as a more determining factor.

In addition to the data on Spanish and Californian slender wild oats and on Spanish *A. sativa*, there is a set of recent works on *Avena* species suggesting adaptive responses to environmental factors in which populations were collected or grown. Table 2 summarizes these data.

The next example relates life cycle characteristics to temperature adaptedness through a physiological mechanism: heat production. Nevo et al. (1992) studied the heat production of accessions of wild *Hordeum spontaneum* and *Triticum dicoccoides* collected in cold and warm places. They found that seedling heat pro-

Table 1. Percentages of *Avena sativa* plants with each multilocus association

	Province				
	BA	TO	SO	ZA	CA
Number of populations	48	22	20	38	20
Badajoz four-locus	73.26	10.22	0.00	8.75	8.29
Badajoz six-locus	69.18	8.44	0.00	8.03	7.32
Soria four-locus	4.08	33.78	70.73	27.68	28.29
Soria six-locus	2.65	9.33	40.49	3.03	21.46
Average temperature	16.8	15.0	10.5	14.8	17.1/11.4
Days of frost	10.7	32.6	92.1	19.6	2.7/50.0

Castellón (CA) is a very diverse province in geography and climate. Climatic data reférs respectively to localities near the Mediterranean Sea and in the inner mountain areas.

Table 2. Data on *Avena* species suggesting adaptive responses to environmental factors

Species	Materials from:	Marker type	
Avena sterilis[1]	World collection	Morphological	Interrelationships between genetic and ecological factors.
Avena byzantina[2] *Avena sativa*	USA and Canada	Quantitative	Large groups that generally correspond to latitude of origin or adaptation.
Avena byzantina[2] *Avena sativa*	USA and Canada	Qualitative (isozymes, proteines, morphol.)	Spring and fall cultivars are genetically different groups.
Avena byzantina[3] *Avena sativa*	USA	RFLP	Spring and fall cultivars separated. Only moderate correlation between genetic distance and coefficient of parentage.
Avena sativa[4]	USA	Production traits	Oat breeding since 1914 primarily has improved oat cultivars to perform (adaptedness) in stressful environments.

[1] Rezai, A. & K.J. Frey, 1988. Euphytica 39: 113–118.
[2] Souza, E. & M.E. Sorrells, 1991. Crop Sci 31: 599–605; 31: 605–612.
[3] O'Donoughue, L.S. et al., 1994. Crop Sci 34: 1251–1258.
[4] Lynch, P.J. & K.J. Frey, 1993. Crop Sci 33: 984–988.

duction was significantly higher at 5° C than at 24° C, that wild barley generated significantly more heat than wild wheat at both temperatures, and that at 24° C the wild barley from warm places generated more heat than accessions from cold ones. They hypothesized that both inter- and intraspecific differences in heat production were adaptive and evolved by natural selection. The intraspecific genetic polymorphism in heat production between wild barley populations presumably enables the steppe and desert populations to complete their annual cycle earlier than the cold mountain populations; thus, early ripening is adaptive in that it allows for avoiding heat and severe drought with desiccation in early summer in the desert. They also hypothesized on the adaptive significance of higher heat production at 5° C: it may enhance growth when it is limited by the ambient temperature. Likewise, the differences between wild barley and wild wheat may largely explain the differences in the geographical distribution of both wild and cultivated gene pools of the two cereals.

Edaphic adaptation

Edaphic factors also play a decisive role in determining plant adaptedness. There are classical examples of how stressing soil components, such as high concentrations of heavy metals, can lead to a rapid adaptive genetic differentiation among plant populations, even among closely located populations of outbreeding species. The ability to colonize these stressful habitats is limited to those species which have the genetic potential to become adapted to such contaminated areas. Thus, for instance, metal tolerance seems to be almost limited to three families (Poaceae, Brassicaceae and Caryophyllaceae) among those found in the United Kingdom. Examples of populations adapted to toxic levels of heavy metals have been described in species of the genera *Agrostis, Anthoxanthum, Deschampsia, Festuca, Holcus, Mimulus, Silene*, among others, and as being due to either polygenes or major genes. Although all these species can evolve metal tolerance, not all of them are found on all contaminated sites, almost certainly because they have other edaphic requirements which prevent them from colonizing certain areas, although their absence is sometimes due to the lack of the appropriate genes for metal tolerance in populations near toxic sites (see Al-Hiyaly et al., 1990; Meharg et al., 1993; Turner, 1994, and references in these papers).

Adaptation to edaphic conditions can also occur rapidly in response to non-stressing conditions. Thus, for instance, *Anthoxanthum odoratum* showed considerable genetic change within populations six years after a new liming treatment had been used (Snaydon & Davies 1982). Another example is *Plantago lanceolata*, an outbreeding species and a natural component of grazed pastures and hay fields with a cosmopolitan distribution in a wide range of habitats. In this species, although highly plastic, genetically determined differences in plant development are maintained and have a great impact on lifetime seed yield at three different grassland sites which differ in soil characteristics, as has been indicated by reciprocal transplant experiments, directed to the study of genetic differentiation and adaptation to local conditions (van Tienderen & van der Toorn, 1991a, b). Soil components can also affect adaptive components through maternal effects since variation in soil nutrient level may change seed quality and seedling survival without necessarily affecting seed mass (see Aarssen & Burton, 1990). Likewise, soil pH is another important environmental factor in plant adaptation and indeed was the single most important factor in the adaptation of seven complete and twelve substituted triticales across Spain. In this man-made species the adaptation to acid and alkaline soils seems to be largely controlled by the single wheat/rye chromosome 2D(2R) substitution, for which both types differ. Complete triticales seem better adapted to acid soils, whereas substituted types are generally more suited to alkaline soils (Royo et al., 1993).

Salinity is one of the main edaphic factors limiting species distribution, and crop cultivation and yield. It is becoming a more and more serious problem for agriculture as more land is becoming salinized owing to inadequate agricultural practices. The possibility of selecting and breeding for enhanced salinity tolerance has received considerable attention over the last decade (Shannon, 1985). Fortunately, some degree of genetic variation to salinity tolerance is present in many crop species and it can be rapidly screened and used to select better cultivars genetically adapted to higher salt concentration (McNeilly, 1990; Al-Khatib et al., 1993).

Edaphic-biotic factors

Edaphic adaptedness not only involves the soil, understood as a set of chemical and geological components, but also the biota component, in particular fungi and bacteria. It is now clear that an understanding of micro-organism effects is of considerable value in increasing the efficiency, reliability, and effectiveness of plant breeding and cultivar evaluation (Funk et al., 1994). The influence of rhizosphere micro-organisms on plant growth and plant competitive ability in natural plant communities is substantial and biotic interactions between roots and associated microorganisms can result in positive or negative impacts on plant productivity. The list includes, not only *Rhizobium* and other nitrogen-fixing microbes, but also other bacteria and mycorrhizal fungi (Chanway et al., 1991). Certain lines of evidence suggest that these micro-organisms represent the limiting factor in the distribution range of many plant species (Chanway et al., 1991). Likewise, it has been found that the heavy metal tolerance of certain plants may vary with mycorrhizal status and the role of ectomycorrhiza is well established in the protection of trees from the effects of high external concentrations of heavy metals (Turner, 1994). Plant growth-promoting rhizobacteria can be very useful in increasing the yield of adapted crop varieties. Pure cultures of *Bacillus* isolated from the rhizosphere of a bread wheat cultivar, promoted root growth when inoculated on seeds of the same cultivar, but did not when a parental wheat cultivar was tested. Some of the *Bacillus* strains increased seedling emergence, and others increased shoot weight and tiller number (Chanway & Nelson, 1990). Soil biota influences Douglas-fir (*Pseudotsuga manziesii*) seedling growth. Seedling biomass of two Douglas-fir ecotypes was greatest in unpasteurized soil collected from the same area as the seed. In addition an ecotype responded specifically to biota present only in the corresponding soil (Chanway & Holl, 1992).

Finally, in understanding the relationships between soil, plants and micro-organisms, we cannot forget the micro-organisms are themselves living beings and therefore subjected to adaptation. Thus McArthur et al. (1988), in research into the genetic variation of the soil bacterium *Pseudomonas cepacia*, indicated that the relationship between genetic diversity and environmental variation suggested a pattern of microgeographical adaptation due to selection.

Isozymes: adaptive roles

In addition to the examples of *Avena* mentioned here and in Allard (this issue), other crop and related wild species in which biochemical and molecular polymorphisms have been correlated with ecogeographical factors are *Hordeum vulgare – H. spontaneum* and *Triticum dicoccoides*. This research, most of it contributed by the groups led by Allard and by Nevo respectively, have demonstrated by using different statistical methods but similarly oriented studies that isozymes, ribosomal intergenic spacer, and even storage seed protein polymorphisms are distributed nonrandomly, that at least part of this variation should be adaptive in response to environmental challenges, and correlated with a range of ecogeographical factors (see Allard et al., 1992; Chalmers et al., 1992; and references in these papers). These grass species, and others such as *Lens culinaris*, are predominantly self-fertilizing species and in many cases multilocus associations were involved in the adaptive evolution. There are also examples of correlation between the distribution of molecular polymorphisms and clinal variation of environmental parameters in cross-fertilizing species, mostly forest trees. Unlike self-fertilizing species in which gene flow among populations is greatly limited, in cross-fertilizing species allozyme frequency clines have been attributed, in addition to adaptive causes, to gene flow between previously isolated populations differing in allele frequencies.

Although it is generally accepted that at least part of the molecular variation should be adaptive, there is no agreement as to how much of the variation is adaptive. Furthermore, the direct relationship between allozymes (and other molecular genetic markers) and adaptation has been, and still is, an uneasy hypothesis. Since standard population genetic theory suggests that most of the molecular genetic variation is adaptively neutral, and that adaptedness results from the accumulation of many alleles, each of small effect (polygenes), and adaptedness due to major genes should be unusual, to admit that a few 'qualitative' Mendelian genes are directly involved in plant adaptation is heterodox. Thus, hitch-hiking and other genetic mechanisms have been invoked as alternative hypotheses to explain the adaptive role of allozymes from the early times of populational studies on plant allozyme variation (Hedrick & Holden, 1979). However, the role of major genes in adaptation has been claimed again recently (Macnair, 1991; Orr & Coyne, 1992) it being pointed out that although the polygenic view of adaptation may be correct, it is not strongly supported by evidence, and that major genes are particularly important in the adaptation to rapid changes in the environment (Macnair, 1991). In such situations, populations must adapt rapidly or become locally extinct, or fail to colonize.

Another major argument against the adaptive role of molecular polymorphisms is that, although significant correlations suggest adaptedness, they cannot establish it unequivocally: correlations do not prove causality by themselves. A significant problem in understanding the direct implications of enzyme polymorphisms in adaptation is the gap between the kind of information obtained by population geneticists, in the form of alternatives in electrophoretic mobility, and by plant biochemists and physiologists, in the form of kinetic and metabolic properties. Protein function and stability are the sum of multiple intramolecular interactions, many of which may be affected when even a single amino acid residue is altered, and we should not forget that, although it is true that differences in electrophoretic mobility do not always imply differences in enzyme properties, the opposite is also true: the lack of electrophoretic differences does not imply the lack of shifts in kinetic properties, as Dubuc et al. (1988) demonstrated for the malic enzyme of populations of *Echinochloa crus-galli* from sites of contrasting climates. An additional problem is that sometimes selection may not act on individual genes but on allelic and non-allelic combinations of them, as was mentioned earlier in this report, in Allard's article (this issue) and in other studies (see references in Allard et al., 1992; Allard et al., 1993). Thus, the effect on adaptedness of allozymes may be different depending upon interactions with other enzymes, making it difficult to ascertain the individual contribution of each gene.

Correlations between enzyme properties and/or alleles and temperature are relevant examples since plant adaptedness along latitudinal and altitudinal tem-

Table 3. Examples of correlation between enzyme variants and environmental factors

Species	Characteristics	Enzyme	Differences in:
Arabidopsis thaliana[1]	Different climates	MDH	Energy of activation.
		GOT	Concentration in cold adapted genotype.
Bromus mollis[2]	Fast vs. slow allele response to flooding and temperature	ADH	Enzyme activity. Dry matter production. Seed germination.
Echinochloa crus-galli[3]	Quebec vs. Mississippi	PPDK	Rate of activation. Cold inactivation.
Echinochloa crus-galli[4]	Quebec vs. Mississippi	PEPC	Thermostability. Energy of activation. K_m, V_{max}/K_m.
		MDH	Energy of activation. K_m, V_{max}/K_m.
		ME	Energy of activation. V_{max}/K_m.
		PPDK	Thermostability.
Lathyrus japonicus[5]	Different latitudes	MDH	Thermal properties.
Triticum[6]	Spring vs. winter	MDH	Thermostability. Energy of activation. V_{max}/K_m.
		GOT	No differences.
Typha latifolia[7]	Different climates	MDH	Thermostability. Activation energy.
Typha[8]	Different distributed species	MDH	Thermostability.
Viola[9]	Different distributed	MDH	V_{max}/K_m.

[1] Simon, J.-P. et al., 1983. Oecologia 60: 143–148.
[2] Brown, A.H.D. et al., 1976. Aust J Biol Sci 29: 389–396.
[3] Simon, J.-P. & M.D. Hatch, 1994. Aust J Plant Physiol 21: 463–473.
[4] Potvin, C., J.-P. Simon, 1990. Evol Trends Plants 4: 98–105.
[5] Simon, J.-P., 1979. Oecologia 39: 273–287.
[6] Simon, J.-P. et al., 1989. Environm Exptl Bot 29: 445–456.
[7] McNaughton, S.J., 1974. Am Nat 108: 616–624.
[8] Liu, E.H. et al., 1978. Am J Bot 65: 214–220.
[9] Simon, J.-P. et al., 1986. J Ecol 74: 19–32.

perature gradients are evident. In particular it has been pointed out that enzymes are tailored through evolutionary time to fit a particular thermal environment (Simon, 1979), that many lesions in plants due to high temperature are due to the inactivation of thermolabile enzymes, and that mutations conferring thermolability tend to accumulate in organisms not subjected to high temperatures (Parsons, 1987). Table 3 shows some examples of differences of enzyme properties mainly correlated with differences in habitat temperature.

One of the clearest signs of differential adaptedness of plant allozyme variants at populational level has been recently contributed by Bergmann & Gregorius (1993). They found that a latitudinal frequency cline observed for two alleles of an IDH locus in European *Abies alba* suggested a correlation between temperature gradient and thermostability of the two allozymes. Using extracts from endosperms of homozygous trees incubated under different temperatures and followed by electrophoresis, it turned out that the allozyme which prevailed in the southern populations distinctly exceeded the alternative allozyme in thermostability in trees from both extremes of the climatic transect. This result strongly supports IDH polymorphism as a result of adaptation to temperature regimes.

The implications of enzyme kinetic properties, thermostability and optimum temperature are relevant in adaptedness and crop production. It has been suggested that the apparent Michaelis K_m can be used to characterize the temperature adaptation of the organism from which the enzyme was extracted. Likewise other parameters such as the ratio V_{max}/K_m and the thermal kinetic windows are important to adaptedness and crop production. Burke et al. (1988) defined the thermal kinetic window (TKW) of optimal enzyme function as the temperature range in which the value of the apparent K_m was within 200% of the minimum apparent *Km* value observed for the enzyme, and Burke (1990) indicated that the TKW is a satisfactory measure of the thermal optimum of cellular metabolism. Temperatures above the TKW impact all aspects of the photosynthesis system in plants. Different species such as cotton or wheat have different thermal kinetic windows. Therefore, the identification of the TKW for each crop species can aid in the interpretation of differential temperature responses of crop species (Burke, 1990). However, it has been argued that an increase in apparent K_m will not necessarily reflect any impairment of enzyme function *in vivo* (Turner et al., 1994).

Restriction fragment length polymorphisms (RFLP) may probably represent a set of genetic markers closer to neutrality than isozymes. However, they have been in some way correlated with the environmental conditions in which species or cultivars grow. In particular, length polymorphism of the ribosomal intergenic spacer (IGS) has been related with a range of ecogeographical factors and/or changes observed following natural or artificial selection in *Avena barbata, Hordeum spontaneum, H. vulgare* and other grass species (see Cluster & Allard, 1995; Chalmers et al., 1992; Saghai-Maroof et al., 1990 for references), and with different salinity in a completely different group of species or subspecies of the genus *Salicornia* (Davy et al., 1990). In these latter species the frequency of individuals with a fragment of 12 kb is strongly associated with the frequency and duration of tidal flooding. Is the IGS always an unselected marker linked to adaptively selected genes, as some authors indicate, or is it sometimes adaptive by itself? Recent studies (Melchinger et al., 1994; O'Donoughue et al., 1994) showed that several tenths of RFLPs clearly separated spring and winter cultivars of European barley (48 cultivars) and American hexaploid oats (83 cultivars). On the contrary only a moderate correlation between genetic distance and the coefficient of parentage or coancestry was found (O'Donoughue et al., 1994; Graner et al., 1994). Both groups argued unrealistic assumptions or inaccuracies in estimating the coefficient as the main cause for this lower correlation. I would like to open an alternative hypothesis. Could adaptive selection to different growth conditions be more important in molding the genotype than ancestry-progeny parentage? The current importance of DNA markers and DNA polymorphisms (RFLP, RAPD, etc.) in basic and applied genetic studies is known. The relevant question from the point of view of this work is: What proportion of this variation at DNA level is adaptive and what is strictly neutral? It is obvious that in a near future DNA markers will serve not only to locate quantitative trait loci but also to identify, isolate and study them.

It is obvious that to complete adaptive puzzles we need a better understanding of the genetic mechanism of adaptation and the physiological processes by means of which adaptedness and plasticity are expressed, which is of considerable urgency given the prospect of possible rapid climatic changes. The alternative of 'migration', that is to cultivate different genotypes or species better adapted to the new conditions of a given area, is possible for most crop species; but the change may be too fast for many wild species, and in particular it can be particularly dramatic for forest trees (Stettler & Bradshaw, 1994).

References

Aarssen, L.W. & S.M. Burton, 1990. Maternal effects at four levels in *Senecio vulgaris* (Asteraceae) grown on a soil nutrient gradient. Am J Bot 77: 1231–1240.

Al-Hiyaly, S.A.K., T.M. McNeilly & A.D. Bradshaw, 1990. The effect of zinc contamination from electricity pylons. Contrasting patterns of evolution in five grass species. New Phytol 114: 183–190.

Al-Khatib, M., T. McNeilly & J.C. Collins, 1993. The potential of selection and breeding for improved salt tolerance in lucerne (*Medicago sativa* L.). Euphytica 65: 43–51.

Allard, R.W., 1988. Genetic changes associated with the evolution of adaptedness in cultivated plants and their progenies. J Hered 79: 225–238.

Allard, R.W., P. García, L.E. Sáenz de Miera & M. Pérez de la Vega, 1993. Evolution of multilocus structure in *Avena hirtula* and *Avena barbata*. Genetics 135: 1125–1139.

Allard, R.W., Q. Zhang, M.A. Shagai Maroof & O.M. Mouna, 1992. Evolution of multilocus genetic structure in an experimental barley population. Genetics 1131: 957–969.

Bergmann, F. & H.R. Gregorius, 1993. Ecogeographical distribution and thermostability of isocitrate dehydrogenase (IDH) allozymes in European silver Fir (*Abies alba*). Biochem Syst Ecol 21: 597–605.

Bradshaw, A.D., 1965. Evolutionary significance of phenotypic plasticity in plants. Adv Genet 13: 115–153.

Burke, J.J., 1990. High temperature stress and adaptation in crops. In: R.G. Alscher & J.R. Cummings (Eds). Stress Responses in Plants: Adaptation and acclimatation mechanisms, pp. 295–309. Wiley-Liss, New York.

Burke, J.J., J.R. Mahan & J.L. Hatfield, 1988. Crop-specific thermal kinetic windows in relation to wheat and cotton biomass production. Agron J 80: 553–556.

Chalmers, K.J., R. Waugh, J. Watters, B.P. Foster, E. Nevo, R.J. Abbott & W. Powell, 1992. Grain isozyme and ribosomal DNA variability in *Hordeum spontaneum* populations from Israel. Theor Appl Genet 84: 313–322.

Chanway, C.P. & F.B. Holl, 1992. Influence of soil biota on Douglas-fir (*Pseudotsuga menziesii*) seedling growth: the role of rhizosphere bacteria. Can J Bot 70: 1025–1031.

Chanway, C.P. & L.M. Nelson, 1990. Field and laboratory studies of *Triticum aestivum* L. inoculated with co-existent growth-promoting *Bacillus* strains. Soil Biol Biochem 22: 789–795.

Chanway, C.P., R. Turkington & F.B. Holl, 1991. Ecological implications of specificity between plants and rhizosphere microorganisms. Adv Ecol Res 21: 121–169.

Cluster, P.D. & R.W. Allard, 1995. Evolution of ribosomal DNA (rDNA) genetic structure in colonial Californian populations of *Avena barbata*. Genetics 139: 941–954.

Davy, A.J., S.M. Noble & R.P. Oliver, 1990. Genetic variation and adaptation to flooding in plants. Aqua Bot 38: 91–108.

Dubuc, M., L. Lapointe & J.-P. Simon, 1988. Kinetic properties of $NADP^+$-malic enzyme in two populations of the C_4 weed species *Echinochloa crus-galli* (L.) Beauv. (Poaceae) from sites of contrasting climates. Funct Ecol 2: 353–362.

Emery, R.N.J., C.C. Chinnappa & J.G. Chmielewski, 1994. Specialization, plant strategies, and phenotypic plasticity in populations of *Stellaria longipes* along an elevational gradient. Int J Plant Sci 155: 203–219.

Erskine, W., A. Hussain, M. Tahir, A. Bahksh, R.H. Ellis, R.J. Summerfield & E.H. Roberts, 1994. Field evaluation of a model of photothermal flowering responses in a world lentil collection. Theor Appl Genet 88: 423–428.

Funk, C.R., F.C. Belanger & J.A. Murphy, 1994. Role of endophytes in grasses used for turf and soil conservation. In: C.W. Bacon & J.F. White (Eds). Biotechnology and Endophytic Fungi of Grasses, pp. 201–209. CRC Press, Boca Raton, Florida.

García, P., F.J. Vences, M. Pérez de la Vega & R.W. Allard, 1989. Allelic and genotypic composition of ancestral Spanish and colonial Californian gene pools of *Avena barbata*: Evolutionary implications. Genetics 122: 687–694.

Graner, A., W.F. Ludwig & A.E. Melchinger, 1994. Relationships among European barley germplasm: II. Comparison of RFLP and pedigree data. Crop Sci 34: 1199–1205.

Hällgren, J.-E. & G. Oquist, 1990. Adaptations to low temperatures. In: R.G. Alscher & J.R. Cumming (Eds). Stress Response in Plants: Adaptation and acclimation mechanisms, pp. 265–293. Wiley-Liss, New York.

Hedrick, P.W. & L. Holden, 1979. Hitch-hiking: An alternative to coadaptation for the barley and slender wild oat examples. Heredity 43: 79–86.

Hoffman, A.A. & P.A. Parsons, 1991. Evolutionary genetics and environmental stress. Oxford Sci Publ, Oxford.

Kato, K. & H. Yokoyama, 1992. Geographical variation in heading characters among wheat landraces, *Triticum aestivum* L., and its implication for the adaptability. Theor Appl Genet 84: 259–265.

Macnair, M.R., 1991. Why the evolution of resistance to anthropogenic toxins normally involves major gene changes: the limits to natural selection. Genetica 84: 213–219.

McArthur, J.V., D.A. Kovacic & M.H. Smith, 1988. Genetic diversity in natural populations of a soil bacterium across a landscape gradient. Proc Natl Acad Sci USA 85: 9621–9624.

McNeilly, T., 1990. Selection and breeding for salinity tolerance in crop species. A case for optimism? Acta Oecol 11: 595–610.

Meharg, A.A., Q.J. Cumbes & M.R. Macnair, 1993. Pre-adaptation of Yorkshire fog, *Holcus lanatus*. L. (Poaceae) to arsenate tolerance. Evolution 47: 313–316.

Melchinger, A.E., A. Graner, M. Singh & M.A. Messmer, 1994. Relationships among European barley germplasm: I. Genetic diversity among winter and spring cultivars revealed by RFLPs. Crop Sci 34: 1191–1199.

Nevo, E., A. Ordentlich, A. Beiles & I. Raskin, 1992. Genetic divergence of heat production within and between the wild progenitor of wheat and barley: evolutionary and agronomical implications. Theor Appl Genet 84: 958–962.

O'Donoughue, L.S., E. Souza, S.D. Tanksley & M.E. Sorrells, 1994. Relationships among North American oat cultivars based on restriction fragment length polymorphisms. Crop Sci 34: 1251–1258.

Orr, H.A. & J.A. Coyne, 1992. The genetics of adaptation: A reassessment. Am Nat 140: 725–742.

Palta, J.W. & G. Simon, 1993. Breeding potential for improvement of freezing stress resistance: Genetic separation of freezing tolerance, freezing avoidance, and capacity to cold acclimate. In: P.H. Li & L. Christersson (Eds). Advances in Plant Cold Hardiness, pp. 299–310. CRC Press, Boca Raton, Florida.

Parsons, P.A., 1987. Evolutionary rates under environmental stress. Evol Biol 21: 311–347.

Pérez de la Vega, M., P. García & R.W. Allard, 1991. Multilocus genetic structure of ancestral Spanish and colonial Californian populations of *Avena barbata*. Proc Natl Acad Sci USA 88: 1202–1206.

Pérez de la Vega, M., L.E. Sáenz de Miera & R.W. Allard, 1994. Ecogeographical distribution and differentiation of multilocus allelic association in Spanish *Avena sativa* L. Theor Appl Genet 88: 56–64.

Platenkamp, G.A.J., 1990. Phenotypic plasticity and genetic differentiation in the demography of the grass *Anthoxanthum odoratum*. J Ecol 78: 772–788.

Roberts, E., R. Summerfield, R. Ellis & A. Qi, 1993. Adaptation of flowering in crops to climate. Outlook Agric 22: 105–110.

Royo, C., A. Rodríguez & I. Romagosa, 1993. Differential adaptation of complete and substituted Triticale. Plant Breeding 111: 113–119.

Saghai-Maroof, M.A., R.W. Allard & Q. Zhang, 1990. Genetic diversity and ecogeographical differentiation among ribosomal DNA alleles in wild and cultivated barley. Proc Natl Acad Sci USA 87: 8486–8490.

Shannon, M.C., 1985. Principles and strategies in breeding for higher salt tolerance. Plant Soil 89: 227–241.

Shonnard, G.C. & P. Gepts, 1994. Genetics of heat tolerance during reproductive development in common bean. Crop Sci 34: 1168–1175.

Simon, J.-P., 1979. Adaptation and acclimatation of higher plants at the enzyme level: Latitudinal variations of thermal properties of NAD malate dehydrogenase in *Lathyrus japonicus* Willd. (Leguminosae). Oecologia 39: 273–287.

Snaydon, R.W. & T.M. Davies, 1994. Rapid divergence of plant populations in response to recent changes in soil conditions. Evolution 36: 289–297.

Stettler, R.F. & H.D. Bradshaw, 1994. The choice of genetic material for mechanistic studies of adaptation in forest trees. Tree Physiol 14: 781–796.

Turner, A.P., 1994. The responses of plant to heavy metals. In: S.M. Ross (Ed). Toxic Metals in Soil-Plant Systems, pp. 153–187. John Wiley and Sons, Chichester.

Turner, L.B., C.J. Pollock & G.O. Edmeades, 1994. Thermal kinetics of glutathione reductase and their relation to thermotolerance in diverse cultivars in maize. Ann Bot 74: 254–250.

van Tienderen, P.H. & J. van der Toorn, 1991a. Genetic differentiation between populations of *Plantago lanceolata*. I. Local adaptation in three contrasting habitats. J Ecol 79: 27–42.

van Tienderen, P.H. & J. van der Toorn, 1991b. Genetic differentiation between populations of *Plantago lanceolata*. II. Phenotypic selection in a transplant experiment in three contrasting habitats. J Ecol 79: 43–59.

Via, S., 1994. The evolution of phenotypic plasticity: What do we really know? In: L.A. Real (Ed). Ecological Genetics, pp. 33–57. Princeton Univ Press, Princeton, New Jersey.

Yamaguchi-Shinozaki, K. & K. Shinozaki, 1994. A novel cis-acting element in an *Arabidopsis* gene is involved in responsiveness to drought, low temperature, or high-salt stress. Plant Cell 6: 251–264.

P.M.A. Tigerstedt (Ed.), Adaptation in Plant Breeding, 39–44, 1997.

Climatic adaptation in subterranean clover populations

Efisio Piano, Luciano Pecetti & Antonio M. Carroni
Istituto Sperimentale per le Colture Foraggere, viale Piacenza 29, 20075 Lodi, Italy

Key words: adaptive traits, annual legumes, climatic variation, Mediterranean, self-regeneration, subterranean clover, *Trifolium subterraneum*

Summary

Long-term persistence and, hence, agronomic success as a pasture of the annual species subterranean clover depend primarily on seed yield and seed survival over seasons. In natural populations, plant characteristics influencing seed setting and formation of seed reserves in the soil are expected to be 'adjusted' to the prevailing environmental conditions of the sites of origin. Knowledge on plant/environment relationships may provide information on adaptive strategies of persistence, and guidelines for selecting adapted varieties to specific conditions. On pure lines from a number of populations such relationships were assessed for flowering time, seed yield, burr fertility, individual seed weight, initial hardseededness, and rate of hardseededness breakdown over summer. Flowering time decreased on decreasing annual rainfall, i.e., on shortening the growing season, as adaptive response to the need of producing adequate seed before the onset of the dry season. Individual seed weight decreased on decreasing rainfall, and increasing temperatures. Hard-seed maintenance over summer was higher in populations from hot and dry environments, where the marked effect of temperature on hardseededness breakdown exerts a strong selective pressure. Within-population variation, assessed on flowering time, was particularly wide, with early genotypes occurring even in populations from long-season environments. The adaptive relevance of maintaining high levels of within-population polymorphism to cope with unpredictable climatic fluctuations is discussed. Number of constituent lines as a measure of the population structure, and intra-population variation were both influenced by altitude and rainfall, tending to decrease as the climatic selective pressure becomes severe, under both low-rainfall, hot conditions and high-elevation, cold-prone environments.

Introduction

The climatic conditions of the Mediterranean region favour annual species, which grow during the cool and wet season and carry through the dry period as seed (Buddenhagen, 1990). Subterranean clover (*Trifolium subterraneum* L.) is a Mediterranean-native, annual, self-regenerating pasture legume characterized by mechanisms of seed burial. Seeds are formed in 'burrs' which, in ordinary conditions, develop underground. Being an annual, its long-term persistence depends primarily on seed yield and seed survival over seasons in the soil. While the former is influenced by different plant characters among which maturity grading is of major importance (Rossiter, 1966; Piano et al., 1993), the latter is governed by germination-regulating mechanisms, such as seed coat impermeability (hardseededness), which enable the constitution of an adequate seed bank in the soil (Quinlivan, 1971).

Because of the species' prevalent inbreeding mating system, natural populations of subterranean clover are composed of clusters, the majority of which are expected to be pure lines (Katznelson, 1974). These populations certainly 'adjust' their adaptive traits to the environments where they naturally grow. As the biological fitness largely corresponds to the agronomic success of the species, i.e., reliability of pasture self-regeneration from season to season (Rossiter, 1966), knowledge on the relationships between population characteristics and climatic features of the environments of origin provides information on adaptive strategies of persistence and guidelines for selecting adapted varieties to given conditions.

This paper refers on the variation among and within natural populations of subterranean clover, and its relationships with the climatic variation of the sites of origin.

Materials and methods

A set of 374 pure lines from 61 collection sites in Sardinia, Italy, were evaluated for a range of bio-agronomic characters during three seasons at Sanluri, south Sardinia, under dense sward and simulated grazing conditions, in micro-plots of about 0.5 m^2 located in two replications of a randomized complete block design. These lines were part of a main collection of 783 lines singled out on the basis of various morpho-physiological and chemical diagnostic traits from 84 natural populations collected in as many Sardinian sites (Piano, 1984).

Initial and residual hardseededness, as percentage of impermeable seed sampled respectively at field maturity (June) and at the end of October, were recorded in the first season. The residual hardseededness was then standardized over the initial one, and the relative ratio expressed the rate of breakdown of hardseededness over summer. Seed yield, both as number and weight in a core sample of 40 cm^2 from each plot; number of seeds per burr; and individual seed weight were recorded in the second season. Flowering time, expressed as number of days from sowing (second half of November) to first inflorescence appearance, was recorded on the whole collection of 783 lines grown in semi-dense rows in separate unreplicated trials in two consecutive seasons.

The population was defined as the set of sympatric lines from each collection site. In the statistical analyses, the average of these lines was assumed as the population mean. Relative to the strains in the bio-agronomic evaluation, differences among populations (sites) were detected by analysis of variance (ANOVA) for each character. Population means were regressed on a range of environmental parameters of the collection sites: altitude; annual rainfall; number of rainy days per year; daily minimum temperature of the coldest month (January); daily maximum temperature of the hottest month (July); and daily maximum temperature of April and October, taken as representative of the intermediate seasons.

Data on flowering time from the whole collection were used for deeper investigation of the variation among and within populations, and of the relationships between this variation and the climatic variation of the sites of origin. A partially hierarchical ANOVA was performed on flowering time, where the variation among populations was tested against the pooled variation among lines within populations (nested factor), and this latter against the experiment error term (line $\times$ season). Flowering time of both individual lines and populations were related with the two main climate-determinant parameters, i.e. altitude and rainfall, by simple and multiple regression analyses. Intra-population ranges of variation (difference between the earliest and latest flowering lines), assumed to be the most representative index of the adaptive 'flexibility' of the populations, and number of lines per population, as a measure of the population structure, were also regressed on site climatic characteristics.

Results and discussion

Significant variation among the 61 populations ($P \leq 0.01$) was found for all the adaptive characters examined in the bio-agronomic evaluation (data not shown).

The relationship between plant characteristics and environmental features of the sites of origin are summarized in Figure 1. Regressions on minimum temperature of January and number of rainy days were omitted as they provided similar information as those on altitude and annual rainfall, respectively. Regressions on temperature of July were also omitted as this parameter proved less discriminant than temperatures in April or October.

The variation for seed yield, both as number and weight, and for number of seeds per burr were not significantly related to the variation of any of the examined climatic parameters. The 'flat' response of seed yield over these parameters suggests that each population in its complex attains a 'threshold' seed production, likely optimal in terms of fitness, independently from the variation of the same climatic parameters. The observed differences of seed yield among sites of origin seems, therefore, influenced by inter-site variation for other environmental features, in particular edaphic characteristics. Number of seeds per burr also proved a stable character across climatic conditions.

Of the other plant characters, flowering time was mainly related with annual rainfall of the sites, while the relationships with altitude and temperatures,

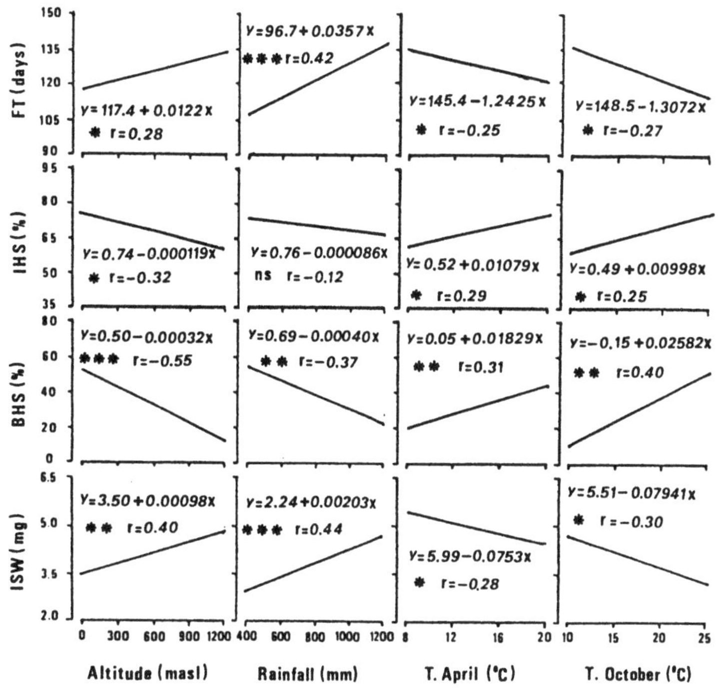

Figure 1. Linear regressionship of population mean values of: i) flowering time (FT); ii) initial hardseededness at plant maturity (IHS); iii) rate of breakdown of hardseededness, measured as maintenance of hard seeds over summer (BHS); and iv) individual seed weight (ISW), on four climate-related parameters of the sites of origin. ns: P > 0.05; *: P ≤ 0.05; **: P ≤ 0.01; and ***: P ≤ 0.001 for both the regression coefficients *b* and the correlation coefficients *r*.

although significant, were lower (Figure 1). The distinct decrease of maturity grading on decreasing annual rainfall, i.e., on shortening of the growing season, confirms that a main adaptive requirement of the species is the need of flowering to start early enough to facilitate burial of the reproductive structures and allow adequate seed setting before the onset of the dry season (Quinlivan & Francis, 1971; Rossiter, 1978).

Contrary to flowering time, the initial proportion of hard seeds at plant maturity seemed mainly temperature-dependent, for initial hardseededness lowered on increasing site altitude and decreasing temperatures (Figure 1). Rate of hardseededness breakdown was significantly influenced by all climatic parameters, particularly by site elevation (Figure 1). The higher the altitude and rainfall, and the lower the temperature levels of the site, the lower the ability of the populations to maintain seed coat impermeability over summer. This latter result reflects the relative impact of natural selective pressures in the different environments. As seed softening is governed by the diurnal fluctuations of temperatures in the soil during summer (Taylor, 1981), in hot and dry sites, where the softening-inducing actions of the environment are marked, the natural selection has favoured genotypes with greater ability to maintain seed coat impermeability. That the ability to maintain hardseededness is greater in environments with high temperature in October (Figure 1) indicates that the length of seed exposure to the softening actions plays a major role as a selective factor, for high temperatures in October are indicative of 'long-summer' sites.

Individual seed weight clearly decreased on decreasing annual rainfall and altitude, and increasing temperature levels of the collection sites (Figure 1). The advantage of smaller seeds in dry and hot environments is probably that they mature faster (Francis & Gladstones, 1974; Pecetti & Piano, 1994), and, due to higher surface/volume ratio, need less water for germination than larger seeds. Examining the physical changes leading to the breakdown of seed coat impermeability, Russi et al. (1992) have also suggested that the larger the seed, the greater is expected the softening stress induced by temperature fluctuations. Indeed, a trend of small-seeded populations to better maintain hardseededness over summer was also found in this

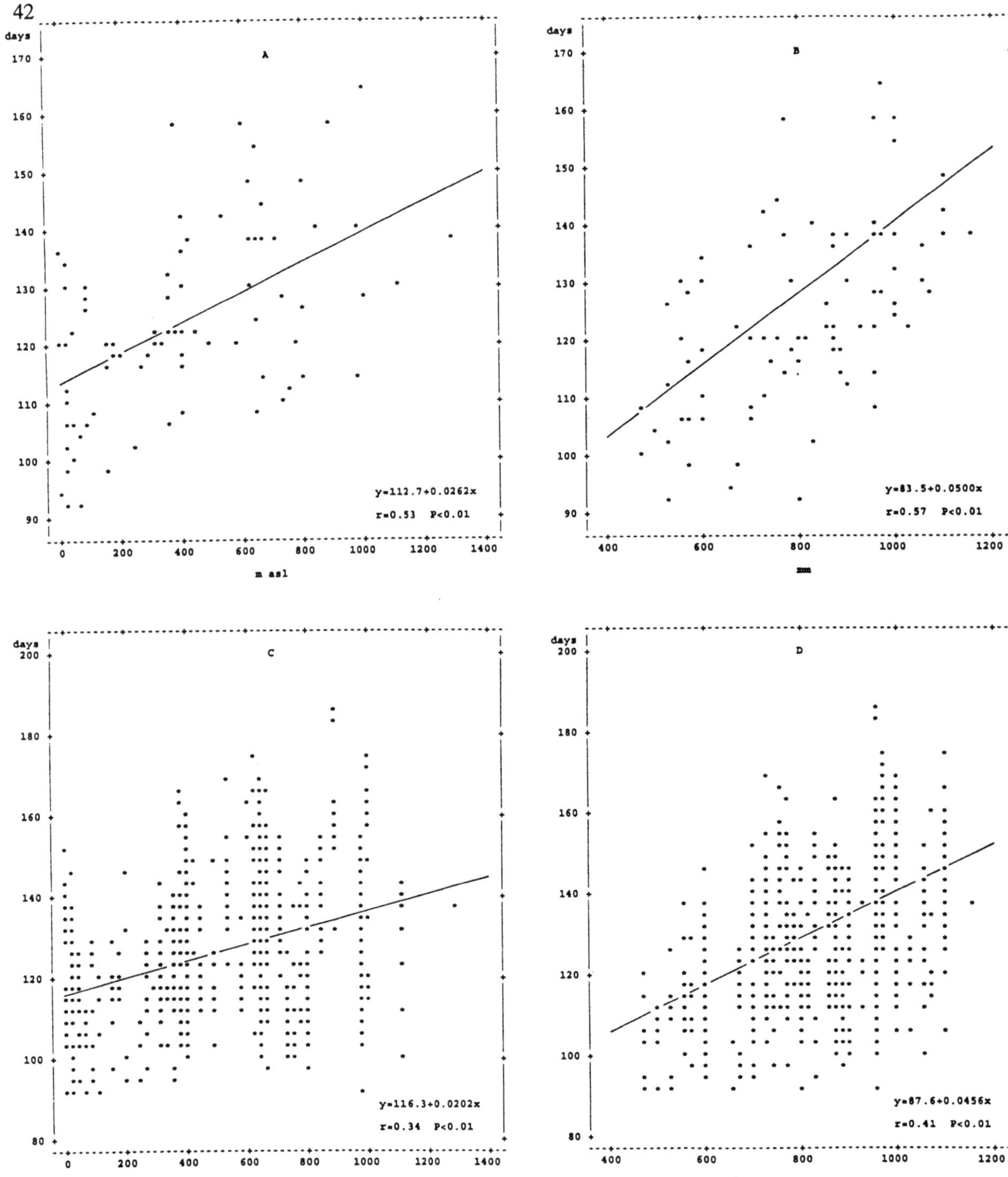

Figure 2. Plot and linear regression of: i) population mean flowering time on altitude (A) and annual rainfall (B) of the sites of origin; and ii) individual genotype flowering time on altitude (C) and annual rainfall (D) of the sites of origin.

study (r = - 0.31, P < 0.01). This would represent an additional advantage of small-seeded populations in dry and hot environments.

The observed variation of seed size, coupled with the stability of both seed yield and number of seeds per burr across climatic conditions, suggests that the other yield component, namely number of burrs, varies in a compensative way relative to the variation of seed size.

Despite the clear relationships observed between adaptive attributes and climatic parameters, the propor-

Table 1. Multiple linear regression parameters of flowering time of individual lines and populations of subterranean clover on altitude and rainfall of collection sites

	Intercept	$b_{Alt.}$	b_{Rain}
Lines	100.0***	0.006*	0.027***
Populations	99.4***	0.008 ns	0.024*

ns: P > 0.05; *: P < 0.05; ***: P < 0.001.

Table 2. Quadratic regression of number of lines per population and intra-population range of flowering time on altitude and rainfall of collection sites, and estimated values of the dependent variables at three different levels of the climatic parameters

Regression equations	Estimated values		
Number of lines per population	a	b	c*
Altitude**: $y = 3.2 + 0.028\ x - 0.0000212x^2$	5.8	11.6	6.8
Rainfall: $y = -34.2 + 0.099\ x - 0.0000527x^2$	4.3	12.2	8.7
Intra-population range (days)	a	b	c
Altitude: $y = 9.1 + 0.062\ x - 0.0000432x^2$	14.9	30.4	21.9
Rainfall: $y = -53.8 + 0.017\ x - 0.0000866x^2$	13.1	28.3	24.5

* Altitude: a = 100 m; b = 900 m; c = 1200 m. Rainfall: a = 550 mm; b = 900 mm; c = 1200 mm.
** Linear and quadratic coefficients of all regressions significant at $P \leq 0.01$.

tion of variation accounted for by these relationships was limited, as inferred by the rather low coefficients of correlation. Environmental variables other than those here examined, particularly edaphic factors and other site-specific situations (e.g., topography), are expected to influence inter-population variation.

Wide inter-genotype and inter-population variation were detected for flowering time in the 783-line collection. Genotype values ranged from 90 to 185 days and population means from 91 to 164 days. In the hierarchical ANOVA both the variance among populations and that among genotypes within populations were highly significant (P < 0.001). The extent of the latter appeared somewhat larger than expected for an inbreeding species. About 45% of the populations had a range among genotypes comprised between 20 and 40 days while some populations showed a range of about 55 days. The variation for intra-population ranges was, at least partly, associated with the number of genotypes singled out per population (r = 0.66, P < 0.001). The latter trait varied from 1 to 26, with an average of 9.2.

Regressions of population mean flowering time on altitude and annual rainfall of the sites of origin confirmed the relationships found with data from the bio-agronomic evaluation, with flowering time linearly increasing on increasing both environmental parameters (Figures 2A and 2B). The results of multiple linear regression analysis of flowering time on annual rainfall and altitude (Table 1) agreed with previous findings (Piano et al., 1993) indicating that the main determinant of the variation in flowering time is the variation in rainfall rather than in altitude and, hence, related temperature levels. Therefore, the observed relationship between altitude and maturity grading seems substantially an indirect effect of the positive correlation existing between site rainfall and elevation (r = 0.78, P < 0.001). The limited influence of altitude *per se* on flowering time is made evident in Fig. 2B by the great spread of mean values of populations from the sites at lowest altitudes.

The wide variation among genotypes originating from similar environmental conditions, relative to both altitude and rainfall, and the better fit of the regression on rainfall than altitude, are evident in Figures 2C and 2D. Overlapping plots do not appear in the graph which, therefore, does not take into account the frequency of the various flowering time values. The densest cloud of plots should have obviously appeared around the fitted line.

Confirming previous observations on germplasm from Sicily (Piano et al., 1993), a main feature which emerges from the distribution of genotype flowering time is that distinctly late lines do not occur in sites with low rainfall, while very early genotypes occur in both dry and wet environments (Figure 2D). Conversely, relatively late genotypes can be found in low-elevation environments, while early strains remain widespread along the altitude range as for rainfall (Figure 2C).

Both the population structure, expressed as the number of constituent lines, and the intra-population variation, expressed as the difference between the earliest and latest line, were substantially influenced by both altitude and rainfall of sites of origin, according to a curvilinear relationship (Table 2). Apparently, the populations tend to reduce their intrinsic variation and simplify their structure as the selective pressure of the climate becomes extreme, approaching to low-rainfall, hot environments at one side, and to high-elevation, cold-prone environments at the other side. The largest variation occurs in populations from intermediate climatic conditions which are rather favourable in the long term, although subject to the wide seasonal climatic fluctuations typical of Mediterranean conditions.

Conclusions

The findings of this investigation have shown how native populations of a peculiar species such as subterranean clover attain a mean adjustment of adaptive features of reproductive significance to the prevailing climatic conditions of the environments where they evolve, in order to ensure reliability of self-regeneration and, hence, persistence from season to season. In this respect, this kind of eco-physiological studies may provide guidelines for selecting varieties adapted to specific environmental conditions. This mean adjustment is, nonetheless, accompanied by the occurrence of relatively high levels of intra-population variation, which represents *per se* a further adaptive strategy of population survival. At very high population density, as generally occurring in Sardinia (1000–2000 seedlings m^{-2}), it has been inferred that even the very low outcrossing rate of 0.15% estimated for the species can produce several hybrids per square metre per year (Marshall & Broué, 1973). The released variation can be subsequently fixed by the inbreeding, and made available to the natural selective pressures. That the wide intra-population divergence is maintained indicates that, in the long run, it confers an adaptive advantage under the varying climatic conditions of the Mediterranean environment. The high level of polymorphism observed for a trait such as flowering time, which has direct and indirect influences on seed yield and, hence, self-regeneration ability of the populations, appears in this context of paramount importance to cope with the unpredictable seasonal climatic fluctuations, and ensure continued persistence. This explains the maintenance of very early variants even in populations from environments with distinctly long growing season. Evidence has been acquired that the population adjustment to the climatic variation also involves the population structure and its inherent level of variation, which tend to diminish in complexity and magnitude, respectively, as the given selective climatic pressure becomes progressively more severe.

Based on the observed wide intra-population diversity, the adoption of appropriate mixtures of selected strains rather than mono-line varieties could be envisaged from a breeding point of view, to ensure reliability of seed setting which is the main prerequisite for the agronomic success of the species.

References

Buddenhagen, I.W., 1990. Legumes in farming systems in Mediterranean climates. In: A.E. Osman, M.H. Ibrahim & M.A. Jones (Eds). The role of legumes in the farming systems of the Mediterranean areas. p. 3–29. Kluwer Academic Publishers, Dordrecht.

Francis, C.M. & J.S. Gladstones, 1974. Relationships among rate and duration of flowering and seed yield components in subterranean clover (*Trifolium subterraneum*). Aust J Agric Res 25: 435–442.

Katznelson, J., 1974. Biological flora of Israel. 5. The subterranean clovers of *Trifolium* subsect. *Calycomorphum* Katzn. *Trifolium subterraneum* L. (*sensu lato*). Isr J Bot 23: 69–108.

Marshall, D.R. & P. Broué, 1973. Outcrossing rates in Australian populations of subterranean clover. Aust J Agric Res 24: 863–867.

Pecetti, L. & E. Piano, 1994. Observations on the rapidity of seed and burr growth in subterranean clover. J Genet & Breed 48: 225–228.

Piano, E., 1984. Preliminary observations on the structure and variability of Sardinian populations of subterranean clover. Genet Agr 38: 75–90.

Piano, E., F. Spanu & L. Pecetti, 1993. Structure and variation of subterranean clover populations from Sicily, Italy. Euphytica 68: 43–51.

Quinlivan, B.J., 1971. Seed coat impermeability in legumes. J Aust Inst Agric Sci 37: 283–295.

Quinlivan, B.J. & C.M. Francis, 1971. The effect of burr burial on the seed of some early maturing subterranean clover cultivars. Aust J Exp Agric Anim Husb 11: 35–38.

Rossiter, R.C., 1966. The success or failure of strains of *Trifolium subterraneum* L. in a Mediterranean environment. Aust J Agric Res 17: 425–446.

Rossiter, R.C., 1978. The ecology of subterranean clover-based pastures. In: J.R. Wilson (Ed.). Plant relations in pastures. p. 325–339. CSIRO, Melbourne.

Russi, L., P.S. Cocks & E.H. Roberts, 1992. Coat thickness and hard-seededness in some *Medicago* and *Trifolium* species. Seed Sci Res 2: 243–249.

Taylor, G.B., 1981. Effect of constant temperature treatments followed by fluctuating temperatures on the softening of hardseeds of *Trifolium subterraneum* L. Aust J Plant Physiol 8: 547–558.

P.M.A. Tigerstedt (Ed.), *Adaptation in Plant Breeding*, 45–48, 1997.

Selection for low temperature tolerance in potato through anther culture

Eva K. Calleberg & Arnulf Merker
Department of Plant Breeding, Swedish University of Agricultural Sciences, Box 7003, S-75007 Uppsala, Sweden

Key words: anther culture, *in vitro* selection, low temperature tolerance, potato

Summary

The main purpose of this study was to use the anther culture system to perform *in vitro* selection for tolerance to low non-freezing temperatures in potato. Anther cultures were incubated at various temperatures (10, 15, 20, 25, 30 °C) for induction of embryos and calli followed by plant regeneration. Anther-derived tuber-bearing clones from clone 73 were analysed for cold tolerance to investigate whether the incubation temperature caused a selection. The criteria for establishing the level of low temperature tolerance included chlorophyll fluorescence, biomass production, and the degree of chlorosis. Although, a correlation between temperature tolerance of the anther-derived clones and the *in vitro* incubation temperature could not be found, significant differences in tolerance could be identified between the anther-derived clones. Furthermore, a significant correlation was found between the temperature-related embryogenic potential (embryo or calli production at 10 °C/embryo or calli production at the optimal temperature) of 6 tetraploids and their low temperature tolerance. This indicates a relationship between gene-expression of the embryogenic microspore and the sporophyte. Accordingly, even though no selection effects could be found in this particular experiment, prerequisites for *in vitro* selection seem to be present.

Introduction

Cultivated potato *Solanum tuberosum* L. subsp. *tuberosum* is adapted for growth in cool climates, but is still only moderately cold tolerant compared with other crops. Kristjansdottir (1991) demonstrated that two common Swedish cultivars showed a decreased biomass production as well as chilling injury at 10 °C/4 °C compared with some Andean clones with expected frost tolerance. Accordingly, in areas where the growth period includes longer periods of low temperatures and occasionally frost, such as alpine and subarctic regions, the cultivated potato is not completely adapted and consequently the vegetative growth is hampered, which may result in a decreased yield. Since the potato is a major food source in certain areas there is interest for extending the temperature tolerance of the potato. Commonly the cultivated potato is tetraploid, which makes the inheritance complicated, and therefore breeding at the dihaploid level would be an alternative strategy (Wenzel, 1979).

The purpose of this study was to use anther culture as a tool for producing dihaploids and to simultaneously select for low temperature tolerance *in vitro* by applying a selection pressure, utilizing the natural genetic variability following meiotic recombination, and a possible gametoclonal variation (Morrison & Evans, 1988). Prerequisites for *in vitro* selection using anther or microspore culture are firstly that the gene expression in the gametophyte is comparable to the expression in the sporophyte, and secondly that the mechanism associated with the stress tolerance largely is determined at the cellular level (Lashermes, 1991). According to Tanksley et al. (1981), Sari-Gorla et al. (1986), and Pedersen et al. (1987) an extensive expression overlap of genes coding for enzymes were found between pollen and sporophyte in tomato, maize and barley respectively. Additionally, since the microspore will be engaged in an embryogenic rather than a normal gametophytic developmental pathway, the correspondence of gene expression between the embryo and the whole plant may be even higher (Lashermes, 1991). Selection for low temperature tolerance on a cellular level appears promising, since tolerance to cold stress is suggested to be due to the composition and changes in membrane structure (for review see Lyons, 1973). Pollen selection for low temperature tolerance was demonstrated by Zamir et al. (1981). They utilized a pollen mixture from the cold tolerant wild species *Lycopersicon hirsutum* and the cultivated chilling-sensitive *L. esculentum* for fertilization of a male sterile *L. esculentum* in a cold environment, resulting in a zygote formation involving preferentially *L. hirsutum* gametes. Furthermore Zamir et al. (1982) and Zamir & Gadish (1987) demonstrated pollen selection for low temperature tolerance using a F_1 hybrid between *L. esculentum* and *L. hirsutum*, while Lyakh & Soroka (1993) similarly demonstrated selection for cold tolerance in maize. Kristjansdottir (1990) noted that low temperature tolerant Andean potato clones maintained their germination capacity better than chilling-sensitive European clones when temperature was decreased. *In vitro* selection using anther culture or microspore culture has been demonstrated for several characters e.g. photosynthetic efficiency in tobacco (Medrano & Primo-Millo, 1985), salt tolerance in barley (Ye et al., 1987), herbicide tolerance in rape-seed (Swanson et al., 1988), freezing tolerance in rape-seed (Orr et al., 1990), and tolerance to *Fusarium* in wheat (Fadel & Wenzel, 1993).

The selection can be performed at different stages of the anther culture procedure and, consequently, different selection units will be involved: microspores, multicellular structures, embryoids or plantlets. The microspore stage offers the largest population of genetically variable individuals, and furthermore an easily manageable system with small spatial demands. In this study we chose to select for cold tolerance by applying low temperatures at the microspore stage and throughout the culture period, until the emergence of macroscopical embryos.

Results

Frequency of anther derived plants produced at various temperatures

In this study four tetraploid potato clones were included, three of them having a background in several tuber bearing wild species (199.13, 201.5, and 201.12) (Landeo, 1980) and one clone having its origin in a seed population from a cross between clone 201.5 and a selfing of cv. Annika (S12) (Kristjansdottir, 1991). After treatment and inoculation of anthers according to Calleberg & Johansson (1993) the cultures were immediately incubated at various temperatures (10, 15, 20, 25, 30 °C). Plants were most frequently produced at 20 °C (e.g. 21 plants from 201.12) and 25 °C (e.g. 36 plants from 199.13) except for clone 73, where the highest number of plants were obtained at 15 °C (20 plants), while the plant production was almost completely inhibited in all clones at 10 and 30 °C. Generally more than 50% of the anther derived clones were dihaploid i.e. with 24 chromosomes, while the remaining clones were doubled and consequently had 48 chromosomes.

Analysis of selection efficiency for low temperature tolerance of anther-derived clones from clone 73

Due to the low number of tuber-producing clones originating from anther cultures of 199.13, 201.5 and 201.12 incubated at the lower temperatures (10 and 15 °C), the selection efficiency for cold tolerance was determined analysing anther-derived clones from clone 73 solely. Totally 24 tuber-producing clones (24 and 48 chromosomes), of which 12 originated from anther cultures incubated at 15 °C, 7 from 20 °C, and 5 from 25 °C, were analysed for low temperature tolerance using chlorophyll fluorescence (Smillie & Hetherington, 1983; Walker et al., 1990; Kristjansdottir, 1993), biomass production (Kristjansdottir, 1989) and the degree of chlorosis. All experiments resulted in a higher degree of low temperature tolerance among plants originating from 2 °C, while an incubation temperature of 15 °C unexpectedly resulted in clones with a lower cold tolerance. Clones originating from 25 °C generally tended to have an intermediate tolerance.

Analysis of the variation of low temperature tolerance within the group of anther-derived clones originating from clone 73

The individual anther-derived clones were analysed for their low temperature tolerance resulting in a significant variation within the group. Clones with a higher level of tolerance than the donor clone 73 could be identified.

Correlation between temperature-related embryogenic potential and low temperature tolerance for 6 tetraploid clones

The embryogenic potential (embryos or calli/100 anthers) at the various incubation temperatures was recorded for 6 tetraploids and the temperature related embryogenic potential (embryo or callus production at 10 °C/embryo or callus production at the optimal temperature) was calculated. When these data were correlated to the low temperature tolerance of each clone a significant positive correlation was found.

Conclusion

The main objective of this study was to investigate whether it was possible to select *in vitro* for cold tolerance by incubating anther cultures of potato at low temperatures. According to the results obtained from analysis of this material no relationship between selection temperature and cold tolerance was found. A limitation in this study, however, was the low number of plants produced in anther cultures incubated at the lower temperatures (mainly at 10 °C), where selection was believed most likely to occur. This made the interpretation difficult and hindered detection of possible selection effects.

Despite the fact that selection effects could not be shown, the results indicate that *in vitro* selection for cold tolerance may still be possible. Of major importance for selection is that the dihaploids show sufficient genetic variation for the character of interest. Since the parents of clone 73 differed significantly in low temperature tolerance according to Kristjansdottir (1991), a genetic variation between the dihaploids was expected, which indeed was verified by this study. Additionally, a correlation between the low temperature tolerance of a clone and its temperature related embryogenic potential was found. Thus, if a clone is highly tolerant to low temperatures, the embryo and callus production of the same clone is higher at the lower incubation temperatures, compared with clones having less cold tolerance. Consequently, it is indicated that the gene-expression for cold tolerance in embryogenic microspores is comparable to the gene-expression in the sporophyte, which is in accordance with the findings of Zamir et al. (1981) and Kristjansdottir (1990), who suggested a relationship between gametophytic and sporophytic gene-expression for cold tolerance after cold-treatment of pollen from tomato and potato respectively.

In conclusion, *in vitro* selection for low temperature tolerance through anther culture of potato should theoretically be possible. However, to obtain a selection system sufficient for an efficient production of cold tolerant plants, further understanding of the embryogenesis in relation to cold stress as well as the *in vitro* culture effects is necessary.

References

Calleberg, E.K. & L.B. Johansson, 1993. The effect of starch and incubation temperature in anther culture of potato. Plant Cell, Tissue and Organ Culture 32: 27–34.

Fadel, F. & G. Wenzel, 1993. *In vitro* selection for tolerance to *Fusarium* in F_1 microspore populations of wheat. Plant Breeding 110: 89–95.

Kristjansdottir, I.S., 1989. Influence of suboptimal temperature on biomass production of potato populations of Andean and European origin. Euphytica 44: 23–35.

Kristjansdottir, I.S., 1990. Pollen germination *in vitro* at low temperatures in European and Andean tetraploid potatoes. Theor. Appl. Genet. 80: 139–142.

Kristjansdottir, I.S., 1991. Low temperature tolerance of biomass and pollen germination in potato clones of Andean and European origin. Euphytica 58: 71–80.

Kristjansdottir, I.S. & A. Merker, 1993. Temperature-related changes in chlorophyll fluorescence and contents of chlorophyll and carotenoids in Andean and European potato clones. Plant Breeding 111: 148–154.

Landeo, J.A., 1980. Development of highland tropic populations. In: Utilization of the Genetic Resources of the Potato III. pp. 54–64. Report of the Planning Conference 1980, CIP, Lima, Peru.

Lashermes, P., 1991. Screening for stress tolerant genotypes via microspore *in vitro* cultures. In: INRA (Ed). Physiology-Breeding of Winter Cereals for Stressed Mediterranean Environments (Montpellier, France, 3–6 July 1989). pp. 461–479. Paris (Les Colloques n° 55).

Lyakh, V.A. & A.I. Soroka, 1993. Influence of low temperature treatment of maize microgametophytes in F_1 in the structure and cold tolerance of resulting populations. Maydica 38: 67–71.

Lyons, J.M., 1973. Chilling injury in plants. Ann. Rev. Plant Physiol. 24: 445–466.

Medrano, H. & E. Primo-Millo, 1985. Selection of *Nicotiana tabacum* haploids of high photosynthetic efficiency. Plant Physiol. 79: 505–508.

Morrison, R.A. & D.A. Evans, 1988. Haploid plants from tissue culture: new plant varieties in a shortened time frame. Biotechnol. 5: 684–690.

Orr, W., A.M. Johnson-Flanagan, W.A. Keller & J. Singh, 1990. Induction of freezing tolerance in microspore-derived embryos of winter *Brassica napus*. Plant Cell Rep. 8: 579–581.

Pedersen, S., V. Simonsen & V. Loeschcke, 1987. Overlap of gametophytic and sporophytic gene expression in barley. Theor. Appl. Genet. 75: 200–206.

Sari-Gorla, M., C. Frova, G. Binelli & E. Ottaviano, 1986. The extent of gametophytic-sporophytic gene expression in maize. Theor. Appl. Genet. 72: 42–75.

Smillie, R.M. & S.E. Hetherington, 1983. Stress tolerance and stress-induced injury in crop plants measured by chlorophyll fluorescence *in vivo*. Plant Physiol. 72: 1043–1050.

Swanson, E.B., M.P. Coumans, G.L. Brown, J.D. Patel & W.D. Beversdorf, 1988. The characterization of herbicide tolerant plants in *Brassica napus* L. after *in vitro* selection of microspores and protoplasts. Plant Cell Rep. 7: 83–87.

Tanksley, S.D., D. Zamir & C.M. Rick, 1981. Evidence for extensive overlap of sporophytic and gametophytic gene expression in *Lycopersicon esculentum*. Science 213: 453–455.

Walker, M.A., D.M. Smith, K.P. Pauls & B.D. McKersie, 1990. A chlorophyll fluorescence screening test to evaluate chilling tolerance in tomato. HortSci. 25 (3): 334–339.

Wenzel, G., O. Schieder, T. Przewozny, S.K. Sopory & G. Melchers, 1979. Comparison of single cell culture derived *Solanum tuberosum* L. plants and a model for their application in breeding programs. Theor. Appl. Genet. 55: 49–55.

Ye, J.M., K.N. Kao, B.L. Harvey & B.G. Rossnagel, 1987. Screening salt tolerant barley genotypes via F1 anther culture in salt stress media. Theor. Appl. Genet. 74: 426–429.

Zamir, D., S.D. Tanksley & R.A. Jones, 1981. Low temperature effect on selective fertilization by pollen mixtures of wild and cultivated tomato species. Theor. Appl. Genet. 59: 235–238.

Zamir, D., S.D. Tanksley & R.A. Jones, 1982. Haploid selection for low temperature tolerance of tomato pollen. Genetics 101: 129–137.

Zamir, D. & I. Gadish, 1987. Pollen selection for low temperature adaptation in tomato. Theor. Appl. Genet. 74: 545–548.

P.M.A. Tigerstedt (Ed.), *Adaptation in Plant Breeding*, 49–58, 1997.

Climatic adaptation of trees: rediscovering provenance tests

Csaba Mátyás
University of Sopron, Hungary, P.O. Box 132, 9401 Sopron, Hungary

Key words: adaptation, climate change, provenance research, phenotypic stability, natural selection, genetic, diversity

Summary

Common garden testing of populations of different origin started with forest trees more than two hundred years ago. Since then, so-called provenance tests have been established with most commercially important species. Beyond the strictly silvicultural goals, the tests offer excellent opportunities to study intraspecific genetic variation patterns and represent probably the most powerful available tool for testing hypotheses of climatic adaptation in trees.

Analysis of adaptive traits (mostly juvenile height growth) in provenance experiments indicate the existence of very effective constraints on adaptedness. The performance of populations plotted against an ecological-climatic factor exhibits a characteristic pattern and can be described by response functions. The population average of a fitness-related trait for a locally adapted population is often significantly lower than that of populations from other environments; usually the ones from milder climate perform better. The phenomenon is interpreted as adaptation lag. Suboptimal adaptation is compensated by a high level of genetic diversity. Molecular genetic studies confirm the high level of allelic and individual genetic diversity in forest trees. A consequence of individual homeostasis, phenotypic stability of populations is usually also high; the sensitivity to environmental changes is generally moderate. Phenotypically stable populations are valuable not only because of a wider range of potential cultivation but specifically because of a greater ability to adjust to unexpected changes. This trait should receive more attention in the future for obvious reasons.

The maintenance of a high within-population genetic variance is favored by the genetic system of the investigated species (effective gene flow, outbreeding, high genetic load, etc.). Random events and long-lasting biotic interactions are further effects impairing the efficiency of natural selection.

In view of expected climate instability, genetic adaptability of forest trees causes serious concern due to their long lifespan compared to the rapidity of expected changes in environmental conditions. The potential of provenance tests to interpret long-term adaptational processes should be utilized to analyze, model and predict response of trees to climate change. Although seldomly appreciated, provenance research might be among the most important contributions of forestry to biological sciences.

Introduction

Due to numerous reasons the domestication of forest trees is still in its beginnings. There are only few genera where monoclonal plantations of selected genotypes are in commercial use (e.g. *Populus, Salix, Cryptomeria*, some Eucalypts). In most cases forests are regenerated with generatively propagated seedling populations; this provides a comparatively wide adaptability to the diverse site conditions encountered in forestry.

Nevertheless the issue of climatic adaptation appears as a crucial problem for forestry in view of the expected climate instability. The longevity of trees make a fast adjustment to changed conditions more difficult than in agriculture. Long-term adaptedness and stability should be therefore of higher concern than possible gains in timber or fibre yield. The fact that breeding of forest trees has not progressed very far yet, offers some advantages. The use of practically wild populations in breeding experiments may offer informations which are valuable far beyond their immediate goal. Genetic variation of wild populations can be analysed to study mechanisms of climate adaptation. Due to the lifespan of trees, climate fluctuations, secular changes and rare events must have shaped the adaptational strategy both on individual and population level. Trees are therefore interesting objects for studying long-term climatic adaptation.

Climatic adaptation of plants involves both the genetic adaptation of populations and the ability of individuals to buffer environmental changes through modifying their phenotypical response. Genetic adaptation is understood as a change in gene frequency, directed toward a theoretical optimum in a given ecological stiuation. The genetic transformation is expected to increase the average fitness of the population. Adaptation should be understood not only as an adjustment to the current conditions, it also includes the ability to adapt to future (unspecified) changes. Covering relatively short periods in time, adaptational changes in the population may be regarded as reversible.

Individual or 'physiological' homeostasis (acclimation or plasticity), appears on population level as phenotyic stability. Although clearly determined by hereditary factors, phenotypic stability does not involve genetic change in the population (therefore the often synonymously used term 'genotypic flexibility' is somewhat misleading). Phenotypic stability or plasticity denotes a change in the expressed phenotype as a function of changing environmental effects. (It is related to the term reaction norm, which, in a narrow sense, describes the reaction to a specific effect). The interpretation of adaptation in natural plant populations is difficult. The safest and simplest is the analysis of climatic adaptation, through observing differences in populations of different origins and relating their response to ecological conditions at their location of origin. To define or even quantify adaptedness is certainly a question of interpretation. Without disregarding conflicting opinions (Tigerstedt, 1992) in this respect, the suggestion of Ayala (1969) is adopted: to assess adaptedness, the ability of the population to transform available nutrients and energy into its own living matter is analyzed. This corresponds to dry matter production, in forestry to volume (timber) yield. (It cannot be denied that within a closed forest stand of high density, juvenile height growth plays a crucial role and is of high adaptive value – less so in open woodlands.) Adaptational differences between populations, manifesting themselves in growth, hardiness and resistance, affect volume production directly and are therefore of critical importance in forestry. Out of the components of yield of a forest stand height growth is under strong genetic control; only this growth trait will be discussed.

Throughout the paper 'adaptive variation' is used for climate-related variation between populations, although it is clear that other forms of variation support adaptation (and fitness) as well; however climate-related variation is the most prudent interpretation of adaptive genetic variation.

The paper deals with phenomena observed on widespread, outbreeding tree species, mostly conifers. Taxa with restricted or fragmented distribution areas may exhibit very different patterns, also depending on their evolutionary past.

Discovering effects of climatic factors on hereditary traits: a retrospective

The awareness that plant development is affected by climatic events can be traced back into antiquity. The understanding however, that widespread species adapt to different ecological conditions through adjusting genetic variation locally, is relatively recent.

Interestingly it were forest trees that attracted attention first. Eighteenth century naval shipyards were seriously concerned about the declining supply of good quality oak and pine timber, and became interested in the question whether seed provenance might influence growth and form of trees in man-made forest plantations. The inspector-general of the French navy, the prominent botanist, Duhamel du Monceau established the first comparative trials in France with Scots pine seed of Baltic origin between 1745 and 1755, however no records of his pioneering work survived (Langlet, 1971).

Seventy years later it was André de Vilmorin, who in the 1820's endeavored to repeat Duhamel's effort and established plantations of Scots pine populations of various origin and other introduced exotic species on his estate Les Barres (Durand, 1984). Vilmorin's results were published after 36 years of observation in 1862, stating differences between geographic provenances and proposing the existence of continuous (clinal) variation (Wright & Bull, 1963). The experiments attracted so much attention in forestry circles, that the posthumus report was soon translated into other languages.

Simultaneously botanists and agronomists began investigating possible links between the origin of populations or cultivated varieties and their phenologic behavior (Schübeler, 1862; Linsser, 1867; De Candolle, 1872; Kerner, 1875 – cit. in Langlet, 1971). In forestry, failures with plantations established from imported seed prompted numerous experiments in the second half of the 19th century, following Vilmorin. Outstanding with respect to their extent and design were the provenance experiments in Russia. Turskiy established between 1877 and 1892 numerous experiments with Scots pine populations near Moscow, which were maintained and reviewed for a century to come (Timofeev, 1973). Later, Ogievskiy established tests with a large number of provenances at 20 different locations – already a network-type experiment (Giertych & Oleksyn, 1981). Similar experiments were outplanted throughout Central and Northern Europe, in Germany, Austria and Sweden (discussed in the excellent paper of Langlet, 1971).

The interest in provenance trials with forest trees was so great, that when the Union of Forest Research Institutions (IUFRO) was founded in 1892, it was among its first tasks to initiate international provenance experiments. The first in a long sequence of IUFRO-coordinated provenance tests with various species was established in 1907 in 9 countries (Giertych, 1979). The scope of some of these tests is impressive: e.g. the international provenance test with Norway spruce initiated in 1962 includes 1100 provenances from Siberia to Iberia; a total of 20 parallel experiments in 14 countries have been established (Weisgerber et al., 1976).

It became evident already around the turn of the century that forest trees harbor a great wealth of intraspecific variability which at least in part reflects the climatic differences within the area of distribution of the species. It was obvious that the morphological and growth differences between the populations may be utilized to improve the quality of cultivated forests (such a proposal may be traced in the archives of the Swedish navy as early as 1759; cit. in Langlet, 1971).

The importance of ecologic factors, especially of climate, in shaping intraspecific variability became widely recognized, however, only after experiments with herbaceous plants yielded similar results. The work initiated by the Carnegie Institution of Washington is presumably the best-known investigation of herbaceous plants collected along altitudinal transects. The common-garden experiments proved the existence of genetic differences within the species. The results published in four volumes (Clausen et al., 1940, 1945, 1948, 1958) became a citation classic in ecological literature, especially the work on *Achillea* (Hiesey et al., 1942, respectively Clausen et al., 1948) and was often used as an illustration of continuous, clinal variation along an altitudinal (climatic) transect, which was not the conclusion of the authors.

Aims and significance of provenance research

Provenance research is the expression used in forestry for the analysis of common garden plantations of tree populations originating from geographically different locations. 'Provenance' means simply the source of a population sample representing a defined area. The provenances (synonymously: sources) originate from seed collected from identified stands or regions which need not be indigenous. (Therefore in tree improvement literature and forest legislation a distinction is made between provenance and origin, the latter meaning the ancestral, original population.)

Due to their size and lifespan, forest trees seem to be very difficult objects for common garden or controlled environment tests. Although growth chamber tests can be carried out barely beyond the early juvenile stage, early growth, developmental and physiology traits correspond satisfactorily to initial field results on the level of provenances, and correlate well with climatic data at the locations of origin (Giertych & Farrar, 1962; Yeatman, 1966; Ledig, 1992). Nursery tests are also frequently used to establish adaptive differences in a geographic area (Rehfeldt, 1988; Sorensen & Weber, 1994).

Provenance tests of trees are time consuming, require large areas and are very costly – nevertheless, as documented in the earlier chapter, provenance testing was conceived more than two hundred years ago and became a standard procedure for a century because of the economic interest in maintaining vitality and growth vigor of forest stands over decades until maturity. Provenance testing has been applied to practically all major forest tree species of the Northern hemisphere and also to numerous tropical species of economic interest. Many of them originate from international cooperation coordinated by IUFRO, FAO or other agencies.

The original intention and principal goal of provenance tests is to identify stands, populations or areas which provide the most desired traits and commercially best results at the test location. In course of time the tests have been used to study quantitative genetic phenomena (e.g. juvenile-mature correlation, ratio of inter- and intrapopulation variation, genotype-

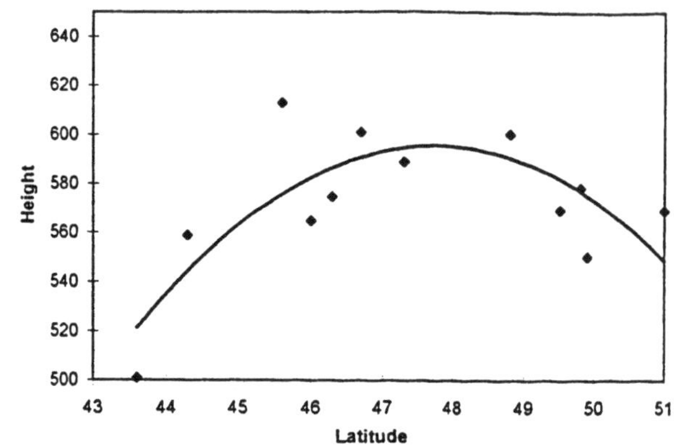

Figure 1. Mean height of 12 jack pine (*Pinus banksiana*) provenances from West Ontario in Swastika, Ont., plotted against the latitude of their origin. The scatter of values indicates a fairly weak adaptation to latitudinal effects in spite of simple topography (R^2 = 0.52, sign. at P = 5% level). The latitude of the test site is 48° 0′N (Mátyás 1987).

environment interactions, etc.). Provenance data also serve as a basis to delineate seed (planting) zones as well as breeding zones (Kung, 1981; Raymond & Lindgren, 1986, 1990). Beyond strongly practice-oriented intentions serving silviculture, the tests offer an excellent opportunity to analyse intraspecific diversity in other than growth traits (allozymic, chemical, physiological, morphological variation) which was exploited also for taxonomic purposes.

Tree breeding research, advancing toward clonal selection, advanced-generation breeding and genetic engineering, regarded until recently provenance tests as a useful but already somewhat redundant starting point. With the increasing understanding of the significance of genetic diversity, provenance tests became interesting as valuable archives of genetic diversity both for breeding and conservation.

The tracing of between-provenance variation probably represents the most powerful available tool for testing hypotheses of climatic adaptation in trees. Instead of analysing genetic changes in subsequent generations, an unthinkable task in forestry, the observed geographic variation is interpreted as an adaptive response to changes in climate conditions. The outplanting of the tests may be regarded as a simulation of environmental change over time and may be modeled (Mátyás & Yeatman, 1987, 1992; Mátyás, 1994; Schmidtling, 1995). This approach offers direct applications in forecasting climate change effects on trees and forests.

Provenance tests covering different habitats may be utilized also for biological site indication. The growth and vigor of populations used as standards characterize the environment, the growth potential and possible risks (Lindgren & Persson, 1995). Compared to common garden experiments with short-lived plants, an advantage might be the long duration of the tests. Effects of annual weather fluctuations and rare anomalies are integrated in the end result. The longevity of the objects themselves binds the organisms stronger to the given environment, than in the case of ephemeral plants. This is not to imply a much closer adaptation to local conditions; a 'perfect' adaptation would result in the long run in the extinction of the species. On the contrary: adaptive strategies seem to include the maintenance of a sufficient level of variability to adjust to continuous and unpredictable fluctuation of conditions during the tree's lifespan and into an obscure future.

These distinct evolutionary and ecological implications make provenance tests important and interesting objects to study beyond direct silvicultural applications. Although provenance research might be among the most important contributions of forestry to biological sciences, up to now its results have failed to capture much attention outside the forestry community. Even the fact that much of the climatic adaptation research has been initiated and studied on forest trees is not generally known.

Genetic variation and climate factors

To demonstrate the effects of climate on intraspecific genetic variation, it is sufficient to plot provenance data against an important climate or geographic factor of the location of origin (Figure 1). Depending on the nature of the observed trait, adaptive genetic variation may exhibit certain geographic patterns; although linked to climate, the various traits may show independent patterns depending on the effect the climate has on that specific trait. Correlations between trait variation and climate parameters may suggest the adaptive pressure exerted on the trait – in a sense demonstrating its adaptive significance (Table 1). Comparing different types of traits, the variation of growth- and phenology-related traits seems to be in closest agreement with cli-

Table 1. Correlation of growth and morphology traits of 35 East European Scots pine populations with climate factors at the location of origin (measured in Kámon, Hungary; from Mátyás 1987)

	Number of frost-free days	Geograph. latitude (photoperiod)	Average July temperature	Annual precipitation
11-year height	0.81	- 0.69	0.62	- 0.14
Height incr., 11th year	0.86	- 0.74	0.69	- 0.17
Branch number, top whorl	0.44	- 0.56	0.41	- 0.18
Branch angle	0.70	- 0.54	0.35	- 0.08
Needle-cast resistance	0.46	- 0.15	0.06	0.29
Bud break	- 0.84	0.72	- 0.62	0.11
Bud set	- 0.82	0.75	- 0.58	0.17
Winter color of needles	0.80	- 0.72	0.75	- 0.29

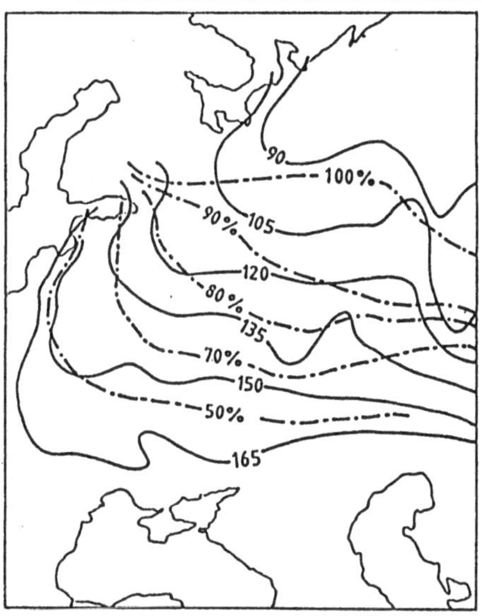

Figure 2. Variation of budset of one year old East European Scots pine (*Pinus sylvestris*) seedlings in early September in Hungary (broken line: percentage of seedlings with buds). The geographic pattern of phenology closely follows thermal conditions (full line: number of frost-free days) (Mátyás, 1981).

matic factors. Morphological and allozyme variation generally do not correlate well with climate.

From a forestry point of view, growth and survival traits are the most important. In temperate and boreal forests, the adaptive variation of growth-related traits (height increment, bud break and formation, hardiness, etc.) are primarily shaped by the thermoperiod (Sarvas, 1972, 1974; Prokazin & Bogachev, 1975; Morgenstern, 1978). Growth and phenology data of provenances correlate best with thermal parameters whereas heat sum, temperature average, maxima and minima yield roughly the same results. This is also in agreement with experiments in controlled environments (Giertych & Farrar, 1962; Yeatman, 1966).

The effect of photoperiod on trees has been also widely investigated, especially because latitude data are usually more precise and easier to establish than thermal parameters. Vaartaja's (1959) classic work led even to the definition of 'photoperiodic ecotypes'. Dormling (1979) was able to prove the importance of photoperiod in case of Norway spruce for budbreak and bud formation, as well as in initiating and terminating dormancy. This seems to be in strong contradiction with the beforesaid. In fact, there is certainly a synergism in photo- and thermoperiodic regulation of growth (Koski & Sievänen, 1985), where the importance of photoperiod seems to increase toward the arctic timberline and high elevation sites (Mátyás, 1981). On the other hand, in most cases there is a strong confounding effect between photoperiod (latitude) and thermal conditions for obvious reasons. The two effects can be successfully separated only in mountainous areas or in anomaly regions, such as Eastern Europe, where – as a consequence of the Gulf stream – isotherms run partially perpendicular to latitude. For this area, the clear dominance of the thermal factor is obvious (Figure 2). The increased importance of photoperiod appears in climatically severe areas receiving less heat. Sarvas proposed the limit for dominant thermal regulation of growth of conifers around 62 ° N. latitude (which corresponds in Scandinavia to roughly 1000 degree-days). More recent investigations in provenance tests support this (Eriksson, 1980; Mátyás, 1981).

The effect of precipitation seems to be minor in affecting climatic adaptation of trees, even if the species is at its limits regarding moisture conditions; moisture-related intraspecific differentiation has not been demonstrated.

Table 2. Ten-year height and survival of some transferred Scots pine provenances in an experiment near Moscow (from Mátyás, 1989)

Provenance	Average height (cm)	Survival (%)	Annual mean temperature	Geographic latitude
	in the Moscow test		at the location of origin	
Moscow (local)	400	87	3.4	56
Minsk	409	88	5.0	54
Brest	435	92	7.0	52
Volynsk	397	65	7.0	50
Donetsk	418	68	8.0	48

Phenotypic stability in tree populations

The most striking feature of provenance tests is the generally low sensitivity of populations to changing environments, even with regard to highly adaptive traits such as height growth. Provenances transferred over large distances into very different environments are able to grow and even compete with the native, local populations (Table 2). This indicates a very high level of individual homeostasis. As a consequence, the phenotypic stability of populations is usually high. Reaction norms of individual populations are usually not well expressed within considerably large geographic areas, and growth differences between adjacent areas are difficult to detect in absence of steep gradients (e.g. mountain slopes). On geographically not fragmented plains the distance between populations with measurable growth differences may exceed 50 to 100 km along ecological gradients.

Stability differences have also been observed among provenances in some coniferous species. Furthermore, stable provenances seem to concentrate in certain 'stable' areas. While the value of stable populations is very great, the evolutionary causes for their appearance are not clear. Phenotypic stability *per se* need not be adaptive. However various theoretical models agree upon the fact that environmental heterogeneity in time favors the evolution of stability by increasing individual homeostasis (Mitton & Grant, 1984; Parsons, 1987; Powers et al., 1991; Scheiner, 1993). Indications that high climatic instability may be responsible for the evolution of phenotypically stable populations, has been found for some coniferous species (Mátyás, 1986).

Phenotypically stable populations are valuable not only because their potential cultivation area is considerably larger than that of unstable populations, but specifically because they have a greater potential to adjust to unexpected environmental fluctuations. Therefore, stability as a trait should receive even more attention in the future.

Adaptive growth patterns and climate: an example of lodgepole pine

The shape and distinctness of between-population adaptive genetic variation depends on the climate gradients. Low lapse rates correspond to less expressed patterns and *vice versa*. An example of pattern changes is presented with data of lodgepole pine (*Pinus contorta*) in western North America.

In a range-wide test of lodgepole pine on Vancouver Island, Ying & Liang (1994) found an unexpectedly narrow band of populations adapted to the local climate. At this test location the area represented by sufficiently adapted populations has a maximum width of only 60 km, however it extends over 3° latitude. The elevational band is even more restricted: only 250 m.

In the rugged mountains of Idaho and Wyoming the species is distributed over more than 2000 m of elevational difference and grows across comparably steep gradients: the frost-free period varies from 120 days to less than 30 (!) days. Instead of expected 'local' strains, a clearcut elevational (thermal) cline can be detected across a topographically complex area, in spite of numerous disjunctions in distribution. Most of the geographic variation found is elevational (Rehfeldt, 1988).

There is no reason to interpret any of the described patterns as local (ecotypic) or non-clinal adaptation. Usually the lack of proper ecological data may create such impressions. E.g. in Idaho, 300 m in elevational change resulted in measurable genetic differences. Although seemingly small, this change equals to 24 days in length of growing season: a very essential difference which clearly has its genetic consequences.

Similarly, the suggested 'very close' adaptation in coastal British Columbia populations is the result of a very steep east-west gradient from very long and mild growing season at sea level to the harsh climate in the nearby Coast Range. Altitudinal differentiation is by no means more explicit than latitudinal; it is the thermal lapse rate which differs.

For the central Sierra Nevada in California, altitudinal and latitudinal lapse rates have been calculated for temperature. The latitudinal gradient is 0.46 °C/lat. degree, the elevational gradient 0.55 °C/100 m. Converted for the conditions on the generalized west slope of the Sierra Nevada 1 °C of average temperature change occurs over 222 km distance in N-S (latitudinal) direction, while over only 5 km along the slope (elevation) – a difference of two orders of magnitudes, in terms of directional selection pressure, which explains the development of narrow bands of similarly adapted populations (Mátyás, 1995).

Genotype-environment interactions measured in tests are relatively weak, but increase generally toward the climatic limits of the species, indicating a growing climatic selection pressure. This phenomenon is observable in numerous conifers near the boreal forest limits (Ericsson, 1980). Lodgepole pine exhibits this effect as well on the northwestern limits of distribution. In the very harsh climate of Northern British Columbia and the Yukon lodgepole pine occurs in a narrow elevational band between 600 and 1200 m, however over a latitudinal range of 10°. Ying & Illingworth (1986) tested sampled populations at two Yukon locations:

	Elev. (m)	Aver. temp (°C)	January	Precip. (mm)*
Whitehorse	663	- 1.0	- 19.0	172
Watson Lake	710	- 3.0	- 23.7	288

At Whitehorse the response differences were remarkably moderate. At Watson Lake in the Rocky Mts., however, the conditions are closer to the climatic tolerance limit of the species: the same populations show strongly accentuated differences (Figure 3).

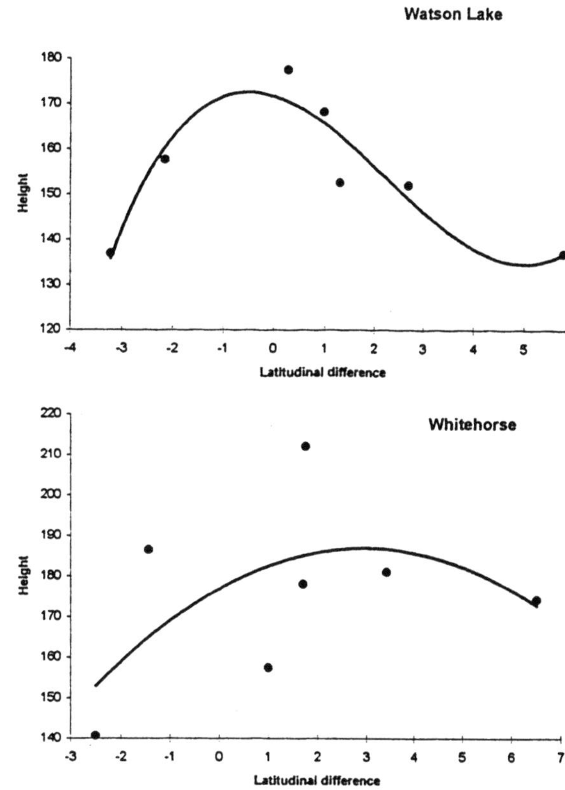

Figure 3. Mean height of 7 lodgepole pine (*Pinus contorta*) provenances from northwestern Canada in two experiments, plotted against the latitudinal difference between the original location and the test site. Negative values indicate southern provenances transferred northward; populations from the North got positive values. While both test sites are close to the northern limit of distributional range of the species, Watson Lake is the harsher location. Note the more accentuated response ($R^2 = 0.65$) at Watson Lake, as compared to Whitehorse (data from Ying & Illingworth, 1986).

Constraints on climatic adaptation

The presented features of adaptive genetic variation in trees illustrate very effective constraints on 'perfect' adaptation. Provenance tests prove the ineffeciency of natural selection to shape a genetic makeup which provides for phenotypes attaining the optimum for a given trait at a given site. The population average of a fitness-related trait for a locally adapted population is often lower than the average of other populations introduced from other environments; a phenomenon termed as adaptation lag (Mátyás, 1991). Suboptimal adaptation has to be compensated for by a sufficiently high level of individual homeostasis. In case of weakly expressed, constrained adaptation to the environment, a high level of diversity has to be expected in trees both on individual and population level.

In fact, genic diversity of trees seems to be relatively high compared to other organisms, this is supported by biochemical marker studies (Hamrick & Godt, 1990). Highest level of allozymic variation is observed in widely distributed, strictly outbreeding, wind-pollinated and wind-dispersed species, such as most conifers and many of the broadleaved tree species. Even though there are observations which support certain correlations between climate factors and frequency patterns of molecular markers (Bergmann & Gregorius, 1993), climatic adaptation seems to have however rather limited effect on the genic diversity indicated by isozymes (Prus-Glowacki, 1991; Savolainen, 1994), resulting in low inter- and high intrapopulation variation. A similar variation pattern is exhibited by several other traits of less adaptive significance, such as certain morphological traits and wood characteristics (fibre length and other anatomic traits, chemical composition, etc.). Interestingly, fertility (cone or seed yield) seem to follow this pattern as well: individual variation in this trait is overwhelming compared to certainly existing, but by far weaker geographic effects.

In case of highly adaptive traits, displaying a clear climate-related variation on the between-population (provenance) level, diversity can be still substantial among individuals. A conspicuous manifestation is the diversity in the length of the growing season; phenophases within one population may differ considerably and may appear in form of polymorphisms (late- and early flushing types). In a Scots pine test, the variation in termination of growth in autumn was more than 70 days within one population (Mátyás 1987). Likewise, a single elevational cline at one location may encompass nearly all the adaptive genetic variation observed within a range of distribution (Rehfeldt, 1988).

The importance of genic diversity itself for the survival and reproductive success of a species or population, although generally recognized, should not be overestimated (Ledig, 1986). Many studies propose a link between protein heterozygozity and consistency of performance (Mitton & Grant, 1984). Considering the limited predictive value of enzymatic diversity on heterozygosity of loci determining adaptive traits, these results cannot be generalized. Homeostasis is not a function of average observed heterozygosity, it is likely regulated both through changing allelic expression and changing interactions among loci in relation with environmental conditions (Scheiner, 1993), which means that it is the genic diversity of certain loci only, which matters – and those may not be evaluated by molecular analytic methods.

The role of the environment in adaptation is generally understood in two ways: first, by setting the fitness function and selecting the fittest individuals, and second, affecting the developmental process of the individual by determining the phenotype. In maintaining the adaptation lag, both the genetic system of the species and random effects play a role, which is seldom recognized. The genetic system of many commercially important temperate-boreal tree species is characterized by very effective gene flow through wind-pollination, an outbreeding mating system with strong mechanisms to prevent inbreeding and a high genetic load (Ledig, 1986; Koski, 1991). These genetic factors act against genetically related neighborhoods and local adaptation. Peculiarities of reproductive ecology (e.g. differential fertility, polyembriony, non-random mating; see Müller-Starck, 1991) prevent further a straightforward action of abiotic selective forces.

Randomness, the occurrence of stochastic events may have a very important role in shaping diversity. Rare and irregularly occurring circumstances are usually disturbances contributing to a high level of temporal heterogeneity on time scales which are difficult to follow. On an evolutionary sclae, the effect of migration, fragmentation, isolation and drift induced by climatic shifts across geologic times, can be demonstrated on the genetic structure of populations and species, which contributes to the presently observed patterns of diversity (Hamrick & Godt, 1990; Ledig, 1986). On the scale of the lifespan of trees, regeneration success and biotic interactions have strongly random components. The possibility for regeneration through appearance of suitable niches is unpredictable, depending on some catastrophic event or mortality of neighbors. Biotic interactions include long-lasting competition with the same and other species, and incidental interactions with consumers, parasites, pathogens, etc. In the entireness of ecosystems, it seems that evolution promotes rather the complexity of biotic interactions at the expense of the proper fitting of the organism-environment relation. The effectiveness of natural selection is further impaired by the random character of accessibility to abiotic limiting factors. It is interesting that heat, the main factor clearly distinguished for shaping climatic adaptation, is the only abiotic factor beside the atmosphere, which can be regarded as conditional, i.e. less affected by competition and other biotic interactions in the ecosystem (on the other hand moisture and light conditions are strongly determined by canopy dynamics).

Aspects of breeding

The concepts of fitness and adaptedness seem to gain more importance in cultivated ecosystems, partially because of concerns that high inputs are affecting the environment and because of projected climate instability. In re-considering sustainability and low-input management, it may be of value to analyze successful adaptive strategies of wild plants, including trees. In forestry, it is proposed to breed in favor of generalists, i.e. phenotypically stable populations. Individual and population diversity should be enhanced by introducing a maximum of different lines which produce the reproductive material for artificial regeneration (Ledig & Kitzmiller, 1992). It seems also worthwhile to increase efforts to understand the genetic mechanisms of high-level homeostasis and to trace naturally stable populations and strains – a valuable natural phenomenon which should be utilized.

References

Ayala, R.J., 1969. An evolutionary dilemma: fitness of genotypes versus fitness of population. Canad. J. Genet. Cytol. 11: 439–456.

Bergmann, F. & H.R. Gregorius, 1993. Ecogeographical distribution and thermostability of isocitrate dehydrogenase (IDH) allozymes in European silver fir (*Abies alba*). Biochem. Syst. Ecol. 21: 597–605.

Clausen, J., D.D. Keck & W.M. Hiesey, 1940, 1945, 1948, 1958. Experimental studies on the nature of species. Vol. 1 to IV. Carnegie Inst. Publ. Nr. 520, 564, 581 and 615, Washington, D.C.

Dormling, I., 1979. Influence of light intensity and temperature on photoperiodic response of Norway spruce provenances. Proc. IUFRO Norway Spruce Breeding Conf., Bucharest, 398–408.

Durand, R., 1984. L'arboretum National des Barres. Paris.

Ericsson, G., 1980. Severity index and transfer effects on survival and volume production of *Pinus silvestris* in Northern Sweden. Stu. For. Suec. Nr. 156.

Giertych, M., 1979. Summary of results of Scots pine height growth in IUFRO provenance experiments. Silvae Gen. 28 (4): 136–152.

Giertych, M. & J.L. Farrar, 1962. A provenance study of jack pine seedlings. Silvae Gen. 11: 111–114.

Giertych, M. & J. Oleksyn, 1981. Summary of results on Scots pine volume production in Ogievskij's pre-revolutionary Russian provenance experiments. Silvae Gen. 30 (2–3): 57–74.

Hamrick, J.L. & M.J. Godt, 1990. Allozyme diversity in plant species. In: A.H.D. Brown, M.T. Clegg, A.L. Kahler & B.S. Weir (Eds). Plant population genetics, breeding and genetic resources. pp. 43–63. Sinauer Ass.

Hiesey, W.M., J. Clausen & D.D. Keck, 1942. Relations between climate and intraspecific variation in plants. Am. Natur. 72.

Koski, V., 1991. Generative reproduction and genetic processes in nature. In: M. Giertych & Cs. Mátyás (Eds). Genetics of Scots pine. pp. 59–72. Elsevier Publ. & Akademiai Kiadó Budapest.

Koski, V. & M. Sievänen, 1985. Timing of growth cessation in relation to the variations in the growing season. In: P.M.A. Tigerstedt, P. Puttonen & V. Koski (Eds). Crop Physiology of Forest Trees. pp. 167–193. Helsinki Univ. Press.

Kung, F.H., 1981. Delineating seed collection zones based on multiplantation provenance tests. Proc. 16th South. For. Tree Impr. Conf. May 22–28, 1981. pp. 83–96. Virginia, USA.

Langlet, O., 1971. Two hundred years of genecology. Taxon 20 (5/6): 653–722.

Ledig, F.T., 1986. Heterozygosity, heterosis and fitness in outbreeding plants. In: M. Soule (Ed). Conservation biology: the science of scarcity and diversity. pp. 77–106. Sinauer Assoc., Sunderland, Mass.

Ledig, F.T., 1992. Genecology: the fitness of the organism and the fitness of the environment. Proc. 12th North Am. For. Biol. Workshop, Aug. 17–20, 1992. pp. 27–49. Sault Ste Marie, Ont., Canada.

Ledig, F.T. & J. Kitzmiller, 1992. Genetic strategies for reforestation in the face of global climate change. For. Ecol. Manage. 50: 153–169.

Lindgren, D. & A. Persson, 1995. Vitalization of results from provenance tests. Proc. XXth IUFRO World Congr., Tampere (in press).

Mátyás, Cs., 1981. Phenological variability of East European Scots pine provenances. Erdészeti Kutat., Budapest, 74: 71–80 (in Hungarian).

Mátyás, Cs., 1986. Improved planting stock in forestry. Akademiai Publ., Budapest (in Hungarian).

Mátyás, Cs., 1987. Adaptation processes in forest tree populations. Dr. Sci. Thesis, Sopron, 195 pp. (in Hungarian).

Mátyás, Cs., 1989. Genetic and ecological constraints of adaptation. Proc. IUFRO Int. Symp. on Forest Genetics, Breeding and Physiology. pp. 79–90. Voronezh, USSR.

Mátyás, Cs., 1991. Adaptation lag: a general feature of natural populations. Proc. Joint Meeting of WFGA and IUFRO Working Parties, Olympia, Wa. (USA) Aug. 20–25, 1990. Pap. Nr. 2.226.

Mátyás, Cs., 1994. Modeling climate change effects with provenance test data. Tree Physiol. 14: 797–804.

Mátyás, Cs., 1995. Climate of the Sierras. Unpubl. report, USDA PSW Expt. Sta., IFG Placerville, 30 pp.

Mátyás, Cs. & C.W. Yeatman, 1987. Adaptive variation of height growth of *Pinus banksiana* populations. Erd. Faip. Egy. Közl. 1–2: 191–197 (in Hungarian).

Mátyás, Cs. & C.W. Yeatman, 1992. Effects of geographical transfer on growth and survival of jack pine populations. Silvae Gen. 43 (6): 370–376.

Mitton, J.B. & M.C. Grant, 1984. Associations among protein heterozygosity, growth rate, and developmental homeostasis. Ann. Rev. Ecol. Syst. 15: 479–499.

Morgenstern, E.K., 1978. Range-wide variation of black spruce. Can. J. For. Res. 8: 463–473.

Müller-Starck, G., 1991. Genetic processes in seed orchards. In: M. Giertych & Cs. Mátyás (Eds). Genetics of Scots pine. pp. 147–162. Elsevier Publ. & Akademiai Kiadó Budapest.

Parsons, P.A., 1987. Evolutionary rates under environmental stress. Evol. Biol. 21: 311–347.

Powers, D.A., T. Lauerman, D. Crawford & L. DiMichele, 1991. Genetic mechanisms for adapting to a changing environment. Ann. Rev. Genet. 25: 629–659.

Prokazin, E.P. & A.V. Bogachev, 1975. Hereditary adaptation of Scots pine to climate factors and possibilities of its assessment and prognosis. In: Genetics, selection and seed production. VNIILM, Pushkino (USSR) (in Russian).

Prus-Glowacky, W., 1991. Biochemical polymorphism. In: M. Giertych & Cs. Mátyás (Eds). Genetics of Scots pine. pp. 73–81. Elsevier Publ. & Akademiai Kiadó Budapest.

Raymond, C.A. & D. Lindgren, 1986. A model of genetic flexibility. In: D. Lindgren (Ed). Provenances and forest tree breeding for high latitudes. pp. 159–177. Kempe Symp. Rapp. 6, Umea.

Rehfeldt, G.E., 1988. Ecological genetics of *Pinus contorta* from the Rocky Mountains (USA): a synthesis. Silvae Gen. 37 (3–4): 131–135.

Sarvas, R., 1972, 1974. Investigations on the annual cycle of development of forest trees. Part I and II. Comm. Inst. For. Fenniae 76.3 and 84.1.

Savolainen, O., 1994. Genetic variation and fitness: conservation lessons from pines. In: V. Loeschke, I. Tomink & S.K. Jain (Eds). Conservation genetics. pp. 27–36. Birkhauser Verl., Basel.

Scheiner, S.M., 1993. Genetics and evolution of phenotypic plasticity. Ann. Rev. Ecol. Syst. 23: 1–14.

Schmidtling, R.C., 1995. Developing a seed transfer model for shortleaf pine. Proc. 8th Bienn. South. Silvic. Conf. Auburn AL, Nov. 1994 (in press).

Sorensen, F.C. & J.C. Weber, 1994. Genetic variation and seed transfer guidelines for ponderosa pine in the Ochoco and Malheur National Forests of Central Oregon. USDA Dor. Serv. Res. Pap. PNW-RP 468.

Tigerstedt, P.M.A., 1992. Why do exotic trees often exceed the yield of endemic trees? In: D. Lindgren (Ed). *Pinus contorta* – from untamed forests to domesticated crop. pp. 60–68. Umea.

Timofeev, V.P., 1973. Growth traits of Scots pine of various origin at the forest experimental fields of the Timiryazev Agric. Academy. Izd.-vo Kolos, Moscow, 2: 130–146 (in Russian).

Vaartaja, O., 1959. Evidence of photoperiodic ecotypes in trees. Ecol. Monogr. 29: 91–111.

Weisgerber, H., W. Dietze, J. Kleinschmidt, J. Racz, H. Dietrich & R. Dimpflmeier, 1976. Ergebnisse des internationalen Fichten-Provenienzversuches '1962. Teil 1: Phenologische Beobachtungen und Höhenwachstum bis zur ersten Freilandaufnahme. Allg. Forst- und Jagdztg. 147: 227–235.

Wright, J.W. & W.I. Bull, 1963. Geographic variation in Scots pine. Silvae Gen. 12: 1–40.

Yeatman, C.W., 1966. Geographic variation in jack pine seedlings. Ph.D. Thesis. 128 pp.

Ying, Ch.C. & K. Illingworth, 1986. Lodgepole pine provenance research in northwestern Canada with particular reference to the Yukon territory. In: D. Lindgren (Ed). Provenances and forest tree breeding for high latitudes. pp. 189–200. Kempe Symp. Rapp. 6, Umea.

Ying, Ch.C. & Q. Liang, 1994. Geographic pattern of adaptive variation of lodgepole pine within the species' coastal range: field performance at age 20 years. Forest Ecol. Manage. 67: 281–298.

P.M.A. Tigerstedt (Ed.), Adaptation in Plant Breeding, 59–67, 1997.

Genetic and physiological mechanisms of plant adaptation

V.A. Dragavtsev
N.I. Vavilov Institute of Plant Industry, St. Petersburg, Russia

Key words: genetic mechanisms, physiological mechanisms, quantative characters, transgressions, heterosis, genotype-environment interactions

Summary

In the period of 1961–1996, as a result team-work with geneticists from Krasnodar and Kharov (A.B. Djakov, P.P. Litun), a new eco-genetical model of organization of quantitative characters was developed. This model differs from the classical G. Mendel one and from K. Mather's model of polygenic inheritance by the following major attributes: (1) redetermination of the spectrum, genes determining genetical variability of a character, when the limiting factor of the environments is changed; (2) module organization of quantitative characters – one resulting and two components; (3) if there are two known values of three essentail characteristics: (a) adaptive properties of genotypes, (b) definite dynamics of lim-factors, (c) genetic parameters of a population, the possibility to determine quantitatively and definitely any third unknown characteristic.

The new model makes it possible to predict a wide series of situations which could not be predicted before – for example, predictions of transgression, heterosis, genotype-environment interaction phenomenon, genotype, genetic and environmental correlations, configurations of Hayman's graphs without organization of diallele crossings, amplitudes of genetic variability in characters, number of genes determining the level and genetic variability of a character, etc. On the basis of this model it would be possible to create the quantitative theory of plant ontogenesis.

From modern plant genetics and plant physiology we know the next multi-levels regulations of adaptive polymorphism in ontogenesis and phytocenosis:

I. Systems of intra-cell regulations

levels of nucleus

1. Blockade and induction of genes activity.
2. Transcription.
3. Transfer m-RNA from nucleus to cytoplasm
4. Jumping genes.
5. Mutations.
6. Changing of chromosomes number (polyploidy, monosomics).

levels of cytoplasm

7. Translation.
8. Activation of protein molecules.
9. Compartmentation.
10. Functioning of proteins.
11. Acting of mitochondrion's and chloroplast's genes

II. Systems of organismal regulations

1. Fields and gradients.
2. Exchange of necessary metabolites (source, sink).
3. Correlations, induced by regulator-metabolites: evocators, inductors.
4. Hormonal regulations.
5. Mixoploidy.

III. Systems of the factors of organism- environment regulations
   1. Systems of direct response for environment (modifications).
   2. Systems of beforehand coordinating of ontogenesis (vernalization, photoperiodism).
   3. Systems of environmental induction of genes activity (heat-shock proteins, Ca^{2+}-calmodulin, transmembran potential, modulators, silentsers, enhansers).
   4. Systems of epigenes (epigenetical switching of operons, giving genotrophs, nicotinotrophs etc.).
   5. Systems of redetermination of genetic formula (spectrum of genes) of quantitative characters.

IV. Systems of phytocenotic level of regulators
   1. Rise of group characters.
   2. Interactions in cenosis (competition, amensalism, commensalism, allelopathy).
   3. Temporal dynamic of phytocenosis.
   Total number of levels – 24.

First important result in this view is: it is impossible to lead to all regulations of adaptive polymorphism in ontogenesis and phytocenosis only to the regulations of the genes level.

Second important result: we must determine which level from 24 have the maximal contributions in the process of settling the adaptive attributes. This knowledge we must have for effective management of plants polygenic adaptivity.

Caxer (1) was the first, who pointed out that Mendel's theory is a model. The list of essential variables of this model is as follows: a chromosome with a linear arrangement of loci on it, alleles (genes), which may be either in homozygous or in heterozygous conditions at each locus, interactions of alleles within a locus (additivity, dominance, overdominance), interactions among loci (epistases of different types). This model described reliably and predicted the behavior of Mendel's 'major' genes, the path of which 'gene-character' was always the only one possible, and practically did not depend upon the fluctuations of environmental limiting factors.

Classical Mendelian genetics, like thermodynamics in a constellation of physical sciences, was always the science of phenomena, i.e. it used the conceptions being far from biochemical and molecular mechanisms of inheritance and hereditary manifestation of ontogenesis. Such concepts of classical genetics as 'gene', 'dominance', 'epistasis', 'pleiotropy' etc., are absolutely phenomenal, and notwithstanding the fast development of molecular genetics, biochemistry, physiology, their mechanisms are not yet known.

Since 1900s genetics of quantitative characters has begun to develop parallel to classical genetics, because the scientists realized that the creation of a powerful breeding theory was impossible without genetics of complex quantitative characters.

Mendelian genetics empirically met with two groups of characters: a) alternative (qualitative), for which the path 'gene-character' is the only one possible and is manifested discretely, and b) quantitative, for which the path 'gene-character' is not clearly single-valued (2).

Unfortunately, G. Mendel's model so strongly influenced the minds of investigators that the creation of a theory and methods of quantitative genetics was based on essential variables of G. Mendel's model. It is enough to study comprehensively the terminology of a diallel analysis by Hayman & Griffing (3) to be convinced that the main concepts (essential variables) of the G. Mendel's model form the basis of algorithms of diallel crosses.

An obvious fact of strong dependence of the level of development of quantitative character upon the limits of an environment, and variations of the genotype ranks for productivity, depending upon environmental fluctuations, so prevented the introduction of refined logics of mendelism into genetics of quantitative characters that the necessity of creating another problem arose, the problem of genotype × environment interaction, but it was moved rather far from the proper methods of a genetical analysis.

In 1984 a new model of organization of a complex quantitative character of plants (and possibly of all living creatures) was described by us (4). It was based on a new set of essential variables, such as redetermination of gene spectra and their number when environmental limiting factors were changed, a module organization of a complex character (two component characters and one resultant character), essential characteristics of a lim-factor itself (whether it has its own variance or not, whether there is a competition for it in phytocenosis or not, the strength and duration of action), the shockmoments produced by lim-factors in ontogenesis and initiation of component characters by the stages of ontogenesis.

The new model was created by us after revealing the phenomenon of redetermination of genetic organization of quantitative characters (5). Before revealing the phenomenon of redetermination of genetic organization of complex character the conception of polygene inheritance by K. Mather (6) predominated in genetics of quantitative characters, the main point of which was as follows: any dimensional character is determined like a qualitative character by traditional 'Mendelian' genes, they are numerous (hence they are named polygenes) and the contribution of each gene into a character is minor. The spectrum of loci and the set of alleles at each locus, determining the level and genetic variability of a character by single values, are stable for each individual of a population. The set of loci, the main effects of which determine additive genetic variability of character, presents a genetic formula of this character (7). Environmental variability of a character influences mainly the activity of polygenes in a stable set of genes of the polygene system of a character. Any fluctuations of the environmental conditions are assumed to cause ecological variability, changing only the activity of loci in the polygene system, constant by its gene set. That's why we do not need to control particular environmental factors, it is quite enough only to dimension ecological variance, to subtract it from phenotypical one and to determine genetic variance.

The conception of K. Mather in combinations with the approaches and methods of R. Fisher (8, 9), S. Wright (10), J. Lush (11) helped to establish the genetic parameters of a population in certain environment at a definite stage of ontogenesis in a definite structure of cenosis. But it failed in predicting the direction and strength of the drift of genetic parameters of a population when changes occurred in environments and cenosis.

The experiments, for example the DIAS program (3), revealed very strong and inexplicable dynamics of genetic parameters. This sharply reduced the possibilities of creating stable breeding strategies on the basis of a polygene theory of organization of quantitative characters.

We have established (4) that the ecologo-genetical formula of a quantitative character is a system of interrelated polygenes and environmental limiting factors.

A situation when in a definite system the changes occur in the set of essential variables, reflecting the essence of its organization, we describe as the redetermination of biological systems.

Besides the former revealed phenomenon of the redetermination of the products of alleles of a heterozygous locus (isoenzymes), determining the development of a character in the dynamics of environmental limiting factors (12), we also detected two earlier unknown phenomena's, causing the loss of stability in the path 'gene products – a quantitative character'.

Phenomenon of changes in the spectrum of loci. This determines the level and genetic variability of a quantitative character in the dynamics of environmental limiting factors. This phenomenon is true for the majority of quantitative characters: from the most complicated resultant ones (for example, the 'mass of grains per plant') up to more simple (for example, the 'synthesis of phytoalexin').

This phenomenon is revealed only in the path 'gene product-character' and is expressed in the fact, that the chains of a phytoalexin synthesis contribute into the productivity of a genotype against the background contaminated with rust, but they cease to determine productivity when rust is absent. Another example: genetical polymorphism of wheat varieties for the duration of the stages of ontogenesis does not contribute into productivity against the background of a constant limit, uniform during the whole period of vegetation, but it acquires significance for breeding if limits are changed by stages of plant development.

The model for polygene inheritance by K. Mather (classical model of quantitative genetics) is deficient, as it cannot explain detail the organization and functioning of genetic formulas of quantitative characters. It disregarded the functional organization of genetic formula and the system: genetic formula – environment.

Phenomenon of changes in the spectrum of modules. A module is an elementary unit of the description of an organization of the system of a quantitative character. It consists of three interrelated characters: onc resultant and two components. For example, the mass of grains per plant is the product of the number of grains per plant and the mass of one grain. A module of lower rank: the number of grains per plant is the product of the number of ears and the number of grains in the ear. The modules are arranged in 'pyramides' of modules. They are two: 1) the pyramid of structural modules, the peak of which denotes the yield of a dry biomass in phytocenosis per unit area, and the foundation of which denotes the products of certain genes or metabolic ways, 2) the pyramid of time modules, its peak denotes the period of time: sowing – harvest, its foundation denotes shorter period of time, significant for yield formation. When environmental limits are changed the component modules either inside a pyramid or between the pyramids change their contri-

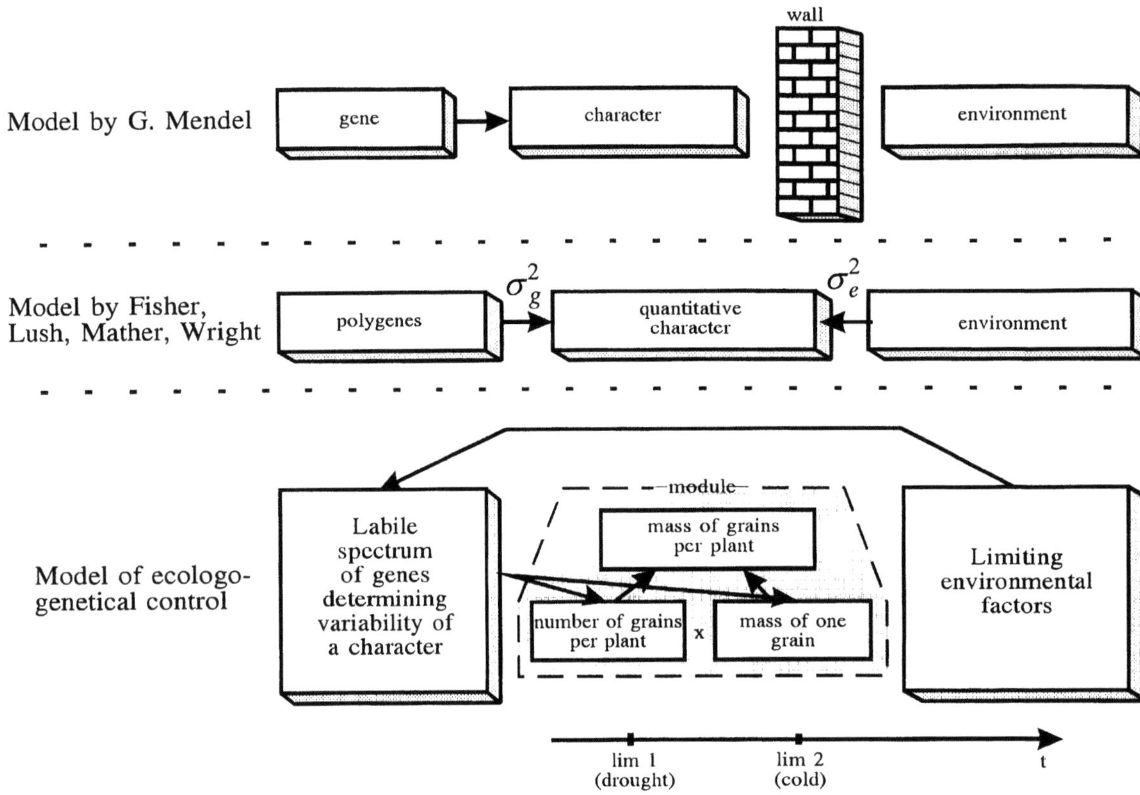

Figure 1. Three models of organizations of quantitative characters.

butions into a final resultant character, this causes the loss of stability in the structure of the hierarchical chain of the modules, 'contributing' to a resultant character.

A module helps to study the elements of the system in its integrity and to evaluate quantitatively the dynamics of processes, kinetics of the state of the system and to measure the contribution of each component of the module into a resultant character.

These two new phenomena together with the previously established mechanism of interrallelic complementation by Finchem (12) is a manifestation of different levels of the general universal biological phenomenon, viz. redetermination of genetic organization of a quantitative character.

The phenomenon of redetermination of genetic organization is displayed at all levels of organization of living matter. For quantitative chacters (when solving genetic and breeding problems) the mechanisms of redetermination at organism, population and cenosis levels acquire special importance.

Therefore, it is impossible to explain genetical organization of a particular quantitative character only on the basis of the parameters of genotypical distributions. A profound analysis of the dynamics of environmental limiting factors, influencing the components of a definitive module at the stages of their development in ontogenesis is required.

So, the model of polygene inheritance by K. Mather, developed on the basis of Mendel's model, appeared to be non-adequate to the real ecologogenetical organization of a quantitative character due to the incomplete set of essential variables, included into the model (only genetical characteristics without environmental limits, without modules, without stages of ontogenesis, without changes in the spectrum of loci).

The phenomenon of redetermination of genetic organization of quantitative characters radically changes former conceptions of genetical organization of complex characters. It becomes evident that a strict stable determination (the so called development on the basis of strict stable programs) does not exist in reality. Proceeding from this postulate there exists an aimed development, the main point of which is a functional organization of an ecologogenetical system of a quantitative character.

Any fluctuations of environmental limiting factors result in regular redetermination of gene spectra, this causes the changes of both, an amplitude of genetic variability of a quantitative character (in certain cases it becomes equal to zero) and the number of loci, determining a mean value and genetic variability of a character.

This phenomenon in genetics is analogous to the effect of dispersion in quantum mechanics when the laws of classical mechanics of motion of particles are broken (13). The effect of dispersion in the physcial micro-world reflects its system organization. The analogous phenomena are typical for the biological macro-world at the level of genetical organization of eucaryotic quantitative characters. But if in quantum mechanics time and motion are reversible and dispersion is stochastic, in biology time and motion are irreversible and monodirectional, and dispersion of genetic information is determined by the functional organization of systems and subsystems. Figure 1 presents a chart of three models of organization of quantitative characters. Figure 2 is a diagrammatic representation of evolution of three branches of genetics: classical, quantitative and molecular.

What are the advantages of the model of ecologo-genetical organization of quantitative characters in the theory of breeding and genetics?

Understanding of the mechanisms and regularities in the reactions of the integral system of an organism (genotype) to the environmental limiting factors. For example, a character of photosynthesis intensity under favorable conditions of growth will be determined by genes for photosynthesis, but against the background of another environmental limit (for example, drought) – by genes for drought resistance, against the background of such an environmental limit as cold – by genes for cold hardiness. If there is a strong polymorphism in a population for genes conferring drought resistance, then against the background of drought a polymorphism for photosynthesis intensity will be displayed. If there is no polymorphism in a population for genes conferring cold hardiness, then against the background of cold a polymorphism for photosynthesis intensity (as well as for any other character of productivity) will disappear.

Understanding of the mechanisms of the development of a quantitative character in ontogenesis. Let us consider a module: the mass of grains per plant is a resultant character (RCh), the number of grains per plant is the first component character (NG), the mass of one grain is the second component character (MG). NG is initiated in spring at a tillering stage (wheat), MG – in autumn in the period of grain forming. If we know the environmental limiting factors in the period of initiation and forming NG and MG and the adaptive properties of a genotype (at least a geographical point of its occurrence) we can easily predict the levels of each component of the module and the level of a resultant character.

Understanding of the mechanisms and prediction of the rate of reaction and homeostasis of productivity of genotypes in different ecological conditions. If one genotype possesses strong genetical systems of drought resistance and cold hardiness, and the other one is with high drought resistance and low winter hardiness, then in two environments: 'drought' and 'cold' the first genotype will manifest homeostasis for the yield and the second will sharply reduce the yield against the background of cold. For predicting the rate of reaction and homeostasis of a quantitative character of a genotype it is necessary to know the lim-factors of the environment in geographical points of its testing and the extent of adaptivity of a genotype to different lim-factors.

Understanding of the mechanisms of drifts of dominance in quantitative characters and predictions of these drifts in different environments. When crossing a coldhardy genotype and a drought resistant genotype and in case of additive unification of their adaptive systems in the F1 plants, the second parent will dominate against the background of cold. A great deal of reliable experimental data on the drift of dominance of quantitative characters can be found in a team monography (3).

Understanding of the nature of ecologically dependent heterosis and prediction of its manifestation under the conditions of ecological or competitive limitation of growth processes. If drought resistant and coldhardy genotypes are crossed, and then the parents and the F1 plants are grown against the background when NG is formed under drought and MG – under cold conditions the F1 plants will surely display heterosis due to the recombination of gene spectra within the components of the module of RCh. In the monograph (3) there are many examples testifying to the regular appearance and disappearance of heterosis. The results of special

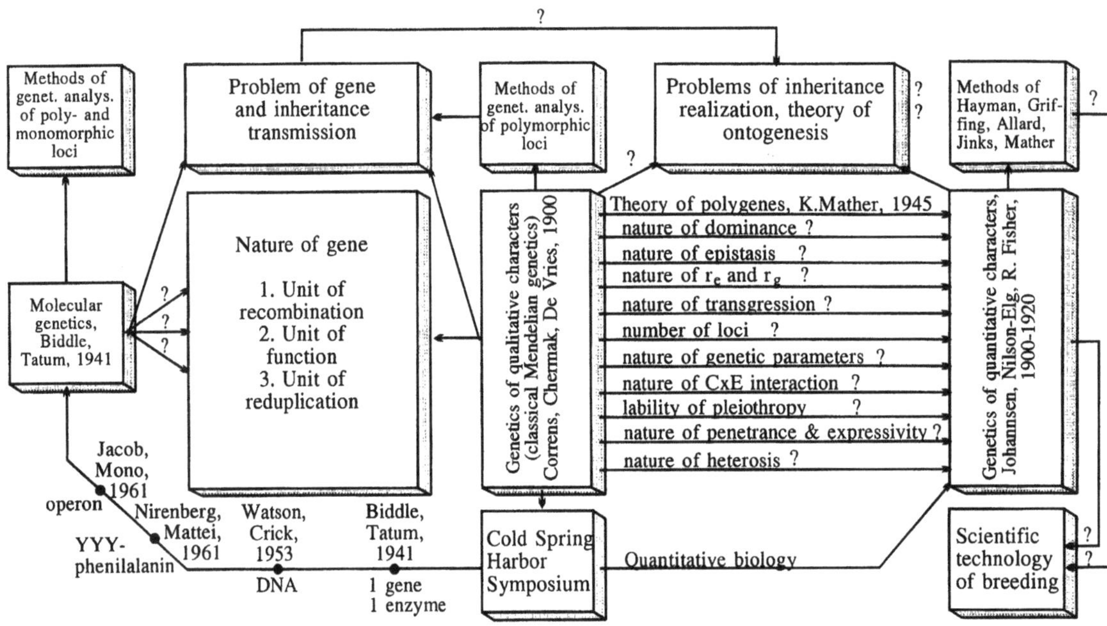

Figure 2. Diagramatic representation of evolution of three branches of genetics: classical, quantitative and molecular.

investigations on the predictions of heterosis are published in the works (8, 15).

Understanding of the nature of drifts and prediction of changes in the signs and values of genotypical, genetical (additive) and ecological correlations. A detailed theory and experimental data on the prediction of genetical and ecological correlations are published in the Reports of Academy of Sciences of USSR, N 3, v. 274, p. 720 and in the Journal 'Genetics', 1986, N 4, p. 616–623.

Controlling the amplitude of genetic variability of a quantitative character in a population by means of changing environmental lim-factors. If the spring wheat variety Saratovskaya 29, which possesses no immunity genes for brown rust, and the variety AS – an analogue of Saratovskaya 29, carrying immunity genes, are grown against the background infested with brown rust, the genetic variance for productivity will be significant. If these genotypes are grown against the background not-infested with rust, the variance for productivity disappears. In the first case we promoted a great genetic variability for productivity, in the second case we reduced it to zero.

Controlling the number of genes, determining the level and genetic variability of a character in a population by changing environmental lim-factors. If in a previous experiment one crosses the variety Saratovskaya 29 and the variety AS, the segregation for productivity against the background, infested with brown rust will be 3 : 1, as the AS variety carries one macrogene of immunity. When grown against the ordinary background (without rust) the divergence in segregation disappears, the distribution of genotypes becomes normal (polygene variant).

Correct choice of characters to be used for selection in particular ecological conditions. The selection for the ratio grain: straw (or harvest index) is known to give good results under relatively favorable conditions for wheat cultivation, when this index is determined by genes of attraction of plastic substances from the straw into the ear. Against the background of drought (Saratov, Tselinograd) where the yield is determined by the genes of drought resistance, the selection for this index gives no results, because any mutation, increasing resistance to drought simultaneously increases the straw mass and the ear mass while the index is left unchanged.

Other examples of the choice of optimum characters for selection see in (3, p. 207).

Artificial 'construction' of ideal foil-characters (characters with zero genetic variance) for express identifi-

cation of genotypes by phenotypes. In this case such an environmental lim-factor is chosen to which the genotypes from a particular set are similarly adaptive (no genetic polymorphism). If such a lim-factor influences the population at the moment when a certain quantitative character is being initiated and formed, the genetic variance of this character will be close to zero.

Understanding the transgression mechanisms and prediction of their appearance in particular hybrid populations (families from diallel crosses). If formation of productivity takes place in a breeding zone against the background of one and the same lim-factor (for example, drought on virgin soils of Kazakhstan) the transgression for productivity can be obtained from the crosses of a drought resistant variety (producing the maximum of a dry biomass per unit area in phytocenosis) and variety, carrying good genes of attraction (affecting the distribution of plastic substances from straw and leaves into an ear). If breeding is carried out in Western Siberia (in the period of NG initiation it is drought, and in the period of MG formation it is cold) the transgression for productivity will be obtained as a result of crossing a drought resistant variety with a cold hardy variety (by combination of two systems of adaptivity), and the like.

Understanding of mechanisms and prediction of the effects of a genotype × environment interaction (G × E). The theory of development and the methods of predicting the G × E effects are available details in the publications (16, 17).

Understanding of the nature of pleiotropy for quantitative characters and prediction of the 'breaks' in pleiotropic complexes in different environmental conditions. If against a background of a long-acting limiting factor in ontogenesis a mutation arises, which reduced adaptivity to this factor, it will also reduce the values of different features, for example, the length of a lower internode and the length of a pod in a pea plant. Against the background of this lim-factor we have a stable phenomenon of pleiotropy. But as soon as we transfer this mutant to another background, where there is no drought in the period of the development of a lower internode and it appears only in the period of pod growth, the 'pleiotropic complex' breaks.

Establishment of new principles in choosing parental pairs for solving three objectives of breeding. a) fast an reliable identification of genotypes by phenotypes in selection, b) obtaining the predicted transgressions, c) breeding for homeostasis.

Theory and practice of improving seed production. If we select the best phenotypes within a cultivar in a year, non-typical for breeding in this zone we significantly worsen the cultivar instead of its improvement.

Working out the general technology of plant breeding, utilizing the systems of artificial climate for studying adaptive properties of genotypes at different stages of their development in different lim-factors.

Possibility, having two known values out of three essential characteristics. a) adaptive properties of varieties, b) definite dynamics of lim-factors, c) genetic parameters of a population – to determine quantitatively and definitely any third unknown characteristic.

Let us consider one of the principle consequences of the ecologo-genetical model.

If fluctuations of environmental limiting factors promote changes in the gene spectrum, determining genetic variability of a quantitative character, it means that in principle it is impossible to obtain stable 'passport' characteristics of the genetical 'structure' of any complex character. So, the expression 'genetics of quantitative characters' is meaningless if it is separated from the definite dynamics of environmental limiting factors.

If it is impossible to obtain stable genetic information about any character influenced by the genotype × environment interaction, it might be not rational to study genetics of character (because they are numbered in thousands). It might be more reasonable to find out what genetic-physiological systems are involved by breeders in the improvement of plant species and then to study a genetic organization of these systems.

Ecologo-genetical model completely rules out the use of marker genes (signals of A.S. Serebrovsky (18), molecular markers (19), RFLP markers (20) *for breeding of quantitative character.* The labile genetic formula of a complex character cannot correlate with any rigid structure of markers even in case of their dense coverage of an entire genom.

In this respect we completely agree with Maletsky (14), notwithstanding the disagreement in arguments.

Possibilities of predicting the configurations of diagrams according to Hayman and the ranks of para-

meters of Hayman without accomplishing the diallel crosses. If we intend to include in the matrix of diallel crosses the Sweden (coldhardy) varieties and the varieties, adapted to the Saratov region (drought resistant) and to carry out the experiment in the Tyumen province (drought in spring influences the character the 'number of grains per plant' and early colds in autumn influence the character the 'mass of one grain'), then by sure we shall obtain the graph of Hayman for the character the 'mass of grains per plant' with a strong effect of 'overdominance' (the regression line will cross X-axis far away from zero point of coordinates).

From Figure 1 (the model of ecologo-genetical control) it follows that if a component character NG is influenced by drought and a character MG – by cold, then *the genetic nature of a resultant character: the 'mass of grains per plant' cannot be described either by the language of traditional mendelian genetics, or by the language of molecular genetics, or by the language of any branches of genetics, existing at present.* The truth is in fact that genetic variability of NG is determinated by genes for drought resistance, MG – by genes for cold hardiness, and RCh by both clusters of genes, accompanied with the drift in time and multiplicative effect (NG multiplied by MG gives RCh). So, the real nature of the initiation and development of RCh can be described only by the language of model of ecologo-genetical control.

This makes doubtful correctness of all previous research works in which the investigators made attempts to describe the genetical 'construction' of quantitative characters by language of traditional Mendelian (or quantitative) genetics, not taking into account their module structures and the dynamics of environmental limiting factors.

Possibility of creating the quantitative theory of plant ontogenesis on the basis of the model of ecologo-genetical control. At present, on the basis of the theory of redetermination of genetic formulas (the genes spectra) of quantitative characters in the dynamics of environmental lim-factors and on the basis of the theory of differential activity of genes (heat-shock proteins, hormonal drifts, calmodulin effects, transmembranous potentials, regulatory genes, etc.) it is possible to predict the ontogenesis ranks of different genotypes in any combinations and successions of environmental lim-factors. In future, while the experiments being equipped with devices and computers, we would be able to proceed from predicting the ranks to the quantitative predictions of the level of polygene characters of different genotypes.

Ecologo-genetical model of organization and control of quantitative characters provides new possibilities for genetic survey of the populations of monopodial coniferous plants. These plants have no intercalary meristems and therefore an annual linear axis increment is fixed on the trunk forever and is not changed during apical growth of the tree. It is evident that individuals with great linear increment in the cold years possess valuable genes for cold hardiness, and the great increment in the drought years testifies to the presence of genes for drought resistance. Individuals with genetic homeostasis are known to have good increments in the years with different lim-factors, i.e. they will be real 'plus-trees', the most promising for vegetative (or possibly seed) propagation.

There are at least 10–15 principle consequences resulting from an ecologo-genetical model, and they are of great help in solving the most important tasks of the theory of plant and animal breeding.

References

1. Кэксер, Г., 1963. Кибернетические модели развития и наследственности/Моделирование в биологии. М.: ИЛ, С.42–64.
2. Малецкий, С.И., 1991. Слитное наследование (новая парадигма). Новосибирск: ИЦИГ СО РАН.
3. Драгавцев, В.А., Р.А. Цильке, Б.Г. Рейтер и др, 1984. Генетика признаков продуктивности яровых пшениц в Западной Сибири. Новосибирск: СО АН.
4. Драгавцев, В.А., П.П. Литун, И.М. Шкель, Н.Н. Нечипоренко, 1984. Модель эколого-генетического контроля количественных признаков растений./Докл. АН СССР, Т.274, N 3. С.720–723.
5. Драгавцев, В.А., Н.В. Утемишева, 1975. К проблеме онтогенетической изменчивости генетико-статистических параметров в растительных популяциях/Генетика. N 11. С.128–140.
6. Mather, K., 1949. The genetical theory of continuous variation./Hereditas, Suppl. V.
7. Игнатьев, М.В., 1933. Количественный анализ действия наследственности и среды/Биологический журнал. Т.2, вып. 45. С.458.
8. Fisher, R.A., 1918. The correlation between relatives on the supposition of mendelian inheritance./Trans. Roy. Soc. Edinburgh, V.52, P.399.
9. Fisher, R.A., 1958. The genetical theory of natural selection. 2-nd. New York.
10. Wright, S., 1952. The genetics of quantitative variability/ Quantitative inheritance, Edinburg, 1950. Ed. London.
11. Lush, J.L., 1949. Heritability of quantitative characters in farm animals./Proc. 8-th Juf. Congress Genetios./Hereditas. Suppl. 1, 356.
12. Финчем, Дж., 1968. Генетическая комплементация. М.: Мир, 1968.
13. Ландау, Л.Д., Е.М. Лифшиц, 1963. Квантовая механика, нерелятивистская теория. М.: Физматгиз, 2-е изд. С.543.
14. Рахман, М., В.А. Драгавцев, 1990. Новые подходы к прогнозированию гетерозиса у растений/С-х. биология. N 1. С.3–12.
15. Драгавцев, В.А., 1992. Новые подходы к прогнозированию гетерозиса. Использование гетерозиса у овощных и бахчевых культур. Л.: ВИР, (Сб. по прикл. бот., ген., и сел. Т.145). С. 9–11.
16. Драгавцев, В.А., А.Ф. Аверьянова, 1983. Переопределение генетических формул количественных признаков пшеницы в разных условиях среды./Генетика. Т.19, N 11. С. 1811–1917.
17. Драгавцев, В.А., А.Ф. Аверьянова, 1983. Механизмы взаимодействия генотип-среда и гомеостаз количественных призначов растений./Генетика. Т.19, N 11. С. 1806–1810.
18. Серебровский, А.С. Методы сигналей, 1970./Генетический анализ. М.: Наука, С.252–273.
19. Stuber, C.W., R.R. Moll, M.M. Goodman, N.E. Shaeffer & B.S. Weir, 1980. Allozyms frequency changes associated with selection for increased grain yield in maize./Genetics. V.95 P.225–236.
20. Godshalk, E. D., M. Lee, K.R. Lamkey, 1990. Relationship of restriction fragment length polymorphisms to a single-cross hybrid performance of maize/TAG, V.80. P.273–280.

P.M.A. Tigerstedt (Ed.), *Adaptation in Plant Breeding*, 69–75, 1997.

Characterization of alfalfa (*Medicago sativa* L.) following *in vitro* selection for salt tolerance

A. Safarnejad[1], H.A. Collin[1], K.D. Bruce[1] & T. McNeilly
Department of Environmental and Evolutionary Biology, University of Liverpool, P.O. Box 147, Liverpool L69 3BX, U.K.; present address: [1] Department of Genetics and Microbiology; University of Liverpool, P.O. Box 147, Liverpool L69 3BX, U.K.

Key words: alfalfa, antioxidant, Southern blotting, proline, salinity, somaclones, lucerne, *Medicago sativa*

Summary

Alfalfa (*Medicago sativa* L.) is a valuable forage crop which is grown in areas of limited rainfall, high temperature and where the land is often salt affected. Seedlings of the commercial variety CUF 101 and the more salt tolerant breeding line CUF101-1S were used as explant material to produce tissue cultures for an *in vitro* screen for salt tolerance. Callus cultures were placed on a regeneration medium containing 0–350 mol m^{-3} NaCl to establish the highest possible salt concentration that would allow buds to regenerate and survive as plantlets. Buds were regenerated in 250 mol m^{-3} NaCl then isolated, grown to maturity and set seed. The response to salt tolerance in the short term was investigated by exposing 14-day-old seedlings to 200 mol m^{-3} NaCl. Measurement of growth, survival, proline concentration and the activity of antioxidant enzymes were made after 14 days exposure to NaCl. One somaclone in particular (6R2IV) compared with the parent line showed increased salt tolerance, greater accumulation of proline and a greater increase in the antioxidant enzyme, glutathione reductase. A potentially important gene is pA9 which is known to be responsible for the production of a proline rich cell wall protein. Polymerase Chain Reaction (PCR) amplification of a region of the pA9 gene indicated that it was present in both tolerant and sensitive lines. Southern blotting has shown that the copy number of the pA9 was present as multiple copies in the tolerant clone compared with a single copy in the parents.

Abbreviations: GR – glutathione reductase

Introduction

The effects of salinity on plant growth have been the subject of intensive research for nearly a century because of the limit to productivity by salt stress in crop plants in arid and semi-arid areas. Alfalfa (*Medicago sativa* L.) is the most important forage crop in these areas and hence it is important that its productivity be increased by an improvement of salt tolerance. A plant breeding approach has only had limited success in improving the salt tolerance of crop species (Epstein et al., 1980). These methods require long term selection and testing and have produced an improved germplasm capable of germinating at high levels of salinity (Dobrenz et al., 1983), but are slow to yield substantial improvement in salt tolerance and growth of crop plants (Winicov, 1991). Selection through tissue culture technology offers an alternative route for improvement to breeding in alfalfa (Arcioni et al., 1990). For example somaclonal variation which represents new variation commonly occurs in many plant species following regeneration from tissue culture (Larkin & Scowcroft, 1981). Somaclonal variation in alfalfa has been used to select for salt tolerance (Bingham & McCoy, 1986). In this technique an *in vitro* selection pressure can be applied to a large population of somatic cells to uncover a desired phenotype so that the number of plants that eventually require field evaluation can be decreased (Hasegawa et al., 1994), thus saving time and space. Selection can be carried out *in vitro*, by culturing either callus pieces, cell suspensions, protoplasts, plantlets, embryos or microspores in the presence of a screening agent (Collin & Dix, 1990).

Although considerable research has been conducted on salt tolerance of alfalfa (Stavarek & Rains, 1984; McCoy, 1987b; Winicov, 1991) in most of this work, salt-tolerant cell lines have not been regenerated, or if plants have been established, they did not inherit the salt tolerance. There has been one report (Winicov, 1991) where whole plant NaCl tolerance was expressed in plants regenerated from NaCl-tolerant cell lines of thc noncommercial variety Regen-S. Winicov & Deutch (1994) made use of biochemical and molecular techniques to characterize new genotype variability. In such a multifactorial response, the basis of tolerance may be related to biosynthesis of intracellular compatible solutes (Delauney & Verma, 1993), the level of antioxidant enzymes (Foyer et al., 1994), or ionic flux (Niu et al., 1993). Winicov et al. (1989) showed that the acquisition of salt-tolerance by callus cells of alfalfa was accompanied by an increased expression of a subset of genes of which some were expressed constitutively, while other were induced. Salt-tolerant alfalfa have been used to characterize a cDNA clone that identified transcripts accumulated specifically in alfalfa roots in response to NaCl (Winicov & Deutch, 1994).

The aim of this work was to screen for increased salt tolerance in the commercial variety of alfalfa, CUF 101, a line CUF101-1S and their regenerants using an *in vitro* screen. The *in vitro* selection was based on the use of differentiating cultures so that selection for salt tolerance would occur during the regeneration process. The further aim was to use biochemical and molecular techniques to distinguish between salt tolerant and sensitive somaclonal and parental alfalfa.

Materials and methods

Plant materials

Seeds of parent CUF 101 were supplied by Cleanseed Pty Ltd (POBox 31, Bongendore NSW, Australia 2621). The CUF 101 progeny, CUF101-1S, which showed moderate salt tolerance was produced at the University of Liverpool (Al-Khatib et al., 1993).

Figure 1. Callus of *M. sativa* shown multiple shoot buds.

Initiation and regeneration of tissue culture

Seeds were sterilised by 0.1% w/v mercuric chloride and 0.1% w/v sodium dodecyl sulphate for 10 min followed by 15% v/v 'Domestos' bleach for 20 min then washed in five changes of sterile distilled water. Surface sterile seeds were placed singly on the surface of 10 ml aliquots of a germination medium containing either Murashige & Skoog (1962), or half strength Schenk & Hildebrandt (1972) medium, with 1% w/v sucrose, no growth regulators and 0.8% Difco Bacto Agar. The media were contained in Universal vials which were maintained at 25 °C under fluorescent light on a 12 h photoperiod. After 8 days, 1 cm length of hypocotyl was placed on the surface of 10 ml BII medium (Saunders & Bingham, 1975) in Universal vials. After two weeks, each complete explant with all its attached callus was transferred to a regeneration medium, Boi2y (Saunders & Bingham, 1975), and incubated as before. After two to five weeks buds were visible on the callus. When larger these shoots were individually transferred to a modified Schenk and Hildebrandt medium (1972) (GS) containing 1% w/v sucrose, no growth regulators and 0.8% Difco Bacto Agar. Well rooted plantlets were transferred to Jiffy-7 peat moss containers (Jiffy products Ltd., Norway) in a greenhouse mist bed for about 10 days, and subsequently potted into a soil-sand-peat moss mixture (3:1:1). Inflorescence were self pollinated by hand then

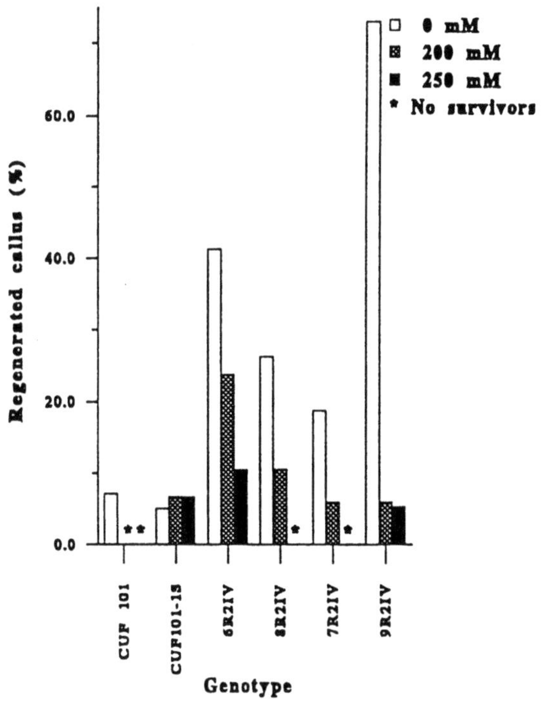

Figure 2. Regeneration in callus during *in vitro* selection of *M. sativa* cv. CUF 101, somaclones (6R2IV, 8R2IV, 7R2IV, 9R2IV) and parent CUF101-1S in presence of salt (NaCl).

covered immediately with non moisture-proof glassine bags for a few days (Sayers & Murphy, 1966). The bags were removed to allow seed pods and seed to mature. The seeds from regenerants then underwent another cycle of tissue culture initiation and regeneration so as to increase the regeneration frequency. Cultures from seeds of CUF 101, CUF101-1S line and 4 regenerants were placed on a regeneration medium containing 0–350 mol m^{-3} NaCl. Shoots on the medium with a high concentration of NaCl (250 mmol m^{-3} were transferred to GS medium for root induction and shoot elongation. The plants were allowed to flower then cross pollinated within plants derived from the same clone to set seed.

Plant growth in NaCl stress

Uniform seeds were surface sterilised in 2% v/v sodium hypochlorite for 2 minutes, rinsed in sterile water and dried at room temperature. For each replicate, 20 seeds were evenly spaced over a raft of alkathene beads in pots containing 300 ml of nutrient solution, modified half strength Rorison solution (Hewitt, 1966) plus 4 mol m^{-3} $CaCl_2$. All the pots were arranged randomly in a tray that was filled with water and placed under clear plastic in a growth room at 25–27° C. After two weeks shoot length and root length were measured and leaves extracted for estimates of proline and antioxidant enzymes. All the equipment used was sterilised to reduce the effect of contaminating organisms on the performance of the seedlings.

Preparation of extracts for proline estimation

Seedling tissue was frozen in liquid nitrogen then when required was homogenized in methanol:chloroform:water (MCW 12:5:1 /V) using 0.2 gram tissue per 2 cm^3 of MCW at room temperature and the proline concentrations estimated by a modification of the method described by Singh et al. (1973). The only modification was that the tubes were heated in a boiling water bath for 65 minutes.

Preparation of extracts for enzyme assays

Fresh seedlings (400 mg) were homogenized in 400 μl 0.1 M K_2HPO_4 (PH 7.0) in a pestle and mortar. The homogenized samples were washed with 2 × 200 μl 0.1 M K_2HPO_4 into Eppendorf tubes and centrifuged for 10 minutes (10000 rpm) in an Eppendorf microfuge. The pellet was resuspended in 800 μl 0.1 M K_2HPO_4 and centrifuged. The combined supernatant was used for the determination of enzyme activity of glutathione reductase (Halliwell & Foyer, 1978).

PCR amplification of isolated DNA

Genomic DNA, extracted from leaflets (Yu & Pauls, 1993) were used in PCR amplification, primers (OMID 1 and OMID 2 representing 200 bp of pA9) were added to 50 mol m^{-3} dNTPs, 2.5 U of Taq DNA polymerase (Boehringer, UK) and 10 μl of 10 × Taq DNA polymerase buffer. Reaction volumes were made to 100 μl with sterile distilled water. Amplification conditions were one cycle of 94 °C for 4 min followed by 30 cycles (94 °C 1 min, 55 °C 1 min, and 72 °C 10 min) and one cycle of 72 °C for 10 min using a Perkin Elmer Cetus 480 thermocycler.

Southern blotting

DNA isolated from leaves (Doyle et al., 1990) was digested with the restriction endonuclease *Eco*RI. DNA (10 μg/lane) was subjected to electrophoresis,

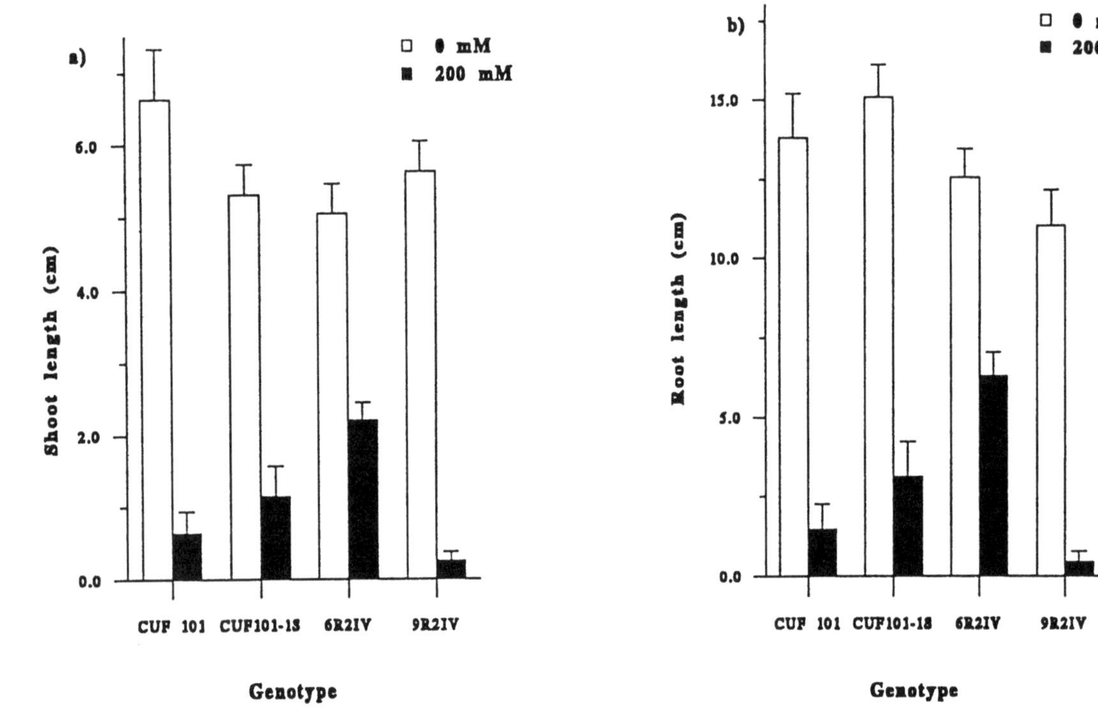

Figure 3. Growth in 200 mol m^{-3} NaCl of 14-day-old seedlings of parental and somaclonal genotypes of *M. sativa*. Somaclones were derived from an *in vitro* screen at 250 mol m^{-3} NaCl. a: Shoot length. b: Root length.

transferred to a nylon membrane, and hybridized to a radioactive probe of pA9-like sequence. Hybridisation protocols were as recommended in the Gene Screen^{TM} Plus instruction manual (DuPont), with the exception that 65 °C washes were used in place of 60 °C in order to detect sequence similarity of greater than 70%. Membranes were exposed to Fuji RX film at - 70 °C and film developed.

Results

In vitro *selection for salt tolerance*

Before differentiating cultures could be used in an *in vitro* selection procedure it was necessary to increase the regeneration capacity of the cv. CUF 101. There was limited regeneration of alfalfa CUF 101 and CUF101-1S in the first tissue culture cycle whereas regeneration in the 2nd tissue culture cycle was enhanced 15 fold (Figures 1 and 2). The percentage of regenerated callus decreased with increase of NaCl in the medium. There were no surviving regenerants of CUF 101 at 200 and 250 mol m^{-3} NaCl and no survivors at 250 mol m^{-3} NaCl for the regenerants 8R2IV and 7R2IV somaclones. One of the somaclones (6R2IV) showed enhanced regeneration in the presence of 200 and 250 mol m^{-3} NaCl in compared with the parent explants (Figure 2).

Growth measurement

Seed from CUF 101, CUF101-1S and from clones that survived the 250 mol m^{-3} *in vitro* screen were grown in 200 mol m^{-3} NaCl for 14 days. Root length was reduced in all genotypes in the presence of 200 mol m^{-3} NaCl. The reduction was 50% of the control in zero NaCl for 6R2IV and 20% of the control in CUF101-1S (Figure 3b). The reduction in shoot length showed the same pattern but with less of a reduction in 6R2IV (40%) and CUF101-1S (20%) compared with their zero controls (Figure 3a).

Proline estimation

Proline accumulation in 14-day-old seedlings of CUF 101, CUF101-1S and 6R2IV exposed to 200 mol m^{-3} NaCl was measured (Figure 4). No difference in proline accumulation was detected in genotypes in the absence of NaCl. At 200 mol m^{-3} NaCl, proline

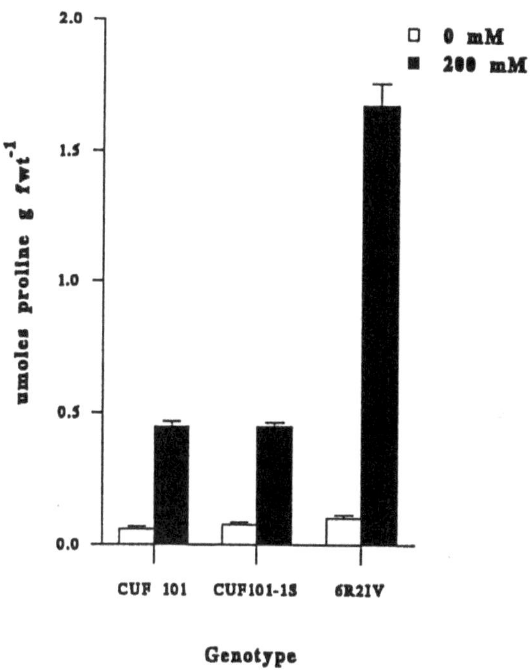

Figure 4. Proline accumulation in 14-day-old seedlings of *M. sativa* cv. CUF 101, somaclone 6R2IV and parent CUF101-1S in response to salinity stress (NaCl).

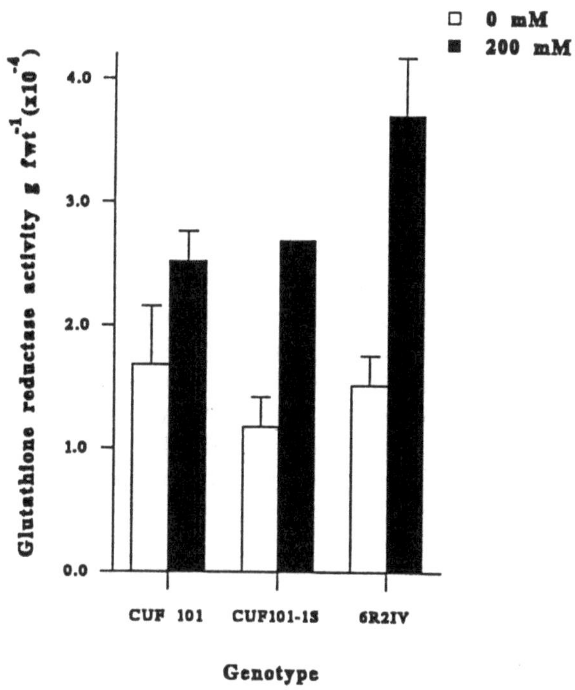

Figure 5. Activity of glutathione reductase in 14-day-old seedling of *M. sativa* cv. CUF 101, somaclone 6R2IV and parent CUF101-1S in response to salinity stress (NaCl).

accumulation increased in all genotypes with the very much larger increase in 6R2IV. Proline accumulation in 6R2IV at 200 mol m^{-3} NaCl increased 16.2 fold compared with 0 mol m^{-3} NaCl and 3.8 fold higher than that measured in CUF 101 and CUF101-1S.

Enzyme assay

The activity of gluthatione reductase antioxidant enzyme in 14-day-old seedlings of parental CUF 101, CUF101-1S and somaclone 6R2IV exposed to 200 mol m^{-3} NaCl was measured. There was no significant difference in GR activity between control of different genotypes, however, GR activity in 6R2IV in 200 mol m^{-3} NaCl increased 246% ± 32% above 0 mol m^{-3} NaCl levels (Figure 5). At 200 mol m^{-3} NaCl, GR activity was 35% higher in 6R2IV than that in CUF 101 and CUF101-1S while CUF 101 increased 148% ± 14% and in CUF101-1S 227% above 0 mol m^{-3} NaCl levels.

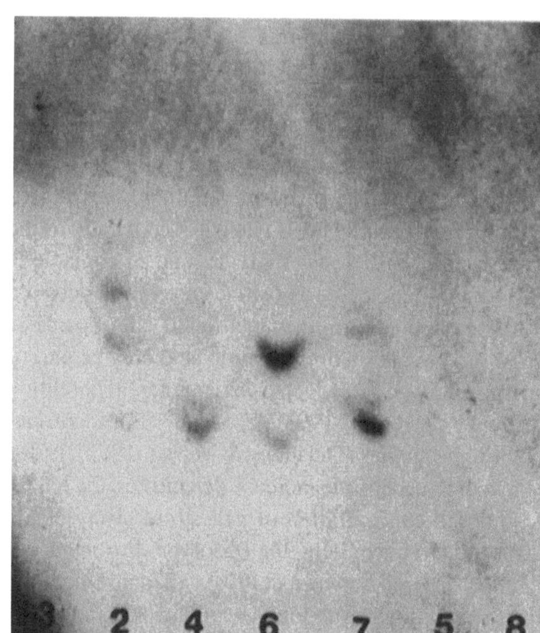

Figure 6. Hybridization of PCR products of *M. sativa* cv. CUF 101 (3), somaclones a [9R21V (2), 6R2IV (6), 7R1 (7), 7R2VO (5), 7R2 (8) and parent CUF101-1S (4).

PCR amplification and Southern Blot analysis

To attempt to distinguish between salt tolerant and sensitive lines of alfalfa, experiments using Polymerase Chain Reaction (PCR) amplification of a region of the pA9 gene were initiated. This indicated that pA9-like sequences were present in tolerant and sensitive lines. To estimate the number of genes corresponding to the pA9 sequence within the alfalfa genome, genomic DNA from tetraploid plants (*M. sativa*, CUF101-1S and somaclone 6R2IV) was digested with the restriction endonuclease *Eco*RI and the resulting fragments were tested for hybridization to pA9-like sequences. One band was observed in CUF101-1S (Figure 6, no. 4) and two bands in somaclone 6R2IV (Figure 6, no. 6), thus indicating the pA9 was present as multiple copies in the tolerant clone compared with a single copy in the parent.

Discussion

The aim of the work was to select for increased salt tolerance in alfalfa by screening differentiating cultures in the presence of high concentration of NaCl. The advantage of this method is that it avoids a separate differentiation stage for selected cells. In addition shoot and root induction and elongation occur in the presence of salt. The results from this study show one of the somaclones (6R2IV) was significantly more tolerant at 200 and 250 mol m^{-3} NaCl level than CUF101-1S.

The data indicated a very strong correlation between proline accumulation, antioxidant enzyme activity and the ability of alfalfa seedling to grow on a nutrient solution with NaCl. The NaCl-induced proline increase in CUF 101, CUF101-1S and in somaclone 6R21V ($\times$ 16) indicated a positive correlation between proline accumulation and adaptation to salt or osmotic stress. Accumulation of this amino acid may be part of general adaptation to adverse environmental condition (Delauney & Verma, 1993). That there is no correlation in some studies (Delauney & Verma, 1993), is due to the fact that a proline increase depends on the length of time and concentration of salt stress (Bray et al., 1991) and as shown here, the genotype. Proline accumulation differed between cultivars adapted to certain growth conditions or regions, as well as within species tolerant to drought or salinity (Heuer et al., 1994). In alfalfa, the most tolerant species exhibited the highest potential for proline accumulation in response to salinity (Heuer et al., 1994) suggesting that it was involved in the osmotic adjustment of salt-stressed alfalfa. This study showed a positive correlation between increased GR activity and salt tolerance in seedlings of alfalfa. It is unknown whether the increase was due to an up regulation of the genes controlling the synthesis of these enzymes or an increased activation of constitutive enzyme pools (Gossett et al., 1994). Other studies have indicated that salt stress increased synthesis of specific proteins in some plants (Hurkman et al., 1989).

Experiments using Polymerase Chain Reaction (PCR) amplification of a region of the pA9 gene indicated that it was present in tolerant and sensitive lines. Southern blotting has been used to determine the copy number of the pA9 in all of these genotypes. It was present as multiple copies in the tolerant clone compared with a single copy in the parents. The hypothesis is that pA9 encodes a protein, related to proline-rich cell wall proteins, which may be important in salt-inducible gene expression in alfalfa (Winicov & Deutch, 1994).

References

Al-Khatib, M., T. McNeilly & J.C. Collins, 1993. The potential of selection and breeding for improved salt tolerance in lucerne (*Medicago sativa* L.). Euphytica 65: 43–51.

Arcioni, S., F. Damiani, M. Pezzotti & E. Lupotto, 1990. Alfalfa, Lucerne (*Medicago* spp.). In: Y.P.S. Bajaj (Ed.), Biotechnology in Agriculture and Forestry, Legumes and Oilseed Crops I, 10: 197–241.

Bingham, E.T. & T.J. McCoy, 1986. Somaclonal variation in alfalfa. In: J. Janick (Ed.) Plant Breeding Reviews, 4: 123–152. AVI, Westport, Connecticut.

Bray, L., D. Chriqui, K. Gloux, D. LeRudulier, M. Meyer & J. Peduzzi, 1991. Betaines and free amino acids in salt stressed vitroplants and winter resting buds of *Populus teichcarpa* × *deltoides*. Physiologia Plantarum 83: 136–143.

Collin, H.A. & P.J. Dix, 1990. Culture systems and selection procedures. In: P.J. Dix (Ed.) Plant Cell Line Selection, procedures and application, pp. 3–18. VCH, Weinham, Germany.

Delauney, A.J. & D.P.S. Verma, 1993. Proline biosynthesis and osmoregulation in plants. The Plant Journal 4: 215–223.

Dobrenz, A.K., M.H. Schonhorst, J.E. Stone, R.K. Thompson, S. Allen & D. Smith, 1983. Registration of AZ-GERM SALT-I nondormant alfalfa germplasm. Crop Sci 23: 807.

Doyle, J.J., J.L. Doyle & L.H. Bailey Hortoriun, 1990. Isolation of plant DNA from tissue. Focus 12: 13–15.

Epstein, E., J.J. Norlyn, G.W. Rush, R.W. Kingsbury, D.W. Kelly, G.A. Cunningham & A.F. Wrona, 1980. Saline culture of crops: a genetic approach. Science 210: 399–404.

Foyer, C.H., P. Descourvieres & K.J. Kunert, 1994. Protection against oxygen radicals: an important defence mechanism studied in transgenic plants. Plant, Cell and Environment 17: 507–523.

Gossett, D.R., E.P. Millhollon, M.C. Lucas, S.W. Banks & M.-M. Marney, 1994. The effects of NaCl on antioxidant enzyme activities in callus tissue of salt-tolerant and salt-sensitive cotton cultivars (*Gossypium hirsutum* L.). Plant Cell Reports 13: 498–503.

Halliwell, B. & C.H. Foyer, 1978. Properties and physiological function of a glutathione reductase purified from spinach leaves by affinity chromatography. Planta 139: 9–17.

Hasegawa, P.M. R.A. Bressan, D.E. Nelson, Y. Samaras & D. Rhodes, 1994. Tissue culture in the improvement of salt tolerance in plants. In: A.R. Yeo & T.J. Flowers (Eds), Soil Mineral Stresses, Approaches to Crop Improvement, Monographs on Theoretical and Applied Genetics 21: 83–125.

Hewitt, E.J., 1966. Sand and water culture methods used in the study of plant nutrition. Comm Agric Bur Tech Comm No. 22.

Heuer, B., 1994. Osmoregulatory role of proline in water- and salt-stressed plants. In: M. Pessarakli (Ed.), Handbook of Plant and Crop Stress, pp. 363–383.

Hurkman, W.J., C.S. Fornari & C.K. Tanaka, 1989. A comparison of the effect of salt on polypeptides and translatable mRNAs in roots of a salt-tolerant and salt-sensitive cultivar of barley. Plant Physiol 90: 1444–1456.

Larkin, P.J. & W.R. Scowcroft, 1981. Somaclonal variation: a novel source of variability from cell cultures for plant improvement. Theor Appl Genet 60: 197–214.

McCoy, T.J., 1987b. Characterization of alfalfa (*Medicago sativa* L.) plant regenerated from NaCl-tolerant cell lines. Plant Cell Rep 6: 417–422.

Murashige, T. & F. Skoog, 1962. A revised medium for rapid growth and bio-assay with tobacco tissue cultures. Plant Physiol 15: 473–573.

Niu, X., M.L. Narasimhan, R.A. Salzman & P.M. Hasegawa, 1993. NaCl regulation of plasma membrane H^+-ATPase gene expression in a glycophyte and a halophyte. Plant Physiol 103: 713–718.

Saunders, J.W. & E.T. Bingham, 1975. Growth regulator effects on bud initiation in callus cultures of *Medicago sativa*. Am J Bot 62: 850–855.

Sayers, E.R. & R.P. Murphy, 1966. Seed set in alfalfa as related to pollen tube growth fertilization frequency, and post fertilization ovule abortion. Crop Sci 6: 365–368.

Schenk, R.U. & A.C. Hildebrandt, 1972. Medium and techniques for induction and growth of monocotyledonous and dicotyledonous plant cell cultures. Can J Bot 50: 199–204.

Singh, T.N., L.G. Paleg & D. Aspinall, 1973. Stress metabolism. Variations in response to water deficit in the barley plant. Aust J Biol Sci 26: 65–76.

Stavarek, S.J. & D.W. Rains, 1984. Cell culture techniques: selection and physiological studies of salt tolerance. In: R.C. Staples & G.H. Toenniessen (Eds) Salinity tolerance in plants. Strategies for crop improvement, pp. 321–334. Wiley, New York.

Winicov, I., 1991. Characterization of salt tolerant alfalfa (*Medicago sativa* L.) plants regenerated from salt tolerant cell lines. Plant Cell Rep 10: 561–564.

Winicov, I. & C.E. Deutch, 1994. Characterization of cDNA clone from salt-tolerant alfalfa cells that identifies salt-inducible root-specific transcripts. J Plant Physiol 144: 222–228.

Winicov, I., J.H. Waterborg, R.E. Harrington & T.J. McCoy, 1989. Messenger RNA induction in cellular salt tolerance of alfalfa (*Medicago sativa*). Plant Cell Rep 8: 6–12.

Yu, K. & K.P. Pauls, 1993. Rapid estimation of genetic relatedness among heterogeneous populations of alfalfa by random amplification of bulked genomic DNA samples. Theor Appl Genet 86: 788–794.

P.M.A. Tigerstedt (Ed.), *Adaptation in Plant Breeding*, 77–80, 1997.

Photoperiod insensitivity gene essential to the varieties grown in the northern limit region of paddy rice (*Oryza sativa* L.) cultivation

Yutaka Okumoto, Katsuyuki Ichitani, Hiromo Inoue & Takatoshi Tanisaka
Laboratory of Plant Breeding, Faculty of Agriculture, Kyoto University, Kyoto 606-01, Japan

Key words: heading time, gene analysis, *Oryza sativa*, photoperiod insensitivity, rice

Summary

The release of extremely early maturing varieties has made it possible to cultivate rice in Hokkaido (NL45-42°) in Japan, the northern limit region of paddy rice cultivation. Until then, rice cultivation in this region has been impracticable due to the climate condition, especially short summer and long-day more than 15 hours during summer. Experiment results confirmed that the success of rice cultivation in this area depends on raising photoperiod insensitivity varieties with short basic vegetative growth period. Moreover, in this study, the genetic factors controlling the photoperiod insensitivity of Hokkaido varieties were analyzed by using 8 kinds of tester lines for three loci, *E1, E2,* and *E3*, controlling photoperiod sensitivity. It was found out that all the varieties examined carry *e1*, a photoperiod insensitivity allele of *E1* locus, but as for the other loci, *E2* and *E3*, the existence of plural alleles were recognized. We have already clarified that almost all the japonica-type varieties grown in Japan (except Hokkaido) and Taiwan carry *E1* bringing about strong photoperiod sensitivity. Accordingly, it can be concluded that *e1* is the gene essential to Hokkaido varieties: the interchange of *E1* with *e1* enabled rice cultivation under long-day condition.

Introduction

Japonica-type varieties of rice are widely cultivated through the temperate zone in Asia. Formerly, rice was a short-day plant well adapted to warm climate; hence, rice cultivation had been impracticable in high-latitude area due to the short summer and long-day more than 15 hours during summer, until extremely early maturing varieties were raised. It was during 1900 to 1930 that such varieties were firstly released and planted in the northern end island, Hokkaido, in Japan (NL 45-42°). The varieties raised for rice cultivation in Hokkaido also enabled rice cultivation even in Hei Long Jiang province (NL53-43°) of China. Previous works on geographical analysis of heading traits in Japanese rice varieties (Hosoi, 1981; Tanisaka et al., 1992) indicated that the extremely early maturing of Hokkaido varieties depends on almost complete elimination of photoperiod sensitivity.

On the other hand, it is known that there exist at least five heading-time loci, *E1, E2, E3, Se-1,* and *Ef-1*, in rice and the latter 4 loci are responsible for the variation of heading time among Japonica-type varieties except Hokkaido ones (Okumoto et al., 1991, 1992a, 1992b): the former 4 loci mainly control the degree of photoperiod sensitivity (Yamagata et al., 1986; Yokoo & Fujimaki, 1971), while the last one controls the period of basic vegetative growth (Tsai, 1976). In this study, first, the degree of photoperiod sensitivity (PS) and the period of basic vegetative growth (BVG) of 63 Japanese varieties including 27 Hokkaido ones were estimated, and then the genotypes for three loci, *E1, E2* and *E3*, of 9 varieties which were selected from the above 27 Hokkaido varieties were determined. On the basis of the results obtained from the above two experiments, the genetic factors controlling the photoperiod insensitivity of Hokkaido varieties were discussed.

Table 1. Genotypes of 8 tester lines used in this experiment

Line	Genotype
EG0	*e1e1e2e2e3e3*
EG1	*E1E1e2e2e3e3*
EG2	*e1e1E2E2e3e3*
EG3	*e1e1e2e2E3E3*
EG4	*E1E1E2E2e3e3*
EG5	*e1e1E2E2E3E3*
EG6	*E1E1e2e2E3E3*
EG7	*E1E1E2E2E3E3*

Materials and methods

Estimation of PS and BVG

Twenty seven varieties cultivated in Hokkaido (NL45-42°) and 36 varieties cultivated in Tohoku (NL41-36°) and South-west (NL36-31°) regions of Japan were grown under two kinds of day-length conditions, a short-day-length (10 h) and a long-day-length (16 h), after 25 days of raising seedlings. Their basic vegetative growth (BVG), which was expressed by the days to heading under 10 h, and photoperiod sensitivity (PS), which was expressed by the difference between days to heading under 16 h and that under 10 h, were estimated.

Identification of the genotype for three loci, E1, E2 *and* E3

Of the above 27 Hokkaido varieties estimated for BVG and PS, 9 varieties were chosen and crossed with eight tester lines for three loci, *E1, E2* and *E3* (Table 1), to determine the genotype for the loci. The F_2 populations were submitted to gene analysis for heading time. In each cross, about 300 F_2 plants were used. All the F_2 plants were grown in a paddy field in Kyoto (NL35°01′) together with their cross parents (ca. 50 plants).

Results

The estimated BVG and PS were analyzed for their genetic variations and their relationships with the heading time under the natural conditions in Kyoto. As seen from Figure 1, all the varieties were clearly divided into three groups according to BVG, PS and the heading time under the natural conditions in Kyoto: a group consisting of the varieties cultivated in Hokkaido (Group H), a group consisting of the varieties cultivated in Tohoku region (Group T) and a group consisting of the varieties cultivated in South-west region (Group S). The average values of BVG in Group H, T and S were 66.0, 76.3 and 70.7, respectively. The difference in BVG among the three groups was not so large, but the varieties of Group H generally showed the least value of BVG. On the other hand, the difference in PS among the three groups was extremely large: the average values of PS of Group H, T and S were 1.8, 19.5 and 70.7, respectively. Thus, the varieties cultivated in higher latitude area tend to show smaller PS than those cultivated in lower latitude area, and most of the Hokkaido varieties showed little response to day-length treatment.

Figure 2 shows the frequency distributions for heading date of the F_2 plants derived from the crosses between three EG lines and two Hokkaido varieties, Ishikari (IS) and Wasefukoku (WF). Moreover, in Table 2, the presence or absence of late transgressive segregants in the F_2 were shown. Under the natural condition in Kyoto, photoperiod-sensitivity alleles, *E1, E2* and *E3*, inevitably bring about delayed heading; hence, the presence of these dominant alleles in a tested variety can be known from the appearance of late transgressive segregants in the F_2. For instance, suppose that a variety carries *E3*, but not either *E1* or *E2*, and late transgressive segregants appear in the F_2 from the crosses with EG0, EG1, EG2, and EG4 (Figure 2). Since these 4 tester lines do not carry *E3*, the appearance of late transgressive segregants, resulting from the joint effect of *E3* with the other late heading-time gene(s), e.g. *E1* or *E2*, will be the proof that this variety surely carries *E3*. This instance comes under the case of WF: in the cross between EG1 (*E1E1e2e2e3e3*) and WF (*e1e1e2e2E3E3*), the late transgressive segregants of EG6 (*E1E1e2e2E3E3*)-type were observed.

As seen from Figure 2 and Table 2, late transgressive segregants did not appear in the F_2 from the crosses of Yukara (YU), Norin 20 (N20), Ishikari (IS), and Kitahikari (KT) with all the EG lines. This indicates that these four varieties carry none of *E1, E2* and *E3*. For Kirara 397 (KR) and Shiokari (SH), late transgressive segregants did not appear only in the crosses with EG lines carrying *E2*, indicating that both varieties carry *E2* but not either *E1* or *E3*. For Eiko (EI) and Wasefukoku (WF), late transgressive segregants appeared in the crosses with EG1, EG2 and EG4. This indicates that both varieties carry *E3*, but not either *E1* or *E2*. For Sasahonami (SA), late transgressive segre-

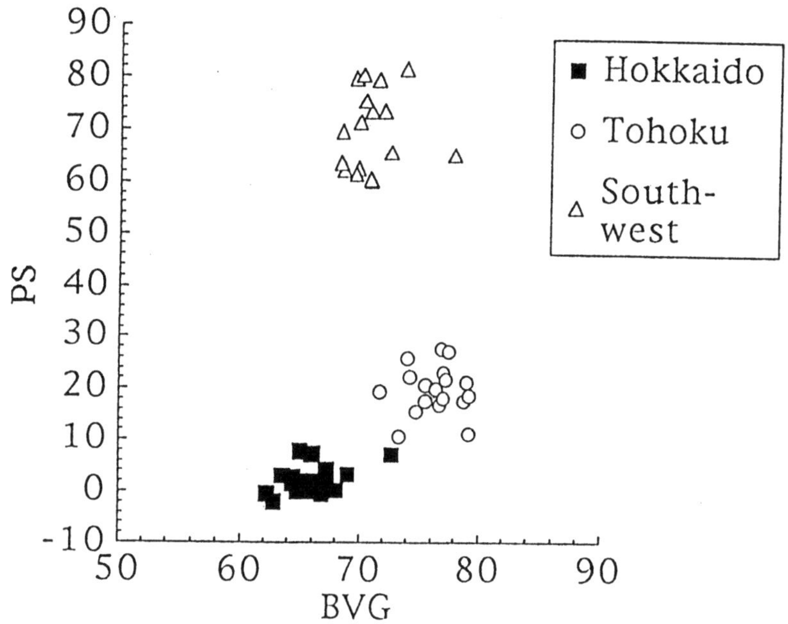

Figure 1. Basic vegetative growth (BVG) and photoperiod sensitivity (PS) of rice varieties cultivated in Hokkaido, Tohoku and South-west regions.

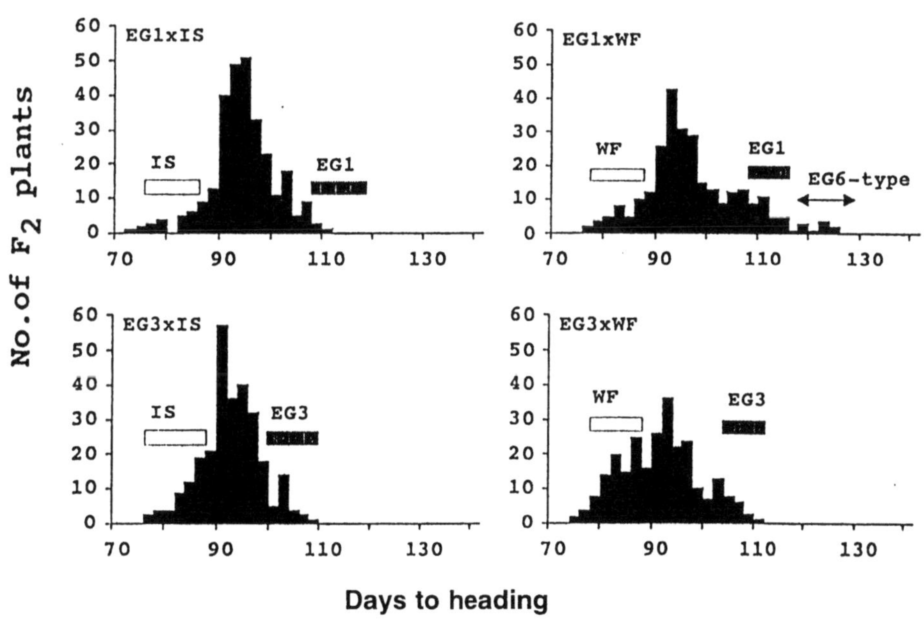

Figure 2. Frequency distributions of the heading date in F_2 populations from the crosses between EG lines and two Hokkaido varieties, Ishikari (IS) and Wasefukoku (WF).

gants appeared in the crosses with EG1, EG2 and EG3, but not in the crosses with EG5 and EG7. It was not obvious whether late transgressive segregants appeared in the cross with EG6 or not, but the results obtained surely indicate that SA carries *E2* and *E3*, but not *E1*.

Table 2. Presence (+) or absence (-) of transgressive segregants towards lateness in the F_2 derived from the crosses between Hokkaido varieties and tester lines

Tester line	Varieties cultivated in Hokkaido[1]								
	YU	N20	IS	KT	KR	SH	EI	WF	SA
EG0	-	nt	nt	-	+	nt	nt	nt	nt
EG1	-	-	-	-	+	+	+	+	+
EG2	-	-	-	-	nt	-	+	+	+
EG3	-	-	-	nt	+	+	-	-	+
EG4	nt	nt	nt	nt	nt	nt	+	+	nt
EG5	-	-	-	nt	-	-	-	-	-
EG6	nt	-	-	nt	nt	nt	nt	-	-
EG7	-	-	-	nt	nt	nt	-	-	-

[1] nt: not tested.

Table 3. Genotypes for 4 photoperiod sensitivity loci of nine Hokkaido varieties

Variety	Genotype[1]	
Yukara (YU)	*e1e1e2e2e3e3*	$Se\text{-}1^e Se\text{-}1^e$
Norin20 (N20)	*e1e1e2e2e3e3*	unknown
Ishikari (IS)	*e1e1e2e2e3e3*	unknown
Kitahikari (KT)	*e1e1e2e2e3e3*	$Se\text{-}1^e Se\text{-}1^e$
Kirara397 (KR)	*e1e1E2E2e3e3*	$Se\text{-}1^n Se\text{-}1^n$
Shiokari (SH)	*e1e1E2E2e3e3*	$Se\text{-}1^e Se\text{-}1^e$
Eiko (EI)	*e1e1e2e2E3E3*	unknown
Wasefukoku (WF)	*e1e1e2e2E3E3*	unknown
Sasahonami (SA)	*e1e1E2E2E3E3*	$Se\text{-}1^n Se\text{-}1^n$

[1] Genotypes for *Se-1* locus was identified by Ichitani et al. (1995).
$Se\text{-}1^e$ and $Se\text{-}1^n$ are photoperiod insensitive allele and photoperiod sensitive allele of *Se-1* locus, respectively.

Discussion

The results of gene analysis were summarized in Table 3. In this Table, the genotype for *Se-1* locus, which was identified by Ichitani et al. (1995), was also given. As seen from Table 3, all the varieties examined did not carry *E1* but *e1*, a photoperiod insensitivity allele of *E1* locus, but as for the other loci, *E2, E3*, and *Se-1*, a photoperiod sensitivity allele was detected in some of these varieties. We have already clarified that almost all the japonica-type varieties grown in regions except Hokkaido carry *E1* bringing forth strong photoperiod sensitivity (Okumoto et al., 1991, 1992a, 1992b). On the other hand, Yokoo et al. (1980) indicated that most of the varieties cultivated in Tohoku region (NL41-36°) carry $Se\text{-}1^e$, a photoperiod insensitive allele of *Se-1* locus, while most of the varieties cultivated in Southwest region (NL36-31°) carry $Se\text{-}1^n$, a second potent allele bringing forth comparatively strong photoperiod sensitivity (Okumoto et al., 1991). This may suggest that most of the Hokkaido varieties (NL45-42°) also do not carry $Se\text{-}1^n$ but $Se\text{-}1^e$ in that the photoperiod sensitivities of these varieties are extremely weak. However, the results by Ichitani et al. (1995) clearly showed that its potent allele $Se\text{-}1^n$ is present in many of the Hokkaido varieties. Accordingly, it can be concluded that *e1* is the gene essential to Hokkaido varieties and that the interchange of *E1* with *e1* enabled rice cultivation in high-latitude area under long-day condition in summer.

References

Chandraratna, M.F., 1953. A gene for photoperiod sensitivity in rice linked with apiculus color. Nature 171: 1162.

Hosoi, N., 1981. Studies on meteorological fluctuation in the growth of rice plants. V. Regional difference of thermo-sensitivity, basic vegetative growth and factors determining the growth duration. Japan J Breed 31: 239–250.

Ichitani, K., Y. Okumoto & T. Tanisaka, 1995. Identification of the genotypes for the photoperiod sensitivity locus, *Se-1*, in extremely early varieties of rice. Breed Sci (in press).

Okumoto, Y., T. Tanisaka & H. Yamagata, 1991. Heading-time genes of the rice varieties grown in south-west region in Japan. Japan J Breed 41: 135–152.

Okumoto, Y., T. Tanisaka & H. Yamagata, 1992a. Heading-time genes of the rice varieties grown in the Tohoku-Hokuriku region in Japan. Japan J Breed 42: 121–135.

Okumoto, Y., A. Yoshimura, T. Tanisaka & H. Yamagata, 1992b. Analysis of a rice variety Taichung 65 and its isogenic early-heading lines for late-heading genes *E1, E2*, and *E3*. Japan J Breed 42: 415–429.

Tanisaka, T., H. Inoue, S. Uozu & H. Yamagata, 1992. Basic vegetative growth and photoperiod sensitivity of heading-time mutants induced in rice. Japan J Breed 42: 657–668.

Tsai, K.H., 1976. Studies on earliness genes in rice, with special reference to analysis of isoalleles at the *E* locus. Japan J Genet 51: 115–128.

Yamagata, H., Y. Okumoto & T. Tanisaka, 1986. Analysis of genes controlling heading time in Japanese rice. In: Rice Genetics I, pp. 339–349, IRRI, Los Baños, Philippines.

Yokoo, M. & H. Fujimaki, 1971. Tight linkage of blast-resistance with late maturity observed in different indica varieties of rice. Japan J Breed 21: 35–39.

Yokoo, M., F. Kikuchi, A. Nakane & H. Fujimaki, 1980. Genetical analysis for heading time by aid of close linkage with blast resistance in rice. Bull Natl Inst Agric Sci Ser D31: 95–126.

P.M.A. Tigerstedt (Ed.), Adaptation in Plant Breeding, 81–85, 1997.

Adaptive properties of *Picea abies* progenies are influenced by environmental signals during sexual reproduction

Øystein Johnsen & Tore Skrøppa
Division of Silviculture, Norwegian Forest Research Institute, Høgskoleveien 12, 1432 Ås, Norway

Key words: seed orchard, phenology, frost hardiness, environmental preconditioning, temperature, photoperiod, Norway spruce, *Picea abies*

Summary

Several independent tests have shown that climate and weather during sexual reproduction influence the adaptive properties of the *Picea abies* progenies. This phenomenon is expressed in seed orchards established by moving parent trees, propagated as grafts, from north to south, from high to low elevation, or from outdoor to indoor greenhouse conditions. The progenies exhibit delayed flushing in the spring, later growth cessation of leader shoots in the summer, delayed bud-set, higher frequency of lammas shoots and delayed development of frost hardiness in the autumn compared to progenies reproduced in the colder native environment. The altered performance is persistent. We have found no effect on progenies of photoperiod and temperature treatments given to the males during meiosis and pollen production. However, when crosses were made in early spring (March), inside a heated greenhouse (short day, high temperature), the progenies are less frost hardy during cold acclimation than progenies from identical crosses performed in late spring (May; long day high temperature) in the greenhouse. The most hardy offspring were from crosses performed under outdoor conditions in May (long day, low temperature). These results indicate that some stages in reproduction, such as female meiosis, pollen tube growth, syngamy and early embryo development, are sensitive to temperature and/or photoperiod which then alter the phenotypic performance of the offspring. The most likely explanation is the existence of a regulatory mechanism affecting the expression of genes controlling adaptive traits. If this is true, it must have implications for the genetic interpretation of provenance differences in Norway spruce.

Introduction

Norway spruce (*Picea abies* (L.) Karst) is the most widely planted conifer tree species in Europe. Seeds for practical plantations are collected in forest stands or in seed orchards after mating between clones that have been selected for superior phenotypic performance. Within the natural range of distribution a great number of climatic races (provenances) have developed (Schmidt-Vogt, 1978). They show clinal variation along latitudinal and altitudinal gradients for traits that describe the annual growth cycle, in particular growth start and cessation and the development of autumn frost hardiness. This variation is thought to reflect climatic adaptation to photoperiod and temperature (Ekberg et al., 1979).

One of the constraints of seed production and breeding in Norway spruce is scarce flowering and poor seed ripening at northern latitudes and high altitudes. An important strategy has thus been to locate seed orchards and clone banks in warm regions. A more frequent, abundant flowering and better seed ripening under favorable climatic conditions are the main reasons for this practice (Werner, 1975; Schmidtling, 1983, 1984). In some cases the vegetatively propagated clones (grafts or cuttings) have been moved long distances from north to south or from high to low elevation, or indoor containerized seed orchards have been tried out as a supplement to outdoor seed orchards (Ross et al., 1986; Philipson, 1990; Johnsen et al., 1994a, 1994b).

The idea to produce seeds in a warm non-native environment, either by clonal transfer to an outdoor seed orchard or by use of heated greenhouses, is based on the assumption that the performance of offspring is not affected by changed environmental conditions. However, several independent studies have shown that the climate and weather during sexual reproduction influence adaptive properties of *Picea abies* progenies. In this paper we give an overview of the main results from these studies. The change in phenology of the reproductive process is discussed in relation to the environmental signals given by temperature and photoperiod. Possible mechanisms are briefly discussed in relation to possible sensitive stages during reproduction.

Sexual reproduction in warm non-native environments

Provenance and half-sib family comparisons

In Norway, seeds for the central and northern part of the country (lat. 63–67° N) are produced in an orchard located in the southern part at lat. 58° N, and seeds for high altitudes are produced in an orchard near the sea level. The parental clones in these orchards are phenotypically selected trees from native stands in the same area where the seedlings from the orchard seeds are to be planted. Both orchards are well isolated from outside pollen sources, and were expected to give rise to seedlings well adapted to the native areas.

When seedlings from the southern seed orchard were grown in nurseries in the northern environment, it was observed that the orchard progenies exhibited a phenology different from seedlings of autochthonous northern forest stands. With these casual observations in mind, Bjørnstad (1981) studied plants from open pollinated seed after cone collection on the selected plus trees in the native stands, and compared them with their half-sibs after controlled pollination (using pollen mix from northern males) in the southern orchard. He found that the seed orchard progenies produced terminal buds 1–4 weeks later in the autumn than their northern half-sibs. Further studies showed that the seed orchard progenies were different from their northern half-sibs in many traits related to phenology. They flushed later in the spring, terminated leader growth later in the summer, they had higher frequencies of lammas shoots, lignified the annual ring later in autumn and were more damaged in artificial freezing tests when plants developed frost hardiness in the autumn (Johnsen, 1989a, 1989b). The difference between pairs of half-sibs persisted even after a clonal propagation with rooted cuttings, and the seed orchard progenies were 15% taller than northern half-sibs after 6 years from seeds (Johnsen, 1989b). Phenological observations in progeny tests with families from the southern seed orchard show that they perform like a more southern provenance even after 17 years from seeds (Edvardsen, 1995). The hypothesis that the observed differences were an effect of phenotypic selection had to be rejected as progenies of the plus trees did not perform differently from progenies of non-selected trees from the same stands (Johnsen & Østreng, 1994).

Similar effects have been observed on progenies after crosses performed at low elevation sites between Norway spruce clones originating from high altitudes (Johnsen, 1988). Seedlings from controlled crosses had an extended shoot elongating period at ages nine and ten years compared with trees from high altitude provenances and had a lower level of autumn frost hardiness than comparable seedlings from natural stands (Skrøppa, 1994).

Results from both field trials and practical plantings indicate that the observed influences of the crossing environment may have practical consequences under extreme climatic situations (Johnsen et al., 1989; Skrøppa, in preparation). The effects may either be positive or negative for survival and vigor of the plantation, depending on how the climatic extremes are related to the annual growth rhythm of the material. The situation seems to be similar to that of a provenance transfer. It will be advantageous under certain conditions (e.g. avoidance of late spring frost damage), but the opposite under other conditions (e.g. tolerance of early autumn frost).

Full-sib comparisons

In order to give conclusive evidence that progenies are influenced by climate and weather during sexual reproduction, more precise comparisons between progenies from the same parents produced in contrasting environments were needed. The same controlled crosses were thus performed to create identical full-sib families under different climatic conditions, both at different latitudes in Finland (Skrøppa et al., 1994), and inside a heated greenhouse vs. outside in a nearby seed orchard (Johnsen et al., 1995). The progenies that were sexually reproduced under warm conditions (southern latitudes or inside the greenhouse) were generally less frost hardy in the autumn than their full-sibs from the cold conditions. From these studies we could conclude that Norway spruce progenies were influenced both by the genetic constitution of their parents and by environment during sexual reproduction. This has also been found in *Pinus sylvestris*, although the effects of the parental environment seem to be less pronounced in this species (Lingren & Wang, 1986; Dormling & Johnsen, 1992; Andersson, 1994; Lindgren & Wei, 1994).

Interaction between photoperiod and temperature and timing of sexual reproduction

A common factor involved in seed production under warm non-native conditions is that the increased heat sum in spring induces an earlier start of the reproductive process at a shorter photoperiod compared with the cooler native environments far north or at high elevation. Especially a north to south transfer accentuates the differences in photoperiod, but even a strict high to low altitude transfer on the same latitude can result in as much as 2 h difference in photoperiod when pollination starts in the first half of May at low elevation compared to four weeks later at high elevations. In addition, a higher temperature sum is normally attained during seed maturation late in summer and early autumn. If this interaction between temperature and photoperiod actually starts the triggering signal which then influence the progeny, a warm seed year with an early flowering could give less hardy progenies than a cool seed year with a late flowering even when seeds are produced from the same seed orchards. One study has indeed shown that seedlings from bulked seed lots of two seed orchards, from a cool (1987) versus a warm (1989) seed year, responded differently to photoperiodic treatments. Seedlings of the cool seed year formed terminal buds at shorter nights in growth chambers and formed buds earlier under natural days in nurseries than did seedlings from the warm seed year (Kohmann & Johnsen, 1994). In this connection it is interesting to note that in the study comparing full-sib pairs from the heated greenhouse and the outdoor seed orchard (Johnsen et al., 1995), the potted grafts were located inside the greenhouse for only three weeks, starting some few days before pollination until the female flowers were no longer receptive. However, due to this treatment, the reproductive process started three weeks earlier in the greenhouse that in the outdoor seed orchard. We thus suspect that an interaction between photoperiod and temperature is the major environmental trigger starting a signal which is then transduced to exert its influence on the progenies by a yet unknown mechanism (see Skrøppa & Johnsen, 1994 for a comprehensive discussion).

Sensitive stages during reproduction

To be able to understand the process we need to know the stages in the reproductive process which are sensitive to the environmental trigger. Possible selective events occurring during the reproduction could be preferential segregation or meiotic drive during formation of pollen and megaspores, megaspore degeneration, competition among 3–7 pollen grains during tube growth, or postmeiotic embryo competition among the 2–4 genetically different embryos in each developing seed (Owens & Blake, 1985). We have tested whether photoperiod and temperature treatments (long and short days/high and low temperature) during male meiosis and pollen maturation (before pollination) could give any differences in progeny performance, but in two independent experiments we found no evidence for altered autumn hardiness related to these pollen treatments (Johnsen et al., 1996 in press). Thus, the process with greatest potential for genetic selection, due to the enormous amount of pollen grains produced, does not seem to be involved as a sensitive stage. In a third experiment we made crosses inside a heated greenhouse in early spring (March) when days were short (14 h), repeated the crosses inside the greenhouse in May when the days were longer (18 h), and in addition made the identical crosses outside the greenhouse in May. No treatments were given to the male parents in this experiment. The progenies reproduced in early spring were less frost hardy than their full-sibs reproduced outdoors in May (Johnsen et al., 1996 in press). This result indicates that somewhere during the time interval from female meiosis, pollen tube growth, syngamy, embryo competition and to early embryogenesis, a signal is transduced, triggered by photoperiod and/or temperature, which then alter the phenotypic performance of the progenies.

A more precise timing of photoperiodic and temperature treatments to the various reproductive stages in the female cones is needed in future experiments to reveal the stages which are sensitive to environmental influence, and to find out whether the phenomenon is triggered only by temperature or photoperiod or an interaction between the two climatic factors. Sarvas

(1968) has related pre- and postzygotic stages in Norway spruce to accumulated heat in day degrees above 5° C. This summer (1995) we timed heat treatments according to these data and we intend to test progenies from these treatments. In addition, ovules have been dissected and fixed throughout the experimental period for light microscopy examinations in order to determine phenological development in relation to the accumulated heat, and to detect potential effects of the heat treatments on anatomical and histological aspects in the ovules. This information will be used in future experiments for more precise timing of treatments to significant events in the reproductive process. If we find the developmental stages in question, a search for mechanisms can hopefully be initiated.

Possible mechanisms

From the data referred to above, it can be concluded that the environmental signal is transduced during sexual reproduction in the female flowers. However, the nature of this particular signal transduction (see Bowler & Chua, 1994) and how it exerts its influence at the molecular level is unknown. Our data indicate that the effect of the parental environments is persistent in the progeny. Thus, the mechanism should to our opinion operate either on a genetic (change in gene frequency caused by selection, or rapid genomic changes) or an epigenetic (gene regulation altered by imprinting) level (see Skrøppa & Johnsen, 1994). The reproductive process offers several possibilities for directed selection. However, the potential effect of gametophytic and sporophytic selection is rather limited due to the low number of pollen grains in each pollen chamber and embryos in a developing seed in spruce (Owens & Blake, 1985). Thus, unless the selection is going in the same direction in a three stage sequence (megaspore degeneration, pollen grain and embryo competition) it can hardly account for the observed effects reported.

Virtually no information is available about rapid genomic changes or epigenetic effects in spruce. Genomic imprinting is, however, being increasingly accepted as a fundamental and widespread process that determines, in ways not predicted by the laws of Mendelian inheritance, whether a particular gene will be expressed or not (Matzke & Matzke, 1993). Interestingly, Meyer et al. (1992) found that environmental factors influenced 35S promoter methylation of a maize A1-gene construct in transgenic petunia and its colored phenotype. While blossoms on field-grown plants flowering early in the season were predominantly red, later flowers on the same plants showed weaker coloration. The reduction of the A1-specific phenotype correlated with methylation of the 35S promoter. Moreover, they found that the stability of pigmentation correlated with the time of seed production. The A1-gene construct was rather insensitive to DNA methylation in progeny from flowers of young parental plants produced early in the season, but became susceptible to methylation within progeny from subsequent later crosses. Similar observations have been made in non-transgenic *Zea mays* where methylation was more pronounced in upper ears and tassels (Federoff & Banks, 1988). So far, we can only speculate that such gene regulation, caused by activation or deactivation of certain genes by environmental conditions during reproduction, regulate the phenotypic expression of adaptive traits in Norway spruce.

Genetic interpretation of provenance variation

Traits related to phenology of Norway spruce provenances are clinally correlated to latitude and altitude. As much as 60–93% of the variation have been explained by these parameters for traits such as bud set in the autumn, cessation of leader growth, duration of the growth period and development of autumn frost hardiness (Dormling, 1979; Skrøppa & Magnussen, 1993; Dæhlen et al., 1995). Our results indicate, regardless of type mechanism involved, that phenotypic provenance variation is not only caused by natural selection among different genotypes within populations, but also directed by environmental signals received by the female parents during sexual reproduction. Both processes seem to change phenology traits of the progeny in the same direction, and may explain why Norway spruce provenances so strongly show clinal variation along latitudinal and altitudinal gradients for traits that describe the annual growth cycle.

Acknowledgements

This research have been supported by The Norwegian Ministry of Agriculture and the Nordic Forest Research Co-operation Committee.

References

Andersson, B., 1994. After-effects of maternal environment on autumn frost hardiness in *Pinus sylvestris* seedlings in relation to cultivation techniques. Tree Physiol 14: 313–322.

Bjørnstad, Å., 1981. Photoperiodical after-effect of parent plant environment in Norway spruce (*Picea abies* (L.) Karst) seedlings. Medd Nor Inst skogforsk 36: 1–30.

Bowler, C. & N.-H. Chua, 1994. Emerging themes of plant signal transduction. The Plant Cell 6: 1529–1541.

Dormling, I., 1979. Influence of light intensity and temperature on photoperiodic response of Norway spruce provenances. In: Proc. of the IUFRO joint meeting of working parties on Norway spruce provenances and Norway spruce breeding. Lower Saxony Forest Res Inst Dept Forest Tree Breeding. Echerode FRG, pp 398–408.

Dormling, I. & Ø. Johnsen, 1992. Effects of the parental environment on full-sib families of *Pinus sylvestris*. Can J For Res 22: 88–100.

Dæhlen, A.G., Ø. Johnsen & K. Kohmann, 1995. Autumn frost hardiness in young seedlings of Norway spruce from Norwegian provenances and seed orchards. (Norwegian with English summary.) Rapp Skogforsk 1/95: 1–24.

Edvardsen, Ø.M., 1995. Frø fra Lyngdal frøplantasje til nordnorsk bruk? (Seeds from Lyngdal seed orchard for use in the north?) MS Thesis, Institute of Forest Sci Agr Univ of Norway. 52 pp.

Ekberg, I., G. Eriksson & I. Dormling, 1979. Photoperiod reactions in conifer species. Holarct Ecol 2: 255–263.

Federoff, N.V. & J.A. Banks, 1988. Is the Suppressor-Mutator element controlled by a basic developmental mechanism? Genetics 120: 559–570.

Johnsen, Ø., 1988. After-effects on progenies from orchard clones moved to non-native environments. In: J. Worrall, J. Loo-Dinkins & D.T. Lester (Eds). Proc. of the Tenth North American Forest Biology Workshop, Vancouver, B.C. July 20–22, 1988, pp 1–11.

Johnsen, Ø., 1989a. Phenotypic changes in progenies of northern clones of *Picea abies* (L.) Karst. grown in a southern seed orchard. I. Frost hardiness in a phytotron experiment. Scan J For Res 4: 317–330.

Johnsen, Ø., 1989b. Phenotypic changes in progenies of northern clones of *Picea abies* (L.) Karst. grown in a southern seed orchard. II. Seasonal growth rhythm and height in field trials. Scan J For Res 4: 331–341.

Johnsen, Ø. & G. Østreng, 1994. Effects of plus tree selection and seed orchard environment on progenies of *Picea abies*. Can J For Res 24: 32–38.

Johnsen, Ø., J. Dietrichson & G. Skaret, 1989. Phenotypic changes in progenies of northern clones of *Picea abies* (L.) Karst. grown in a southern seed orchard. III. Climatic damage in a progeny trial. Scan J For Res 4: 343–350.

Johnsen, Ø., G. Haug, O.G. Dæhlen, B.S. Grønstad & A.T. Rognstad, 1994a. Effects of heat treatment, timing of heat treatment, and gibberellin $A_{4/7}$ on flowering in potted grafts of *Picea abies*. Scan J For Res 9: 333–340.

Johnsen, Ø., O.G. Dæhlen, G. Haug, B.S. Grønstad & A.T. Rognstad, 1994b. Female cone abortion and full seed production in an indoor seed orchard with potted grafts of *Picea abies*. Scand J For Res 9: 329–332.

Johnsen, Ø., T. Skrøppa, G. Haug, I. Apeland & G. Østreng, 1995. Sexual reproduction in a greenhouse and reduced autumn frost hardiness of *Picea abies* progenies. Tree Physiol 15: 551–555.

Johnsen, Ø., T. Skrøppa, O. Jünttila & O.G. Dahlen, 1996. Influence of the female flowering environment on autumn frost hardiness of *Picea abies* progenies. Theor Appl Genet (in press).

Kohmann, K. & Ø. Johnsen, 1994. The timing of bud-set in seedlings of *Picea abies* from seed crops of a cool versus a warm summer. Silvae Genet 43: 328–332.

Lindgren, D. & Q. Wang, 1986. Are genetic results influenced by the environment during seed maturation? In: R.J. Weir (Ed.), Proc. of the IUFRO – Conference on Breeding Theory, Progeny Testing and Seed Orchards. 13–17 Oct. 1986, Williamsburg, VA, USA. N.C. State University, Raleigh, pp. 192–199.

Lindgren, D. & R.-P. Wei, 1994. Effects of maternal environment on mortality and growth in young *Pinus sylvestris* field trials. Tree Physiol 14: 323–327.

Matzke, M. & A.J.M. Matzke, 1993. Genomic imprinting in plants: Parental effects and transinactivation phenomena. Ann Rev Plant Physiol Plant Mol Biol 44: 53–76.

Meyer, P., F. Linn, I. Heidmann, Z.A.H. Meyer, I. Niedenhof & H. Saedler, 1992. Endogenous and environmental factors influence 35S promoter methylation of maize a1-gene construct in transgenic petunia and its colour phenotype. Mol Gen Genet 231: 345–352.

Owens, J.N. & M.D. Blake, 1985. Forest tree seed production. Information Report PI-X-53 Petawawa National Forestry Institute Chalk River Ontario.

Philipson, J.J., 1990. Prospects for enhancing flowering of conifers and broadleaves of potential silvicultural importance in Britain. Forestry 63 (3): 223–240.

Ross, S.D., A.M. Eastham & R.C. Bower, 1986. Potential for container seed orchards. In: Conifer Tree Seed in the Inland Mountain West. Symposium Proceedings. USDA For Serv Gen Tech Rep INT-203. pp 180–186.

Sarvas, R., 1968. Investigations on the flowering and seed crop of *Picea abies*. Comm Inst For Fenn 67 (5): 1–84.

Schmidtling, R., 1983. Rootstock influences flowering, growth and survival of loblolly pine grafts. For Sci 29: 117–124.

Schmidtling, R., 1984. Planting south of origin increases flowering in shortleaf (*Pinus echinata* Mill.) and Virginia pines (*Pinus virginiana* Mill.) Silvae Genet 33: 140–144.

Schmidt-Vogt, H., 1978. Genetics of *Picea abies* (L.) Karst. Annales Forestales 7: 145–186.

Skrøppa, T., 1994. Growth rhythm and hardiness of *Picea abies* progenies of high altitude parents from seeds produced at low elevation. Silvae Genet 43: 95–100.

Skrøppa, T. & S. Magnussen, 1993. Provenance variation in shoot growth components of Norway spruce. Silvae Genet 42: 111–120.

Skrøppa, T. & Ø. Johnsen, 1994. The genetic response of plant population to a changing environment: the case for non-Mendelian processes. In: T.J.B. Boyle & C.E.B. Boyle (Eds), Biodiversity, Temperate Ecosystems, and Global Change, NATO ASI Series, Vol I 20, pp 183–189. Springer-Verlag, Berlin Heidelberg, 1994.

Skrøppa, T., T. Nikkanen, S. Routsalainen & Ø. Johnsen, 1994. Effects of sexual reproduction at different latitudes on performance of the progeny of *Picea abies*. Silvae Genet 43: 297–303.

Werner, M., 1975. Location, establishment and management of seed orchards. In: R. Faulkner (Ed.) Seed Orchards. For Comm Bull (U.K.) 54: 49–57.

P.M.A. Tigerstedt (Ed.), *Adaptation in Plant Breeding*, 87–94, 1997.

The role of selection on the genetic structure of pathogen populations: Evidence from field experiments with *Mycosphaerella graminicola* on wheat

Bruce A. McDonald, Christopher C. Mundt[1] & Ruey-Shyang Chen[2]
Department of Plant Pathology and Microbiology, Texas A&M University, College Station, TX 77843-2132, U.S.A.; [1] Department of Botany and Plant Pathology, Cordley Hall 2082, Oregon State University, Corvallis, OR 97331-2902, U.S.A.; [2] present address: Department of Plant Pathology, National Chung Hsing University, Taichung 402, Taiwan

Key words: coevolution, DNA fingerprints, *Mycosphaerella graminicola*, RFLPs, *Septoria tritici*

Summary

Coevolution refers to reciprocal genetic changes that occur in two or more ecologically interacting species. In agricultural ecosystems, we are especially concerned with the genetic response of pathogen populations to resistant cultivars produced by plant breeding programs. It would be useful to be able to predict whether disease resistance is likely to be durable or ephemeral before a cultivar is widely grown. Though it may not be possible to predict durability in advance, knowledge of the genetic structure of pathogen populations may prove useful for making predictions about the rate at which pathogens adapt to resistant varieties. Much has been learned about the genetic structure of populations of obligate fungal pathogens such as rusts and mildews, which have become paradigms for plant pathology. We have focused our effort on the population genetics of the less known, non-specialized, necrotrophic pathogens, such as the Septorias of small grains. Our approach has been to use DNA fingerprinting and RFLP analysis to conduct field experiments that elucidate how populations of fungal pathogens adapt in agroecosystems. Our results suggest that mating system may have a greater impact than natural selection on the genetic structure of populations of *Mycosphaerella graminicola* (anamorph *Septoria tritici*).

Introduction

Coevolution has been defined as 'reciprocal genetic changes that occur in two or more ecologically interacting species' (Futuyma & Slatkin, 1983). Thus coevolution represents a very broad category of potential interactions among plants, animals, insects and microbes. In this paper, we will refer to the specialized case of coevolution where a plant host and a pathogenic fungus interact. We will only consider coevolution as it occurs in agricultural pathosystems, ignoring coevolution that happens in natural pathosystems. This very narrow focus simplifies our consideration of coevolution because plant breeders control half of the coevolutionary interaction.

In agricultural systems, we are not studying coevolution as it was originally defined. Plant breeders largely determine which resistance genes are present in the host plant, so the plant population does not exhibit the reciprocal response to changes in pathogen virulence seen in natural systems. However, since the pathogen population is forced to respond to any changes made in the host population, evolution of the pathogen population is likely to reflect the man-directed evolution of the cultivated host. The question that is most relevant to 'coevolution' in agricultural ecosystems is, how does the pathogen population respond to genetic changes in the host population? In other words, how does the pathogen adapt to the host? The rationale for studies of coevolution in agricultural ecosystems is that we hope to develop new knowledge that can be used to manage resistance genes in agricultural ecosystems. The goal is to minimize the evolutionary potential of the pathogen population, leading to stable, economically acceptable levels of disease.

Based on decades of observing epidemics of rusts on cereals, most plant pathologists and plant breeders came to the conclusion that the genetic composition of the host population is the primary determinant of the genetic structure of a corresponding pathogen population. Plant pathologists have long known that most major resistance genes have a finite lifespan in agricultural systems. After introducing host genotypes containing a single major resistance gene, it was common to find that the pathogen population experienced a rapid increase in the frequency of a corresponding virulence gene, causing the resistance gene to lose its effectiveness. The natural interpretation of this 'boom and bust' cycle was that selection, through the gene-for-gene interaction (Flor, 1956; Staskawicz et al., 1995), was the major driving force behind pathogen evolution. Mutations for virulence occurred within a clonal lineage of the rust or mildew, and selection increased the frequency of the virulent mutant until it reached the threshold for economic damage. This scenario has largely been confirmed using molecular markers for leaf rust on wheat (Kolmer et al., 1995) and powdery mildew on barley (Brown & Wolfe, 1990) and it also appears to also be the case for rice blast (Zeigler et al., 1995) and potato late blight (Goodwin et al., 1995).

Several experiments have been conducted to measure selection in obligate biotrophs such as rusts and mildews (e.g. Groth & Ozman, 1994; Kolmer, 1993; Chin & Wolfe, 1984). The biggest experiments have been conducted by farmers who grow one resistant cultivar over a large area (Stakman & Christensen, 1960). Investigators usually find that there is rapid selection for particular virulence genes in the rust or mildew population that correspond to resistance genes present in the host. These observations led to the present paradigm that selection is the dominant force governing the genetic structure of populations of pathogenic fungi. But few experiments have been conducted with the generalist, non-obligate pathogens that do not appear to conform to Flor's gene-for-gene hypothesis.

In 1989, we began a series of field experiments to learn about the population genetics of the wheat leaf blotch pathogen, *Mycosphaerella graminicola* (anamorph *Septoria tritici*). Through these experiments, we learned that 1) field populations of *M. graminicola* have a high level of gene and genotype diversity worldwide (McDonald et al., 1995), 2) the sexual stage has a major impact on the genetic structure of these populations (Chen & McDonald, 1996), 3) populations exhibit little evidence for genetic drift over time, suggesting that effective population sizes are large (Chen et al., 1994), 4) populations separated by long distances can be genetically very similar, suggesting significant levels of gene flow among these populations (McDermott & McDonald, 1993; McDonald et al., 1995). In this paper, we will present the results of an experiment that explored the effect of different host populations on the genetic structure of populations of *M. graminicola*.

M. graminicola is best known as a foliar pathogen on wheat but it also infects several other grasses (Brokenshire, 1975), hence it appears to be more of a generalist pathogen than a host-specialized pathogen. It is a disease problem on wheat worldwide, especially in wet and cool climates. Though some researchers have proposed that the wheat-*M. graminicola* pathosystem conforms to a gene-for-gene interaction (Eyal et al., 1985), the inheritance of virulence in the pathogen has not been determined because no one has made a controlled cross with this fungus. It appears that the expression of virulence in this system is very sensitive to environmental conditions, hence it has proven difficult to reproduce disease symptoms using the same differentials at different research stations. *M. graminicola* produces two types of spores during its life cycle. The asexual spores (pycnidiospores), which have a long, filiform shape, are spread by splash-dispersal and have limited potential for long-distance movement. These were thought to be the primary reproductive spore for epidemic progression in wheat fields (King et al., 1983). They also were considered to be the major source of primary inoculum until recently. The sexual spores (ascospores) have a short, ellipsoid shape that is suitable for long-distance transmission. Several studies (Shaw & Royle, 1989; Schuh, 1990), including our own (McDonald & Martinez, 1990a), suggested that the ascospores were the primary source of inoculum that initiated an epidemic. We also provided evidence that the sexual stage has a major impact on the genetic structure of these populations (Chen & McDonald, 1996). The sexual stage has been found in most countries where investigators have looked for it, but it is difficult to differentiate from other fungal fruiting bodies that colonize wheat straw between growing seasons. We have proposed that the sexual ascospores are the primary mechanism for gene flow among populations on a regional basis (Boeger et al., 1993), though we consider it likely that infected seed may be more important for gene flow on a global scale (McDonald et al., 1995).

We have developed and utilized four categories of DNA-based genetic markers in *M. graminicola*.

1. Restriction fragment length polymorphism (RFLP) markers based on hybridization of anonymous DNA probes to homologous, selectively neutral nuclear loci provide moderately polymorphic, codominant markers that can be used much like isozymes (McDonald & Martinez 1990b). Individual RFLP loci have been used to measure gene diversity (Boeger et al., 1993; McDonald & Martinez, 1990a) and genotype diversity (Boeger et al., 1993; Chen et al., 1994), to assess genetic similarity among individuals (McDonald & Martinez, 1991b), to determine the relative contributions of sexual and asexual reproduction (Chen & McDonald, 1996), to infer gene flow (Boeger et al., 1993), and to measure the genetic stability of populations over time (Chen et al., 1994). 2. DNA fingerprints are based on hybridization of anonymous DNA probes to families of repetitive DNA sequences that are dispersed throughout the nuclear genome (McDonald & Martinez, 1991b). DNA fingerprints have been used to measure genotypic diversity (Chen et al., 1994; Chen & McDonald, 1996), to monitor the movement of individual clones over time and space (Chen et al., 1994), and to demonstrate the distribution of genetic variation on a microgeographical scale (Boeger et al., 1993; McDonald & Martinez, 1990a). 3. Mitochondrial DNA (mtDNA) haplotypes are determined by hybridizing purified, labeled total mtDNA to the mtDNA fraction of the genome present in a Southern blot of total DNA digested with a restriction endonuclease. Though the mtDNA has proven to be much less variable than nuclear DNA, we have used mtDNA haplotypes to locate a potential center of diversity (and the potential center of origin) of *M. graminicola* (McDonald et al., 1995). 4. Electrophoretic karyotypes generated by pulsed field gel electrophoresis were used to physically map RFLP loci to particular chromosomes. This linkage information was needed for analyses of linkage disequilibrium among RFLP loci. The karyotype data also showed that high levels of karyotype variation may exist within field populations of pathogenic fungi and demonstrated duplications of RFLP loci in natural populations (McDonald & Martinez, 1991a).

We conducted a field experiment in Corvallis, Oregon in 1989–1990 to study the effect of host genotype on the genetic structure of *M. graminicola* populations. We began the experiment with a series of hypotheses based on what was known about *M. graminicola* in 1989. We expected that the experimental plots would be inoculated by a cloud of ascospore inoculum, representing the maximum possible genetic diversity for a population of *M. graminicola*. Following the initial infection, we postulated that selection would increase the frequencies of pathogen genotypes that were best adapted to each host genotype, leading to a reduction in overall genetic diversity over time. We assumed that the fungus would reproduce largely through production of asexual spores, leading to a decrease in genotypic diversity, an increase in gametic disequilibrium, and changes in allele frequencies over time. Finally, we expected that disruptive selection in heterogeneous host populations would lead to a correlation between the genetic diversity of a host population and the genetic diversity of the corresponding pathogen population.

Materials and methods

Field plot design

In October of 1989, four wheat varieties that differed in resistance to *M. graminicola* were planted in pure stands and in all possible two-, three-, and four-way variety mixtures (15 treatments total) in a randomized complete block design with three replications. Gene was considered a resistant variety, Madsen was moderately resistant, and Stephens and Malcolm were susceptible. The field site was fallow in the year prior to initiating the experiment. Individual plots in the field measured 1.5 × 6.1 m in size.

Fungal collections

Natural infection by *M. graminicola* was distributed uniformly through the field by mid-January. We believe that the initial infection came from ascospores originating from stubble in wheat fields ~ 80 km distant from the field site. The plots were sampled on March 6 and May 30, 1990, hereafter referred to as early- and late-season, respectively. Twenty infected leaves were chosen randomly along two transects running along the length of each plot. The average distance between collections was 50 cm within each plot. The infected leaf tissue was air dried at room temperature for two weeks prior to making isolations. Only one isolation was made from each infected leaf. A total of 149 isolates were obtained from early in the season and 562 isolates were obtained from late in the season. Sample sizes were less than 900 isolates for both collections because some leaves did not possess viable fungal fruiting bodies and limitations in funding restricted the number of isolates that could be assayed.

DNA extraction, probes, and hybridization

DNA was extracted from each isolate using a CTAB extraction protocol described previously (McDonald & Martinez, 1990b). Purified DNA (4 μg) was digested individually with the restriction enzymes *Pst*I or *Xho*I. DNA fragments were separated on 0.8% agarose gels and then transferred to nylon membranes using the alkaline transfer method (Reed & Mann, 1985) as recommended by the manufacturer (BioRad, Hercules, CA).

Development of anonymous DNA probes used for RFLP analysis was described previously (McDonald & Martinez, 1990b). The source of these probes was a partial *Sau*3A digestion of total DNA from *S. tritici*. DNA fragments ranging from 500–2500 bp in size were size-fractionated before cloning into a pGEM4 (Promega, Madison, WI) plasmid vector. Based on an initial screen, 10 probes that hybridized to single loci, and one probe that hybridized simultaneously to two loci located on different chromosomes (McDonald & Martinez, 1991a, b) were used. Together these probes hybridized to loci on 12 of the 18 chromosomes in the *M. graminicola* genome (McDonald & Martinez, 1991a; B.A. McDonald unpublished results). Two probes (pSTL40 and pSTL70) that hybridized to dispersed, repetitive DNA sequences were also used in this study. The repetitive probes were shown previously to be useful for DNA fingerprinting (McDonald & Martinez, 1991b). Probes were labeled by nick translation following the manufacturer's recommendations (BRL, Gaithersburg, MD).

Analytical methods

Each probe-enzyme combination was defined as a different RFLP locus. DNA fragments or combinations of fragments with different sizes were treated as alleles at each RFLP locus. Only *Pst*I was used to digest DNA from the late-season collection in order to lower the cost of collecting data. As a result, there were larger sample sizes for all RFLP loci that used *Pst*I than for RFLP loci that use *Xho*I. Isolates having the same multilocus haplotype (i.e., having the same alleles at each of the single RFLP loci) were compared using DNA fingerprints. Isolates having the same multilocus haplotype and DNA fingerprint were assumed to be individual members of the same clone (McDonald & Martinez, 1991b). Based on frequencies of alleles at individual loci, and frequencies of individual fragments in DNA fingerprints, the probability that two individuals would have the same DNA profile by chance was less than 10^{-9} on average in this population (McDonald & Martinez, 1991b; McDonald unpublished). To determine whether repeat sampling of the same clone would affect measures of association among loci, a second 'clone-corrected' data set was constructed by including only one representative of each clone for each analysis of gametic disequilibrium.

The measure of genotypic diversity (Stoddart & Taylor, 1988) was based on the number of isolates with different multilocus haplotypes and DNA fingerprints in each collection. Only isolates with the full complement of single-locus RFLP data and fingerprints were included in this analysis. Genotypic diversity in each collection was measured using the formula

$$\hat{G} = \frac{1}{\sum_{x=o}^{N} \{(f_x) \cdot (x/N)^2\}}$$

where N is the sample size and f_x is the number of genotypes observed x times in the sample. The maximum possible value for Ĝ, which occurs when each individual in the sample has a different genotype, is the number of individuals in the sample. To compare Ĝ in collections with different sample sizes, we divided Ĝ from each collection by N to calculate percentage of maximum possible diversity that was obtained (Chen et al., 1994). A t-test was used for statistical comparisons between the normalized measures of Ĝ (Chen et al., 1994).

Gametic disequilibrium between pairs of loci was calculated using the following methods suggested by Weir (1990). A test for significance of the disequilibrium coefficient across all alleles for each pair of loci was formulated with the chi-square test statistic

$$\chi_T^2 = \sum_{u=1}^{k} \sum_{v=1}^{l} \frac{n\hat{D}_{uv}^2}{\tilde{P}_u \tilde{P}_v}$$

for all pairs of alleles *u* and *v*, which were present at loci that had *k* and *l* alleles total, respectively. *n* was the number of individuals in the sample and $\hat{D}_{uv}$ was the maximum likelihood estimator for the coefficient of disequilibrium between alleles *u* and *v*. The observed allele frequencies for the loci were $\tilde{P}_u$ and $\tilde{P}_v$, respectively. This chi-square statistic had (*k*-1)(*l*-1) degrees of freedom for each pair of loci. A test for the significance of the disequilibrium coefficient between each pair of alleles at two loci was formulated with the chi-square statistic

Table 1. Frequencies of the one or two most common alleles at each RFLP locus for three pure-stand host treatments at the end of the growing season. The number of alleles found in the total collection of 711 isolates is shown in parentheses. Most of the alleles were found only once in the sample. Data for the most resistant host, Gene, are not shown because sample sizes were too low for a meaningful comparison

RFLP locus (# alleles total)	Allele	Madsen (MR) (N = 34)	Stephens (S) (N = 57)	Malcolm (S) (N = 56)	Significance
pSTS192A (9)	1	1.00	0.95	0.88	NS
pSTS192B (6)	1	1.00	1.00	0.98	NS
pSTS14 (4)	1	0.68	0.88	0.73	NS
	2	0.32	0.11	0.25	
pSTS2 (12)	1	0.65	0.69	0.64	NS
	3	0.27	0.17	0.20	
pSTL10 (20)	1	0.74	0.67	0.73	NS
	3	0.18	0.31	0.14	
pSTS196 (17)	1	0.91	0.88	0.82	NS
	2	0.00	0.02	0.05	
pSTL2 (8)	1	0.44	0.39	0.54	NS
	2	0.53	0.52	0.38	
pSTL53 (13)	1	0.53	0.46	0.52	NS
	3	0.15	0.14	0.20	
pSTS199 (16)	1	0.27	0.34	0.34	NS
	2	0.53	0.57	0.36	
pSTS197 (11)	1	0.44	0.70	0.59	NS
	2	0.50	0.23	0.30	
pSTL31 (10)	1	0.77	0.63	0.71	NS
	2	0.21	0.18	0.20	
pSTS43 (9)	1	0.59	0.68	0.59	NS
	2	0.24	0.26	0.32	

$$\chi^2_{uv} = \frac{n\hat{D}^2_{uv}}{\tilde{P}_u(1-\tilde{P}_u)\tilde{P}_v\tilde{P}_v}$$

This chi-square statistic had one degree of freedom. As before, all alleles present at frequencies of less than 10% and 20% in the late- and early-season populations, respectively, were pooled into a single category. Pairs of alleles that showed a significant ($p < 0.05$) departure from random expectations were tested further with Fisher's exact test of independence (Weir, 1990). The two-tailed exact test for independence between pairs of alleles at different loci was calculated with the computer program FISH 6, version 1.001 (Engels, 1988). In each exact test, all alleles except for the tested allele pairs were pooled into a single category. The level of significance used for the exact test was $p < 0.05$.

Results

A high level of genetic diversity was found both early and late in the growing season. On average, 11 different alleles were found for each RFLP locus across the 711 isolates assayed (Table 1). Nei's measure of gene diversity averaged across all loci was H = 0.41 early in the season and H = 0.40 late in the season. There was no change in genotypic diversity over the course of the growing season. DNA fingerprints differentiated 120 different genotypes among the 129 isolates tested from the early season collection. The measure of genotypic diversity was 87% of its maximum value. In the late season collection, 497 unique DNA fingerprints were found among 544 isolates tested. Genotypic diversity was at 83% of its maximum value. There was no statistical difference between the genotypic diversity of early- and late-season populations. The most common genotype was found twice in a plot from early in the season and twice in the same plot late in the season,

Table 2. Nei's average gene diversity (H) across 12 RFLP loci for *M. graminicola* populations sampled from 4-way, 3-way, 2-way mixtures and pure stands of wheat at the end of the growing season

	4-way mixture N = 53	3-way mixtures N = 170	2-way mixtures N = 190	Madsen N = 34	Malcolm N = 56	Stephens N = 57
H	0.45	0.42	0.44	0.40	0.46	0.40

showing that asexual reproduction occurred over the course of the growing season.

The genetic structures of early- and late-season populations were virtually identical. There was no evidence for population subdivision among early and late season populations. G_{st} averaged 0.006 across all loci. There were no significant differences in frequencies of *M. graminicola* RFLP alleles between host varieties at the end of the growing season (Table 1). This suggested that there was not selection for particular RFLP alleles corresponding to specific host genotypes. There also was no correlation between the genetic diversity present in the host and pathogen populations (Table 2). Gene diversity in each pathogen population was high regardless of the complexity of the host population. No significant multilocus associations were detected among RFLP loci. The populations appeared to be random mating for both early and late season populations.

Discussion

We began this experiment with the assumption that selection played a major role in the genetic structure of *M. graminicola* populations. We set about to falsify the null hypothesis that the genetic composition of the pathogen population would not change over the course of the growing season as a result of selection. We expected that the pathogen genotypes best adapted to each host would increase in frequency on that host. We were surprised to find no evidence for adaptation to any of the host genotypes over the growing season. We did not detect any increase in the clonality of the pathogen population over time, and we found no evidence of selection for particular fungal clones in any plot or host treatment. We conclude that there was no evidence that different host genotypes caused selection for particular pathogen genotypes in this experiment.

We consider two hypotheses to explain the lack of response by the pathogen population in this experiment. The first hypothesis is that the effects of selection are minor compared to other evolutionary forces such as drift, gene flow, or mating system. A previous experiment that compared the genetic structure of a population in this location over a three year period led us to conclude that drift is insignificant (Chen et al., 1994). A comparison of the population from Oregon with populations from California and Israel indicated that the populations showed little evidence for population subdivision (Boeger et al., 1993; McDonald et al., 1995). We interpreted this as evidence for gene flow among these populations, and estimated that movement of ~ 12 individuals each generation would be necessary to account for their high level of similarity. Because it is impossible to determine if these *M. graminicola* populations are at drift/migration equilibrium, we cannot conclude that our estimate of gene flow, Nm = 12, is due strictly to migration. If we assume that the populations are at equilibrium, however, the large effective size of *M. graminicola* populations makes it unlikely that this degree of gene flow would have a significant impact on the genetic structure of the *M. graminicola* population described in this experiment.

Another equally valid hypothesis is that our experimental design was not adequate to detect selection if it occurred. It is certain that we made an incorrect assumption about the biology of *M. graminicola* at the start of our experiment. We assumed that the primary infection would be due to the ascospores (products of sexual reproduction) and that subsequent infection would be largely from the pycnidiospores (products of asexual reproduction). Thus, we anticipated that gene and genotype frequencies would change as a result of selection for particular pathogen clones on different host varieties. This experiment was designed to detect changes in frequencies of particular clones. It now appears that sexual reproduction played a much greater role in infection throughout the growing season then was previously recognized (Chen & McDonald, 1996). Populations of *M. graminicola* in Oregon are characterized by high levels of genotypic diversity distributed on a very fine spatial scale (Chen et al., 1994). In the current and previous studies, we found that nearly every isolate of *M. graminicola* had a unique genotype, unless samples were taken in very close spatial proximity. This high degree of genotypic diversity is consistent with field observations (Mundt unpublished) that the sexual stage of *M. graminicola* is common in Oregon, and that populations established from autumn ascospore infections are large, perhaps on the order of hundreds of genotypes per sq m.

If sexual reproduction is common in *M. graminicola* then putative genes conditioning specific pathogenicity to host cultivars are likely to be distributed among a large number of pathogen genotypes. Under this scenario, the ability to detect changes in frequency of pathogen genotypes would be limited, and our experimental design was inadequate to detect selection. In fact, despite the absence of evidence for selection based on RFLPs, greenhouse studies of isolates sampled from Oregon suggest that host selection can have a significant impact on pathogenicity of *M. graminicola* (Ahmed et al., 1996). The pathogenicity of isolates sampled from California and Oregon differed significantly (Ahmed et al., 1995), despite the similarity in allele frequencies as measured by RFLPs.

The anonymous RFLP loci used in this experiment were chosen to represent a random cross section of the nuclear genome. The results from this experiment offer additional evidence that they are selectively neutral, and probably not tightly linked to genes that are under selection. Thus, the nuclear RFLP loci may not have been appropriate for studying the process of pathogen evolution in this experiment. We presently are conducting a field competition experiment that takes into account our new knowledge of the biology of *M. graminicola*.

Acknowledgements

This work was supported by the Texas Agricultural Experiment Station (Hatch project # 6928), the National Science Foundation (Grant # DEB-9306377), and the USDA National Research Initiative Competitive Grants Program (Grant # 93-37303-9039).

References

Ahmed, H.U., C.C. Mundt & S.M. Coakley, 1995. Host-pathogen relationship of geographically diverse isolates of *Septoria tritici* and wheat cultivars. Plant Pathol 44: 838–847.

Ahmed, H.U., C.C. Mundt, M.E. Hoffer & S.M. Coakley, 1996. Selective influence of wheat cultivars on the pathogenicity of *Mycosphaerella graminicola* (anamorph *Septoria tritici*). Phytopathology 86: 454–458.

Boeger, J.M., R.S. Chen & B.A. McDonald, 1993. Gene flow between geographic populations of *Mycossphaerella graminicola* (anamorph *Septoria tritici*) detected wit RFLP markers. Phytopathology 83: 1148–1154.

Brokenshire, T., 1975. The role of graminaceous species in the epidemiology of *Septoria tritici* on wheat. Plant Pathol 24: 33–38.

Brown, J.K.M. & M.S. Wolfe, 1990. Structure and evolution of a population of *Erisyphe graminis* f.sp. *hordei*. Plant Pathol 39: 376–390.

Chen, R.S. & B.A. McDonald, 1996. Sexual reproduction plays a major role in the genetic structure of populations of the fungus *Mycosphaerella graminicola*. Genetics 142: 1119–1127.

Chen, R.S., J.M. Boeger & B.A. McDonald, 1994. Genetic stability in a population of a plant pathogenic fungus over time. Mol Ecol 3: 209–218.

Chin, K.M. & M.S. Wolfe, 1984. Selection on *Erisyphe graminis* in pure and mixed stands of barley. Plant Pathol 33: 535–546.

Engels, B., 1988. FISH6. Version 1.001. University of Wisconsin Genetics Department, Madison, WI 53706.

Eyal, Z., A.L. Scharen, M.D. Huffman & J.M. Prescott, 1985. Global insights into virulence frequencies of *Mycosphaerella graminicola*. Phytopathology 75: 1456–1462.

Flor, H.H., 1956. The complementary genic systems in flax and flax rust. Advanc Genet 8: 29–54.

Futuyma, D.J. & M.S. Slatkin, 1983. Coevolution. Sinauer Associates, Sunderland, MA.

Goodwin, S.B., L.S. Sujkowski & W.E. Fry, 1995. Rapid evolution of pathogenicity within clonal lineages of the potato late blight disease fungus. Phytopathology 85: 669–676.

Groth, J.V. & E.A. Ozmon, 1994. Contrasting effects of asexual reproduction and random mating on changes in virulence frequency in a field collection of *Uromyces appendiculates*. Phytopathology 84: 566–569.

King, J.E., R.J. Cook & S.C. Melville, 1983. A review of Septoria diseases of wheat and barley. Ann Appl Biol 103: 345–373.

Kolmer, J.A., 1993. Selection in a heterogeneous population of *Puccinia recondita* f.sp. *tritici*. Phytopathology 83: 909–914.

Kolmer, J.A., J.Q. Liu & M. Sies, 1995. Virulence and molecular polymorphism in *Puccinia recondita* f.sp. *tritici* in Canada. Phytopathology 85: 276–285.

McDermott, J.M. & B.A. McDonald, 1993. Gene flow in plant pathosystems. Ann Rev Phytopathol 31: 353–373.

McDonald, B.A. & J.P. Martinez, 1990a. DNA restriction fragment length polymorphisms among *Mycosphaerella graminicola* (anamorph *Septoria tritici*) isolates collected from a single wheat field. Phytopathology 80: 1368–1373.

McDonald, B.A. & J.P. Martinez, 1990b. Restriction fragment length polymorphisms in *Septoria tritici* occur at a high frequency. Curr Gen 17: 133–138.

McDonald, B.A. & J.P. Martinez, 1991a. Chromosome length polymorphisms in a *Septoria tritici* population. Curr Gen 19: 265–271.

McDonald, B.A. & J.P. Martinez, 1991b. DNA fingerprinting of the plant pathogenic fungus *Mycosphaerella graminicola* (anamorph *Septoria tritici*). Exp Mycol 15: 146–158.

McDonald, B.A., R.E. Pettway, R.S. Chen, J.M. Boeger & J.P. Martinez, 1995. The population genetics of *Septoria tritici* (teleomorph *Mycosphaerella graminicola*). Can J Bot 73 (supplement): S292–S301.

Schuh, W., 1990. Influence of tillage systems on disease intensity and spatial pattern of Septoria leaf blotch. Phytopathology 80: 1337–1340.

Shaw, M.W. & D.J. Royle, 1989. Airborne inoculum as a major source of *Septoria tritici* (*Mycosphaerella graminicola*) infections in winter wheat crops in the UK. Plant Pathol 38: 35–43.

Stakman, E.C. & J.J. Christensen, 1960. The problem of breeding resistant varieties. Plant Pathology: An Advanced Treatise 3: 567–624.

Staskawicz, B.J., F.M. Ausubel, B.J. Baker, J.G. Ellis & J.D.G. Jones, 1995. Molecular genetics of plant disease resistance. Science 268: 661–667.

Stoddart, J.A. & J.F. Taylor, 1988. Genotypic diversity: estimation and prediction in samples. Genetics 118: 705–711.

Weir, B.S., 1990. Genetic Data Analysis. Sinauer Associates, Sunderland, MA.

Zeigler, R.S., L.X. Cuoc, R.P. Scott, M.A. Bernardo, D.H. Chen, B. Valent & R.J. Nelson, 1995. The relationship between lineage and virulence in *Pyricularia grisea* in the Philippines. Phytopathology 85: 443–451.

P.M.A. Tigerstedt (Ed.), *Adaptation in Plant Breeding*, 95–101, 1997.

Diversity among Finnish net blotch isolates and resistance in barley

J. Robinson & M. Jalli
Agricultural Research Centre of Finland, Institute of Crop and Soil Science, Plant Breeding Research Section, 31600 Jokioinen, Finland

Key words: barley, net blotch, *Pyrenophora teres*, resistance, *Hordeum vulgare*

Summary

Seedlings of a differential barley (*Hordeum vulgare* L.) series (21 genotypes) and of six check genotypes were used in the greenhouse to assess variation in virulence among 20 single-spore isolates of the net blotch pathogen, *Pyrenophora teres* Drechs. f. *teres* Smedeg., collected from various sites in Finland. The experiment was run twice and symptom expression was recorded on the first three leaves. Analysis of second leaf symptom scores from Run 1 indicated differences in virulence between isolates, all of which were pathogenic, and differential resistance among the barleys. The virulence of *P. teres* isolates appeared to be conditioned by the host barley from which the isolate derived; the average virulence of isolates collected from a susceptible host was greater than that of isolates collected from a less susceptible host. Results from Run 2 were similar regarding resistance within the barleys, but variation in virulence among the *P. teres* isolates was not consistent with that from Run 1. CI 9819 carries duplicate genes for resistance to *P. teres* (*Rpt1b* and *Rpt1c*), and CI 7548 possesses *Rpt3d*. Both genotypes were highly resistant to all isolates. The *Rpt1a* gene of Tifang (CI 4407) conferred resistance to all isolates in Run 2, but only to about half of the isolates in Run 1. The checks, including two of the symptomatically most resistant Nordic barley genotypes, were universally susceptible during these stringent tests. No selective pressure has been placed on Finnish isolates of *P. teres* through previous deployment of major resistance genes, and it is speculated that any variation in virulence among isolates is likely to be due to a combination of evolutionary forces including, natural selection, random genetic drift and gene flow.

Introduction

Net blotch of barley (*Hordeum vulgare* L.), caused by *Pyrenophora teres* Drechs. f. *teres* Smedeg., is an economically important disease worldwide (Shipton et al., 1973), and particularly in Finland. With the exception of a report on its distribution in Finland (Mäkelä, 1975) results of research on it in Finland have not appeared in English.

The interactions between host plant and *P. teres* and symptom development are greatly influenced by the environment (Shipton et al., 1973), making it difficult to assess the degree of pathogenic differentiation among *P. teres* isolates. Specific pathotypes have, however, been reported from many countries (See Steffenson & Webster, 1992a). The purpose of this study was to assess diversity of virulence among Finnish isolates of *P. teres* from a range of sites and identify useful sources of resistance for use in a resistance breeding programme.

Materials and methods

Net blotch isolates

Leaves of barley plants with net blotch symptoms were collected from 11 sites in Finland during summer 1994, and the host plant from which the leaves were taken was recorded (Table 1). Leaf tissue with lesions was surface sterilized in 50% ethanol for 15 s and in 2% NaOCl for 30 s, and then rinsed with distilled water and placed on 2.3% lima-bean agar (LBA, 23 g Bacto-Lima Bean Agar, 1 L water) in sterile Petri dishes. The

Table 1. Origins of *Pyrenophora teres* isolates used in two experiments

Isolate	Location	Grid reference	Host barley
1	Ylistaro	62° 56′N, 22° 29′E	Arve
2	Ylistaro	62° 56′N, 22° 29′E	Pohto
3	Mikkeli	61° 40′N, 27° 13′E	Arve
4	Pälkäne	61° 20′N, 24° 13′E	Arve
5	Pälkäne	61° 20′N, 24° 13′E	Pohto
6	Sotkamo	64° 06′N, 28° 20′E	Arve
7	Sotkamo	64° 06′N, 28° 20′E	Pohto
8	Tohmajärvi	62° 14′N, 30° 19′E	Arve
9	Tohmajärvi	62° 14′N, 30° 19′E	Pohto
10	Jokioinen	60° 49′N, 23° 30′E	Arve
11	Vihtavuori	62° 20′N, 26° 10′E	Arve
12	Vihtavuori	62° 20′N, 26° 10′E	Pohto
13	Mietoinen	60° 38′N, 21° 51′E	Arve
14	Mietoinen	60° 38′N, 21° 51′E	Pohto
15	Ruukki	64° 40′N, 25° 05′E	Arve
16	Ruukki	64° 40′N, 25° 05′E	Pohto
17	Maaninka	63° 09′N, 27° 02′E	Arve
18	Maaninka	63° 09′N, 27° 02′E	Pohto
19	Peipohja	61° 17′N, 22° 14′E	Arve
20	Peipohja	61° 17′N, 22° 14′E	Pohto

Table 2. Mean squares and their statistical significance for selected factors from the ANOVA for symptom scores on 27 barley genotypes artificially infected with 20 isolates of *Pyrenophora teres*, during two experiments (Runs 1 and 2) conducted in the greenhouse, on the second and third seedling leaves

Factor	df	Run 1 Leaf 2	Run 1 Leaf 3	Run 2 Leaf 2
		MS[a]	MS	MS
Replicate (R)	1	197.53***	182.99***	3.04ns
Isolate (I)	19	217.19***	132.92***	251.10***
error a (RI)	19	11.46	16.38	4.56
Genotype (G)	26	716.24***	802.00***	315.63***
GI	494	10.57***	14.35**	7.33***
error b (GR[I])	520	4.46	11.03	3.29
Regressions				
Joint	1	3103.91***	3039.31***	2050.98***
Slope	19	2.16ns	2.53ns	15.85***
Intercept	19	36.19***	17.31***	62.75***
error	500	1.66	2.08	1.21
Time[b]	2[c]	713.34***	652.49***	517.25***
error c	1080[d]	2.31	8.30	2.10

[a] ***, significant at $P \leq 0.001$; **, significant at $P \leq 0.01$; ns, not significantly different ($P > 0.05$).
[b] interactions with time not included
[c] df = 1 for Run 2 Leaf 2
[d] df = 540 for Run 2 Leaf 2.

Petri dishes were kept under near UV light and a 12 h photoperiod at 18° C for 14 days. A spore suspension was made by flooding Petri dishes with 100 ml of sterile water and 0.75 ml of the suspension was spread on LBA. Cultures were incubated under near UV light and a 12 h photoperiod at 18° C. For each experiment the same original stock of conidia was used to produce new inoculum.

Differential genotypes

The following differential genotypes were used: 1) Tifang CI 4407, possesses the *Rpt1a* gene 2) Canadian Lakeshore CI 2750, 3) Atlas CI 4118, 4) Rojo CI 5401, 5) Coast CI 2235, 6) Manchurian CI 739, 7) Ming CI 4797, 8) CI 9819, possesses *Rpt1b* and *Rpt1c* genes 9) Algerian CI 1179, 10) Kombar CI 15694, 11) CI 11458, 12) CI 5791, 13) Harbin CI 4929, 14) CI 7584, possesses the *Rpt3d* gene 15) Prato CI 15815, 16) Manchuria CI 2330, 17) CI 5822, 18) CI 4922, 19) Cape CI 1026, 20) Beecher CI 6566, 21) Rika CI 8069. These genotypes were suggested by Steffenson & Webster (1992a) to be a useful standard series for researching response to *P. teres* infection. Six Nordic 6-row spring barley genotypes, which had shown various degrees of symptomatic response to *P. teres* infection in the field (Robinson, unpublished), were also included: 22) Arve, 23) Agneta, 24) Artturi, 25) H6221, 26) Pohto, 27) WW7977.

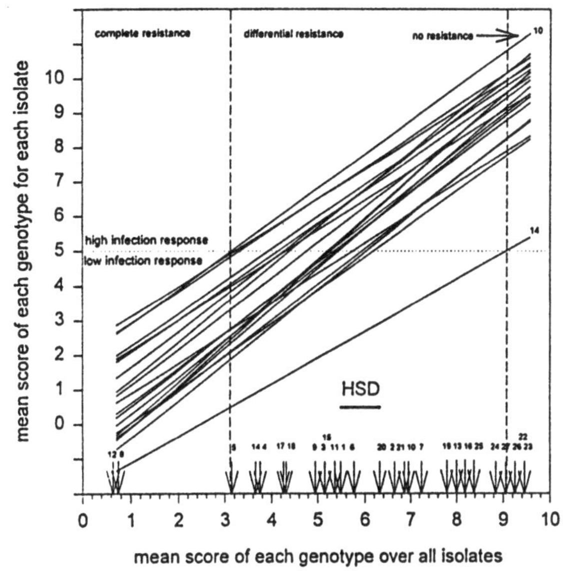

Figure 1. Regression lines showing the relationship between mean symptom scores of each barley genotype for each isolate of *Pyrenophora teres* and the overall mean symptom score for each barley genotype over all *P. teres* isolates, for the second leaf during the first run (data are means from two replicate blocks and three times). Means for the barley genotypes are indicated by arrows on the abscissa; numbers refer to genotype designations given in Materials and methods. A Tukey's HSD ($\alpha = 0.05$) is given for separating significantly different genotype means. *P. teres* isolates 10 and 14 (see Table 1) are indicated.

Inoculation and scoring

After two weeks conidia were collected by flooding the Petri dishes with distilled water and scraping the culture into a beaker of distilled water. Density of conidia was assessed in a haemocytometer and the concentration adjusted to 2×10^4 conidia per ml.

Five viable barley seeds were sown per pot (6 cm diam.) containing nutrient-supplemented peat, and were placed in a greenhouse at 18–22° C, 12 h photoperiod. The pots were arranged in a split plot design with two replicate blocks. Isolates were assigned to main plots and barley genotypes to sub-plots (PROC PLAN, SAS Institute, 1989). Approximately two weeks after planting, when seedlings were at GS12/GS13 (Tottman & Makepeace, 1979), relative humidity in the greenhouse was raised to 100%, and main plots were inoculated with conidial suspension at a rate of 0.5 ml per pot. A control set of genotypes, which received no inoculum, was included in each block. High humidity was maintained for two days, by when the first net blotch symptoms were evident. Infection response on the first three leaves was recorded, using the 10 point scale of Tekauz (1985), for single plants harvested from each pot. Scores were taken from the first three leaves of three seedlings per pot over three days during Run 1 and from the first two leaves of two seedlings per pot over two days during Run 2. Scores $\geq$ 5.0 were categorized as high infection responses (HIR) and those < 5.0 low infection responses (LIR).

Data analysis

The data for Run 1 and Run 2 were analysed independently as split-split plot designs (PROC ANOVA, SAS Institute, 1989), and regressions (Finlay & Wilkinson, 1963) plotted according to Sato & Takeda (1993) using Sigma Plot software (Jandel Scientific, 1993). Significantly different means were separated with Tukey's HSDs ([α = 0.05], SAS Institute, 1989). Heterogeneity of slopes and intercepts for regression lines was established using PROC GLM (Littell et al., 1991).

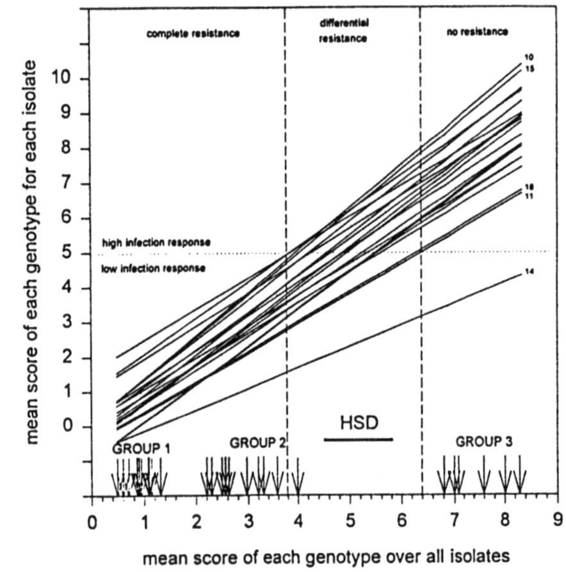

Figure 2. Regression lines showing the relationship between mean symptom scores of each barley genotype for each isolate of *Pyrenophora teres* and the overall mean symptom score for each barley genotype over all *P. teres* isolates, for the third leaf during the first run (data are means from two replicate blocks and three times). Means for the barley genotypes are indicated by arrows on the abscissa. Group 1 contains genotypes 5, 8, 12, 1, 14, 18, 4, 9, 3, 15 and 6; Group 2: 16, 13, 11, 20, 17, 2, 19, 10, 7 and 21; Group 3: 22, 23, 24, 25, 26 and 27 (numbers refer to genotype designations given in Materials and methods). A Tukey's HSD ($\alpha = 0.05$) is given for separating significantly different genotype means. *P. teres* isolates 10, 11, 14, 15 and 18 (see Table 1) are indicated.

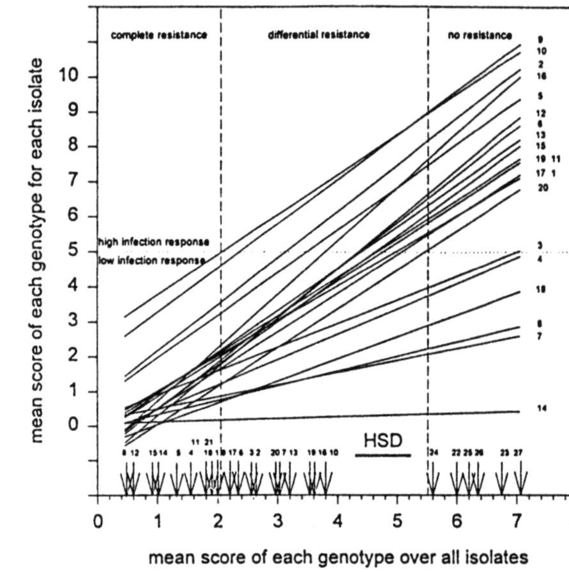

Figure 3. Regression lines showing the relationship between mean symptom scores of each barley genotype for each isolate of *Pyrenophora teres* and the overall mean symptom score for each barley genotype over all *P. teres* isolates, for the second leaf during the second run (data are means from two replicate blocks and two times). Means for the barley genotypes are indicated by arrows on the abscissa; numbers refer to genotype designations given in Materials and methods. A Tukey's HSD ($\alpha = 0.05$) is given for separating significantly different genotype means. All *P. teres* isolates (see Table 1) are indicated.

Results

Net blotch symptoms developed well during both runs, lesions becoming apparent within two days of inoculation. For both runs the entire range of scores was evident during symptom assessment although the artificially induced epidemic was more severe during Run 1 than Run 2. The disease did not extent to the third leaf in Run 2, other than on the Nordic genotypes. During both runs and for the three seedling leaves, genotypes 8 (CI 9819) and 12 (CI 5791) were virtually symptomless. During both runs isolate 14 of *P. teres* was very weakly pathogenic, and isolate 10 on both occasions, on average, exhibited the highest degree of pathogenicity.

Selected mean squares and their statistical significances for ANOVAs and regression analyses are given in Table 2. Data for Run 2 were transformed with (x + 0.5) and analysed, but this did not alter the conclusions reached from analysis of untransformed data. Initially, analyses were done on complete data sets including two (Run 2) and three leaves (Run 1). There were significant differences in the degree of symptom expression – the second seedling leaf was most affected by the disease and the third leaf the least. In these analyses the first order interactions between leaf and the other factors were generally statistically significant, but very small in comparison with the main effects, and not useful to present, with the exception of the leaf x genotype interaction for Run 1 ($F = 20.3$, $P = 0.0001$, df = 52, 1080). This significant interaction reflects the differences in third seedling leaf susceptibility between genotypes, the Nordic genotypes being much more susceptible than the remainder.

From Table 2 it is apparent that there were significant differences between isolates, genotypes and their interactions. The disease increased with time, as evidenced by the statistically significant time component. The isolate x time interaction ($F = 3.24$, $P = 0.001$, df = 38,1038) is accounted for by isolates 3, 8, 14, 15, 18 and 20 initiating a much lower increase in symptom expression than the other isolates between the second and third scorings. The significant genotype x isolate interactions indicate that differential reactions occurred. In the reduced analyses, for specific leaves, the first order interactions with time were also statistically significant, but were again small, and are omitted from Table 2. However, the significant genotype x time interaction for Run 1 was principally the result of genotypes 2, 8 and 12 expressing less severe symptoms on the third assessment than on the second. The second order interactions with time were not meaningful, and the third order interaction was not significant. It was felt that a large genotype by isolate matrix detailing the HIR/LIR reactions would not be useful given that each isolate exhibited a unique pattern of response on the differential genotypes and these patterns were not consistent between runs. Therefore, regression plots and analyses were employed to provide a more accessible summary of the interactions. Plots for Run 1 Leaf 2, Run 1 Leaf 3 and Run 2 Leaf 2 are represented by Figures 1–3 respectively. Means for the genotypes are indicated on the abscissas with arrows and a Tukey's HSD is provided to separate statistically significant means. The data provided in Table 2 indicate that there were statistically significant differences in intercept for the 20 lines in each of the three plots, but that during the first run, for leaves 2 and 3, there were no statistically significant differences in slopes.

Results of ANOVAs from further reduced data sets, omitting isolates 3 and 10, which were only collected from a single host cultivar (Table 1), indicated significant differences in virulence associated with site of collection ($F = 14.66, P = 0.0001$, df = 8,17) and influence of host cultivar ($F = 34.64, P = 0.0001$, df = 1,17) in Run 1, but only for site in Run 2 ($F = 17.10$, $P = 0.0001$, df = 8,17). An interaction between site and host cultivar in Run 1 ($F = 12.10$, $P = 0.0001$, df = 8,17 Run 2; $F = 0.09$, $P = 0.7698$, df = 1,17) was accounted for by the behaviour of isolates 14 (virtually non-pathogenic) and 13. The effect of site was not consistent between runs, but the effect of host cultivar on symptom expression in Run 1 was consistent for leaves 1, 2 and 3, and the data are graphed in Figure 4. Regression lines for Arve and Pohto differed in slope only for Leaf 3 ($F = 13.37, P = 0.0006$, df = 1,50), but in intercept for all three leaves ($P < 0.0001$, df = 1,51).

Discussion

From the point of view of practical barley breeding, the results of these experiments indicate that any barley from the differential series contains useful levels of resistance to *P. teres* that might be transferred to Nordic genotypes: all were significantly less affected by *P. teres* than the six Nordic genotypes during both runs (confirmed with data from a third and fourth run; data not given), considering symptom scores on the first (data not shown), second and third leaves. Genotypes, moreover, ranked very similarly considering symptom expression on the first three seedling leaves. It appeared that the Nordic genotypes did not contain any of the resistance genes carried by the members of the differential series. However, it was shown that large differences exist between the infection responses of the Nordic genotypes in the field, that is, differences in quantitative resistance to *P. teres*, where epidemics are invariably much milder than those created during these experiments and where plants grow to maturity (Robinson, unpublished).

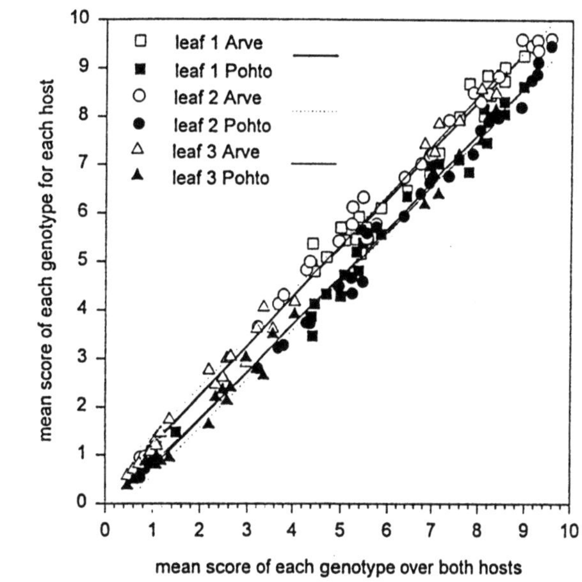

Figure 4. Regression lines showing the relationship between mean symptom scores for both barley host genotypes (Arve and Pohto) for all isolates of *Pyrenophora teres* and the overall mean symptom score for each barley genotype over both hosts, for the first run and the first three seedling leaves. Individual data items are means from two replicate blocks, three times and nine sites.

Steffenson & Webster (1992a), using the differential series employed in these studies, reported the existence of 13 *P. teres* pathotypes among a collection of isolates from California. They established that ten of the differentials were resistant to all of the pathotypes and each differential was resistant to six or more pathotypes. The results reported here indicate that under a severe artificial epidemic (Figure 1) only genotypes 12 (CI 5791) & 8 (CI 9819), both of Ethiopian origin and both in the group of completely resistant genotypes reported by Steffenson & Webster (1992a), were resistant to all *P. teres* isolates, but that symptom scores on the third leaf (Figure 2) placed all genotypes except one (Rika, CI 8069) in the resistant (LIR) category. During the second run (Figure 3) ten genotypes fell into the resistant category. Interestingly, of the 10 universally resistant genotypes reported by Steffenson & Webster (1992a) seven occur in Group 1 (Figure 2) and seven in the 'completely resistant' area of Figure 3 (Tifang, CI 9819, Coast, CI 5791, CI 7584, Rojo and Algerian) and their reaction to the Finnish *P. teres* isolates represents a similarity to the Californian situation. However, Tifang was reported to give an incompatible response by Steffenson & Webster (1992a) and was used in further studies on quantitative resistance to *P. teres* as the completely resistant check genotype (Steffenson & Webster, 1992b). Although there is not universal agreement on allelic and locus relationships, using the designations of Bockelman et al. (1977), Tifang carries the *Rpt1a* gene; CI 7584, *Rpt3d*, and CI 9819, *Rpt1b* and *Rpt1c*. In these experiments Tifang did not exhibit an incompatible response, in Run 1 it was resistant to about half of the isolates, but in Run 2 to all. In Run 1 (Leaf 2) the mean value for its response over all isolates was 5.4, very similar to that (5.3) reported by Sato & Takeda (1993) in their investigations on Japanese isolates of *P. teres*. CI 5791 has been generally reported to be resistant (e.g. Khan, 1969; Tekauz & Buchannon, 1977; Tekauz & Mills, 1974). CI 9819 has also been shown to be resistant on a wide geographic scale, most recently in Japan (Sato & Takeda, 1993).

The patterns of genotype reaction to the 20 *P. teres* isolates are consistent for the second and third seedling leaves in Run 1 (Figures 1 and 2), there being one principal group of regression lines in each instance, with only the regression line for isolate 14 separate from the main group. The lack of virulence of isolate 14 is no doubt associated with phenotype of the isolate in culture: in the Petri dish it was white, fast-growing and produced few conidia while all other isolates were dark, slow-growing and produced abundant conidia. Shipton et al. (1973) reported on such phenotypic differences.

During the second run (Figure 3) the situation was similar to that of the first except that the isolates form a main group of fourteen, with five isolates initiating an LIR on all genotypes, and isolate 14 being virtually non-pathogenic. Isolates 3, 8, 14, 15, 18 and 20, during the first run, were those contributing to the significant genotype x time interaction, weakening in effect, in comparison with the other isolates, with the progression of time, and thereby reflecting their relatively weak degree of pathogenicity (Figure 3). The strict division of pathogenicity of isolates and resistance of genotypes is somewhat arbitrary. For both runs all 20 isolates had unique patterns of HIR/LIR, but they were not consistent between runs (nor for a third and fourth run, data not given) and it is therefore not possible to designate the isolates as different pathotypes. Variation in virulence between Finnish isolates appears inconsistent under stringent conditions although differences appear to be magnified under less severe conditions. Sato & Takeda (1993, 1994) suggested that there was little difference in virulence between Japanese isolates of *P. teres*.

It is apparent, despite adoption of a standard differential series, that comparison and interpretation of results from the same laboratory from different times, and between laboratories, is rendered very difficult by the variable nature of the host-pathogen interaction. This means that although a single score might be used to determine whether an infection response is high or low, the reaction of the barley genotype to any one isolate is largely determined by the severity of the epidemic. Thus, in Run 1 for the second leaf, with the exception of genotypes 12 and 8, no members of the differential series were resistant to all isolates. During the second run however, several genotypes were resistant to all isolates. Three statistically distinct groups of genotypes formed, considering symptoms recorded on the third leaves during Run 1, and all except one genotype were resistant to all isolates of *P. teres*.

The data given in Figure 4 strongly indicate that under conditions of severe epidemic, at least, there was a significant effect on virulence of the isolate associated with the host plant from which it was taken. Under field conditions Arve is considerably less resistant to *P. teres* than Pohto (Robinson, unpublished). Monitored on the first, second and third seedling leaves it is apparent that isolates from Arve were significantly more virulent than those from Pohto. This may represent some form of host specialization or conditioning response. McDonald et al. (1989) reported that the average virulence of *Rhynchosporium secalis* (Oud.) Davis isolates collected from a more resistant host barley population was greater than that of isolates from a less resistant population. This is exactly contrary to the situation reported here, but illustrates that the host plant from which pathogen isolates derive can affect subsequent virulence. Steffenson & Webster (1992a) point out that in their own study they were unable to make any associations between pathogenicity of *P. teres* isolates and original host in the absence of data on the latter.

In Finland net blotch is typically sporadic, and it might therefore be inferred that cycles of extinction and recolonization occur. It is not known what contribution the sexual cycle of *P. teres* makes to the genetic structure of the populations, but recolonization most probably occurs via transport and growing of infected seed. As the pathogen has not been exposed to major resistance genes in the past, and therefore not been subjected to any selective pressure, it would seem unlikely that the coevolution of host and pathogen has been along the lines of the gene-for-gene model (Thompson & Burdon, 1992), but that natural selection, gene flow and random genetic drift have modelled the population structure (McDermott & McDonald, 1993). Molecular markers have been used to assist in determination of the uniqueness of the genetic structure of *P. teres* populations from Canada, Germany and the U.S.A. (Peever & Milgroom, 1993) and it is hoped that in the future they might be used similarly with Finnish isolates of *P. teres*.

Acknowledgements

We thank H.E. Vivar, ICARDA/CIMMYT barley program, Mexico, for providing seeds used during these experiments. T. Hovivuori and A. Kedonperä (MTT; Finland) are thanked for their technical assistance, and E. Nissilä (Boreal Plant Breeding, Finland) is acknowledged for his useful discussion with the authors.

References

Bockelman, H.E., E.L. Sharp & R.F. Eslick, 1977. Trisomic analysis of genes for resistance to scald and net blotch in several barley cultivars. Can J Bot 55: 2142–2148.

Finlay, K.W. & G.N. Wilkinson, 1963. The analysis of adaptation in a plant-breeding programme. Aust J Res 14: 742–754.

Jandel Scientific, 1993. Sigma Plot User's Manual, San Rafael, CA, USA.

Khan, T.N., 1969. Inheritance of resistance to net blotch in barley. I. Factors affecting the penetrance and expressivity of gene(s) conditioning host resistance. Can J Genet Cytol 11: 587–591.

Littell, R.C., R.J. Freund & P.C. Spector, 1991. SAS System for Linear Models, Third Edition, Cary, NC.

Mäkelä, K., 1975. Occurrence of *Helminthosporium* species on cereals in Finland in 1971–1973. J Sci Agric Soc Fin 47: 181–217.

McDermott, J.M. & B.A. McDonald, 1993. Gene flow in plant pathosystems. Annu Rev Phytopathol 31: 353–373.

McDonald, B.A., J.M. McDermott, R.W. Allard & R.K. Webster, 1989. Coevolution of host and pathogen populations in the *Hordeum vulgare-Rhynchosporium secalis* pathosystem. Proc Natl Acad Sci USA 86: 3924–3927.

Peever, T.L. & M.G. Milgroom, 1994. Genetic structure of *Pyrenophora teres* populations determined with random amplified polymorphic DNA markers. Can J Bot 72: 915–923.

SAS Institute Inc., 1989. SAS/STAT User's Guide, Version 6, Fourth Edition, Volume 2, Cary, NC.

Sato, K. & K. Takeda, 1993. Pathogenic variation of *Pyrenophora teres* isolates collected from Japanese and Canadian spring barley. Bull Res Inst Bioresour Okayama Univ 1: 147–158.

Sato, K. & K. Takeda, 1994. Sources of resistance to net blotch in barley germplasm. Bull Res Inst Bioresour Okayama Univ 2: 91–102.

Shipton, W.A., T.N. Khan & W.J.R. Boyd, 1973. Net blotch of barley. Rev Plant Pathol 52: 269–290.

Steffenson, B.J. & R.K. Webster, 1992a. Pathotype diversity of *Pyrenophora teres* f. *teres* on barley. Phytopath 82: 170–177.

Steffenson, B.J. & R.K. Webster, 1992b. Quantitative resistance to *Pyrenophora teres* f. *teres* in barley. Phytopath 82: 407–411.

Tekauz, A., 1985. A numerical scale to classify reactions of barley to *Pyrenophora teres*. Can J Plant Pathol 7: 181–183.

Tekauz, A. & K.W. Buchannon, 1977. Distribution of and sources of resistance to biotypes of *Pyrenophora teres* in western Canada. Can J Plant Sci 57: 389–395.

Tekauz, A. & J.T. Mills, 1974. New types of virulence in *Pyrenophora teres* in Canada. Can J Plant Sci 54: 731–734.

Thompson, J.N. & J.J. Burdon, 1992. Gene-for-gene coevolution between plants and parasites. Nature 360: 121–125.

Tottman, D.R. & R.J. Makepeace, 1979. An explanation of the decimal code for the growth stages of cereals, with illustrations. Ann Appl Biol 93: 221–234.

P.M.A. Tigerstedt (Ed.), *Adaptation in Plant Breeding*, 103–108, 1997.

Interaction of insect digestive enzymes with plant protein inhibitors and host-parasite coevolution

Alexander V. Konarev
All-Russian Institute for Plant Protection (VIZR), Podbelskogo 3, St. Petersburg-Pushkin, Russia

Key words: insects, digestive enzymes, cereals, inhibitors, insect-plant coevolution

Summary

A complex of new approaches was used to study the insect-plant coevolution by using cereal pest digestive α-amylases and proteinases and their proteinaceous inhibitors in cereals. During evolution, plants can weaken the destructive affects of insect hydrolases at the expense of inhibitors in various ways, including (1) increasing the inhibitor activity and heterogeneity, (2) increasing the complexity of an inhibitor set and (3) producing highly specific insect enzyme inhibitors and bifunctional amylase/proteinase inhibitors. Insects, in turn, can decrease the influence of inhibitors by (1) increasing digestive enzyme activities, (2) by modifying a set of related activities of various digestive hydrolases, (3) by decreasing enzyme sensitivity to inhibitors and (4) by destroying inhibitors in guts by proteinases.

Abbreviations: IEF – isoelectric focusing; TI – trypsin inhibitor; CI and C/SI – chymotrypsin and chymotrypsin/subtilisin inhibitors; InsAI12 – insect amylase inhibitor with molecular weight value of about 12 kDa

Introduction

Amylases and proteinases play a key role in the digestion of plant starch and proteins by insects. Various α-amylase and proteinase inhibitors present in seeds and vegetative organs of plants (Buonocore et al. 1977; Boisen 1983; Konarev 1982a, 1987, 1992, 1994; Garcia-Olmedo et al. 1987; Ryan 1990) are able to act as limiting regulators of phytophage number and are involved with pest enzymes in host-parasite coevolution at the molecular level. Some direct and indirect evidences for this will be viewed here using an example of pest digestive enzymes and their inhibitors in cereals.

Digestive enzymes of cereal pests

Insect species damaging cereal grains or grain products differ strongly in the spectrum of various enzymes produced, their isoelectric patterns and by the relative activities of various hydrolases in guts (Konarev 1981, 1986a, 1992; Konarev & Fomicheva 1991a). For example, the guts of larvae and the adults of the yellow mealworm beetle *Tenebrio molitor* L. contain trypsin, chymotrypsin-like and cysteine proteinases. Cysteine proteinases prevail in isoelectric patterns of both *Tribolium confusum* Duv. and *Tr. castaneum* Herbst., although some weak trypsin- and chymotrypsin-like components are also present. In contrast, trypsin-like proteinases were only detected in the guts of the lesser grain borer *Rhyzopertha dominica* F. The trypsin-like enzyme predominated in guts of *Oryzaephilus surinamensis* L. over the other proteinases. The chymotrypsin-like proteinases and other proteinases were detected in isoelectric spectra of proteins in guts of the weevils *Sitophilus granaria* L. and *S. oryzae* L. The insect species also differ significantly in gut α-amylase patterns.

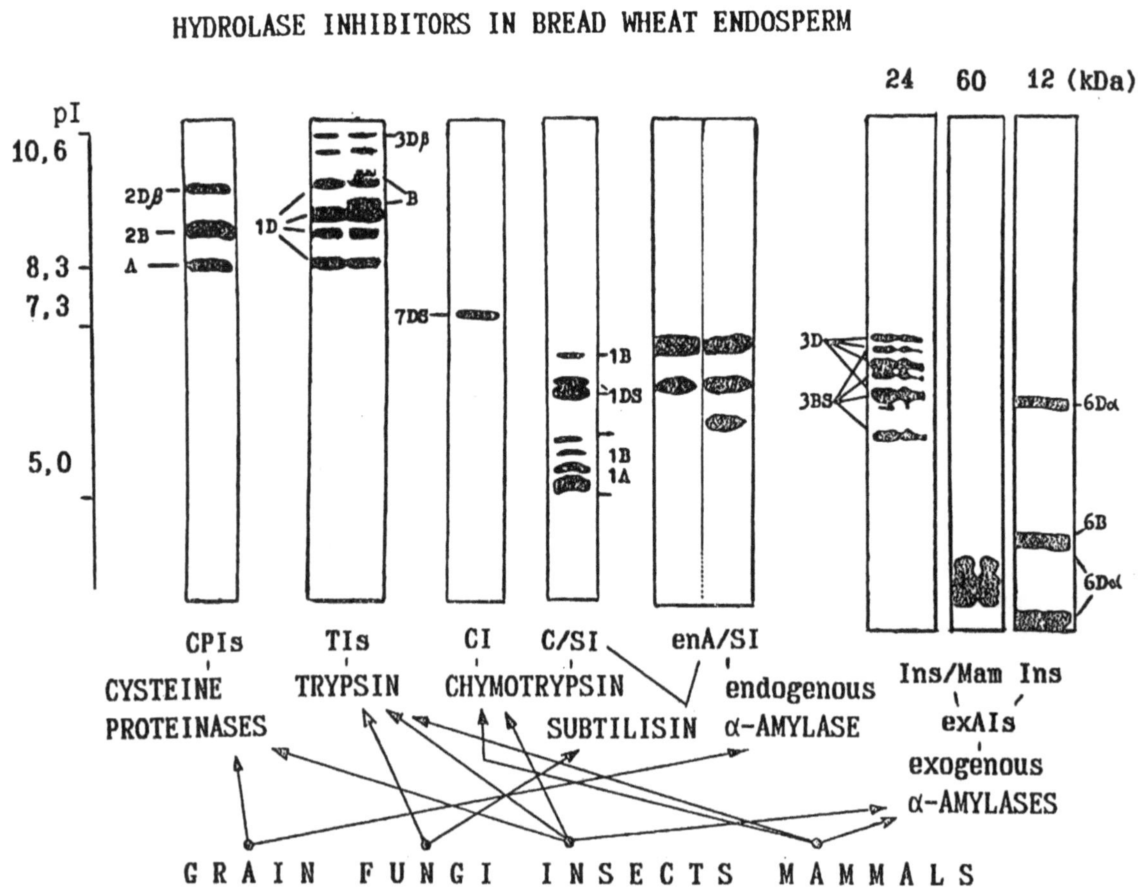

Figure 1. Hydrolase inhibitors in endosperm of bread wheat (*Triticum aestivum* L.). Isoelectric focusing of water-soluble grain proteins followed by the detection of inhibitors. CPIs – cysteine proteinase inhibitors, TIs – trypsin inhibitors, CI – chymotrypsin inhibitor, C/SIs – chymotrypsin and subtilisin inhibitors, enA-SI – endogenous amylase and subtilis in inhibitor, exAIs – exogenous amylase inhibitors. A, B, 1D, . . . – genome and chromosome control of inhibitors.

Insect digestive hydrolase inhibitors in cereals

Universal methods have been developed to detect inhibitors of various α-amylases and proteinases in plant protein patterns separated by isoelectric focusing (IEF), electrophoresis and thin layer gel-filtration (Konarev 1982a, 1986a, 1987, 1992, 1994) (Figure 1). A number of different trypsin inhibitors (TIs) which are active against trypsin-like proteinases from insects, mammals and fungi are localized in wheat endosperm, embryos and leaves (Konarev 1986a, 1987, 1992, 1994; Konarev & Fomicheva 1991). The interrelation between genome composition and the structure of system of endosperm TIs may be seen in the scheme of evolution of this system in genus *Triticum* L. (Figure 2). The scheme is based on the results of analytical IEF of inhibitors and on data about the phylogeny of wheats (Konarev 1983). The TIs in *T. aestivum* L. (A^uA^uBBDD) are represented by the two groups of components differing in values of isopoints (pI), molecular masses and genetic control (Konarev 1987, 1992, 1994). The similarity of bread wheat TIs (12 kDa) controlled by chromosome *1D* with TIs from *Ae. tauschii* Coss. (considered to be a donor of genome *D* to wheat) is evident. The other subsystem of TIs (19 kDa) is similar to TIs of tetraploid wheats from evolutionary group *T. turgidum* L. (A^uA^uBB). The *T. turgidum* L. TIs are close in their turn to TIs of *Aegilops* L. species from section *Sitopsis* Jaub. et Spach. which are considered as probable donors of genome *B*. Very active TIs (about of 14 kDa) are characteristic of all diploid wheat (*AA*) and of tetraploids of the *T. timopheevii* Zhuk. group

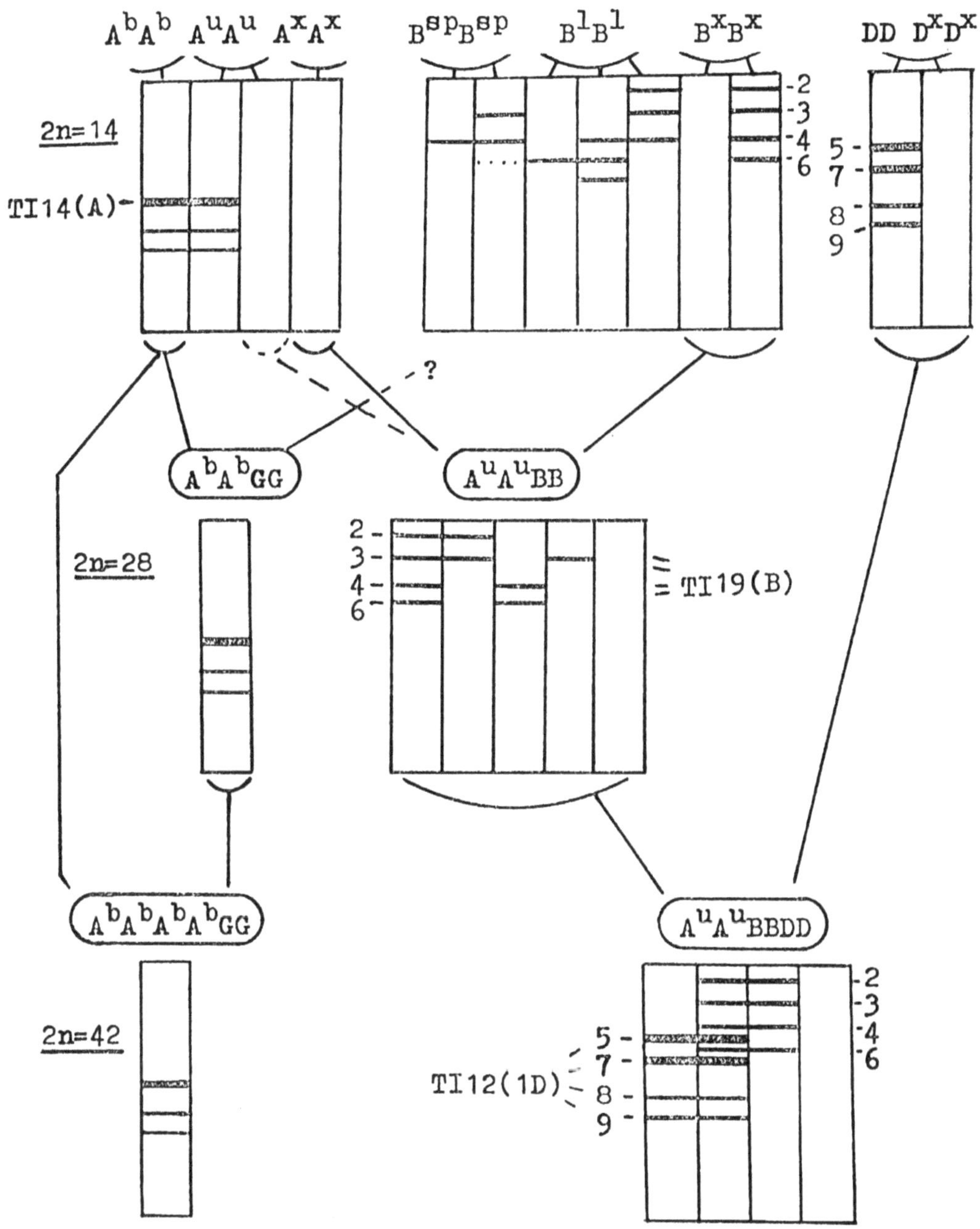

Figure 2. The evolution of the endosperm trypsin inhibitor (TI) system in genus *Triticum* L. The scheme based on the results of the isoelectric focusing of endosperm proteins in pH interval 5–11 followed by the detection of TIs, SDS-PAGE of TIs purified by affinity chromatography etc. and schemes of wheat phylogeny. TI(12/14/19) – TIs fractions with corresponding molecular mass values (in kDa). A^b, A^u, B^1, . . . – genome formulae; *1D* – chromosome control; A^x, B^x, D^x – genomes of assumed (but not found) diploid forms.

(A^bA^bGG), but these are lacking in *T. turgidum* and *T. aestivum*. Thus, at least three types of endosperm TIs occur in *Triticum* species.

Species and varieties of polyploid wheats having genomes *B* and *D* are highly variable in the composition and activity of endosperm TIs. The existence of contrasting forms has led to the establishment of a relationship between the TI activity and the resistance of wheat varieties to *R. dominica* which feed on wheat endosperm and have the trypsin-like digestive proteinases predominating in their guts (Konarev 1986a, 1987, 1994; Konarev & Fomicheva 1991a). It can be supposed that other insects cope more easily with TIs because of the presence of chymotrypsin-like, cysteine and other proteinases.

There are two types of insect chymotrypsin-like proteinase inhibitors in endosperm of wheats and realted cereals: chymotrypsin/subtilisin inhibitors (C/SIs) and chymotrypsin inhibitors (CIs). C/SIs, which are active also on elastase, were found in all of the studied *Triticaea* Dum. and in many other cereals as well. It is the most heterogeneous and variable inhibitor system in the wheat endosperm. Its polymorphism can be even higher than that of the prolamins. The bread wheat C/SIs are controlled by chromosomes *1B* and *1D*. The components characteristic of tetraploid wheats and *Ae. tauschii* are summed up in the C/SI spectra of *T. aestivum* (Konarev 1986b, 1989, 1994).

Native CI from bread wheat (*T. aestivum*) has the molecular mass of about 26 kDa, a pI value of about 7.5 and is not active towards subtilisin or elastase. It is controlled by genome *D* and is absent from other wheats. The identical CI components were found in *T. aestivum, Ae. tauschii* and *S. cereale* (Konarev 1992b).

Cysteine proteinase inhibitors (CPIs) were revealed in the IEF spectra of endosperm protein of wild and cultivated species of wheat, barley, rye and other *Triticeae* cereals (Konarev 1984, 1986a, 1994). The same components of endosperm proteins inhibit the cysteine proteinases of insects and germinated wheat grain as well as ficin and papain. The CPIs in various cereal groups differ in pI values and in other characteristics. The composition of CPIs are specific on the whole for di-, tetra- and hexaploid wheats. *2B* and *2D* chromosomoes of *T. aestivum* are involved in the control of CPIs. The wheat CPIs have a molecular mass of about 11–12 kDa, a value close to rice CPIs (Abe et al. 1987). The CPIs can be viewed both as regulatory and defence (against insects) proteins in cereals. It is important that digestive cysteine proteinases are characteristic of insects, and are not found in the digestive system of mammals. Furthermore, the mixtures of cysteine and serine proteinase inhibitors exhibit synergistic toxicity towards the insects (Oppert et al. 1993).

Independent systems of inhibitors of endogenous and exogenous α-amylases are present in the endosperm of *Triticeae* cereals (Konarev 1982b, 1992, 1994; Mundy et al. 1983). Exogenous amylases inhibitors from wheat may be conventionally divided into insect α-amylase inhibitors (12 kDa) and inhibitors of both insect and mammals α-amylases (24 and 60 (48) kDa) (Buonocore et al. 1977; Konarev 1982a, 1992, 1994; Garcia-Olmedo et al. 1987). α-Amylase inhibitors with a molecular mass value of about 12 kDa (InsAI12) were representative of 22 genera from 12 tribes of *Pooideae* and *Bambusoideae* subfamilies. Inhibitors of insect and mammal α-amylases (24 kDa) were revealed only in *Triticeae* and *Bambusiodeae* cereals. The 24 kDa inhibitors of *Hordeum vulgare* L. are active mainly against insect α-amylases. Inhibitors of the 60 kDa group were detected in wheat, rye and barley species (Konarev 1992, 1994).

The IEF banding patterns of insect α-amylase inhibitors are specific for a few diploid wheats (the majority samples of them have no α-amylase inhibitors), for the evolutionary groups of tetraploid wheats *T. timopheevii* and *T. turgidum*, for hexaploid *T. aestivum* and for different *Aegilops* species. Intraspecific variation of tetra- and hexaploid wheats by α-amylase inhibitors is expressed in the presence or absence of InsAI12 components, controlled by the chromosome *6B* in *T. aestivum* (Konarev 1982a, 1994). In samples of tetraploid wheats the presence or absence of InsAI12 components is accompanied by variation in inhibitor activity. In *T. aestivum* samples the variability of activity of insect amylase inhibitors is reduced by presence of low variable inhibitors with molecular mass value of 24 kDa. It is of interest that correlations between resistance to grain pests and insect α-amylase inhibitor activity are found just among tetraploid wheat varieties (Yetter et al. 1979).

Bifunctional inhibitors of insect α-amylases and trypsin were found in *H. bulbosum* L. (12 kDa, pI values of about 7.3) (Konarev & Fomicheva 1991b; Konarev 1992). A similar inhibitor was found earlier by Richardson et al. (1987) in maize. Such inhibitors may be considered as the next evolutionary step in the formation of efficient defence protein systems after specific inhibitors of insect α-amylases, as they are able to block digestive trypsin-like proteinases too. The protection of amylase inhibitors from inactivation by proteinases can significantly increase their defensive role towards insects. In contrast to other pests, insects which are able to digest wheat amylase inhibitors do not respond to the presence of such inhibitors in diet (Gatehouse et al. 1986).

The digestive α-amylases of different insect species vary strongly in their sensitivity to cereal inhibitors. It is thought that insects which usually feed on grain and its products possess high sensitive amylases (Buonocore et al. 1977). On the other hand, salivary gland and gut α-amylases of a bug, *Eurygaster integriceps* Puts., which is highly specialized in feeding on cereal endosperm, are many hundreds times less sensitive to wheat 12 kDa and 24 kDa inhibitors then α-amylases of less specialized bugs and beetles (Konarev 1981, 1992). It may be suggested that *E. integriceps* have adapted to inhibitors by decreasing the sensitivity of their α-amylases to the inhibitors.

Conclusion

The general tendency of the evolution of insect and other exogenous hydrolase inhibitor systems in the genus *Triticum* consists of an increase in their heterogeneity and in their activity, along with a growth of the ploidy level of their hosts. Alloploidy as a whole expands the adaptive possibilities of plants partly at the expense of the integration of protein and enzyme systems controlled by different genomes. As a result, the range of physiological conditions in which the systems retain activity can be enlarged (Konarev 1983). That can be true also for inhibitor systems. Thus, during evolution, plants can weaken a destructive affect of insect hydrolases by involving the inhibitors in the following ways: (a) by increasing the inhibitor activity in plant tissues; (b) by developing a heterogenous set of inhibitors possessing activities to various enzymes; (c) by forming highly specific inhibitors of insect enzymes; (d) by producing bifunctional inhibitors which are active both to amylases and proteinases; (e) by increasing the complexity of inhibitors with different biochemical properties through the production of isoinhibitors or through allopolyploidy. In their turn, the insects likewise can decrease the influence of inhibitors by: (a) increasing the activity of their digestive enzymes; (b) modifying a spectrum or the relative activities of various digestive hydrolases; (c) destroying the inhibitors in guts via proteinases; (d) decreasing the sensitivity of their enzymes to the inhibitors.

Acknowledgements

I am grateful to Dr. R. Fido and Dr. J. Bailey (Long Ashton Research Station, UK) for helpful discussion and advices. This work was supported by the Russian Fund for Fundamental Investigations (grant No 93-04-21290).

References

Abe, K., H. Kondo & S. Arai, 1987. Purification and characterization of a rice cysteine proteinase inhibitor. Agric. Biol. Chem. 51: 2763–2768.

Boisen S, 1983. Protease inhibitors in Cereals. (Occurrence, properties, physiological role and nutritional influence). Acta Agriculture Scandinavica 33: 371–381.

Buonocore, V., T. Petrucci & V. Silano, 1977. Wheat protein inhibitors of α-amylase. Phytochemistry 16: 811–820.

Garcia-Olmedo, F., G. Salcedo, R. Sanchez-Monge, L. Gomez, J. Royo & P. Carbonero, 1987. Plant proteinaceous inhibitors of proteinases and α-amylases. Oxford Surv. Plant Mol. Cell Biol. 4: 275–334.

Gatehouse, A.M.R., K.A. Fenton, I. Jepson & D.J. Pavey, 1986. The effects of α-amylase inhibitors on insect storage pests: Inhibition of α-amylase *in vitro* and effect on development *in vivo*. J. Sci. Food Agric. 37: 727–734.

Konarev, Al.V., 1981. Specificity of wheat grain albumins (α-amylase inhibitors) toward enzymes of Sunn pest. Bull. VIR 114: 30–32 (in Russian).

Konarev, Al.V., 1982a. Component composition and genetic control of insect α-amylase inhibitors from wheat and aegilops grain. Doklady VASKhNIL 6: 42–44.

Konarev, Al.V., 1982b. Identification of inhibitors of endogenous and exogenous α-amylases among wheat proteins. Bull. VIR 118: 11–12 (in Russian).

Konarev, Al.V., 1984. Identification of inhibitors of thiol proteinases of insects and grain in wheat and other cereals. Doklady VASKhNIL 10: 13–15.

Konarev, Al.V., 1986a. Analysis of proteinase inhibitors from wheat grain by gelatin replicas method. Biolhimiya 51: 195–201.

Konarev, Al.V., 1986b. Relationship and variability of proteinase and amylase inhibitors in wheat and related cereals. Selkochozaystvennaya Biologia 3: 46–52 (in Russian).

Konarev, Al.V., 1987. Variability of trypsin-like proteinase inhibitors in wheat and related cereals. Selkochozaystvennaya Biologia 5: 17–24 (in Russian).

Konarev, Al.V., 1989. The evolutionary variability and genetic control of chymotrypsin-subtilisin inhibitors in wheat. Doklady VASKhNIL 5: 8–10.

Konarev, Al.V., 1992. Systems of hydrolase inhibitors in cereals: organization, functions and evolutionary variability. D. Sci. Thesis. Institute of Biochemistry, Moscow.

Konarev, Al.V., 1994. Variability of hydrolase inhibitors and the problems of evolution, immunity and breeding of cereals. In: V.G. Konarev (Ed). Molecular Biological Aspects of Applied Botany, Genetics and Plant Breeding. (Supplement to 'Theoretical basis of plant breeding', Vol. 1), pp. 83–94. VIR, Sankt-Petersburg.

Konarev, Al.V. & Yu.V. Fomicheva, 1991a. Cross analysis of the interaction of α-amylase and proteinase components of insects with protein inhibitors from wheat endosperm. Biokhimiya 56: 628–638.

Konarev, Al.V. & Yu.V. Fomicheva, 1991b. Hydrolase inhibitors and coevolution of cereals and harmful pests. Abstr. IX All-Union Conference on Plant Immunity to Deceases and Pests, pp. 270–271, Minsk.

Konarev, V.G., 1983. Plant proteins as genetic markers. Kolos, Moscow.

Liang, G., G. Brookhart, G.H. Feng, G. Reeck & K.J. Kramer, 1991. Inhibition of digestive proteinases of stored grain coleoptera by oryzacystatin, a cysteine proteinase inhibitor from rice seed. FEBS Lett. 278: 139–142.

Mundy, J., I. Svendsen & J. Hejgaard, 1983. Barley α-amylase/subtilisin inhibitor. I. Isolation and characterization. Carlsberg Res. Commun. 48: 81–90.

Oppert, B., T.D. Morgan, G. Culbertson & K.S. Kramer, 1993. Dietary mixtures of cysteine and serine proteinase inhibitors exhibit synergistic toxicity toward the red flour beetle *Tribolium castaneum*. Comp. Biochem. Physiol. 105C: 379–385.

Richardson, M., S. Valdes-Rodriguez & A. Blanco-Labra, 1987. A possible function for thaumatin and TMW-induced protein suggered by homology to a maize inhibitor. Nature V. 327, No. 6121: 423–434.

Ryan, C.A., 1990. Protease inhibitors in plants: gens for improving defenses against insects and pathogens. Annual Rev. Phytopathol. 28: 425–449.

Yetter, M.A., R.M. Saunders & H.P. Boles, 1979. α-Amylase inhibitors from wheat kernels as factors in resistance to postharvest insects. Cereal Chem. 56: 243–244.

P.M.A. Tigerstedt (Ed.), *Adaptation in Plant Breeding*, 109–117, 1997.

Adaptation of wheat rusts to the wheat cultivars in former Czechoslovakia

Pavel Bartoš, Eva Stuchlíková & Renata Hanušová
Research Institute of Crop Production, 161 06 Praha-Ruzyně, Czech Republic

Key words: rusts, virulence genes, wheat, resistance genes, Czech Republic, Slovak Republic, *Triticum aestivum*, *Thalictrum species*

Summary

In former Czechoslovakia virulence of rusts attacking wheat was studied since the sixties. Since the same time genes for resistance in the registered cultivars were identified. The role of *Berberis* and *Thalictrum* as alternate hosts for stem rust and leaf rust, respectively, was investigated as well. Determined changes of virulence in rust populations could only partially be ascribed to changes of resistance genes in the grown cultivars. Unnecessary genes for virulence had no negative effect on the fitness of the pathogen. All tested samples of aeciospores from barberries attacked rye, not wheat. None of *Thalictrum* species occurring in the Czech and Slovak Republics was found to host wheat leaf rust. However, the sexual stage of wheat stem rust and wheat leaf rust could be induced on *Berberis vulgaris* and *Thalictrum speciosissimum*, respectively. General epidemiological conclusions are drawn from the results and experience of the last 35 years.

Introduction

Flor's (1956) gene-for-gene hypothesis and application of it by Person (1959) established a base for genetic studies on adaptation of biotrophic pathogens to their host cultivars. As the hypothesis is based on correspondence between virulence genes on the one side and resistance genes on the other side, studies concerned with adaptation of biotrophic pathogens to their host cultivars were usually limited only to specific genes for virulence and specific genes for resistance. Virulence genes in the pathogen are necessary, to match the corresponding resistance genes. However, the adaptation of the pathogens to their hosts is influenced by a complex genetic information of the pathogen and the host. Presence of genes for virulence in the pathogen population cannot be explained only by genes for resistance in the grown cultivars and by selection pressure, as was very well documented by Mac Key (1992). Results of our race surveys lead to similar conclusions.

Material and methods

Virulence of single pustule isolates of leaf and stem rust from the most widespread cultivars was studied on standard differentials every year since the sixties. Standard methods were used for race determination (Browder, 1971; Stakman et al., 1962). Field samples of yellow rust were tested irregularly whenever obtained from the wheat growers. Resistance genes in the grown cultivars were postulated according to their reactions to a set of rust isolates. In most cases results were confirmed by genetic analysis of the hybrids. To study genetic variation of rusts due to recombinations, sexual stage of leaf rust and stem rust was induced by exposing wetted straw with telia over *Thalictrum speciosissimum* Loefl. and *Berberis vulgaris* L., respectively. Straw with telia of leaf rust was collected at three different distant locations. In stem rust, telia formation of race 21 (isolate G69) was induced in the greenhouse. Straw with telia was kept outdoors over winter and exposed above alternate hosts in the greenhouse in February–March. Transfer of nectar with pyknospores was carried out using Flor's method. Susceptible cultivar Little Club was inoculated with the developed

Table 1. The most important leaf rust races identified in the Czech and Slovak Republics in the years 1966–1993 (%)

Race Years	77	14	1	14 SaBa	77 SaBa	61	61 SaBa	53 SaBa	2 SaBa	Other	T*
1966–1969	34.5	34.5	–	–	–	–	–	–	–	31.0	165
1970–1973	47.6	23.2	6.5	5.3	2.4	–	–	–	–	15.0	168
1974–1976	5.4	2.7	1.3	1.3	87.2	–	–	–	–	2.1	149
1977–1980	6.1	8.0	3.6	–	50.2	22.0	–	–	–	10.1	163
1981–1983	–	9.0	–	1.3	3.9	39.4	41.3	–	–	5.1	155
1984–1986	–	7.9	–	6.0	4.7	16.7	42.9	4.7	–	17.1	317
1987–1990	1.6	3.0	–	3.4	4.6	4.6	42.8	23.4	12.4	4.2	368
1991–1993	2.0	–	–	1.0	0.7	9.6	51.0	4.5	13.8	17.4	290

T* – Total number of isolates.

aeciospores. Random single pustule isolates were tested on all (stem rust) or several (leaf rust) standard differentials. Aeciospores collected from barberries from two distant locations (32 samples) were used for inoculation of wheat, rye, barley and oats in separate greenhouse experiments.

Leaf rust

Results

Leaf rust races

The most common leaf rust race till 1968 was race 14. In this race virulence to Lr26 was found in early seventies and designated by the suffix SaBa (= virulence on cv. Salzmünder Bartweizen). Unlike in races that prevailed later, virulence to Lr26 never significantly predominated over avirulence in race 14. In 1969 race 77 prevailed and was dominant till 1979. First it was avirulent to Lr26, later virulence to Lr26 began to spread in race 77. Race 77SaBa accounted for 7.6% of all analyzed samples in 1973, whereas in 1975 it was already 92.9%. In 1980 race 61 predominated. Also this change was very fast. In 1979 the proportion of race 61 was only 5.7% whereas in 1980 already 61.1%. Race 61SaBa prevailed in 1982. The proportion of other races varied, however none of them overpassed the level of 10% except race 53SaBa and 2SaBa. The proportion of race 53SaBa suddenly increased in 1987 to 19% and reached its peak of 32% in the next year. Race 2SaBa overpassed 10% in 1987 and again in the period 1990–1992, reaching its peak of 33% in 1990. In the years 1993–1994 race 61SaBa accounted again for over 90% of all tested isolates. Data on races and virulence genes in the prevailing races are presented in Tables 1 and 2, respectively. Detailed results were published earlier, the last report with citation of previous papers being Bartoš et al. (1994).

Genes for leaf rust resistance

The most common genes in wheat cultivars registered in the Czech and Slovak Republics are Lr3 and Lr26 and combinations of them. In several cultivars resistance genes have not been determined (Table 3). Detailed results of genetic studies were published earlier (Bartoš et al., 1969; Bartoš et al., 1994 – citations of previous papers are included).

Sexual stage of leaf rust

None of two endemic *Thalictrum* species (*T. minus* L and *T. aquilegiifolium* L.) was infected after exposure of germinating telia on wheat straw over the *Thalictrum* plants. Under the same experimental conditions *T. speciosissimum* Loefl. = *T. flavum* subsp.*glaucum* (Desf.) Batt. formed abundantly aecia. Virulence combinations obtained from aeciospores are summarized in Table 4. Eight of 16 possible combinations of virulence were obtained from 97 cultures derived from aeciospores. In the year 1969 when straw with telia was collected, 8 races were determined from field urediospore samples. Virulence pattern of most aecial cultures corresponded to 4 of them. Virulence pattern of the rest of aecial cultures represented 4 other combinations of virulence genes not found in nature in 1969. Detailed results were published earlier (Bartoš, 1970).

Table 2. Genes for virulence in the most widespread leaf rust races and genes for resistance in the most common Czech and Slovak wheat cultivars (1966–1994)

Years	Race Cultivars	Resistance (R) – virulence (p) genes Lr1 (Mal)	Lr2a (Web)	Lr2b + B (Car)	Lr2c + B (Brevit)	Lr2c (Loros)	Lr3 (Med)	Lr11 (Huss)	Lr26 (SaBa)
1966–	14				p	p		p	
1973	77	p	p	p	p	p	p	p	
	Kaštická os., Fanal, Qualitas, Jubilar, Pavlovická 198								
	Mironovská 808						R		
1974–	77SaBa	p	p	p	p	p	p	p	p
1980	61				p	p	p	p	
	Slavia, Jubilar, Zora								
	Mironovská 808, Jubilejná, Iljičovka, Grana						R		
	Aurora						R		R
	Kavkaz, Solaris								R
1981–	61				p	p	p	p	
1986	61SaBa				p	p	p	p	p
	Vala, Regina, Zdar, Košutka								
	Amika						R		R
	Iris								R
1987–	61SaBa				p	p	p	p	p
1994	53SaBa							p	p
	2SaBa						p	p	p
	Zdar, Regina, Ilona								
	Viginta, Hana						R		
	Danubia, Selekta								R
	Sparta						R		R

Discussion

Replacement of race 14 avirulent to Lr3 by race 77 virulent to that gene in the leaf rust population was probably due to the introduction of cultivars Mironovskaya 808 and Bezostaya 1, both possessing Lr3, in 1966. Mironovskaya 808 was the leading cultivar in the late sixties and early seventies.

Cultivars Aurora and Kavkaz both registered in 1972 were first cultivars possessing Lr26. Virulence to Lr26 predominated in race 77 (designated as 77SaBa) in the mid seventies.

As mentioned above, spread of virulence to Lr3 and later to Lr26 in the leaf rust population can be explained by resistance genes of the grown cultivars. For sudden increase of race 77 and later 77SaBa in former Czechoslovakia large areas of cultivars with Lr3 and Lr26 in eastern and southeastern Europe seem to be also responsible. In the early sixties race 77 was the most widespread one in Rumania (Stewart et al., 1966, 1967), Hungary (Bócsa, 1966), Yugoslavia (Boškovič, 1965), Ukraine (Voronkova, 1969). Nover et al. (1964) identified race 77 in samples from Germany, Bulgaria, Finland, The Netherlands, Norway and former Czechoslovakia which shows occurrence of this race also in western and northern Europe.

Due to the wide distribution of cultivars possessing Lr26, particularly Kavkaz, Avrora and derived cultivars, race 77 virulent to Lr26 gained importance in the whole eastern and middle Europe. In the former USSR virulence to Lr26 in race 77 was described by Voronkova as early as in 1964.

With the decrease of race 77 race 61 first avirulent and later virulent to Lr26 gained more importance not only in former Czechoslovakia but also e.g. in Poland (Dwurazna pers. comm., 1978), Italy (Pasquini et al., 1979), Germany (Unger pers. comm., 1985), Hungary (Szunics, 1991).

Table 3. Genes for resistance to leaf and stem rust in winter wheat (*Triticum aestivum* L.) cultivars registered in former Czechoslovakia and in the Czech Republic (1964–1994)

Cultivar	Registered	Sr	Lr	Cultivar	Registered	Sr	Lr
Chlumecká 12	1919–1970	-	-	Vala	1980–1989	29	-
Slovenská 200	1946–1967	-	-	Baranjka (YUG)	1981–1985	*	*
Kaštická osinatá	1954–1973	-	-	Košútka	1981–	+	+
Pavlovická 198	1956–1969	-	-	Odra	1981–1989	-	3, +
Košutská	1956–1968	-	-	Regina	1982–	-	-
Diana I	1960–1969	-	-	Super Zlatna (YUG)	1982–1985	*	*
Fanal (GDR)	1960–1969	-	-	Iris	1983–	31, 11, +	26
Iva	1962–1969	-	-	Sabina	1983–1988	31	26
Qualitas (GDR)	1963–1973	-	-	Zdar	1983–	+	-
Lada	1964–1971	-	-	Danubia	1984–	31	26, +
Draga	1965–1972	-	-	Mara	1984–1990	29	3, +
Mironovskaya 808 (USSR)	1966–1993	Tmp	3	Mironovskaya nizkoroslaya (USSR)	1984–1987	31	26
Bezostaya 1 (USSR)	1966–1973	5, ++	3	Viginta	1984–	5, +, +	3
Belocerkovskaya 198 (USSR)	1967–1972	+	3	Agra	1985–	31, +	3, 26
Diana II	1967–1972	-	-	Hana	1985–	29	3
Jubilar (FRG)	1969–1979	-	-	Roxana	1985–1991	31	26, +
Poros (GDR)	1969–1972	-	-	Selekta	1985–	31	26
Yubileynaya 50 (USSR)	1971–1986	5, +	3+	Branka	1988–1993	31	3, 26, +
Oska	1971–1977	-	-	Sparta	1988–	31	3, 26
Zora	1971–1979	-	-	Ilona	1989–	11	-
Kavkaz (USSR)	1972–1976	31, 5	26	Sofia	1990–	31	3, 26
Aurora (USSR)	1972–1976	31, 5	26, 3	Vlada	1990–	+++	++
/Grana(P)/	1975–1983	-	3	Livia	1991–	31, 11	26
Ilyitchovka (USSR)	1975–1985	5, ++	3	Blava	1992–	+	+
Lena	1975–1979	+	-	Vega	1992–	-	3
Sava(YUG)	1976–1981	+	*	Torysa	1992–	29	+
Slavia	1976–1991	29	-	Simona	1991–	+	-
Solaris	1976–1986	5, 31	26	Senta	1991–	31	3, 26
Zlatna Dolina (YUG)	1976–1981	*	*	Samanta	1993–	+	3
Hela	1979–1988	29	-	Sida	1993–	31	26
Istra	1979–1985	31, 5	3, 26, +	Asta	1994–	+	3
Juna	1979–1985	5	3	Bruta	1994–	+	-
Kormoran (FRG)	1979–1982	-	-	Mona	1994–	31	3, 26
Mironovskaya ulutchshennaya (USSR)	1979–1982	5, +, +	3	Rexia	1994–	+	+
				Siria	1994–	-	+
Mirela	1979–1984	-	-	Trane (D)	1994–	31	26
Amika	1980–1986	5, 31	3, 26				

- no specific resistance, + undetermined gene, * not tested.

Race 61 is avirulent to Lr1. Prevailing avirulence to Lr1 was found in many western European countries by Felsenstein & Fischbeck (1994).

High incidence of race 61, later 61SaBa as well as outbreaks of races 53SaBa and 2SaBa can be hardly attributed to changes of genes for resistance in the grown cultivars. Race 2SaBa is virulent to Lr3 and Lr26 like races 61SaBa and 77SaBa. However, race 53SaBa is virulent only to Lr26 (considering Lr genes in the registered cultivars in former Czechoslovakia). Thus it could not attack 8 out of 17 registered cultivars of winter wheat at the time of its maximum incidence in 1988. Relatively high incidence of the two above mentioned races is most likely to be ascribed to their higher fitness at the given time, in the given area and given varietal structure. It is difficult to decide whether the incidence of races common in the whole Europe is of monotopic or polytopic provenience.

Table 4. Virulence pattern of aecial leaf rust cultures obtained on *Thalictrum speciosissimum* Loefl.

16 possible combinations of reactions on genes for resistance				Number of aecial isolates with corresponding reactions	Corresponding race found in the year of telia collection
Lr1	Lr2a	Lr11	Lr3		
R	R	R	R	20	–
R	R	R	S	2	84
R	R	S	R	13	14
R	S	R	R	–	–
R	R	S	S	24	–
R	S	S	R	1	–
R	S	R	S	–	–
R	S	S	S	9	–
S	R	R	R	–	–
S	R	R	S	–	–
S	R	S	R	–	–
S	S	R	R	–	–
S	R	S	S	–	–
S	S	S	R	–	–
S	S	R	S	2	122
S	S	S	S	26	77

Table 5. The most important stem rust races identified in the Czech and Slovak Republics in the years 1966–1993 (%)

Years	Race						Other	T*
	21	14	34	11	20	1		
1966–1969	52.4	39.3	–	–	–	–	8.3	84
1970–1973	22.0	5.1	2.2	11.3	1.1	20.9	37.4	177
1974–1976	24.2	2.0	49.5	9.1	–	1.0	14.2	99
1977–1980	5.1	2.8	34.8	29.2	–	–	28.1	178
1981–1983	–	4.0	73.0	23.0	–	–	–	100
1984–1986	–	1.2	52.1	–	5.3	2.4	39.0	169
1987–1990	–	–	37.2	50.9	1.0	–	10.9	102
1991–1993	–	–	76.9	23.1	–	–	–	13

T* – Total number of isolates.

Results obtained in experiments with *Thalictrum* indicate that endemic *Thalictrum* species are not likely to contribute to the variation and adaptation of leaf rust in the Czech Republic. Other combinations of virulences were revealed in aecial isolates in addition to some of those identified also from the field urediospore samples. This indicates the potential importance of sexual cycle on the variability.

Comparison of the number of necessary virulence genes to match resistance genes in the grown cultivars (Table 2) shows presence of many unnecessary virulence genes. Obviously, they do not diminish fitness of the pathogen. Mac Key (1992) explains presence of unnecessary virulence genes by their pleiotropic functions.

Data on leaf rust resistance genes in the registered cultivars (Table 3) show that resistance genes Lr3 and Lr26 appeared in the registered cultivars not only before virulence to these genes predominated but also in the present-day cultivars. We presume that the resistance gene Lr3 is linked with genes having positive effect on some agronomically important traits or has positive pleiotropic effect itself. Positive effect of the translocation 1BL/1RS on the yield has been proved although not in all experiments (Bartoš, 1993).

Table 6. Reactions of race 21 of stem rust (isolate G69) and of cultures originated by autogamization on barberry

Rust isolate	Differentials												Number of cultures
	Little Club	Marquis	Reliance	Kota	Arnautka	Mindum	Spelmar	Kubanka	Acme	Einkorn	Vernal	Khapli	
21 (G69)	4	4	0	4	3–4	4	4	3	3	;1–2	;	;	
A	4	1–2+	0	;1	;	;	;	;	;	4	;	;	28
B	4	2	;1–2	;1–2	;	;	;	;	;	3	;1	;	5
C	4	1–2	0	1–2	;	;	;	;	;	3±	3±	;	4
D	4	1–2	4	;1–2	;	;	;	;	;	3	;	;	4
E	4	1–2	0	1–2	0;	;	;	;	;	4	1–2	;	4
F	4	2–3	2	2	;	;	;	3	4	;1	;1	;	3
G	4	;1	0	;	0	;	;	;	;	4	;	;	2
H	4	1	0	;	;	;	;	;	;	;1	;1	;	2
I	4	2	3	1–2	;	;	;	;1	;	4	1–2	;	2
J	4	4	0	3	3	4	4	4	;1	;1	;1	;	2
K	4	4	0	3	3	4	4	2–3	2–3	1–2	;1	;	2
L	4	4	3	3	3	4	4	3	4	4	;1	;	2
M	4	3	3	3	3	3	3–4	3	3	1	;1	;	1
N	4	4	4	4	3	4	4	4	1–2+	;1	;1	;	1
O	4	3–4	2	3–4	3–	1–2	;1	4	4	;1	1	;	1
P	4	4	0	;	;	;	;	;	;	3	;	;	1

Stem rust

Results

Stem rust races

The most common stem rust race in the late sixties was race 21, followed by race 14. Whereas race 21 remained important until 1976, proportion of race 14 dropped in the seventies and early eighties to and below 5% and it was no more found since late eighties. Both races are avirulent to Sr5. The most important change in the stem rust population was increase of virulence to Sr5 due to races 56, 11 and 34 in 1972 the year of a stem rust epidemic. However, races avirulent to Sr5 still remained widespread, the most common ones being 1 and 21. In the mid-seventies virulence to Sr5 represented by races 34 and 11 distinctly prevailed over avirulence, represented mainly by race 21. In the late seventies races 11 and 34 formed almost 2/3 of the stem rust population and have remained the most important races till now.

Results are summarized in Table 5. Detailed results were published earlier, the last paper being Bartoš et al. (1994) with citations of previous reports.

Genes for stem rust resistance

The most common Sr genes in the Czech and Slovak Republics are Sr31, Sr29, Sr5, Sr11. Several genes have not been determined yet. Results are summarized in Table 3, detailed results of genetic studies were published earlier (Bartoš et al., 1970; Bartoš et al., 1994 – citations of previous papers are included).

Sexual stage of stem rust

Reactions of isolates obtained after selffertilization of race 21 on barberry are summarized in Table 6. Most isolates had less genes for virulence than the original uredial culture of race 21. However, one isolate was identical with race 11 virulent to Sr11. Aeciospores collected in the nature on barberries caused infection on rye, not on wheat, barley or oats (Bartoš & Valkoun, 1980).

Discussion

Races 14 and 21 prevailing in former Czechoslovakia in the sixties belonged to the most important races in the whole Europe as well as in Israel (Santiago & Pais, 1963; Massenot, 1965; Gerechter-Amitai, 1966;

Šebesta & Bartoš, 1969). In the same period race 34 that prevailed in former Czechoslovakia with race 11 later in the seventies, was already an important race in southeastern Europe, e.g. in Rumania (Stewart et al., 1966, 1967), former Yugoslavia (Kostič, 1966) and also in Israel (Gerechter-Amitai, 1966).

As stem rust does not overwinter in our country, adaptation of stem rust population to cultivars with Sr5 must have occurred in the countries where inoculum originates from i.e. very likely southeastern Europe. Cv. Bezostaya 1 was the leading cultivar there. In former Czechoslovakia virulence to Sr5 was found already in 1962 i.e. 4 years earlier than the first cultivar with Sr5 was registered.

The change from prevailing avirulence on Sr5 to prevailing virulence in stem rust population in early seventies was influenced by cultivars possessing Sr5. In former Czechoslovakia they were represented mainly by cultivars introduced from the former USSR, Bezostaya 1 (1966), Yubileynaya (1971), Aurora (1972), Kavkaz (1972), Ilyitchovka (1975). Cultivar Bezostaya 1 and other cvs. from the former USSR were donors of Sr5 to many cultivars developed in eastern and southeastern Europe. In former Czechoslovakia the most important of them was cv. Viginta (1984).

In the sixties and seventies another very important cultivar in former Czechoslovakia as well as in some other middle and eastern and southeastern European countries was the Ukrainian cv. Mironovskaya 808 possessing SrTmp. As this gene governs medium resistance adaptation of stem rust observed in the seventies may have occurred by increased agressiveness or appearance of specific virulence to SrTmp.

Virulence to Sr31 has never been found inspite of large areas of cultivars with 1BL/1RS translocation particularly in eastern and southeastern Europe. Unlike in the case of Lr26 which is closely linked with Sr31 no adaptation to Sr31 has been observed till now.

No adaptation was registered to Sr29 possessed by important Czech cultivars Slavia (1976) and Hana (1985) and also Mara (1984) and Torysa (1992).

Virulence to Sr11 possessed by cvs. Iris (1983) and Ilona (1984) has been sporadic till now. It was first found in mid sixties i.e. before cultivars with Sr11 were registered.

The function of the alternate host of wheat stem rust in adaptation of the pathogen to resistant cultivars in the Czech Republic is questionable. All aecial isolates from the nature tested in our experiments were virulent only to rye not to wheat. Barberry bushes were obviously infected with stem rust from couch grass (*Elytrigia repens*) growing below them. Presence of wheat straw with stem rust telia very close to barberry seems to be necessary for successful infection. This situation is very rare in the Czech Republic. However, inoculation experiment showed that alternate host would contribute to the adaptation, if functional. Selfing of race 21 showed heterozygosity at many loci for virulence.

Yellow rust

Incidence of yellow rust in former Czechoslovakia was irregular since the sixties, having only two economically important outbreaks. The first one was in the early-sixties in connection with yellow rust epidemics in western Europe. As in western Europe, virulence to cv. Heines VII was characteristic for yellow rust population. Of Czech cultivars high-yielding cv. Diana released in 1960 was so damaged that no cultivars susceptible at adult stage were allowed to be registered since that time. This regulation was highly effective: the next economically important outbreak of yellow rust followed as late as in 1977–1978. In the mid-seventies susceptible cultivars Sava and Zlatna Dolina from former Yugoslavia were temporarily introduced because of lack of Czech early cultivars. Within two years yellow rust spread to such a degree that they almost stopped to be grown.

Since that time there was no yellow rust epidemic. Virulence to Yr9 was found in 1983 although some data suggested earlier occurrence. Out of many cultivars possessing Yr9 only cvs. Danubia and Iris turned susceptible. Other cultivars with Yr9 possess additional effective genes for resistance.

As yellow rust can overwinter in the Czech Republic and as it often appeared in neighbouring countries non-adaptation to the grown resistant cultivars may be ascribed to the very low level of inoculum making adaptation difficult.

Conclusions

1. Virulence changes in the wheat rusts populations, as surveyed in former Czechoslovakia since 1962, can be only partially attributed to the adaptation to resistance genes in the grown cultivars.
2. Unnecessary virulence genes seem to have no negative effect on the fitness of races possessing them.

3. Incidence of some resistance genes ineffective already for a long time in recently released cultivars is probably due to positive effect of those genes or closely linked genes on agronomically important traits (e.g. resistance genes on translocation 1BL/1RS). We presumed that incidence of some unnecessary virulence genes may have similar reasons (positive effect on fitness of the pathogen).
4. Changes in the predominating races were fast. Airborne inoculum from southern and south-eastern Europe may have caused these shifts. Sporadically occurring races are presumed to be of local origin.
5. Similar leaf rust races as in former Czechoslovakia were often described in Hungary, Ukraine, Poland and former GDR. Similar stem rust races were found in many southeastern countries. The highest similarity in rust races was observed with Hungary. This may confirm the eastern European rust pathway both for leaf rust and stem rust.
6. Alternate host of stem rust, *Berberis vulgaris*, being mostly attacked with rye stem rust, is of little importance for stem rust adaptation to wheat. However, it can facilitate the adaptation of the pathogen by production of sexual recombinants if accidentally attacked with wheat stem rust.
7. Several *Thalictrum* species occur in the Czech and Slovak Republics. However, none of them has been found yet as a host of wheat leaf rust. Sexual process with Czech leaf rust isolates was induced on *Thalictrum speciosissimum* L. ex Loefl. (= *T. flavum* subsp. *glaucum* (Desf.) Batt.) from Canada. Preadaptation to this host is interesting but of little practical importance.
8. Virulence in leaf rust and stem rust to resistance genes in the grown cultivars was usually found before resistant cultivars were registered. It increased with the increasing area of corresponding resistant cultivars. The pathogen was usually preadapted.
9. Mutation for virulence to Lr26 appeared in various races. However, it prevailed in the most widespread races. Genetic background conditioning high fitness seems to be important for good adaptation of mutants.
10. Adaptation of yellow rust to the grown resistant cultivars was very limited. Obviously, when the disease level in an area is low adaptation can be very slow.

References

Bartoš, P., 1993. Chromosome 1R of rye in wheat breeding. Plant Breeding 63 (10): 1203–1211.

Bartoš, P., G.J. Green & P.L. Dyck, 1970. Reactions to stem rust and genetics of stem rust resistance in European wheat varieties. Can J Bot 48: 1439–1443.

Bartoš, P., D.J. Samborski & P.L. Dyck, 1969. Leaf rust resistance of some European varieties of wheat. Can J Bot 47: 543–546.

Bartoš, P., E. Stuchlíková & R. Hanušová, 1994. Genetics of stem and leaf rust resistance of winter wheat cultivars Ilona, Livia, Blava and Torysa. Genetika a Šlechtění 30 (2): 123–132 (orig. in Czech).

Bartoš, P. & J. Valkoun, 1980. Sexual variability of stem rust (*Puccinia graminis*). Ochrana Rostlin 16 (3): 181–190 (orig. in Czech).

Bócsa, E., 1966. Physiologic specialization of brown and black rust in wheat in Hungary. Proc Cereal Rusts Conferences, Cambridge 1964: 321–323.

Boškovič, M., 1965. Identifikacija fizioloških rasa *Puccinia recondita* Rob. ex Desm. f.sp. *tritici* (Eriks.) u Yugoslavji Savrem Poljopriv 4: 1–84.

Browder, L.E., 1971. Pathogenic specialization in cereal rust fungi, especially *Puccinia recondita* f.sp. *tritici*: concepts, methods of study, and application. Tech Bull 1432, USDA.

Felsenstein, F.C. & G. Fischbeck, 1994. Erste Ergebnisse zur großräumigen Differenzierung der Virulenz des Weizenbraunrosts gegenüber den Resistenzgenen Lr19, Lr2a, Lr2b, Lr3Bg, Lr3Ka, Lr15, Lr25 und Lr28. Vortr Pfl-Züch 28: 171–174.

Flor, H.H., 1956. The complementary genic systems in flax and flax rusts. Adv Genet 8: 29–54.

Gerechter-Amitai, Z.K., 1966. Physiologic races of *Puccinia graminis triticiin* Israel in 1965. Robigo 18: 9–10.

Gerechter-Amitai, Z.K. & J. Wahl, 1966. Studies on the physiologic specialization of wheat stem rust in Israel, 1961–1963. Proc Cereal Rusts Conferences, Cambridge 1964: 162–171.

Kostić, B., 1966. Physiologic races of *Puccinia graminis* var. *tritici* Erikss. et Henn. in Southeastern Yugoslavia. Proc Cereal Rusts Conferences, Cambridge 1964: 154–161.

Mac Key, J., 1992. Use of near-isogenic lines in analysing race-specific pathogen interrelations. Norwegian Journal of Agricultural Sciences. Suppl 7: 63–76.

Massenot, M., 1965. Races physiologiques de *Puccinia graminis tritici* isolées à Grignon au cours des années 1962, 1963 et 1964. CR Acad Sci Paris 51: 442–445.

Nover, I., M. Klinkowski & I. Simon, 1964. Ergebnisse der in Halle seit 1945 durchgeführten Arbeiten über den Weizenbraunrost (*Puccinia triticina* Erikss. = *P. recondita* Rob. ex Desm.) Zeitschrift für Pflanzenkrankheiten (Pflanzenpathologie) und Pflanzenschutz 71: 74–89.

Pasquini, M., M.A. Gras & G. Zitelli, 1979. Virulence genes present in populations of *Puccinia recondita* f.sp. *tritici* and *Puccinia graminis* f.sp. *tritici* in Italy during 1977 and 1978. Cereal Rusts Bull 7 (1): 8–15.

Person, C., 1959. Gene-for-gene relationships in host: parasite systems. Can J Bot 37: 1101–1130.

Santiago, J.C. & A.L. Pais, 1963. Data obtained from the European wheat stem rust nursery during the period 1960–1963. Mimeographed. Plant Breeding Station Elvas, Portugal.

Stakman, E.C., D.M. Stewart & W.Q. Loegering, 1962. Identification of physiological races of *Puccinia graminis* var. *tritici*. US Dep Agric ARS E617: 53.

Stewart, D.M., E. Radulescu & F. Negulescu, 1966. Some recent studies on cereal rusts in Roumania. Rev Roum Biol Ser Bot 11: 209–214.

Stewart, D.M., E. Radulescu & F. Negulescu, 1967. Physiologic races of *Puccinia recondita* and *P. graminis* in Roumania in 1964 and 1965. Plant Dis Rep 51: 358–362.

Szunics, L., S. Manninger & L. Szunics, 1991. A búzalisztharmat, levélrozsda és szárrozsda rasszösszetétele. Martonvásár 2: 5–6.

Šebesta, J. & P. Bartoš, 1969. Physiologic specialization of wheat leaf rust in Czechoslovakia in the years 1964 and 1965. Ochrana Rostlin 5: 147–154 (orig. in Czech).

Voronkova, A.A., 1969. Biotypes of race 77 of the causal agent of leaf rust in Kubany and reaction of some wheat cultivars to them. Mikologia i fitopatologia 3 (5): 429–431 (orig. in Russian).

P.M.A. Tigerstedt (Ed.), *Adaptation in Plant Breeding,* 119–133, 1997.

Intergenotypic interactions in plant mixtures

Roy Turkington
Botany Department, University of British Columbia, Vancouver, BC, V6T 1Z4, Canada

Key words: plant mixture, intergenotypic interaction, pasture, *Lolium perenne*, perennial ryegrass, *Trifolium repens*, white clover

Summary

Pastures provide a good model system to study intergenotypic plant interactions because most pastures are mixtures of several sown genotypes and species and the conditions necessary for genetic change are present: (i) an extremely variable biotic and abiotic environment at the local scale, (ii) wide genetic variation, (iii) intense competition so that strong selection is likely, and (iv) long term continuity enables selection to be continuous and cumulative. These together provide a system in which (v) the theoretical outcome of intergenotypic interactions may be readily tested and maximizes the probability of detecting patterns. Interactions between the species will have both ecological and evolutionary consequences, both of which may be of interest to plant breeders. These patterns are detectable at different scales but particular attention will be focused at the level of the individual genotype and especially with mixtures of the grass *Lolium perenne* and the legume *Trifolium repens*. Competition experiments between *L. perenne* and *T. repens* are plentiful, but where specific hypotheses about adaptation at the genotype level have been tested, in all cases *T. repens* grows best when planted with its natural *L. perenne* neighbor – a reciprocal effect in the *L. perenne* has only been reported once. This specific *T. repens* – *L. perenne* neighbor recognition may be mediated through soil microorganisms, particularly *Rhizobium leguminosarum* biovar *trifolii* and *Bacillus polymyxa*.

Introduction

Some of the most important influences impacting an organism arise from its relationship with other organisms, at the same, or at different, trophic levels. This is especially applicable to plants which, for the most part, are relatively immobile and are forced to deal with their neighbors which are often situated in close promixity and at high density. Ever since Charles Darwin, agronomists, ecologists, and evolutionary biologists have been interested in the ways by which species interact and the consequences of these interactions. The Darwinian tradition has focused on biotic interactions, indeed, he wrote that there is a '. . .deeply-seated error of considering the physical conditions of a country as the most important (for its inhabitants); whereas it cannot be disputed that the nature of the other species with which it has to compete, is as least as important, and generally a far more important element of success.' Agronomists have been primarily concerned with plant competition, and grazing, and have focused research effort on maximizing biomass production and on breeding or selecting for competitive ability. Ecologists have been more interested in how interactions cause organisms to specialize, or adapt, to one another and to the physical environment, and how this impacts abundance, distribution and the organization and structure of communities. Evolutionary biologists have focused attention on how the interactions themselves evolve, the nature and extent of speciation, and whether perhaps the interactions have coevolved. Perhaps not enough effort has been made to cross-fertilize these disciplines, and this of course is not necessarily easy, because the disciplines do have different, although somewhat overlapping, goals. The topic I was specifically assigned to address is intergenotypic interactions in plant mixtures. Taking a little liberty with the title, I will attempt to address some of the areas of common interest to the agronomist, the ecologist and the evolutionary biologist. I will do this by focusing my attention on managed (grazed) temperate pasture systems where grasses and legumes are the major plant association.

Some ecological principles

John Hill (1990) has well articulated the problem before us when he says that 'coexistence is crucial to the success of any breeding programme designed to raise the productivity of grass/legume pastures...' Various basic ecological assumptions are implicit in this statement. First, one of the preconditions for coexistence is that factors are present which limit the expression of dominance. In pastures this limitation will usually operate through grazing which will act to debilitate potential dominants (Turkington et al., 1993). Second that neighboring species are competing. Competition is one of the conceptual backbones of ecological thinking but it is still poorly understood. It is notoriously difficult to demonstrate, probably because avoidance of it, in nature, is advantageous in most cases. However, in the crowded conditions typical of pastures it is difficult to believe that competition is not occurring. Third, is the assumption that if species have persisted together for a prolonged period it can be taken as axiomatic that some ecological distinction must exist between them. These differences may have always been present, or as is most commonly argued, there has been some degree of population differentiation by one or more of the species, a differentiation driven by competition. Fourth, that this differentiation permits coexistence, and continued interaction may result in coevolution, although Connell (1980) has argued why coevolution between competitors should not be expected to happen frequently. Finally, that coevolved populations will have a higher biomass production than non-coevolved populations. To varying degrees these assumptions have been tested experimentally in ecological studies, and as expected, there is not always a consistent unambiguous result, although a fairly clear consensus and theoretical framework is emerging (Thompson, 1994).

Intergenotypic interactions – their context in the pasture

Pastures provide us with a good model system to study intergenotypic interactions because most pastures are mixtures of several sown genotypes and species and the conditions necessary for genetic change are present: (i) an extremely variable biotic and abiotic environment at the local scale, (ii) wide genetic variation, (iii) intense competition so that strong selection is likely, and (iv) long term continuity enables selection to be continuous and cumulative. These together provide a system in which (v) the theoretical outcome of intergenotypic interactions may be readily tested and maximizes the probability of detecting patterns. I will consider each of these briefly.

Variable biotic and abiotic environment

The visual impression of pastures being quite uniform is misleading. The only meaningful measure of environmental variation is that which is seen from what I (Turkington, 1975) and Harper (1978) have called the 'plant's-eye-view'. Species with differing growth forms will experience the local environment in entirely different ways. For example, the tillering habit of *Lolium perenne* ensures that genets stay relatively fixed in position and ensures that the individual experiences a relatively constant spatial environment where it is dependent on a strictly local supply of resources. *Trifolium repens* in contrast, has an exploring phenotype that can spread through a grassland by up to 25 cm annually. The whole plant axis therefore samples a diversity of soil microenvironments (Snaydon, 1962) and of neighbors (Burdon, 1980), both spatially and temporally. It is already well established that *T. repens* varies in its aggressiveness to different species of grass (Chestnutt & Lowe, 1970; Turkington & Harper, 1979b; Frame & Newbould, 1986; Harris, 1987), and presumably also in its response to them. MacArthur & Levins (1964) distinguished between environmental variation in which the period of fluctuation is less than the lifespan of the plant and that in which the period is greater. They called this fine-grained and coarse-grained respectively. Thus most tillering pasture grasses will 'see' a coarse-grained environment and rhizomatous or stoloniferous species will 'see' a fine-grained environment.

But the mosaic of vegetation patches that make up a pasture also has a temporal dynamic (Thorhallsdottir, 1990a). Populations of grasses are distributed as a patchwork, corresponding to patterns of past disturbances, colonization, heterogeneity in the physical environment and interspecific interactions (Parish & Turkington, 1990a, 1990b). As plants grow, and patches expand, they inevitably meet and interact with other patches which they may invade, or by which they may be invaded. Plants will die and the gaps produced will be invaded by the same, or a different species (Grubb, 1977). In temperate grasslands dominated by perennials, the dominant species mostly colonize small disturbances by vegetative spread. Growth and death of plants, disturbance, and colonization, all contribute to the dynamics of the pasture community. These dynamics then impose their own structure on the community. As patches meet and interact, their structure will change as they expand or are invaded (Thorhallsdottir, 1990b; Turkington, 1993). Patches may change in age and size distributions, densities, relative abundances, and relative proportions of genotypes, which may in turn alter the local resource levels. These changes in patches and resources will in turn affect the outcomes of interactions. Consequently, there will be a patch dynamics of interactions and a patch dynamics of species distributions, abundances, and composition (Thompson, 1985, 1988). So what is an apparently uniform system to the human observer, is a dynamic and variable system when seen from a plant's-eye-view. This level of variation can clearly impose an array of local directional selection forces which are likely to dominate the fitness of individuals and promote microevolution.

Genetic variation

Trifolium repens is a highly variable species commonly showing considerable differences within and between populations in a wide range of morphological and physiological characters (Turkington & Burdon, 1983; Burdon, 1983; Frame & Newbould, 1986; Harris, 1987). *Trifolium repens* is polymorphic for petiole length (Kerner von Marilaun, 1895), date of first flowering (Davies & Young, 1967), nitrogen fixing ability (Connolly et al., 1969), susceptibility for infection by *Rhizobium* (Mytton, 1975), leaf marks and leaf size (Cahn & Harper, 1976a, 1976b; Evans & Turkington, 1988), relative growth rate (Burdon & Harper, 1980; Trathan, 1983), cyanogenic glycosides (Dirzo & Harper, 1982) and their consequent palatability to some herbivores (Burgess & Ennos, 1987), and stolon length and flower production (Aarssen & Turkington, 1985c). Various studies of intraspecific diversity have been conducted in populations of *T. repens* from an old (approx. 100 years) pasture in North Wales. Burdon (1980) examined 50 plants and showed that they usually differed from each other in several characters and on average differed significantly in 3.3 characters. Cahn & Harper (1976a) identified an average of between 3 and 4 clones per 10 cm × 10 cm square, while Trathan (1985) using isoenzyme analysis, identified 48–50 genotypes per square meter. Gliddon & Trathan (1985) estimated the genetic variation within seven sub populations of *T. repens* and *L. perenne*. All 15 polymorphic loci studied in *T. repens* and 21 out of 23 in *L. perenne* showed significant heterogeneity of gene frequency, indicating that there was genetic differentiation between samples. The level of diversity found within this single population is of a magnitude more commonly encountered in comparisons between populations taken from distinctly different environments. Unfortunately, such detailed data are not available for any of the companion grasses but Aarssen & Turkington (1985c) showed extensive variation for some characters in *L. perenne* and *Holcus lanatus* from pastures in British Columbia. In addition, Breeze et al. (1965) showed that the rate of tillering in *L. perenne* can be significantly altered by selection within clones; the response depends on the age of the clone and is genotype-dependent. This is presumably due to some type of somatic mutation. Kelley & Clay (1987) used two grasses, *Danthonia spicata* and *Anthoxanthum odoratum*, from a 40-year-old field in North Carolina to show that naturally co-occurring genotypes of these species differed in interspecific competitive ability. The competitive performance of a given genotype often depended on the genotypic identity of the competing species, especially in *D. spicata*.

Competition and selection

There seems to be little doubt that competition occurs in pastures. Donald (1963) claims 'intense competition' and 'acute competitive stress' and Snaydon (1978) states that 'intense competition usually occurs within the stand (pasture)...' Almost invariably, when species are experimentally removed, there is a positive growth response in remaining species. However, Fowler (1981, 1982) only detected generally diffuse competition and weak competitive relationships among species in a 30-year-old pasture in North Carolina. The problem generally seems not to be one of demonstrating that competition is happening, but to demonstrate that it is having a structuring effect on the community (Turkington & Mehrhoff, 1990).

There is abundant evidence that genetic changes occur rapidly in both sown and natural populations and it is mostly assumed that these changes are due to competition-driven mortality. Brougham & Harris (1967) have shown that the genetic composition of a sward of *L. perenne* may change by as much as 40%

within 4 months of sowing. The most comprehensive series of studies was that undertaken by Charles and coworkers from 1961–1973 (e.g. Charles, 1961, 1972) and perhaps as few as 10% survive the establishment year. They demonstrated that the genetic composition of a number of species, including *L. perenne* and *T. repens*, changes substantially within a few years. This shift was mostly caused by high mortality in the establishment year but the data from older swards showed further marked changes in populations in established grasslands. Changes were reported in biomass production, flowering date, plant height and tillering. Ennos (1981) demonstrated a significant change in the frequency of linamarase-containing phenotypes of *T. repens* during the transition from the seed to established plants in populations exposed to selection at three different sites.

In older pastures, although the level of variation is still apparently quite high, it is nevertheless declining. McNeilly & Roose (1984) documented a decline from about 40 different genotypes of *L. perenne* per 0.25 m^2 in a 10-year-old pasture to 5 in a 40-year-old pasture. The reason for the death of these individuals and the loss of variation may be related to many factors such as grazing, competition, pathogens, and variable abiotic factors, and it is not clear what role competition plays in this sorting process. These cautions also apply to the survey by Aarssen & Turkington (1985c) who collected 100 clones each of *H. lanatus*, *L. perenne* and *T. repens* from four pastures ranging from 4 to 67-years old. Although it did not occur for all characters, many characters, especially for *L. perenne* and *T. repens*, showed a decline in mean value and variance with increasing pasture age.

Long-term persistence and stability

There is little evidence for seedling establishment in pastures except on molehills (Jalloq, 1975), gopher mounds (Peart, 1989), decayed dung pats (Parish & Turkington, 1990b), and on ant hills (King, 1977). However, this small amount of new recruitment may be important in maintaining genetic diversity within populations (Soane & Watkinson, 1979). Pastures are therefore primarily perennial systems and it is this that ensures that selection will be continuous and cumulative, and provides the opportunity to assess the long-term consequences to community structure.

It is very difficult to get long term data on any one system even though White (1985) reports the census (not repeated) of plants from about a dozen European grasslands from the last century. Few as they are, such data have become available through a number of recent papers (Silvertown, 1980; Silvertown et al., 1994; Dodd et al., 1994, 1995). These studies were based on the Park Grass Experiment at Rothamsted, in England, on plots established in 1856, and provide a superb opportunity to assess stability or equilibrium in plant communities. Initial changes in botanical composition, in response to different fertilizer treatments, was rapid, but within 50 years no further significant changes were detectable in the proportions by weight of grasses, legumes and other species in seven of nine plots. Dodd et al. (1995) call this a dynamic equilibrium because there was year-to-year variation caused by the vagaries of yearly weather conditions, nevertheless there was a range of long-term equilibria in community composition. A similar stabilizing of floristic composition has been demonstrated by Aarssen & Turkington (1985a) using pastures of different age and stringing them together as a time sequence (also see Turkington & Mehrhoff, 1990).

The theoretical outcome of intergenotypic interactions

When a pasture is established with two or more species that are genetically variable, their relationship may change in at least two ways. First, the different populations may adapt separately and independently to the local environment and the response will be intensified by intrapopulation competition. Secondly, the populations may adapt to each other or coevolve and the response here will be intensified by interpopulation, or interspecific, competition (Antonovics, 1978). Various consequences of interspecific competition have been proposed such as character displacement and character convergence. But the most noteworthy feature of such studies is the variability of the results. Many fit the expectation of improved performance through divergence. Just as frequently, competitive performance declines or remains unchanged (Antonovics, 1978). The conclusion of most theoretical studies (e.g. MacArthur & Levins, 1964, 1967; Lawlor & Maynard Smith, 1976) is that populations will tend to diverge in their ecological requirements so that they increasingly use different resources and so reduce or avoid competition. In the ecological literature this theme has become nearly axiomatic even though most of the evidence is circumstantial with little empirical support, but compelling (Grant, 1972; Arthur, 1982; Schoener, 1983; Benton, 1987). Clear evidence in support of the hypothesis is provided by Allard & Adams (1969) with barley and by Joy & Laitinen (1980) with *Trifolium pratense* and *Phleum pratense*. Alternative outcomes of competition have been proposed, such as the evolution of interference mechanisms (Aarssen, 1983; Hairston et al., 1987), and convergence in resource use (Ghilarov, 1984). Nevertheless, the expectation of interspecific, or intergenotypic, competition, is divergence.

Intergenotypic interactions – consequences in the pasture

There are two major methods by which to increase biomass productivity (outside of fertilizing). First, is a straightforward systematic search for good combinations of species or cultivars that are high yielding. Often this has lead to attempts to identify characteristics, and to their subsequent inclusion in breeding programs, of the components of the high-yielding mixtures (Hill & Michaelson-Yeates, 1987; Hill, 1990; Collins & Rhodes, 1990; Annicchiarico & Piano, 1994). Conventional variety improvement programs endeavor to identify superior i.e. higher yielding, genotypes on the basis of pure stand performance. The general conclusion from these studies was stated in Donald's (1963) landmark paper and have been restated elsewhere (Trenbath, 1974; Haynes, 1980; Zannone et al., 1983, 1986) and have not really needed upgrading, that is, that the yield of a mixture (legumes excluded) will usually be less than the higher yielding pure component and higher than the lower yielding pure component, and the total yield may be either greater than, or less than, the mean yield of the two pure components. In terms of yield there seems to be little benefit from mixing grass forage species. Trenbath (1974) describes a few exceptions (Rhodes, 1968, 1970). Such yield advantages of genotypes in binary mixtures may occur when the components are of contrasting growth habit (Rhodes, 1970) or have different seasonal growth patterns (Zannone et al., 1983). Most of these studies have been short-term (usually single generation) and made with random pairs of cultivars or genotypes. There is some evidence that coadaptation can occur among genotypes of cereals (Allard & Hansche, 1964; Allard & Adams, 1969) but these studies ran for up to 18 generations of selection and the Allard and Adams composite cross was derived from 31 varieties of barley.

Secondly, and of more direct interest to me are the ecological and evolutionary consequences of species interactions. Although it is a false premise that competitive ability of individuals and productivity of the community are synonymous, it may not be so errant to argue that success (fitness of individuals) and their productivity are often synonymous; this may be less true in pastures where grazing is a dominant force of selection. Nature selects for success, and if this can be translated as productivity, perhaps nature is already doing the systematic selection which we try to mimic in a simplified fashion in field trials, and natural selection is probably more effective than artificial selection. In our field and pot experiments the measures of performance used most often are dry weight and tiller production, measures of interest to the agronomist.

It would seem likely that coexisting genotypes, if they continue to coexist, should coevolve so that they increase their yield in each other's presence i.e. have ecological combining ability (Harper, 1967). It is unlikely that random selections of genotypes grown in mixture for one, or very few, generations, will generate any type of synergistic effect resulting in increased yield. This requires a shift to using multigeneration selection trials, and genotypes grown in mixture, or, deliberately selecting target combinations of coexisting individuals from natural situations where coevolution is assumed to have already happened. This way, both components of the mixture are exposed to the combined effects of both intra- and inter-genotypic competition (Hill & Michaelson-Yeates, 1987; Hill et al., 1987). It is odd that *T. repens* and grasses are usually bred separately in spite of the fact that *T. repens* is almost universally grown in mixture with grasses. Besides, a coevolved mixture will probably form a more stable combination of genotypes. If, however, the components of a mixture complement each other nutritionally then over-yielding will not be unexpected (de Wit et al., 1966) and it is not surprising that mixtures of grasses and legumes provide the most striking and repeatable examples.

Phenotypic plasticity will act to buffer some of the effects of natural selection and thus act to minimize genetic variation. Within any population there is an interplay of genetic diversity and phenotypic plasticity. Both are a means by which a population may adjust to its environment. Plasticity is particularly crucial in allowing fairly rapid and reversible adjustment to environmental change. Turkington (1989b) has argued that in more 'uniform' pastures we may expect to find more evidence of microevolution, or multiple adaptive optima, than in more 'variable' pastures where more of the variation detected will be phenotypic. However, the consequences of a patchy environment may be increased phenotypic plasticity and/or population differentiation (presumably having a genetic basis) among habitats. In the pasture context where environmental heterogeneity is high, where genetic variability within populations is high, and where competition is intense, there are a number of mechanisms by which a population can apparently become locally specialized. Local differentiation may have occurred by natural selection of genotypes leaving most progeny. This is unlikely because of the low rate of recruitment of seedlings into pastures. If this were the case, given the extremely fine scale at which this differentiation occurs (Aarssen & Turkington, 1985b), the intensity of natural selection would need to be extremely high. Alternatively, and much more likely, the differentiation may have occurred by spatial reassortment of the mobile population of genotypes, such that each genotype is most likely to be found in microenvironments that are most favourable to it. This will be especially so in rhizomatous and stoloniferous species. A third hypothesis is that there is no genetic basis to the apparent differentiation and the local specialization is a result physiological carryover of a plastic response (e.g. Evans & Turkington, 1988; Turkington, 1989b). Whatever the mechanism, the result would be that genotypes of one species found with genotypes of another particular species are specifically 'adapted' and have the optimal growth habits for coexisting. Arguably this should result in an increased yield of the mixture. Genetic change by non-random differential mortality among genotypes is potentially a very important process of adaptation in plant populations.

Ecological and evolutionary consequences

The range of responses by plants to the challenges of their ever-changing environment is quite remarkable. Once we have figured out why the plant population does this or that, it confounds us with yet another ecological or evolutionary solution to a problem that nature has thrown at it. We can detect responses at different scales and some of these may be of a purely academic rather than agronomic interest. I will first of all consider the larger scale (between species) patterns, then within-species patterns, and finally at the individual genotype level.

Between-species patterns

In most of the studies considered here, the specific hypothesis being tested was that the presence or absence of coexistence in the history of a population would influence its fitness in interspecific competition. Evans et al. (1985, 1989) sampled coexisting individuals of *L. perenne* and *T. repens* and from four long-established pastures in Switzerland, France, and Italy, and a 7-year-old pasture in southern England. *Trifolium repens* was grown in two-species mixtures either with its coexisting *L. perenne* or with one of two other *L. perenne* companions with which it had not previously coexisted. By the end of the second harvest year, those mixtures based on coexisting populations yielded 22% more on average than the other mixtures. After the fourth harvest year, overall, *T. repens* yields were significantly higher in the coexisting mixture. In addition, for all five populations the average annual yield of *T. repens* was highest when they were grown in mixture with their coexisting grass. While they detected no significant differences between the grasses, a pattern seems to be emerging (Table 1) where there is a benefit to the grass to be grown with its coexisting clover. On average, coexisting mixtures outyielded the remaining mixtures. Menchaca & Connolly (1990) also examined species interference in *L. perenne* and *T. repens* and their results are based on five sequential harvests after about 3 months. Initial growth of *L. perenne* was vigorous and its influence on *T. repens* was great but it declined dramatically with time. The influence of *T. repens* on *L. perenne* was initially small but it had an enhancing effect in later harvests, presumably due to the increased soil fertility caused by nitrogen fixation. Using the relative resource total (RRT) they showed that there was initially no benefit from mixing but there was a large benefit by the final harvest; this resulted because *T. repens* developed from being a competitor to a net facilitator over time, while *L. perenne* remained a competitor of *T. repens* throughout. In the final two harvests, *L. perenne* individuals performed better in mixture with *T. repens* than in pure stand at the same density. Turkington & Jolliffe (1996) followed the dynamics, over one growing season, of *L. perenne* and *T. repens* in mixtures varying in density and frequency, and using plants from different ages of pastures. Interactions between *Lolium* and *Trifolium* are complex, and change dramatically with pasture age, harvest date, and density. Over successive harvests, the two species' view of each other, and of themselves, changed in a dynamic way. In addition, individuals of

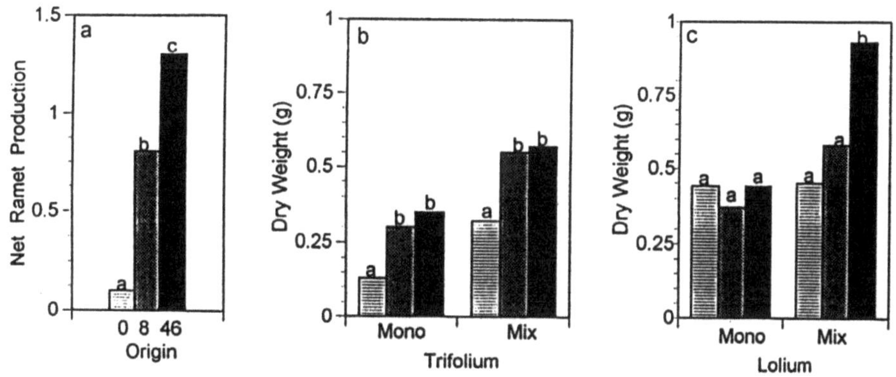

Figure 1. Average growth of *Trifolium repens* (Net ramet production, NRP (A) and dry weight (B) and *Lolium perenne* (dry weight (C)) collected from three pastures of age 0 (seed) (▤), 8 (▩) and 46 (■) years and (a) planted into undisturbed swards in 8- and 46-year-old pastures, (b, c) planted as mixtures and as monocultures into grazed patches in the two pastures. Within a graph, bars sharing a common letter are not significantly different ($p < 0.05$). (From Mehrhoff & Turkington, 1990).

Table 1. Dry matter yield (kg ha^{-1}) of *Lolium perenne* and *Trifolium repens* when grown in mixtures. Mixtures based on *T. repens* growing with its coexisting *L. perenne* and with other *L. perenne* companions. Data are from 1984 and from a mean annual yield from 1981–1984. (From Evans et al., 1989)

Origin of *T. repens*	*L. perenne* companion	1984			Mean 1981–1984		
		T. repens	*L. perenne*	Mixture	*T. repens*	*L. perenne*	Mixture
Switzerland	1	1093	1531	2624	1673	2640	4313
	2	682	1596	2278	1880	3139	5019
	Coexisting pop	1176	2176	3352	2680	3069	5749
Switzerland	1	1600	1380	2980	3853	2578	6431
	2	1171	1742	2913	4016	2958	6974
	Coexisting pop	1616	2823	4439	4312	3237	7549
Italy	1	1674	1618	3292	3820	2605	6425
	2	1110	1956	3066	3626	3038	6664
	Coexisting pop	1803	1625	3428	4110	2791	6901
France	1	1371	2245	3616	3156	3334	6490
	2	1392	2384	3776	3275	3711	6986
	Coexisting pop	1409	2698	4107	3615	3278	6893
England	1	1835	1564	3399	3825	2709	6534
	2	1796	2257	4053	4316	2832	7148
	Coexisting pop	1648	1405	3053	4368	2761	7129
Overall means	1	1515	1668	3183	3265	2773	6038
	2	1230	1987	3217	3423	3136	6559
	Coexisting pop	1531	2145	3676	3817	3027	6844

species collected from different ages of pasture perceived each other in different ways, and never in a reciprocal manner. An analysis of some component of final yield would conceal these dynamic interactions and would lead to erroneous conclusions concerning the behaviour of the components in the mixture. Generally however, *L. perenne–T. repens* mixtures used resources more effectively and were more productive than their corresponding monocultures. Chirwa (1985) collected coexisting pairs of *L. perenne* and *T. repens* from a pasture in Surrey and from a Swiss Alpine pasture, and grew them in all four combinations of *L. perenne* and *T. repens*. He did not detect a leading diagonal effect where the productivity of mixtures based on coexisting components is higher than others, but in three of the four mixtures (including both coexisting pairs) he detected a reduction in the interspecific pressures exerted by both components. He suggests that they may have developed 'tolerance' to one another. Martin & Harding (1981) posed a similar question using *Erodium cicutarium* and *Erodium obtusiplicatum*, common plant species in Californian annual grasslands. They also showed that total seed output and total reproductive rates of coexisting mixtures were higher than those of non-coexisting mixtures. McNeilly (1981) showed evidence of ecotypic differentiation of *Poa annua* with respect to competitive ability against *L. perenne*.

Mehrhoff & Turkington (1990, 1996) propagated individuals of *L. perenne* and *T. repens* from seed and from two different-aged (8 and 46-year old) pastures. The material was then transplanted back into the two pastures as monocultures and as *L. perenne-T. repens* mixtures. The experiments were planted in spring, grazed during summer and fall, allowed to grow until May, then harvested. *T. repens* showed a significant age-related pattern in growth where plant material from older pastures outyielded that from younger pastures both when planted into cleared plots without competitors and into the dense sward (Figure 1a). This indicates that microevolution can apparently occur quite rapidly. In addition, yield per plant for both species was higher in mixture than in monoculture, and the highest mixture yield came from the plant material collected from the oldest pasture (Figure 1b, 1c). It is of interest that the age-related increases in *L. perenne* growth were evident only when grown in mixture with *T. repens*, not in monoculture. In another series of studies, two-species mixtures of *L. perenne* grown with *D. glomerata, H. lanatus*, and *T. repens* were planted in a common garden. Genets of each species were collected from different pastures ranging in age from 0 (freshly sown) to 46 years. For each of the three species combinations, total ramet production was significantly greater in mixtures from older pastures than from younger pastures.

Within-species patterns

Turkington & Harper (1979b) collected ramets of *T. repens* from within patches dominated by each of four different grasses. The *T. repens* populations were grown in all possible combinations with the four species of grass in greenhouse flats containing a standard potting compost. Each clover population tended to grow best with the grass from which it had originally been sampled, i.e., a principal diagonal effect (see Luescher & Jacquard, 1991). A similar result was reported when the same design was done in the field although a local soil influence was also noted. It is evident that the different grasses impose different constraints on the growth of *T. repens* and that the *T. repens* populations has differentiated into subpopulations defined by the identity of the grass neighbor. This experimental design could not rigorously test (i) if the shift was genetic, or merely a plastic response to changing conditions, or (ii) if competition really was the driving force. Neither claim was made in the paper (Turkington & Harper, 1979b) although strongly implied (see Snaydon, 1985), and a subsequent study by Evans & Turkington (1988) highlights the problem.

From an old pasture in British Columbia, Evans & Turkington (1988) collected 100 ramets of *T. repens* from each of four neighborhoods dominated by *D. glomerata, H. lanatus, L. perenne* and *Poa compressa*. These were transplanted into a common garden, without competitors, and later scored for twelve morphological characters. For ten of the characters, a significant proportion of the variation between sampled ramets was accounted for by the identity of the neighboring grass species with which the *T. repens* ramet had been growing in the pasture (Table 2). These same clovers were grown for two more years in the common garden and were then scored for morphological characters as before. In no case was a significant proportion of the variation in characters now found to be due to the previous grass neighbors (Table 2). The clear implication of this study is that divergence patterns in *T. repens* morphology have a plastic rather than a genetic basis.

As a consequence of these apparently conflicting results, Turkington (1989a) constructed a much more complicated and rigorous form of this design, following the protocol of Connell (1980). Specifically he was asking (i) has population divergence happened, (ii) does it have a genetic basis, and (iii) was it caused by competition. Again, a home-site advantage was demonstrated for *T. repens* i.e. divergence had occurred. This was the first to actually demonstrate a genetic shift in the *T. repens* populations in response to competition from their grass neighbors.

Between-genotype patterns

Given Darwin's emphasis on the fate of individuals as the cornerstone of natural selection, it is surprising that we know so little about how the events occurring at this scale are relevant to influencing community structure.

Aarssen & Turkington (1985b) reduced the questions of the previous section from investigating species to investigating genotypes. They collected neighboring pairs of genotypes of *L. perenne* and *T. repens* from four different locations in a 40-year-old pasture

Table 2. Summary of analyses of variance for measured morphological characters of *Trifolium repens* from neighborhoods dominated by *Dactylis glomerata* (D), *Holcus lanatus* (H), *Lolium perenne* (L), and *Poa compressa* (P). The 1982 measurements were made shortly after the original collections, and the 1984 measurements were made on the same plant material four generations (2 years) later. The multiple range tests are on the means for measured characters classified by neighborhood type from which the ramets of *T. repens* were collected; means are ranked from smallest to largest, and underlined sets of means are not significantly different ($p > 0.05$). (From Evans & Turkington, 1988)

	1982		Multiple	1984	
	Percent variation accounted for	Significance	range test	Percent variation accounted for	Significance
Root weight	4.8	**	LD HP	0.0	NS
Shoot weight	4.8	**	LD HP	0.0	NS
Total weight	5.3	**	LD HP	0.0	NS
Primary stolon number	5.7	**	DL HP		Not measured
Total stolon number	2.7	*	DL HP	0.0	NS
Internode number	0.7	NS			Not measured
Primary stolon length	0.7	NS		0.17	NS
Secondary stolen length	2.0	*	LD PH		Not measured
Total stolen length		Not measured	–	0.0	NS
Internode length	2.4	*	LD HP	0.0	NS
Petiole length	20.2	**	LD HP	0.0	NS
Leaf weight	11.8	**	–	0.0	NS
Leaf length	19.6	**	–	0.0	NS

in British Columbia. All 16 possible interspecific combinations of *L. perenne* and *T. repens* genotypes were planted together in pots, the mixture clipped regularly for 1 year, and the cumulative yield of each of the two species recorded. For each pair, the yield of *T. repens* was generally highest when grown in mixture with its natural *L. perenne* neighbor, although a similar pattern was not detected for the grass (cf. Evans et al., 1989). The natural neighbor mixtures did not differ significantly in total yield from pairs of nonnatural neighbors (Figure 2). These results were more-or-less mimicked by Lüscher et al. (1992) using five genotypes each of the same two species. For each of the five natural combinations the percentage of clover was higher than the overall non-neighboring mean. Averaged over six harvests, the diagonal yield of *T. repens* was about 10% higher than that of the average off-diagonal yield (Figure 3). The diagonal yield of *L. perenne* was about 10% lower and total yield about the same as their respective off-diagonal averages.

Gliddon & Trathan (1985) did a rather elegant field experiment using *T. repens* and *L. perenne* transplants from four sites in the old (100 years) pasture in North Wales. They clearly showed that both *T. repens* (especially for number of survivors) and *L. perenne* (for dry weight) are highly locally specialized (Figure 4). A final competition experiment using plant material from two of the sites showed a significant leading diagonal effect.

A final example concerns the work of Collins & Rhodes (1990). They grew six *T. repens* varieties with three *L. perenne* companions: (a) the coexisting *L. perenne* collected from the same location as the *T. repens*, (b) the variety Ba 9462 bred specifically to be compatible with *T. repens*, and (c) S23, a variety of *L. perenne* that has been used to evaluate clover varieties during breeding programs. For all the *T. repens* the greatest *T. repens* yield, and the total mixture yield, was highest in mixtures with the coexisting grass than with Ba 9462 (Table 3). It is notable that all *T. repens*

Table 3. Annual dry matter yield (t.ha^{-1}) of *Trifolium repens* populations in mixture with *Lolium perenne*. (From Collins & Rhodes, 1990)

Yield (t.ha^{-1})				
T. repens population	*L. perenne* companion	*T. repens*	*L. perenne*	Total
Ac 3441	Coexisting	4.6	3.5	8.1
	S. 23	1.2	2.1	3.3
	Ba 9462	3.4	2.8	6.2
Ac 3160	Coexisting	8.4	3.1	11.5
	S. 23	6.4	2.2	8.6
	Ba 9462	7.9	3.2	11.1
Ac 4512-13-14	Coexisting	7.5	3.2	10.7
	S. 23	6.0	2.5	8.5
	Ba 9462	6.8	2.7	9.5
Ac3449-53	Coexisting	7.0	3.4	10.4
	S. 23	5.5	2.7	8.2
	Ba 9462	6.4	3.4	9.8
Menna	Coexisting	7.9	2.8	10.7
	S. 23	6.4	2.3	8.4
	Ba 9462	7.5	2.9	10.3

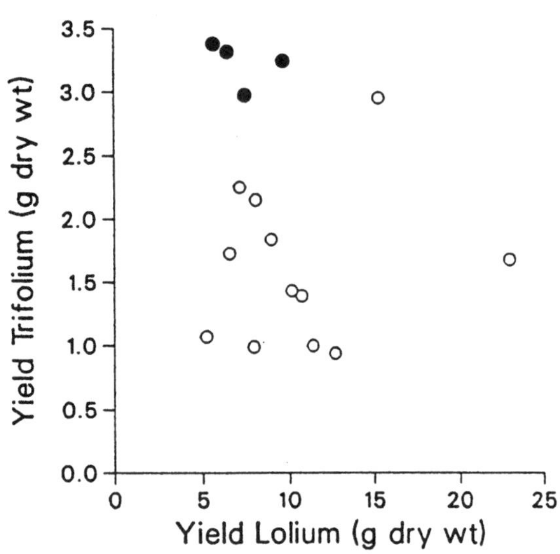

Figure 2. Yield of *Trifolium repens* and *Lolium perenne* when genotypes of the two species were grown in mixture. Mixtures that were collected as natural neighboring combinations in the field are designated by (●) and mismatched combinations by (○) (Data from Aarssen & Turkington, 1985b).

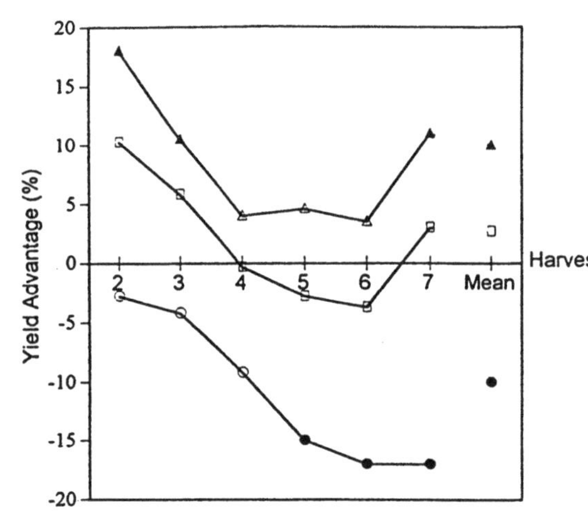

Figure 3. Comparative yield of *Trifolium repens* (Δ), *Lolium perenne* (○), and their mixture (●), over six successive harvests. The yield advantage is a comparative measure of the percentage increase or decrease in yield of natural neighboring combinations compared with the mean yield of all mismatched combinations. Filled symbols indicate a significant neighbour effect ($p < 0.05$) (From Lüscher et al., 1992).

yielded more with Ba 9462 than with S23 suggesting that it may be possible to breed for more general compatibility in grasses. It must be noted however, that some studies have reported that not all *T. repens* grow best with their coexisting grass neighbors (e.g. Collins & Rhodes, 1989).

Competition experiments between *L. perenne* and *T. repens* (at different scales) are plentiful, but where

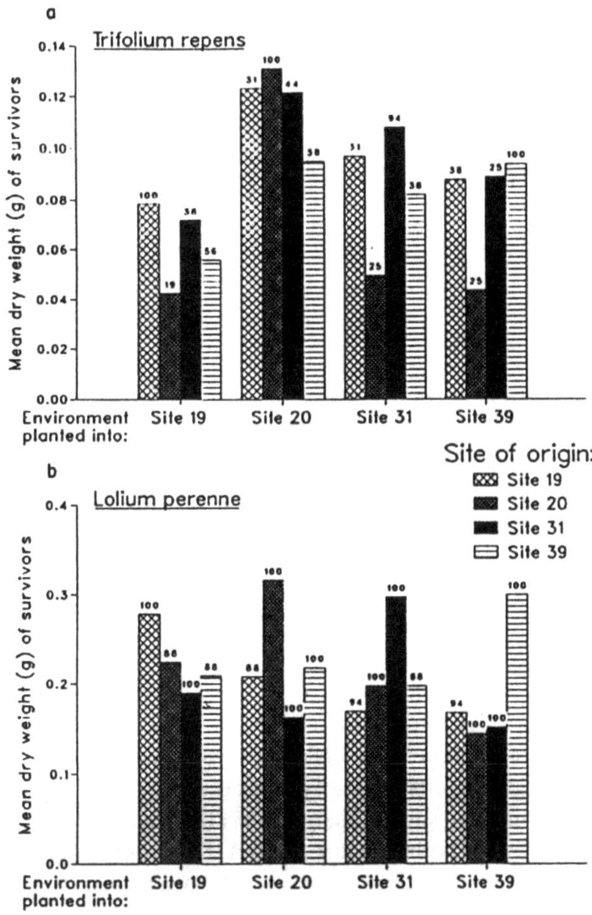

Figure 4. Mean dry weight of survivors of (a) *Trifolium repens* and (b) *Lolium perenne* collected from four sites and replanted into their site of origin and also transplanted into the other three sites. Numbers on top of bars are percentage survival values. (Data from Gliddon & Trathan, 1985).

a specific hypothesis about adaptation has been tested (Turkington & Harper, 1979b; Aarssen & Turkington, 1985b; Evans et al., 1985, 1989; Gliddon & Trathan, 1985; Chanway et al., 1989; Turkington, 1989a), in all cases *T. repens* grew best when planted with its natural neighbor yet a reciprocal effect in the *L. perenne* has only been reported once (Gliddon & Trathan, 1985).

Is there a mechanism?

Watkin Williams (1970), in his opening address to a White Clover Research symposium, sponsored by the British Grassland Society, stated that 'Another major gap in our understanding. . . is the precise relationship between the clover plant *together with its symbiont* (emphasis added), and co-habiting grass species in a mixed sward.' In spite of the plea, and

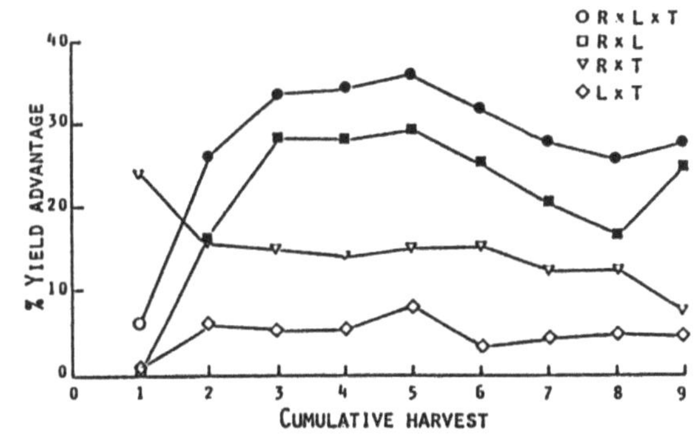

Figure 5. The influence of *Lolium perenne* genotype and *Rhizobium leguminosarum* biovar *trifolii* strain on the dry weight yield of *Trifolium repens*. The yield advantage is a comparative measure of the percentage increase or decrease in yield of natural neighboring combinations compared with the mean yield of all mismatched combinations. Filled symbols indicate a significant yield advantage ($p < 0.05$). (From Chanway et al., 1989).

considering that one entire section of the symposium (almost 80 published pages) was devoted to the white clover/Rhizobium symbiosis, there have been few attempts to include the associated microorganisms in these investigations, and almost nothing that I know of, to do it in the field. A number of more recent studies has attempted to partially remedy this situation (Chanway et al., 1988, 1989, 1991; Holl et al., 1988; Turkington et al., 1988; Thompson et al., 1990). Specifically Chanway et al. (1989) hypothesized that if individual genotypes of *L. perenne* and *T. repens* had become specifically adapted to one another, then the identity of the *Rhizobium* genotypes should be matched to the *L. perenne-T. repens* association. The objectives of the work were to (i) assess the effect of *Rhizobium* on the genotype specificity between individual *L. perenne* and *T. repens* plants collected from a permanent pasture, and (ii) to determine the effectiveness of *Rhizobium* strains when tested with coexistent or unrelated *T. repens* host genotypes grown in species mixture with *L. perenne*. We used the term 'homologous' to describe cuttings of *L. perenne* and of *T. repens*, and *Rhizobium* isolated from *T. repens* root nodules, all collected from the same point in the pasture. The experiment consisted of growing these three components (n = 3 for each factor) in a factorial design. To corroborate the patterns observed by Aarssen & Turkington (1985b), we would have predicted a yield advantage to ramets of *T. repens* when growing with matched *L. perenne* tillers; this effect was detected but it was not significant (Figure 5) i.e. specificity between *L. perenne* and *T. repens*

Table 4. Effect of *Bacillus polymyxa* inoculation on plant dry weight (mg) in a 50:50 (by numbers) mixture of *Lolium perenne* and *Trifolium repens* as genotypic homology is increased. (From Chanway et al., 1988)

Degree of homology	*Lolium perenne* Inoculation -	+	% change	*Trifolium repens* Inoculation -	+	% change
Partial[b]	1631	1502	- 9	688	845	+ 23[a]
Substantial[c]	1856	1820	- 2	721	1025	+ 43[a]
Total[d]	1624	1876	+ 16	619	985	+ 59

[a] Significant at $p < 0.05$.
[b] Plant populations comprised mixtures of three genotypes each of the two species, one of which was homologous to the *Bacillus* inoculum strain.
[c] The yield species was homologous to *Bacillus*; the neighboring plant population comprised a mixture of three genotypes, one of which was homologous to the *Bacillus* inoculum strain.
[d] The yield species was homologous to both the *Bacillus* and the neighboring plant population.

is lost if the coexisting *Rhizobium* is not present, a result not corroborated by Lüscher et al. (1992). When *Rhizobium* from homologous sites is used to inoculate the mixture (Figure 5), the effect becomes very pronounced, with homologous groups outyielding nonhomologous groups by up to 35%. However, a *T. repens* yield advantage of up to 30% can also be achieved when only homologous *Rhizobium-L. perenne* combinations are used, regardless of the identity of the *T. repens*. This indicates that it is the specific associations of *Rhizobium-L. perenne* genotypes that have the greatest impact on *T. repens* yield, even though the *Rhizobium* is symbiotic with *T. repens*.

The work also considered a grass-associated bacterium, *Bacillus polymyxa* (Chanway et al., 1988). It is clear from the data that the genotypic match between plant species and bacterial inoculant is critical to the direction and magnitude of the growth response. As the level of homology was increased between participating species, yield of *L. perenne* increased from slightly below those of the uninoculated controls to 16% ($p > 0.5$) greater under conditions of total homology (Table 4). However, data for the white clover clones reflected a significant yield increase in response to inoculation and increasing homology from 23% with partial homology, to 42% with substantial homology and finally 59% under total homology. These effects appear to be truly novel in an agricultural context and may represent the discovery of a significant component of the variation observed in plant growth promotion by rhizosphere bacteria. Exploitation of these relationships could contribute to the development of inoculants which consistently and effectively promote crop growth.

A final comment

A number of the studies I have described in this paper were done in a pasture in North Wales when it was about 100 years old (Peters, 1980). The pasture has large areas dominated by single grasses; the *L. perenne*-dominated site is about 150 m^2 (Turkington & Harper, 1979a; Thorhallsdottir, 1990a). Here, during its lifetime, the individual *T. repens* will not extend its stolons through many different environments, but will sample its environment in a coarse-grained manner, and may spend many generations coexisting with one species of grass. In contrast, many of the other studies I described were done in an old pasture in British Columbia which was 35–45 years-old during the studies. This pasture has a similar botanical composition to the Welsh pasture but has very different patterns of species distribution. The pasture is a complex mosaic of grass patches where most patches are probably less than 1 m^2 and closer to 0.25 m^2 (Evans & Turkington, 1988). Here, an individual *T. repens* will extend its stolons through many different neighborhoods and sample the different biotic environments in a fine-grained way. Thus, one might expect to detect genetically based microevolutionary changes in the *T. repens* population in the Welsh pasture in response to different coarse-grained patches. In contrast, the British Columbia pasture population of *T. repens* might be expected to respond in a plastic manner to the fine-grained mosaic environment.

References

Aarssen, L.W., 1983. Ecological combining ability and competitive combining ability in plants: Toward a general evolutionary theory of coexistence in systems of competition. Am Nat 122: 707–731.

Aarssen, L.W. & R. Turkington, 1985a. Vegetation dynamics and neighbour associations in pasture-community evolution. J Ecol 73: 585–603.

Aarssen, L.W. & R. Turkington, 1985b. Biotic specialization between neighbouring genotypes in *Lolium perenne* and *Trifolium repens* from a permanent pasture. J Ecol 73: 605–614.

Aarssen, L.W. & R. Turkington, 1985c. Within-species diversity in natural populations of *Holcus lanatus, Lolium perenne*, and *Trifolium repens* from four different-aged pastures. J Ecol 73: 869–886.

Allard, R.W. & J. Adams, 1969. Population studies in predominantly self-pollinating species. XIII. Intergenotypic competition and population structure in barley and wheat. Am Nat 163: 621–644.

Allard, R.W. & P.E. Hansche, 1964. Some parameters of population variability and their implications in plant breeding. Adv Agron 16: 281–325.

Annicchiarico, P. & E. Piano, 1994. Interference effects in white clover genotypes grown as pure stands and binary mixtures with different grass species and varieties. Theor Appl Gen 88: 153–158.

Antonovics, J., 1978. The population genetics of mixtures. In: J.R. Wilson (Ed.), Plant Relations in Pastures, pp 233–252. CSIRO, Melbourne.

Arthur, W., 1982. The evolutionary consequences of interspecific competition. Adv Ecol Res 12: 127–187.

Benton, M.J., 1987. Progress and competition in macroevolution. Biol Rev 62: 305–338.

Breese, E.L., M.D.Haywood & A.C. Thomas, 1965. Somatic selection in perennial ryegrass. Heredity 20: 367–379.

Brougham, R.W. & W. Harris, 1967. Rapidity and extent of changes in genotypic structure induced by grazing in a ryegrass population. NZ J Ag Res 10: 56–65.

Burdon, J.J., 1980. Intra-specific diversity in a natural population of *Trifolium repens*. J Ecol 68: 717–735.

Burdon, J.J., 1983. Biological flora of the British Isles, *Trifolium repens*. J Ecol 71: 307–330.

Burdon, J.J. & J.L. Harper, 1980. Relative growth rates of individual members of a plant population. J Ecol 68: 953–968.

Burgess, R.S.L. & R.A. Ennos, 1986. Selective grazing of acyanogenic white clover: Variation in behaviour among populations of the slug *Deroceras reticulatum*. Oecologia 73: 432–435.

Cahn, M.A. & J.L. Harper, 1976a. The biology of the leaf mark polymorphism in *Trifolium repens* L. I. Distribution of phenotypes at a local scale. Heredity 37: 309–325.

Cahn, M.A. & J.L. Harper, 1976b. The biology of the leaf mark polymorphism in *Trifolium repens* L. II. Evidence for the selection of leaf marks by rumen fistulated sheep. Heredity 37: 327–333.

Chanway, C.P., F.B. Holl & R. Turkington, 1988. Genotypic coadaptation in growth promotion of forage species by *Bacillus polymyxa*. Plant Soil 106: 281–284.

Chanway, C.P., F.B. Holl & R. Turkington, 1989. Effect of *Rhizobium leguminosarum* biovar *trifolii* genotype on specificity between *Trifolium repens* L. and *Lolium perenne* L. J Ecol 77: 1150–1160.

Chanway, C.P., R. Turkington & F.B. Holl, 1991. Ecological implications of specificity between plants and rhizosphere microorganisms. Adv Ecol Res 21: 121–169.

Charles, A.H., 1961. Differential survival of cultivars of *Lolium, Dactylis* and *Phleum*. J Br Grassld Soc 16: 69–75.

Charles, A.H., 1972. Ryegrass populations from intensively managed leys. III. Reaction to management, nitrogen application and *Poa trivialis* L. in field trials. J Agric Sci 79: 205–215.

Chestnut, D.M.B. & J. Lowe, 1970. Agronomy of white clover/grass swards: a review. Occ Sym Br Grassld Soc 6: 191–213.

Chirwa, R.M., 1985. Inter and intra-specific competition in perennial ryegrass-white clover mixtures. MSc Thesis, University of Wales, Aberystwyth, Wales.

Collins, R.M. & I. Rhodes, 1989. Yield of white clover populations in mixture with contrasting perennial ryegrasses. Grass For Sci 44: 111–115.

Collins, R.P. & I. Rhodes, 1990. The basis and significance of compatibility in grass/clover mixtures. Herba 3: 30–32.

Connell, J.H., 1980. Diversity and the coevolution of competitors, or the ghost of competition past. Oikos 35: 131–138.

Connolly, V., C.L. Masterson & D. Conniffe, 1969. Some genetic aspects of the symbiotic relationship between white clover (*Trifolium repens* L.) and *Rhizobium trifolii*. J Theor Appl Gen 39: 206–213.

Davies, W.E. & N.R. Young, 1967. The characters of European, Mediterranean and other populations of white clover (*Trifolium repens* L.). Euphytica 16: 330–340.

Dirzo, R. & J.L. Harper, 1982. Experimental studies on slug-plant interactions. III. Differences in the acceptability of *Trifolium repens* to slugs and snails. J Ecol 70: 101–118.

Dodd, M., J. Silvertown, K. McConway, J. Potts & M. Crawley, 1994. Stability in the plant communities of the Park Grass Experiment: the relationships between species richness, soil pH and biomass variability. Phil Trans Roy Soc B 346: 185–193.

Dodd, M., J. Silvertown, K. McConway, J. Potts & M. Crawley, 1995. Community stability: a 60-year record of trends and outbreaks in the occurrence of species in the Park Grass Experiment. J Ecol 83: 277–285.

Donald, C.M., 1963. Competition among crop and pasture plants. Adv Agron 15: 1–118.

Ennos, R.A., 1981. Detection of selection in populations of white clover (*Trifolium repens* L.). Biol J Linn Soc 15: 75–82.

Evans, D.R., J. Hill, T.A. Williams & I. Rhodes, 1985. Effects of coexistence in the performances of white clover-perennial ryegrass mixtures. Oecologia 66: 536–539.

Evans, D.R., J. Hill, T.A. Williams & I. Rhodes, 1989. Coexistence and the productivity of white clover-perennial ryegrass mixtures. Theor Appl Gen 77: 65–70.

Evans, R.C. & R. Turkington, 1988. Maintenance of morphological variation in a biotically patchy environment. New Phytol 109: 369–376.

Fowler, N., 1981. Competition and coexistence in a North Carolina grassland. II. The effects of the experimental removal of species. J Ecol 69: 843–852.

Fowler, N., 1982. Competition and coexistence in a North Carolina grassland. III. Mixtures of component species. J Ecol 70: 77–92.

Frame, J. & P. Newbould, 1986. Agronomy of white clover. Adv Agron 40: 1–87.

Ghilarov, A.M., 1984. The paradox of the plankton reconsidered; or, why do species coexist. Oikos 43: 46–52.

Gliddon, C. & P. Trathan, 1985. Interactions between white clover and perennial ryegrass in an old permanent pasture. In: J. Haeck & J.W. Woldendorp (Eds), Structure and Functioning of Plant Populations, 2, pp. 161–169. North-Holland, Amsterdam.

Grant, P., 1975. The classical case of character displacement. Evol Biol 8: 237–337.

Grubb, P.J., 1977. The maintenance of species richness in plant communities: the importance of the regeneration niche. Biol Rev 52: 107–145.

Hairston, N.G., K.C. Nishikawa & S.L. Stenhouse, 1987. The evolution of competing species of terrestrial salamanders: Niche partitioning or interference? Evol Ecol 1: 247–262.

Harper, J.L., 1967. A Darwinian approach to plant ecology. J Ecol 55: 247–270.

Harper, J.L., 1978. Plant relations in pastures. In: J.R. Wilson (Ed.), Plant Relations in Pastures, pp. 3–14. CSIRO, Melbourne.

Harris, W., 1987. Population dynamics and competition. In: M.J. Baker & W.M. Williams (Eds), White Clover, pp. 203–275. Commenwealth Agricultural Bureau, Wallingford, UK.

Haynes, R.J., 1980. Competitive aspects of the grass-legume association. Adv Agron 33: 227–261.

Hill, J., 1990. The three Cs - competition, coexistence and coevolution – and their impact on the breeding of forage crop mixtures. Theor Appl Gen 79: 168–176.

Hill, J., K. Mather & P.D.S. Caligari, 1987. Analysis of competitive ability among genotypes of perennial ryegrass. II. Effect upon dry matter production. Euphytica 36: 109–115.

Hill, J. & T.P.T. Michaelson-Yeates, 1987. Effects of competition upon the productivity of white clover-perennial ryegrass mixtures. Seasonal trends. Plant Breed 99: 251–262.

Holl, F.B., C.P. Chanway, R. Turkington & R. Radley, 1988. Growth response of crested wheatgrass (*Agropyron cristatum* L.), white clover (*Trifolium repens* L.), and perennial ryegrass (*Lolium perenne* L.) to inoculation with *Bacillus polymyxa*. Soil Bio Biochem 20: 19–24.

Jalloq, M.C., 1975. The invasion of molehills by weeds as a possible factor in the degeneration of reseeded pastures. I. The buried viable seed population of molehills from four reseeded pastures in West Wales. J Appl Ecol 12: 643–657.

Joy, P. & A. Laitinen, 1980. Breeding for coadaptation between Red Clover and Timothy. Hankkija's Seed Publ 13. Hankkija Plant Breeding Inst, Finland.

Kelley, S.E. & K. Clay, 1987. Interspecific competitive interactions and the maintenance of genotypic variation within two perennial grasses. Evolution 41: 92–103.

Kerner von Marilaun, A., 1895. The Natural History of Plants. Vol. 2. Blackie, London.

King, T.J., 1977. The plant ecology of ant-hills in calcareous grasslands. II. Succession on the mounds. J Ecol 65: 257–278.

Lawlor, L.R. & J. Maynard Smith, 1976. The coevolution and stability of competing species. Am Nat 110: 79–99.

Luescher, A. & P. Jacquard, 1991. Coevolution between interspecific plant competitors? TREE 6: 355–358.

Lüscher, A., J. Connolly & P. Jacquard, 1992. Neighbour specificity between *Lolium perenne* and *Trifolium repens* from a natural pasture. Oecologia 91: 404–409.

MacArthur, R.H. & R. Levins, 1964. Competition, habitat selection, and character displacement in a patchy environment. Proc Nat Acad Sci USA 51: 1207–1210.

MacArthur, R.H. & R. Levins, 1967. The limiting similarity, convergence and divergence of coexisting species. Amer Nat 101: 377–385.

Martin, M.M. & J. Harding, 1981. Evidence for the evolution of competition between two species of annual plants. Evolution 35: 975–987.

McNeilly, T., 1981. Ecotypic differentiation in *Poa annua*: Intrapopulational differences in response to competition and cutting. New Phytol 88: 539–547.

McNeilly, T. & M.L. Roose, 1984. The distribution of perennial ryegrass genotypes swards. New Phytol 98: 503–513.

Mehrhoff, L.A. & R. Turkington, 1990. Microevolution and site-specific outcomes of competition among pasture plants. J Ecol 78: 745–756.

Mehrhoff, L.A. & R. Turkington, 1996. Growth and survival of white clover (*Trifolium repens* L.) transplanted into patches of different grass species. Can J Bot (in press).

Menchaca, L. & J. Connolly, 1990. Species interference in white clover-ryegrass mixtures. J Ecol 78: 223–232.

Mytton, L.R., 1975. Plant genotype × *Rhizobium* strain interactions in white clover. Ann Appl Biol 80: 103–107.

Parish, R. & R. Turkington, 1990a. The influence of dung pats and molehills on pasture composition. Can J Bot 68: 1698–1705.

Parish, R. & R. Turkington, 1990b. The colonization of dung pats and molehills in permanent pastures. Can J Bot 68: 1706–1711.

Peart, D.R., 1989. Species interactions in a successional grassland. III. Effects of canopy gaps, gopher mounds and grazing on colonization. J Ecol 77: 267–289.

Peters, B., 1980. The demography of leaves in a permanent pasture, Ph.D. thesis, University College of North Wales, Bangor, Wales.

Rhodes, I., 1968. Yield of contrasting ryegrass varieties in monoculture and mixed culture. J Brit Grassl Soc 23: 156–158.

Rhodes, I., 1970. The production of contrasting genotypes of perennial ryegrass (*Lolium perenne* L.) in monocultures and mixed cultures of varying complexity. J Brit Grassl Soc 25: 285–288.

Schoener, T.W., 1983. Field experiments on interspecific competition. Am Nat 122: 240–285.

Seaton, A.P.C. & J. Antonovics, 1967. Population interrelationships. I. Evolution in mixtures of *Drosophila* mutants. Heredity 22: 19–35.

Silvertown, J., 1980. The dynamics of a grassland ecosystem: botanical equilibrium in the Park Grass Experiment. J Appl Ecol 17: 491–504.

Silvertown, J., M. Dodd, K. McConway, J. Potts & M. Crawley, 1994. Rainfall, biomass variation, and community composition in the Park Gras Experiment. Ecology 75: 2430–2437.

Snaydon, R.W., 1962. Microdistribution of *Trifolium repens* and its relation to soil factors. J Ecol 50: 133–143.

Snaydon, R.W., 1978. Genetic changes in pasture populations. In: J.R. Wilson (Ed.), Plant Relations in Pastures, pp. 253–272. CSIRO, Melbourne.

Snaydon, R.W., 1985. Aspects of the ecological genetics of pasture species. In: J. Haeck & J.W. Woldendorp (Eds), Structure and Functioning of Plant Populations, 2, pp. 127–152. North-Holland, Amsterdam.

Soane, I.D. & A.R. Watkinson, 1979. Clonal variation in populations of *Ranunculus repens*. New Phytol 82: 557–573.

Thompson, J.D., R. Turkington & F.B. Holl, 1990. The influence of *Rhizobium leguminosarum* biovar *trifolii* on the growth and neighbour relationships of *Trifolium repens* and three grasses. Can J Bot 68: 296–303.

Thompson, J.N., 1985. Within-patch dynamics of life histories, populations; selection over time in small spaces. In: S.T.A. Pickett & P.S. White (Eds), The Ecology of Natural Disturbance and Patch Dynamics, pp. 253–264. Academic Press, Orlando.

Thompson, J.N., 1988. Variation in interspecific interactions. Ann Rev Ecol Syst 19: 65–87.

Thompson, J.N., 1994. The Coevolutionary Process. University of Chicago Press, Chicago.

Thórhallsdóttir, T.E., 1990a. The dynamics of a grassland community: a simultaneous investigation of spatial and temporal heterogeneity at various scales. J Ecol 78: 884–908.

Thórhallsdóttir, T.E., 1990b. The dynamics of five grasses and white clover in a simulated mosaic sward. J Ecol 78: 909–923.

Trathan, P., 1983. Clonal interactions of *Trifolium repens* and *Lolium perenne*. Ph.D. thesis, University of Wales, Bangor, Wales.

Trenbath, B.R., 1974. Biomass productivity of mixtures. Adv Agron 26: 177–210.

Turkington, R., 1975. Relationships between neighbours among species of permanent grassland (especially *Trifolium repens* L.). Ph.D. thesis. University of Wales, Bangor, Wales.

Turkington, R., 1989a. The growth, distribution and neighbour relationships of *Trifolium repens* in a permanent pasture. V. The coevolution of competitors. J Ecol 77: 717–733.

Turkington, R., 1989b. The growth, distribution and neighbour relationships of *Trifolium repens* in a permanent pasture. VI. Conditions effects by neighbours. J Ecol 77: 734–746.

Turkington, R., 1993. Effect of propagule source on competitive ability of pasture grasses: spatial dynamics of six grasses in simulated swards. Can J Bot 72: 111–121.

Turkington, R. & J.J. Burdon, 1983. Biology of Canadian weeds. 57. *Trifolium repens*. L. Can J Pl Sci 63: 243–266.

Turkington, R., F.B. Hall, C.P. Chanway & J.D. Thompson, 1988. The influence of microorganisms, particularly *Rhizobium*, on plant competition in grass-legume communities. In: A.J. Davy, M.J. Hutchings & A.R. Watkingson (Eds), Plant Population Ecology, pp. 343–366. Blackwell, Oxford, England.

Turkington, R. & J.L. Harper, 1979a. The growth, distribution and neighbour relationships of *Trifolium repens* in a permanent pasture. I. Ordination, pattern and contact. J Ecol 67: 201–218.

Turkington, R. & J.L. Harper, 1979b. The growth, distribution and neighbour relationships of *Trifolium repens* in a permanent pasture. IV. Fine scale biotic differentiation. J Ecol 67: 245–254.

Turkington, R. & P.A. Jolliffe, 1996. Interference in *Trifolium repens — Lolium perenne* mixtures: short-term and long-term relationships. J Ecol (in press).

Turkington, R., E. Klein & C.P. Chanway, 1993. Interactive effects of nutrients and disturbance: an experimental test of plant strategy theory. Ecology 74: 863–878.

Turkington, R. & L.A. Mehrhoff, 1990. The role of competition in structuring pasture communities. In: J.B. Grace & D. Tilman (Eds), Perspectives on Plant Competition, pp. 307–340. Academic Press, San Diego.

deWit, C.T., G.P. Tow & G.C. Ennik, 1966. Competition between legumes and grasses. Agric Res Rep 687: 1–30, Wageningen.

White, J., 1985. The census of plants in vegetation. In: J. White (Ed.), The Population Structure of Vegetation, pp. 33–38. Junk Publishers, Dordrecht.

Williams, W., 1970. White clover in British Agriculture. Occ Sym Br Grassld Soc 6: 1–10.

Zannone, L., L. Assemat, P. Rotili & P. Jacquard, 1983. An experimental study of intraspecific competition within several forage crops. Agronomie 3: 451–459.

Zannone, L., P. Rotili, R. Paoletti & C. Scotti, 1986. Experimental studies of grass-legume associations. Agronomie 6: 931–940.

P.M.A. Tigerstedt (Ed.), *Adaptation in Plant Breeding*, 135–142, 1997.

Co-adaptation between neighbours? A case study with *Lolium perenne* genotypes

T. McNeilly[1] & M.L. Roose[2]
[1] *Department of Environmental and Evolutionary Biology, University of Liverpool, P.O. Box 147, Liverpool, U.K.;*
[2] *Department of Botany & Plant Science, University of California, Riverside, California, U.S.A.*

Key words: Lolium perenne, perennial ryegrass, co-adaptation

Summary

Co-adaptation between neighbouring genotypes of *Lolium perenne* was examined in a pot experiment over an 18 month period. Competitive interactions in pure and mixed stands were examined using 9 genotype pairs (established as neighbours from electrophoresis data), 7 from a 10 year-old, and 2 from a 40+ year-old pasture. Combinations of each pair of neighbours grown together, and each grown with 5 randomly chosen alien (non-neighbour) genotypes sampled from the same pastures were also assessed. Seven harvests of green matter produced above 20 mm were made, and 50–135 kg N ha^{-1} was administered after each harvest. Data for dry weight at each harvest for pure and mixed stands, and Relative Yield Totals (RYT) were obtained for neighbours and aliens in each treatment.

In 6 of the 9 pairs of neighbours grown as pure stands, the yields of neighbours did not differ significantly at the final harvest. In 8 of the 9 pairs grown as mixtures of pairs of neighbours, the yield of one of the neighbours was significantly reduced and one neighbour had been eliminated. Over the last 5 harvests, yields from (alien + neighbour) mixtures had the greatest dry matter yields. No suppression of one component by the other was recorded for this (alien + neighbour) combination.

Based upon RYT estimates, neighbours did not yield more when grown together than when grown separately with aliens. There was no relationship between original genotype size in their fields of origin and dry matter yield in this experiment.

This experiment provides no evidence for he evolution of co-adaptation between surviving neighbours in the two pastures from which these individuals were sampled. The genotypic architecture of pasture would seem to be the product of random and/or competitive exclusion of individual genotypes, rather than the evolution of co-adapted groups of individuals through time. The consequences of extreme depletion of genetic variation in the first two years after establishment, the very low rates of seedling establishment in established pastures, and the impact of the grazing animal would suggest that there is little likelihood of or potential for, co-adaptation evolving between neighbours.

Introduction

As much as 90% of the seed sown to establish pasture grassland in the United Kingdom is lost during the establishment year (Charles, 1961). In genetically diverse seeds mixtures this can result in selection pressures which can rapidly bring about marked changes in the genetic structure of surviving populations (Charles, 1961, 1964, 1966; Rossiter, 1966), although many of the deaths are apparently random (Charles, 1961). Such selective changes result in populations adapted to the agronomic practices to which the pastures are subjected – grazing management, fertilizer treatment, species composition of mixture sown (Clements et al., 1983).

The number of genotypes decreases rapidly during the establishment phase, particularly in the early years. At least 37 to 49 perennial ryegrass genotypes were recorded from 0.25 m^2 plots in three 10 year old *Lolium perenne* dominated swards in the United Kingdom (McNeilly & Roose, 1984). By contrast only 5 within a 0.25 m^2 area were found in a pasture known to be at least 40 years old. If adaptations to the general agronomic features of the environment occur in the relatively short term, as reported by Charles (1961, 1964, 1966), Brougham et al. (1960) and Rossiter et al. (1966), the continued decline in genotype numbers might be expected to result in the further evolution of groupings of co-adapted neighbouring genotypes which, as well as possessing general adaptation to agronomic factors, would possess specific adaptation (*sensu* Burdon, 1980) to neighbours of the same or different species. There is some evidence for the evolution of such co-adapted groups in barley (Adams & Allard, 1969), and in *Drosophila* (Seaton & Antonovics, 1967), and within long established natural pastures containing white clover and perennial ryegrass (Evans et al., 1985). There is however no overall consistent body of experimental evidence for the evolution of co-adaptation in plants (Antonivics, 1978; Snaydon, 1978).

Lolium perenne L. is a tufted perennial with a generally poor ability for clonal spread (Hubbard, 1968), and in common with most pasture grasses, rarely grows from seed in established grasslands (Harper, 1977). Hence it is possible that neighbouring individuals may have been neighbours for many years, and there may be the potential in old established pastures for the selection of co-adapted neighbours.

The work described in this paper examines interactions in pure and mixed stands, between selected pairs of neighbouring genotypes of *L. perenne* from pastures 10 years old and at least 40 years old at sampling (McNeilly & Roose, 1984), to assess whether co-adaptation between individual neighbour pairs has occurred.

Materials and methods

Plant material

Nine pairs of neighbouring genotypes (termed neighbours a and b) were chosen for investigation.

Seven pairs (numbered 1–7) were from a sown 10 year old pasture, the remaining two pairs (numbered 8 and 9) were from a permanent pasture known to be at least 40 years old. Each pair of neighbours was propagated from tillers originally collected 5 cm apart on a grid within a 0.25 cm^2 quadrat in the field of origin. Neighbours has distinct isozyme genotypes (McNeilly & Roose, 1984). Material of these 18 genotypes (nine pairs of neighbours) was multiplied in John Innes (JI) potting compost over winter to provide a minimum of 150 tillers per genotype.

Five plants – termed aliens – were sampled at random within both the 10 and 40 year old pastures. Tillers of the two sets of 5 aliens were propagated using the same procedures as for the neighbour pairs.

Experimental design

The experiment was set up in early spring in 17.5 cm diameter plastic pots containing equal parts by volume of a mixture of JI potting compost and horticultural peat, and after a two week establishment period in a glasshouse, the pots were placed in the open at the University Botanic Gardens. The experiment ran for 18 months.

Pots were planted with uniform sized unrooted tillers of the appropriate genotypes, and the few tillers which died were replaced with equivalent spare material. Each pair of neighbouring genotypes was planted as follows:

(i) Pure stand 10 tillers per pot Neighbour a
(ii) Pure stand 10 tillers per pot Neighbour b
(iii) Mixed stand 5 tillers Neighbour a + 5 tillers, 1 from each of the 5 aliens
(iv) Mixed stand 5 tillers Neighbour b + 5 tillers, 1 from each of the 5 aliens
(v) Mixed stand 5 tillers Neighbour a + 5 tillers Neighbour b

In each pot the tillers were arranged equidistant from each other in four rows. In the pots containing mixtures of genotypes (iii, iv, and v), the two components occupied alternate rows.

Pure stands of the groups of 5 alien genotypes from the two pastures were also planted, having 10 tillers per pot, two from each genotype, planted at random.

Each of the nine pairs of neighbours was thus represented by 5 pots (i–iv above), and there were two pots of aliens in pure stands, giving a total of 47 pots per replicate. There were four fully randomised replicate blocks; the position of each pot was re-randomised after each harvest.

Harvests

There were 7 harvests, at each of which plant material above 20 mm was cut, over dried at 60° C for 7 days, and weighed, the components of mixtures being harvested separately. Harvests were made 3, 4, 5, 6,

15, and 17 and 18 months after planting. After each harvest all pots were given 50 kg ha^{-1} N equivalent, and 135 kg ha^{-1} N in the early spring between the 6 and 15 month harvests, all from a 17:17:17 (N; P_2O_5; K_2O) fertilizer.

Data presentation

The experiment was designed to assess whether selection for co-adaptation between neighbouring genotypes in the two swards sampled. De Wit & Van den Bergh (1965) proposed the concept of Relative-Yield-Total (RYT) to assess competitive interactions in competition experiments based upon the replacement series method developed by De Wit (1960). The use of RYT has been discussed by Snaydon (1991), in the comparison of competition experiments using additive or replacement experimental designs. The experiment described in this paper considers data derived from a replacement design from which RYT values have been obtained, the objective being to assess possible differences in yield obtained from mixtures of (possibly co-adapted) neighbouring individuals as opposed to mixtures of each of each pair of neighbours with a set of non-neighbours termed aliens. For such an objective, the use of RYT to provide a standard means of performance comparisons of pairs of genotypes of the same species is considered adequate.

RYT values were calculated from the dry weight yield data for each of the harvests made, where, for pure stands and mixture of A and B,

$$RYT = \frac{\text{Yield of A in mixture}}{\text{Yield of B in mixture}} + \frac{\text{Yield of A in pure stand}}{\text{Yield of B in pure stand}}.$$

If RYT does not differ significantly from 1, the demands of the two components of the mixture are the same, hence competition will be severe. RYT values in excess of 1 imply different resource requirements and exploitation by the components, a situation expected where co-adaptation has evolved between neighbouring individuals. RYT values which are less than 1 imply that the environment is used less efficiently by the mixtures than by the two components in pure stands.

RYT values were calculated for each pair of neighbours when they were grown in mixture together, and for each of the neighbouring individuals when grown in mixtures with the 5 random alien individuals.

Results

Data for relative yield totals for each pair of neighbours grown together, for each of the two neighbours grown separately with aliens at each of the seven harvests, and means over all pairs of neighbours, are given in Figures 1 and 2 respectively.

The results of the analysis of Relative Yield Totals (RYTs) clearly lead to the conclusion that neighbours do not yield more when grown together than when each of the two neighbours is grown separately with a random sample of non-neighbours from the pasture from which the neighbours were collected. RYT estimates were greater than 1.0 for 111 of the 189 combinations (7 harvests × 3 mixtures × 9 pairs of neighbours), and less than 1.0 for only 39. Of the RYTs greater than 1.0, 30% (slightly less than the expected 33%) involved combinations of neighbours. When averaged over all pairs, RYTs for the mixtures of neighbouring individuals, and for individual neighbours grown in mixtures with aliens were greater than 1.0 at all harvests, and significantly greater in 14 of the 21 cases. Analysis of variance (Table 1) showed that only at harvest 5 there was a significant difference between RYTs of neighbours in competition with each other and with aliens, and in this case the mixture of neighbours had a smaller RYT than the (neighbour + alien) mixtures.

When yields were averaged over all neighbours (Table 2), the aliens initially had the lowest yields, both alone in pure stands, and when grown in mixture with either of the two neighbouring individuals. However, the combination (aliens + a neighbour) had the highest DM yields after harvest 2. Aliens when grown in pure stand were not particularly high yielding, and did not differ significantly from pure stands of neighbours except at harvests 1 and 3 when neighbours outyielded the aliens. Yields of neighbours grown with aliens were similar to those of pure stands of neighbours, except at harvest 6 when it was significantly less. The high yield of the aliens competing with individual neighbours does not appear to result from suppression of the neighbour component of the mixture.

The mixture of neighbours was also a high yielding treatment in the majority of harvests, being greater than that of the neighbours in pure stand at all harvests except harvest 2, and significantly greater at harvest 4, and for the sum of yields across harvests. In general mixtures had higher yields than pure stands, but mixtures of two neighbouring individuals did not yield more than mixtures of (a neighbour + aliens).

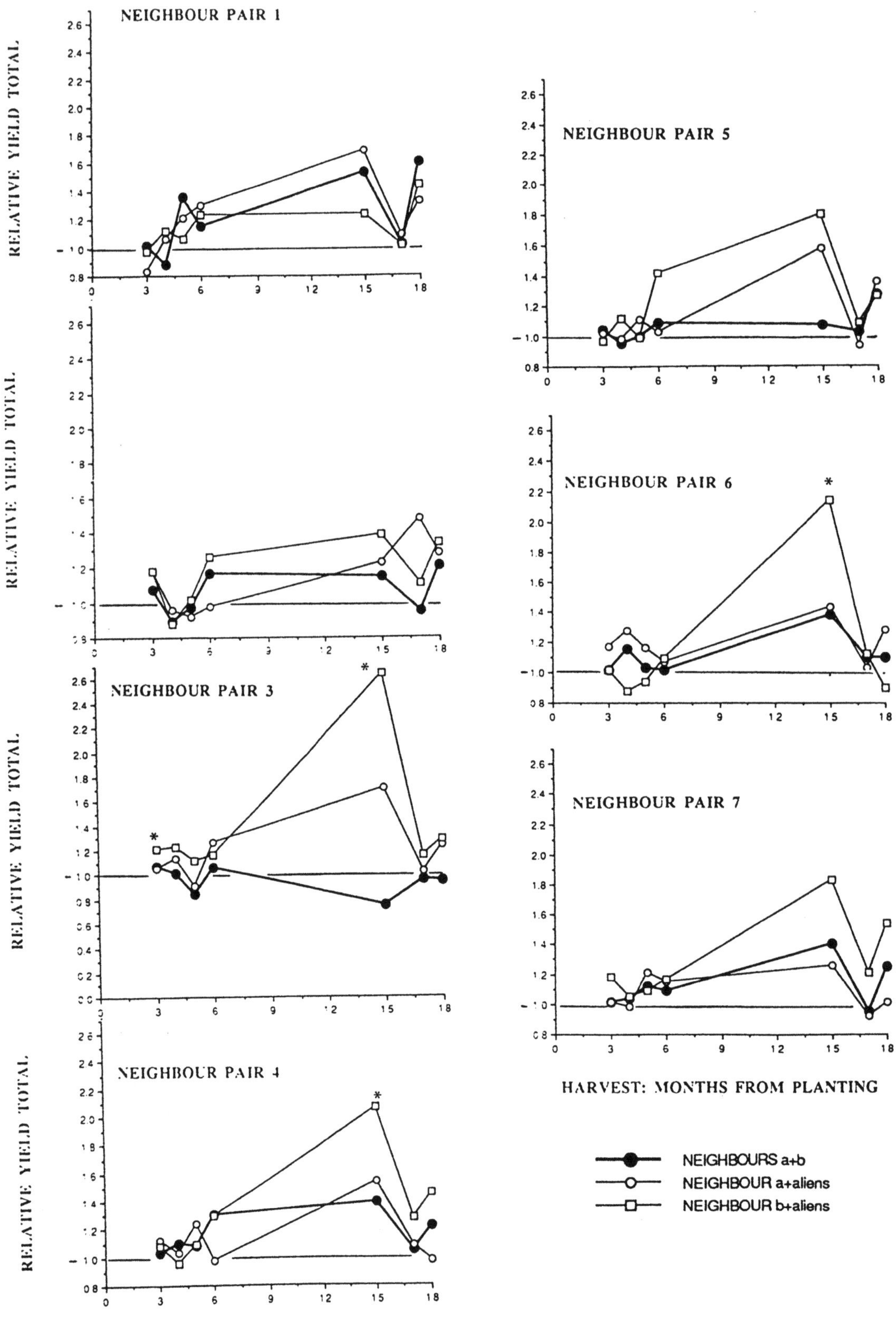

Figure 1. Relative yield totals for pairs of neighbours grown together, and for each neighbour grown separately with aliens over seven harvests.
* denotes RYT significantly different of form 1 at $p<0.05$.

NEIGHBOUR PAIR 8
FROM 50+ YEAR OLD PASTURE

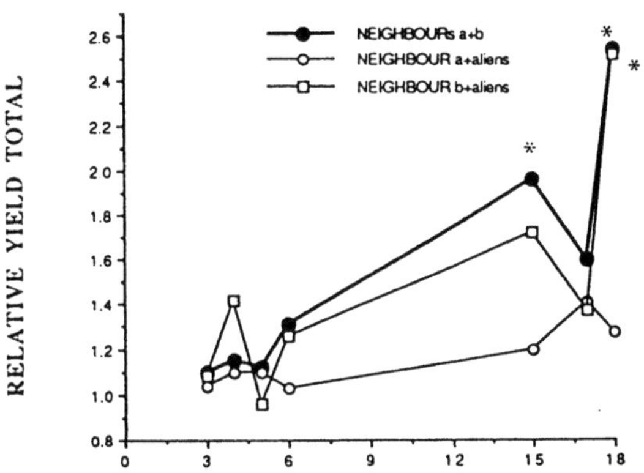

HARVEST: MONTHS FROM PLANTING

NEIGHBOUR PAIR 9
FROM 50+ YEAR OLD PASTURE

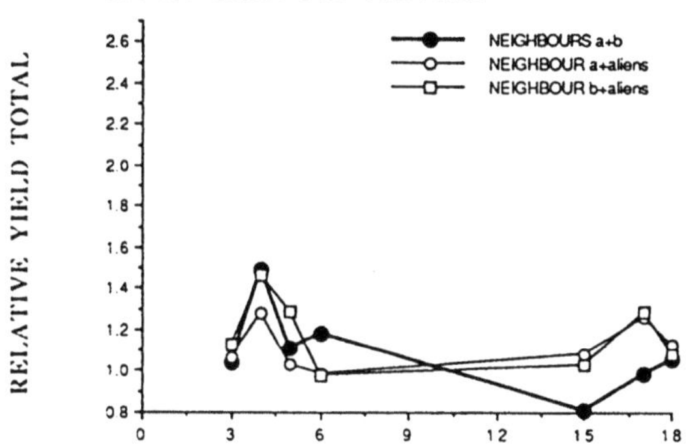

HARVEST: MONTHS FROM PLANTING

Figure 1. Continued.

The neighbouring individuals used in this study differed in area occupied in the pastures from which they were sampled (McNeilly & Roose, 1984), and it is therefore possible to compare performance in this experiment with size in the wild. Neighbours a and b of pair 9, and neighbour b of pair 8 were large in the pasture, occupying at least 830 cm^2 in a pasture that had not been ploughed for at least 40 years, and probably 100–150 years since ploughing (personal communication, owner of field). Neighbouring pairs 1–7 were sampled from a 10 year-old pasture, and from this pasture the largest clones used in this study were (neighbour a) of pair 2, (neighbour b) of pair 3, and (neighbour a) of pair 5, each of which covered areas varying from 200–300 cm^2. In this experiment, plant size was clearly limited by pot size, but DM yield in pure stand at harvest 1, and DM yield in mixture with aliens at harvest 7 provides a crude measure of growth rate and persistence. The correlation coefficients between area occupied in the field of origin and these two measures were not significant ($r = -0.6$, $r = 0.13$ respectively).

Discussion

This study was undertaken to examine the possibility that groups of genotypes may have co-adapted through time in pastures dominated by perennial ryegrass, and that this would be revealed by competitive interactions between them. There can be little doubt that within the populations of perennial ryegrass examined, there must be very considerable niche overlap between genotypes, and considerable selection pressures might be thought likely to operate to promote the evolution of co-adaptation between neighbours. In assessing the findings from this experiment, it should be remembered that it was carried out in pots, which will have inevitably removed the possibility of root based mechanisms which may be a considerably important component of successful co-existence in the field.

The data of Evans et al. (1985) clearly showed that a degree of co-adaptation at the population level between perennial ryegrass and white clover (*Trifolium repens*) had evolved in old Alpine pastures in Switzerland, Italy, and France. This is in contrast to the data presented here for co-habiting neighbouring pairs of perennial ryegrass individuals, which clearly do not give any evidence of the evolution of co-adaptation between neighbours. Thus, whilst at the final harvest yields of neighbours in pure stand differed from each other in only 3 from 9 pairs of neighbours, in 8 of the 9 pairs of neighbours grown together, the yield of one of the neighbours was significantly reduced, and one neighbour, (neighbour b in pair 3) had been eliminated. In a similar experiment, Aarsen & Turkington (1985) also found little evidence supporting the view that co-adaptation had evolved in a 40 year-old pasture based upon competition experiments with pairs of individuals from 5 coexisting species, including *T. repens*, the species which they found to be most sensitive to the nature of its associates. It seems likely that the ability of sward components to reach some degree of co-adaptation as shown for the *L. perenne*/*T. repens* pastures examined by Evans et al. (1985), does not exist in the pastures dominated by *L. perenne* examined in this study, despite the possibility that the progressive elimination of genotypes throughout the life

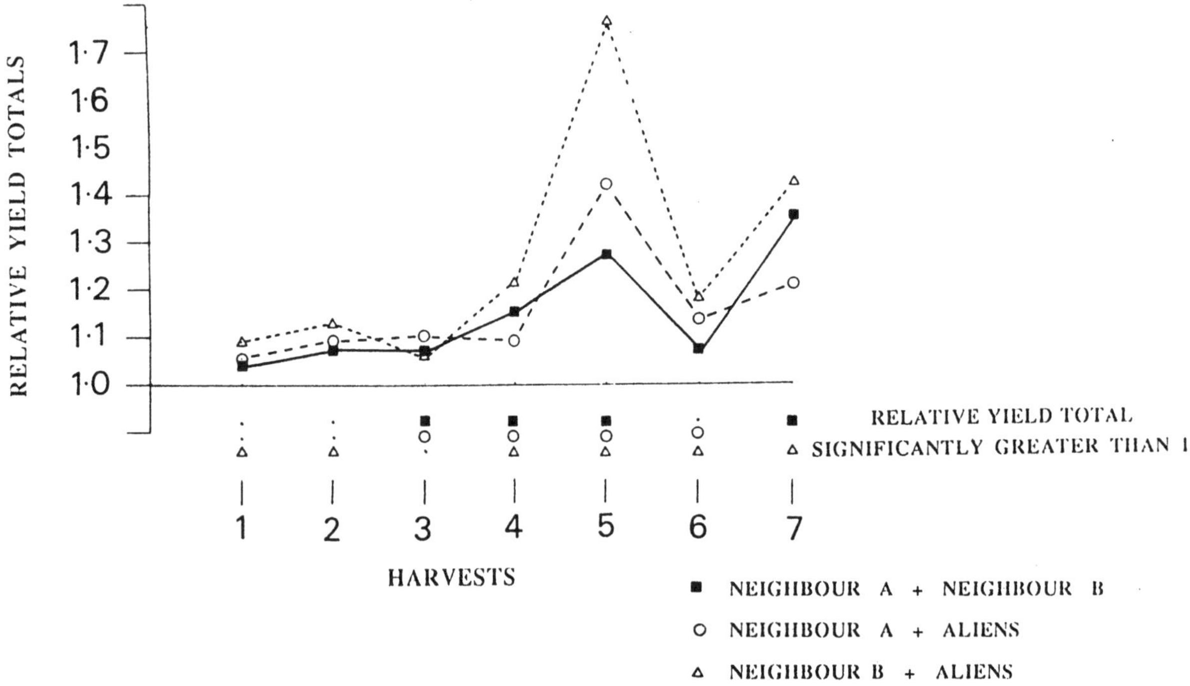

Figure 2. Means of all pairs of neighbours grown together, and for each neighbour grown separately with aliens over seven harvests.

Table 1. Mean squares from analysis of variance of Relative Yield Totals over 7 harvests for 9 pairs of *L. perenne* neighbours grown in 3 combinations, (i) neighbour a with neighbour b; (ii) neighbour a with 5 alien genotypes; (iii) neighbour b with (the same) 5 alien genotypes

Source of variation	Mean square by harvest df	H1	H2	H3	H4	H5	H6	H7	Total
Blocks	3	0.43	1.39	0.09	0.73	0.45	0.40	2.27	0.02
Neighbours (N)[1]	8	0.04	0.26	0.09*	0.05	0.65	0.23	1.16*	0.05*
Combinations[2] (C)	2	0.02	0.03	0.01	0.13	2.30**	0.10	0.44	0.10*
Interaction (N × C)	16	0.02	0.13	0.05	0.06	0.55	0.08	0.32	0.03
Error	78	0.03	0.13	0.04	0.07	0.33	0.11	0.55	0.02

Significant differences: * at $p < 0.05$, ** at $p < 0.01$, otherwise differences not significant.
[1] Neighbours compares mean dry weights of pooled data for the nine pairs of neighbours.
[2] Combinations compares mixture of (neighbour a + neighbour b); (neighbour a + aliens); (neighbour b + aliens).

of the permanent pasture might have resulted in such a situation. In this context it is interesting that it is the *T. repens* component of the Alpine populations that is the major source of yield advantage in them. The work of Turkington & Harper (1979) has also shown the remarkable ability of *T. repens* to co-adapt to particular grass species including *L. perenne*. It may thus be that the perennial ryegrass component of such mixtures does not have the same potential to evolve genotypes which are compatible with white clover as vice versa.

The lack of any evidence for the evolution of coadaptation between the co-existing pairs of genotypes examined here could be attributed to insufficient time for such evolution to have occurred. This might be the case for the 10 year old pasture, despite the considerable reduction in genotype numbers during its early life as shown by the study of Charles (1961, 1964) a process which continues through later life of similar pastures, (McNeilly & Roose, 1984). As stated earlier, the permanent pasture, known to be at least 40 years

Table 2. Overall mean dry matter yields (g 0.024 m^{-2}) of (a) Pure stands of 10 plants per pot of (i) 9 pairs of neighbours; (ii) of 5 alien non-neighbours (2 plants per alien = 10 plants per pot), and (b) Mixtures (i) of two neighbours (the components of each pair of neighbours), and (ii) of mixtures of each neighbour with the 5 alien non-neighbours. Values in rows not followed by the same letter are significantly different at $p > 5\%$

Harvest	Pure stands of		Mixed stands of		
	Neighbours 10 per pot	Aliens 10 per pot*	Aliens + Neighbours a	Neighbour a + Neighbour b	Aliens + Neighbours b
1	4.25ab	3.12a	3.07a	4.34b	4.62b
2	2.54ab	2.22a	2.30ab	2.48ab	2.58b
3	1.89bc	1.65a	2.12d	2.03cd	1.94bcd
4	2.03ab	2.25bcd	2.57d	2.34cd	2.14abc
5	2.94ab	2.40a	4.88c	3.62b	2.78a
6	2.19b	2.20b	3.21c	2.33b	1.76a
7	1.56ab	1.73b	2.26c	1.83b	1.30a
Total	17.11c	15.57ab	20.26d	18.08d	16.96bc

Significance tests not strictly valid; means * based on many fewer replications than other treatments (4 vs 36).

old, may be at least 100–150 years old (owner's personal communication), and in it, McNeilly & Roose (1984) recorded only 7 perennial ryegrass genotypes in 1 metre square, the individuals being large and therefore presumably of considerable age. Yet genotype pairs from this pasture showed no evidence of consistently greater yields from mixed stands, or of balanced component yields, than neighbour pairs from the 10-year-old pasture. Elimination of one of the genotypes from one of the pairs of neighbours seems likely to have occurred had the experiment continued, despite the fact that these clones must, from their size, have co-existed in the wild for a considerable time. Since rates of movement of *L. perenne* plants estimated from McNeilly & Roose (1984) are small, and there is general agreement that the potential for seedling establishment in established grassland is small (Harper, 1977), the origin of any co-adapted individuals must lie within the gene pool of the seed used to establish the pastures examined. The two ryegrass varieties S23 and S24 sown to produce the 10-year-old pasture are significantly diverse in a number of important agronomic traits. The seed rates used to establish the 10-year-old pasture had the potential to produce 425 seedlings per 0.25 m^{-2}, although observations of similar pastures 5 months after establishment revealed a mean seedling density of 175 m^{-2}, seedlings being of a size at which inter-genotype competition would have been negligible (McNeilly, unpublished data). For the evolution of co-adapted neighbours within such a system, we must assume that within this relatively small number of genotypes there must exist sufficient variation in characters of co-adaptive significance to allow natural selection to bring about adaptive changes in population structure. We must further assume that those characters which promote fitness during sward coalescence are those which enhance the fitness of competing adult plants in the mature sward. This seems unlikely, and it has been argued that different characters or character combinations may be of importance at different stages in the life of the plant (Aarsen, 1983). The possible or even probable lack of appropriate genetic variation in characters/characteristics of importance in the evolution of co-adaptation within neighbouring groups, and the high probability that characters promoting seedling and adult fitness do not correlate, suggests that the evolution of co-adapted groups of neighbours is not likely to be a general outcome amongst pasture components.

The situation is made even more complex under grazing management as a consequence of the effects of defoliation, treading, and the deposition of dung and urine by grazing animals (Curll & Wilkinson, 1983). Over a 10 year period, Petersen et al. (1956) concluded that 93–94% of a cattle grazed sward would have been defaecated upon at least once, and 15% would have been defaecated upon four times. These factors would cause the elimination of whole or parts of individual plant genotypes, and would create considerable heterogeneity in soil nutrient status and organic matter, all of which would seem likely to reduce the potential for evolution of co-adapted groups of genotypes, since these effects will be random spatially and temporally.

From analysis of the data of Van den Bergh (1968) and Hill (1974) on competition and co-operation between binary mixtures of cultivars and genotypes it can be concluded that no clear and consistent advantages of mixtures over performance in pure stand of the better component of mixtures occurs; the occasional advantages which did occur were the product of transitory density dependent effects.

The data presented here about competitive interactions between individuals clearly identified as neighbours, suggest that mixtures show some advantage over pure stands (RYT > 1 for 111 out of 189 mixtures). They do not, however, provide any evidence of co-adaptation between neighbours, and support the view of Snaydon (1978) that studies of natural mixed populations have not provided significant evidence of co-adaptation between grass species genotypes. It is possible however that these findings may reflect the consequences of selection for increased ability to persist, and for a greater degree of phenotypic diversity, which results in somewhat reduced niche overlap in general between surviving individuals.

References

Aarsen, L.W., 1983. Ecological combining activity and competitive combining ability in plants; towards a general evolutionary theory of coexistence in plants. Am Nat 122: 707–731.

Aarsen, L.W. & R. Turkington, 1985. Competitive relations among species from pastures of different ages. Can J Bot 63: 2319–2325.

Allard, R.W. & J. Adams, 1969. Population studies in predominantly self-pollinating species XII. Intergenotypic competition and population structure in barley and wheat. American Nat 103: 621–645.

Antonovics, J., 1978. Population genetics of mixtures. In: J.R. Wilson (Ed), Plant Relations in Pastures, pp. 233–252. CSIRO, Melbourne, Australia.

Brougham, R.W., 1960. The effects of frequent hard grazings at different times of the year on the production and species yields of a grass/clover pasture. N Z J Agric Res 2: 1232–1248.

Burdon, J.J., 1980. Intra-specific diversity in a natural population of *Trifolium repens*. J Ecol 68: 717–735.

Charles, A.H., 1961. Differential survival of grass cultivars of *Lolium, Dactylis* and *Phleum*. J Brit Grassl Soc 16: 69–75.

Charles, A.H., 1965. Differential survival of plant types in swards. J Brit Grassl Soc 20: 198–204.

Charles, A.H., 1966. Variation in grass and clover populations in response to agronomic selection pressures. Proc X International Grassl Cong pp 625–629. Helsinki, Finland.

Clements, R.J., M.D. Hayward & D.E. Blyth, 1983. Genetic adaptation in pasture plants. In: J.G. McIvor & R.A. Bray (Eds), Genetic Resources of Forage Plants, pp 101–115. CSIRO, Melbourne, Australia.

Curll, M.L. & R.J. Wilkins, 1983. The comparative effects of defoliation, treading, and excreta on a *Lolium perenne-Trifolium repens* pasture grazed by sheep. J Agric Sci 100: 451–460.

Evans, D.R., J. Hill, T.A. Williams & I. Rhodes, 1985. Effects of coexistence on the performance of white clover-perennial ryegrass mixtures. Oecologia 66: 536–539.

Harper, J.L., 1977. Population Biology of Plants. London Academic Press.

Hill, J.L., 1974. Methods of analysing competition with special reference to herbage plants iii. Monoculture v binary mixtures. J Agric Sci 83: 57–65. Hubbard, C.E., 1978. Grasses. Penguin Books, U.K.

McNeilly, T. & M.L. Roose, 1984. The distribution of perennial ryegrass genotypes in swards. New Phytol 98: 503–513.

Petersen, R.G.K., H.L. Lucas & W.W. Woodhouse, 1956. The distribution of excreta by freely grazing cattle and its effect on pasture fertility. 2 Effect of returned excreta on the residual concentration of some fertilizer elements. Agron J 48: 444–449.

Rossiter, R.C., 1966. The success or failure of strains of *Trifolium subterraneum* in a Mediterranean environment. Aust J Agric Res 17: 425–446.

Seaton, A.P.C. & J. Antonovics, 1967. Population inter-relationships I. Evolution of *Drosophila* mixtures. Heredity 22: 253–269.

Snaydon, R., 1978. Genetic changes in pasture populations. In: J.R. Wilson (Ed.), Plant Relations in Pastures. pp 253–269. CSIRO, Melbourne, Australia.

Snaydon, R., 1991. Replacement or additive designs for competition studies? J Appl Ecol 28: 930–946.

Turkington, R. & J.L. Harper, 1979. The growth distribution and neighbour relationships of *Trifolium repens* in a permanent pasture. IV Fine scale biotic differentiation. J Ecol 67: 245–254.

Van den Bergh, J.P., 1968. An analysis of yields of grasses in mixed and pure stands. Versl Landb Onderz Wageningen, 714.

de Wit, C.T., 1960. On competition. Versl Landb Onderz Nederland 66: 1–82.

de Wit, C.T. & J.P. van den Bergh, 1965. Competition between herbaceous plants. Neth J Agr Sci 13: 212–221.

P.M.A. Tigerstedt (Ed.), Adaptation in Plant Breeding, 143–148, 1997.

Breeding for yield, in mixtures of common beans (*Phaseolus vulgaris* L.) and maize (*Zea mays* L.)

Maria Jose de Oliveira Zimmermann
CNPAF/EMBRAPA, Cx. P. 179, 74001-970, Goiânia, Goiás, Brazil

Key words: common bean, correlations, intercropping, maize, selection gains, selection methods, yields, *Phaseolus vulgaris*, *Zea mays*

Summary

Despite the growing industrialization, technification and transformation that is happening in the agriculture around the world, and despite that agricultural research has always concentrated its effort on sole crops, multiple cropping systems have historically been important for common bean production in tropical countries. The reasons for this fact, are economical and social, as well as biological. Bean breeders have always been questioned on their work, because the development of new varieties is usually done in sole crop, but the varieties are grown in either systems. This paper addresses a set of questions that are usually presented to the breeders, in light of the evidence obtained from many trials conducted in Brazil and in the U.S.A.: Will the genotypes bred for sole crop conditions, perform well when grown in intercrop; How different should a genotype be, for cultivation in intercropping compared to genotypes developed for sole crop conditions; Is there a need for special breeding programs for intercropping and How could a breeding program focus the question of multiple (associated) cropping?

Introduction

The growing of two or more crop species simultaneously during all or part of their life cycle is called intercropping. It is assumed that crop competition will occur during all or part of the crop growth (Andrews & Kassam, 1976). When more than one species are grown in the same area in a year, many different spacial and temporal patterns of overlappings may occur. It is used to call multiple cropping as a general term to designate all these types of arrangements. Since for common bean there are two types of intercrop (simultaneous planting and relay crop), the words 'mixed crop' are synonym of intercrop.

The common bean (*Phaseolus vulgaris* L.) is an annual crop, grown in most tropical countries. In tropical America and parts of Africa, the most traditional form of cultivation for bean is in intercrop with maize, coffee, sorghum, cassava, sugar cane or some other crops, where subsistence agriculture is practiced.

The most frequent species that is found intercropped with bean is maize (*Zea mays* L.). It is estimated that about 60% of the bean production in Latin America comes from intercrop, be it simultaneous intercrop or relay crop.

Due to tradition and to the fact that in the maize-bean associations, maize does not suffer much with the competition and bean is really an additional crop from the same area, intercrop will always surely keep some of its importance. The area decrease which is happening will probably continue until an equilibrium point is reached, below which no further decrease is expected.

The increasing concern on agricultural sustainability also favors the maintenance of intercropping systems, because: a) It is an efficient soil conservation practice due to the increased ground cover that it provides, and to the exploitation of different soil layers due to the different depth of the root systems of the two species and b) It is a safer and more stable system of agricultural production than sole crop, for small farms, where capital is limited and labor is available. If one crop fails, the other can still give some yield.

Despite the importance of mixed cropping for bean production in areas where bean is an important food crop, research and extension services have concentrated their efforts on sole crop. The obtention of improved cultivars was and is always done in sole crop. Even for data analysis and interpretations of more simple variety testing trials, the available methods are not very adequate (Austin & Marais, 1987; Balasubramaniam & Sekayange, 1990; Roman & Martinez, 1988).

Since in the bean-maize intercropping, bean is the species that suffers most with the competition, it is from the point of view of a bean breeder that this paper will try to focus. Bean breeders have always been questioned about their work where it relates to intercrop. The main questions that this paper will address in the following chapters, are: 1. Will the genotypes bred for sole crop conditions, perform well when grown in intercrop? 2. How different should a genotype be, for cultivation in intercropping compared to genotypes developed for sole crop conditions? 3. Is there a need for special breeding programs for intercropping? and 4. How could a breeding program focus the question of multiple cropping?

Will the genotypes bred for sole crop conditions, perform well when grown in intercrop?

There are many papers reporting on correlations among yields of bean cultivar in intercrop and sole crop conditions. In most cases they are significant, with high values and positive sign, meaning that the relative behavior of the cultivars and breeding lines is similar in the two systems (Antunes & Teixeira, 1982; Chagas & Aquino, 1981; Davis & Garcia, 1983; Francis et al., 1978a, 1978b; Furtado et al., 1992; Ramalho et al., 1983; Santa-Cecilia & Ramalho, 1982; Serpa & Barreto, 1982; Vieira & Aidar, 1984; Vieira & Ramos, 1992). Despite that, some significant genotype by cropping systems interactions have also been reported (Francis et al., 1978a; Paniagua, 1977). The interactions have been interpreted as an indication that not all genotypes that are good in intercrop were also good in sole crop and that the correlations may not be the whole story.

Using data from different experiments that had been reported in Brazil, Hamblin & Zimmermann (1986) calculated how successful was the selection, applied in sole crop, for the identification of the best performers for intercrop. They determined the percentage of selected genotypes in one system (sole crop) which coincide with the ones that would be selected in the alternate system, discarding the coincidence that would happen at random.

Selection efficiency formula used by Hamblin & Zimmermann (1986) is

$$Se\% = \frac{Y - A}{X - A} \times 100$$

Where: X = number of genotypes that were selected in sole crop. Y = number of genotypes that coincide, if selection was practiced in intercrop. A = number of coincidences expected to occur by random chance. It is assumed to be proportional to the selection intensity (i) and is calculated as: A = (i) X.

Table 1, using data from Hamblin & Zimmermann (1986), plus some more recent data, depicts the selection efficiencies for some different trials that were conducted in Brazil. In most cases, despite the high correlations, selection efficiency (selection coincidence) was lower than 50%. In some of the few cases that it was higher, when there were disease evaluation data available, yield and disease reaction data were highly correlated in the two cropping systems, indicating that the coincidence was more due to the differential disease reactions of genotypes than to general adaptations to the cropping systems.

The answer to the first question is that germplasm that was developed for sole crop may sometimes be good for mixed crop, but that is not always necessarily true. Disease resistance traits, confer advantage that is important for cultivars in both systems, but physiological adaptation characteristics may be specific.

How different should a genotype be, for cultivation in intercropping compared to genotypes developed for sole crop conditions?

There are some papers relating to traits that are conducive to adaptation to mixed cropping for bean and maize. Bean, being the species generally more affected by cultivation with maize, has been more studied in relation to traits that may improve adaptation to intercrop. Davis & Woolley (1993) mentioned that the traits required for intercropping, in bean and maize, are those that increase the complementarity between species. In the case of bushy beans, which are very weak competitors, traits that relate to a slightly increased competitivity would be required. Another suggestion would be the selection for more efficient patterns of resource use for both species.

Table 1. Selection efficiency % for mixed crop, considering a selection of the highest yielding 33% of cultivars in sole crop

# Cultivars	# Selected	# Alternate environment	# Expected by chance	Selection efficiency %	Correlation	References
19	6	4	2	50	0.88**	Francis et al., 1978a
20	6	2	2	0	0.51*	Francis et al., 1978a
20	6	3	2	25	0.81**	Francis et al., 1978b
20	6	1	2	- 25	0.41ns	Francis et al., 1978b
59	20	13	7	46	0.66**	Antunes & Teixeira, 1982
64	21	13	7	43	0.54**	Antunes & Teixeira, 1982
49	16	14	5	82	0.84**	Vieira & Aidar, 1984
40	13	7	4	33	0.65**	Santa-Cecilia & Ramalho, 1982
40	13	11	4	78	0.98**	Santa-Cecilia & Ramalho, 1982
40	13	5	4	11	0.64**	Ramalho et al., 1983
49	16	13	5	73	0.84**	Vieira & Ramos, 1992
49	16	13	5	73	0.81**	Vieira & Ramos, 1992

*, ** = Significant at the probability levels of 5 and 1% respectively.
ns = not significant.

Zimmermann et al. (1984b) studied the relative importance of some traits in segregating populations of common bean, in intercrop and sole crop. The same interactions that occurred for varieties were also reported for segregating populations and harvest index of beans was negatively related to yield in sole crop and positively in intercrop. All other reported phenotypic and genotypic correlations with grain yield (number of pods per plant, seeds per pod and 100 seed weight), were in the same direction in both systems. Such correlations indicate that it is important, when selecting plants for intercrop, to avoid a correlated reduction in harvest index in order not to obtain plants with excessive vegetative growth. They also show clearly that bean benefit from some increase in competitiveness for cultivation in intercrop.

Path coefficient analysis for simultaneous intercrop and sole crop (Zimmermann et al., 1984b), showed that the importance of some yield components for total grain yield changed with the system, with 100 seed weight having a larger direct effect in intercrop than the other components, while number of pods per plant had the larger direct effect in sole crop.

Evaluation of bean lines and selection for competitive ability can be done even in bean-bean mixtures. Guazzelli (1975) found differences for competitive ability even among small seeded, black bean lines. The same author selected plants for competitive ability in four populations and obtained lines that differed from one another for such trait (Guazzelli, 1976). Later they were tested for yield under relay crop, simultaneous intercrop and sole crop. Lines with higher competitive ability showed better performance in mixed cropping systems than the unselected lines (Vieira & Aidar, 1984; Guazzelli & Kluthcouski, 1988).

Competitive ability is related to vegetative growth. The economic product of a crop such as bean is the seed (reproductive growth). Selection for competitive ability may mean, to some extent, a decrease in the economical efficiency of the plant. These results clearly suggest that the intercropping environment is not being fully exploited by the current cultivars and some gains ought to be expected also from improvement in the management of the crops.

Considering that even for sole crop, there is not a defined ideotype for common bean, the answer to the question is still unclear. There are only a few indications of traits to pay attention to, in order to avoid some undesirable correlated responses of selection.

Is there a need for special breeding programs for intercropping?

This question has been addressed in a few papers in Brazil and other countries, for maize and for bean. For maize, results from Mahajan et al. (1990), suggest that selection may be done in sole crop conditions provided that a few features are considered when making the selections.

In the case of bean, Zimmermann et al. (1984a, 1984b, 1985) addressed the problem by performing

Table 2. Selection efficiency % of two selection criteria: Mean yield (M) and sole crop selection (S)

Source	Selection criteria	Test environment		
		Sole crop	Intercrop	Alternate
Antunes & Teixeira (1982)				
Year 1. Simultaneous int.	M	86	57	–
	S	–	–	43
Year 1. Relay crop	M	57	28	–
	S	–	–	14
Year 2. Simultaneous int.	M	86	57	–
	S	–	–	43
Year 2. Relay crop	M	43	86	–
	S	–	–	29
Santa-Cecilia & Ramalho (1982)				
Site 1. Simultaneous int.	M	78	66	–
	S	–	–	33
Site 2. Simultaneous int.	M	89	89	–
	S	–	–	78
Vieira & Ramos				
Simultaneous intercrop	M	82	73	–
	S	–	–	54
Relay crop	M	82	73	–
	S	–	–	63

genetic studies and selection gains under intercrop and sole crop conditions in segregating populations. Zimmermann et al. (1985) detected a greater number of significant genetic effects in intercrop than in sole crop for grain yield and harvest index of three common bean crosses. Standard unit heritabilities (Zimmermann et al., 1984a) for grain yield were larger in intercrop than in sole crop, indicating that selection gains may be more easily achieved for this system than for sole crop. Realized selection gains for the three populations, calculated for direct selection in each system, indirect gains for each system if selection was practiced in the alternate cultivation pattern and gains in each system for selection practiced based on the average of both, showed that indirect selection for either cultivation system was the one that gave the worse results of all. Selection based on the mean of both systems resulted, in most cases, in similar gains to direct selection. Direct selection for each system, as it was expected resulted in generally higher gains. Similar results are also shown here, with selection gains for mixed cropping being larger when selection was practiced based on the mean of both systems than selection made in sole crop only (Table 2).

A genotype that is indicated for cultivation in any area in Brazil, must have adequate performance in both systems. Selection for either system separated, would possibly result in indications of different cultivars for each system in a region. It is also hard to find any Brazilian state that has more than 500,000 ha in intercrop (number suggested by Davis & Woolley, 1993 as a number that would justify a full fledged breeding program).

The answer to the question is that, in most cases, it does not justify to establish a special breeding program for intercrop.

How could a breeding program focus the question of multiple cropping?

It has been previously suggested by Fyfe & Roges (1965) and by Harper (1967), that if two species are to be grown together, the best will be to breed both crops simultaneously. The same conclusion was achieved by Hamblin & Zimmermann (1986). Hamblin et al. (1976), had suggested a method, based on diallel design, that is difficult to apply. Geraldi (1983) used a 'diallel' approach and he was able to identify combinations that maximized economical returns. The problem with this and other 'diallel' methodologies, is the size of experiments needed to test all combinations of lines.

Davis & Woolley (1993) suggested that, the best approach would be to select each crop species, in early generations, intercropped with a representative of the other species and they also suggested the utilization of single hill plots (consisting of two maize and two bean plants) instead of larger plots. They also consider important to make 'on farm selection' for such system.

For Brazil, due to the importance of disease resistance for bean, it is still recommended that early generation selections be done in sole crop, for such traits with test of improved lines under the two cropping systems. A release will be done only for the lines that perform well in both.

Many different approaches may be tried to obtain genetically improved genotypes for mixed cropping, but many of them even though theoretically perfect, are awkard, difficult to obtain good evaluations, extremely large designs and of little applicability. Generally, selection should be from, as early generation as feasible, in the cropping system that the crop is to be grown.

Conclusions

Genotypes bred for sole crop may not be as good for mixed crop cultivation. Interactions among cultivars and cropping systems do occur. There is no clear answer, even for bean only, how different should a genotype be for one system compared to the other, because there is no defined ideotype for bean for either system. Traits like disease resistance are important for both. Full fledged breeding programs, specially devoted to intercrop are recommended only when that system occupies a very large area in the region. There is not a single answer to how a breeding program should be for mixed cropping. The preferred approach in Brazil is, to make early generation selections in sole crop for qualitative traits, and to make yield tests in the system or systems that the crop will be grown in the region.

References

Andrews, D.J. & A.H. Kassam, 1976. The importance of multiple cropping in increasing world food supplies. In: I.R. Papendick, P.A. Sanchez & G.B. Triplett (Eds), Multiple Cropping, pp 1–11. A.S.A. Special Publ 27, Amer Soc Agron, Madison, WI, USA.

Austin, M.N. & J.N. Marais, 1987. Methods of presenting intercropping results and preliminary results with *Zea mays* and *Phaseolus vulgaris*. S Afr J Plant Soil 4 (1): 1–6.

Antunes, I.F. & M.G. Teixeira, 1982. Produtividade de genotipos de feijão em monocultivo e no cultivo associado com milho nas épocas das águas e da seca em Goiânia, GO. EMBRAPA. Anais 1 RENAFE. 83–88.

Chagas, J.M. & A.R.L. Aquino, 1981. Yield of climbing bean cultivars under two cropping systems. Ann Rep Bean Improv Coop 24: 56–57.

Davis, J.H.C. & S. Garcia, 1983. Competitive ability and growth habit of indeterminate beans and maize for intercropping. Field Crops Res 6: 59–75.

Davis, J.H.C. & J.N. Woolley, 1993. Genotypic requirement for intercropping. Field Crops Res 34: 407–430.

Francis, C.A., M. Prager, D.R. Laing & C.A. Flor, 1978a. Genotype × environment interactions in bush bean cultivars in monoculture and associated with maize. Crop Sci 18: 237–242.

Francis, C.A., M. Prager & D.R. Laing, 1978b. Genotype × environment interactions in climbing bean cultivars in monoculture and associated with maize. Crop Sci 18: 242–246.

Furtado, M.R., A.A. Cardoso, C.D. Cruz, C. Vieira & L.A. Pedernelli, 1992. Linhagens de feijão obtidas no monocultivo e no consórcio com a cultura do milho e seu comportamento nesses sistemas de cultivo. Revista Ceres, U.F.V. Brazil 223: 276–289.

Fyfe, J.L. & H.H. Rogers, 1965. Effects of varying variety and spacing on yields and composition of mixtures of lucerne and tall fescue. J Agr Sci (Cambridge) 64: 351–359.

Geraldi, I.O., 1983. Método de Análise Estatística para Combinacão de Cultivares em Consórcio. Dr Thesis, ESALQ, USP, Piracicaba SP, Brazil. 120 pp.

Guazzelli, R.J., 1975. Competicão Intergenotípica em Feijão (*Phaseolus vulgaris* L.): Estimacão da Capacidade Competitiva. M Sc Thesis, ESALQ, USP Piracicaba, SP, Brazil 62 pp.

Guazzelli, R.J., 1976. Competicão Intergenotípica em Feijão (*Phaseolus vulgaris* L.): Estimacão de Parâmetros Genéticos. Dr Thesis, ESALQ, USP Piracicaba, SP, Brazil. 88 pp.

Guazzelli, R.J. & J. Kluthcouski, 1988. Avaliacão de Germoplasma de Feijão (*Phaseolus vulgaris* L.) para o Consórcio Simultâneo com Milho. EMBRAPA-CNPAF. Comunicado Técnico 21F, 19 pp.

Hamblin, J., J.G. Rowell & R. Redden, 1976. Selection for mixed cropping. Euphytica 25: 97–106.

Hamblin, J. & M.J.O. Zimmermann, 1986. Breeding common bean for yield in mixtures. In: J. Janick (Ed.) Plant Breeding Reviews 4: 245–272. AVI Publ Co Westport, Conn.

Harper, J.L., 1967. A Darwinian approach to plant ecology. J Ecol 55: 247–270.

Mahajan, V., M. Singh, B.K. Sharma & C.S. Patel, 1990. Identification of characters to breed varieties of maize for intercropping with dry bean. Crop Improv 17 (2): 150–154.

Paniagua, C.V., 1977. Identification and stability analysis of traits important to yield of beans in associated culture. PhD Thesis, Michigan State University, East Lansing, MI. 76 pp.

Ramalho, M.A.P., A.C. Oliveira & J.C. Garcia, 1983. Recomendacões para o Planejamento e Análise de Experimentos com as Culturas de Milho e Feijão Consorciadas. EMBRAPA-CNPMS. Documentos 2. 74 pp.

Roman, A. & R. Martinez, 1988. Metodologia para evaluar progenies F_5 a partir de selecciones individuales F_4 de fríjol voluble en el sistema de relevo con maíz. Agronomia Colombiana, V: 8–16.

Santa-Cecilia, F.C.S. & M.A.P. Ramalho, 1982. Comportamento de cultivares de feijão em monocultivo e em associacão com milho. Ciência e Prática 6: 45–54.

Serpa, J.E.S. & A.C. Barreto, 1982. Competicão de cultivares de feijão em consorciacão com o milho nas micro regioes homogêneas 123 e 130 do estado do Sergipe. EMBRAPA-UEPAE Aracaju. Pesquisa em Andamento n. 6. 6 pp.

Vieira, R.F. & H. Aidar, 1984. Avaliacão de cultivares e linhagens de feijão em consórcio com o milho e em monocultivo. EMBRAPA-CNPAF Pesquisa em Andamento n 51. 7 pp.

Vieira, R.F. & J.A.O. Ramos, 1992. Avaliacão de cultivares de Feijão no Consórcio com Milho em Goianira, Goiás. Revista Ceres 39: 491–506.

Zimmermann, M.J.O., A.A. Rosielle & J.G. Waines, 1984a. Heritabilities of grain yield of common bean in sole crop and in intercrop with maize. Crop Sci 24: 641–644.

Zimmermann, M.J.O., A.A. Rosielle, J.G. Waines & K.W. Foster, 1984b. A heritability and correlation study of grain yield, yield components and harvest index of common bean in sole crop and intercrop. Field Crops Res 9: 109–118.

Zimmermann, M.J.O., A.A. Rosielle, K.W. Foster & J.G. Waines, 1985. Gene action for grain yield and harvest index of common bean grown in sole crop and intercrop with maize. Field Crops Res 12: 319–329.

P.M.A. Tigerstedt (Ed.), *Adaptation in Plant Breeding*, 149–152, 1997.

Breeding components for mixture performance

J. Hill
Department of Agricultural Sciences, Royal Veterinary and Agricultural University, Copenhagen, Denmark; Present address: Broneirian, Llanbadarn Road, Aberystwyth, Dyfed SY23, 1HB, Wales, U.K.

Key words: breeding mixture components, coevolution, ecological combining ability

Summary

The problems posed by breeding components for use in binary mixtures are discussed. Breeding and selection strategies designed to overcome some of these problems are outlined.

The problems

Before a breeder embarks upon a breeding and selection programme he would, as a matter of routine, define his objectives. In particular what would be the likely end-use of the potential cultivar; what, in other words, would be its agricultural niche. Defining his objectives allows the breeder to choose a base population and breeding and selection strategies appropriate for their fullfilment. Thus, for example, breeding for disease or drought resistance would require that potential varieties are selected in environments prone to these biotic and abiotic stresses. Yet, when it comes to the breeding of crop varieties destined for use in mixtures, rarely if ever is one component of the mixture evaluated in the presence of the other component(s). Rather is the breeding process aimed at identifying superior genotypes on the basis of their performance in pure stands. It is only afterwards that the putative components are brought together in the hope, often misplaced, that a superior mixture will result. One reason for the failure of this haphazard process is that yielding ability, which is required for high monoculture performance, is not necessarily the same as competitive ability (Sakai, 1961), which is the attribute determining performance in mixtures. Competitive ability may be defined as the ability of one component to obtain limited resources when grown in mixtures with another component, compared to its ability to secure those same resources when grown in pure stands (modified from Snaydon, 1991). Monoculture performance is not necessarily therefore a reliable guide to behaviour in mixtures, and indeed, yielding and competitive ability may even be inversely related. Moreover, in monocultures only intra-component competition is present, whereas in mixtures both intra- and inter-component competition exists. Stresses induced by intra-component competition are expected to be the greater because similar spatial and temporal demands are being made upon the available environmental resources (Mather, 1961; Caligari, 1980).

So what are the requirements of those breeding and selection strategies designed to produce varieties destined for use in mixtures, such as the grass/clover mixtures encountered in temperate pastures, or the mixed cropping systems employed in many developing countries? Clearly the main pre-requisite is to conduct such programmes in the presence of the other component, if we consider just binary mixtures, since only then is it likely that genotypes will 'possess biological characters conducive to synergism' (Allard & Adams, 1969). Allard & Adams were able to substantiate this view from their own work with barley mixtures: selection in such mixed populations they observed 'favours the survival of genotypes that are good competitors and at the same time good neighbours,' thereby improving what is often referred to as the ecological combining ability (Harper, 1967; Allard & Adams, 1969) of the components concerned. By analogy with the well-known genetic concept of combining ability, ecological combining ability may be divided into general and specific. And, again by analogy with Sprague & Tatum's (1942) original definitions of the genetic concept, the general ecological combining ability (g.e.c.a.) of a competitor may be defined as the average performance of all mixtures based on that competitor, while specific combining ability (s.e.c.a.) relates to those mixtures which perform significantly better or worse than expected on the basis of the average of its constituent components (Hill, 1990). Thus the yield of a binary mixture between the ith competitor and the jth associate (Y_{ii}) can be written as

$$Y_{ij} = \mu + c_i + a_j + c_i a_j + e_{ij}$$

where μ is the overall mean, c_i is the g.e.c.a. of the ith competitor, a_i is the g.e.c.a. of the jth associate, $c_i a_i$ is the s.e.c.a. of the ith competitor and the jth associate, and e_{ii} is an error term. Note that this model measures neither the competitive ability of the individual components nor the relative magnitudes of intra- and inter-component competition as such. This latter aspect of competition in particular requires an entirely different experimental approach, which has been reported in detail elsewhere in the literature.

Breeding and selection strategies

Two broad breeding strategies can be envisaged. In the first, attention is focussed on one of the components only. For the sake of convenience this will be referred to as the passive approach. In the second both components may be improved using alternating cycles of selection. This will be termed the active approach. The passive approach lends itself to a North Carolina 2 mating design, in which a number of genotypes from one component could be assessed in all possible binary combinations against a set of testers or associates drawn from the other component (Figure 1). Ideally testers should not be strong competitors, otherwise differences between the genotypes under test may go undetected. Equally the testers should not be so weak as to offer no competitive stress to the genotypes. Material selected on the basis of its general ecological combining ability may then be polycrossed and re-tested against the associates in further cycles of selection. Specific ecological combining ability would also be revealed by this mating design.

In the active approach, the roles of tester and tested are reversed in alternate cycles of selection (Figure 1), but apart from this the basic strategy is the same as for the passive approach. Both a North Carolina 2 or a diallel mating design could be used. The former design would be appropriate for binary mixtures based on different species, such as grass/clover, while the latter is suited to the evaluation of different genotypes or cultivars of the same species. If a diallel design is employed it should preferably be based on Griffing's experimental method 4 (Griffing, 1956), as this would avoid any complications arising from estimates obtained from both monocultures and mixtures.

Furthermore, a diallel should be reserved for the later phases of any breeding programme, by which time the number of components has been whittled down to manageable proportions and when specific ecological combining ability is being sought.

Coevolution and ecological combining ability

So far no mention has been made of coevolution. It is axiomatic that species or genotypes must coexist before they can coevolve. According to Janzen (1980) coevolution occurs when a trait of one species evolves in response to a trait of another species, which trait has itself evolved in response to the trait in the first species. The process is both reciprocal and specific. If only one species diverges, coevolution *sensu stricto* has not occurred (Futuyma & Slatkin, 1983). Convincing experimental evidence of coevolution is hard to find (Snaydon, 1985), as opportunities for independent plant species to coevolve will depend upon their exploitation of the available niches. Similarly, Rhodes (1981) suggested that improved ecological combining ability may be achieved by increasing both the temporal and spatial compatibility of the components. Both coevolution and ecological combining ability depend therefore on the availability and exploitation of suitable niches, such that the competitive stresses between the components are somehow avoided, minimized or tolerated (Hill et al., 1986). The distinction between them is that the former is a process which occurs naturally and often passes unnoticed, while the latter is an attribute which can be bred for, given the appropriate strategy.

In determining whether coevolution has occurred there is a danger of applying agronomic rather than genetic criteria. Coevolution is primarily concerned with improving the genetic fitness of the participants, not with maximizing their agricultural productivity. As Harper (1978) states 'many of the processes that favour members of one species over another in respect of leaving more descendants than their neighbours, may lead in a direction quite opposite to that of agronom-

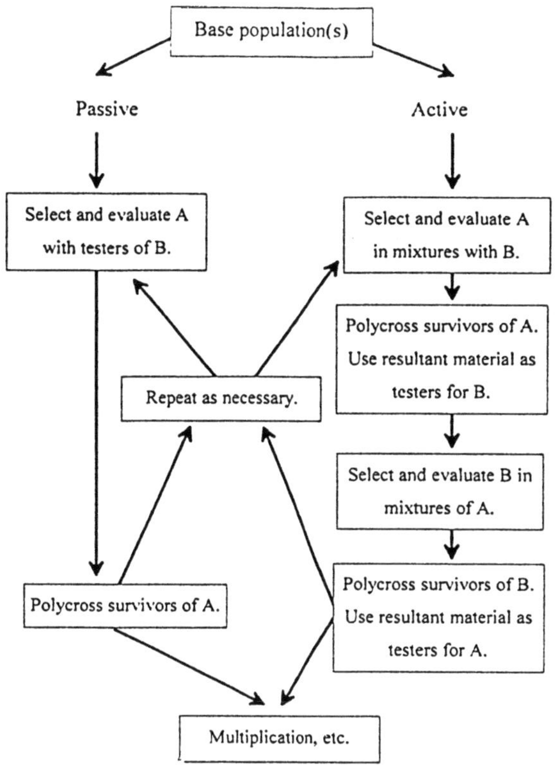

Figure 1. Outline of the active and passive breeding and selection strategies for improving ecological combining ability (after Hill, 1990).

ic optimisation.' Although the evolutionary and plant breeding concepts of coevolution may not be synonymous, it could be argued that artificial selection may enforce coevolution for agronomic traits, since these traits could be related to fitness under these conditions. Greater synonymity between these two concepts may also result if the character concerned is itself related to fitness, such as seed yield, as opposed to a vegetative character such as dry matter production. Thus, although Aarssen & Turkington (1985), Evans et al. (1989) and others observed increased yields from grass/clover mixtures based on components which had previously coexisted, it comes as no surprise to discover that not all such mixtures behave in this way (Collins & Rhodes, 1989). Nevertheless, mixtures based on components having a history of coexistence offer a stopgap until material developed by the strategies suggested earlier becomes available.

The art – or science – of plant breeding could perhaps be construed as the ability to re-assort in decades what it has taken nature centuries to build, in order to accumulate as many agronomically desirable traits as possible into a single variety. It is no easy task; developing varieties specifically for use in mixtures will be no easy task, but then whoever said plant breeding was easy?

Acknowledgement

I am indebted to Dr. E.L. Breese for reading and commenting upon this script.

References

Aarssen, L.W. & R. Turkington, 1985. Biotic specialization between neighbouring genotypes in *Lolium perenne* and *Trifolium repens* from a permanent pasture. J Ecol 73: 605–614.

Allard, R.W. & J. Adams, 1969. Population studies in predominantly self-pollinating species. XIII. Intergenotypic competition and niche differentiation in barley and wheat. Am Nat 103: 621–645.

Caligari, P.D.S., 1980. Competitive interactions in *Drosophila melanogaster*. Heredity 45: 219–231.

Collins, R. & I. Rhodes, 1989. Yields of white clover populations in mixture with contrasting perennial ryegrasses. Grass Forage Sci 44: 111–115.

Evans, D.R., J. Hill, T.A. Williams & I. Rhodes, 1989. Coexistence and the productivity of white clover-perennial ryegrass mixtures. Theor Appl Genet 77: 65–70.

Futuyma, D.J. & M. Slatkin, 1983. Coevolution. Sinauer Associates, Sunderland.

Griffing, B., 1956. Concept of general and specific combining ability in relation to diallel crossing systems. Aust J Biol Sci 9: 463–493.

Harper, J.L., 1967. A Darwinian approach to plant ecology. J Ecol 55: 247–270.

Harper, J.L., 1978. Plant relations in pastures. In: J.R. Wilson (Ed.), Plant Relations in Pastures, pp 3–14. CSIRO, East Melbourne.

Hill, J., 1990. The three C's – competition, coexistence and coevolution – and their impact on the breeding of forage crop mixtures. Theor Appl Genet 79: 168–176.

Hill, J., K. Mather, R.M. Chirwa & P.D.S. Caligari, 1986. The analysis of competitive ability in forage crops. In: M.J. Kearsey & C.P. Werner (Eds), Proc. Sixth Meeting Eucarpia Section Biometrics in Plant Breeding, University of Birmingham, UK, 28 July–1 August, 1986, pp 193–204.

Janzen, D.H., 1980. When is it coevolution? Evolution 34: 611–612.

Mather, K., 1961. Competition and co-operation. Symp Soc Expl Biol 15: 264–281.

Rhodes, I., 1981. The physiological basis of variation in the yield of grass/clover mixtures. Br Grassl Soc Occas Symp 13: 149–161.

Sakai, K.I., 1961. Competitive ability in plants: its inheritance and some related problems. Symp Soc Exp Biol 15: 245–263.

Snaydon, R.W., 1985. Aspects of the ecological genetics of pasture species. In: J. Haeck & J.W. Woldendorp (Eds), Structure and Functioning of Plant Populations. 2. Phenotypic and genotypic variation in plant populations, pp 127–152. North Holland, Amsterdam.

Snaydon, R.W., 1991. Replacement or additive designs for competition studies? J Appl Ecol 28: 930–946.

Sprague, G.F. & L.A. Tatum, 1942. General vs specific combining ability in single crosses of corn. J Am Soc Agron 34: 923–932.

P.M.A. Tigerstedt (Ed.), *Adaptation in Plant Breeding*, 153–159, 1997.

Pines beyond the polar circle: Adaptation to stress conditions

Outi Savolainen
Department of Biology, University of Oulu, PO Box 400, FIN 90571 Oulu, Finland

Key words: adaptation, DNA-markers, marginal populations, *Pinus sylvestris*

Summary

Plant populations in extreme conditions differ from those in more favorable conditions in many respects. They have developed special physiological adaptations for withstanding e.g. extreme temperatures or drought. On the other hand, low survival and fertility may cause smaller effective population sizes, which in turn influences patterns of genetic variation. Scots pine has the widest range of all pine species, extending in the north far beyond the arctic circle. Close to the tree limit, the growing season is short and survival is low. Very northern populations also have lower and more irregular male and female flowering and seed production than southern ones. The consequences of these conditions on genetic variation have been examined. All Finnish populations have similar patterns of variation at allozyme loci and nuclear DNA markers: high variation but no differentiation between populations. Northern populations seem to have slightly fewer recessive deleterious alleles than southern ones. There is, however, strong differentiation between populations with respect to adaptation to the very short growing season. In common garden experiments, northern populations cease growth and set buds much earlier and develop frost tolerance earlier than southern populations. The genetic basis for this high differentiation in bud set can be explored by using dense maps of molecular markers. The implications of these patterns of variations on conservation of genetic resources and tree breeding will be discussed.

Introduction

Distribution limits of plants are probably often set by constraints of the physical environmental (e.g. Woodward, 1995). Close to the distribution borders, plants live in marginal conditions, where populations can barely survive. Such environmentally limiting conditions can be due to extreme temperatures, to drought, or to edaphic conditions. Some classical studies of stress imposing environments include those on the heavy metal tolerance of *Anthoxanthum tenuis* (Bradshaw, 1972). Many later studies have been made by comparing physiological properties of plants in different environments, for trees e.g. Abrams (1994).

The environmental conditions that cause stress are diverse, but the populations do nevertheless share some evolutionary features. The extreme conditions often require specific ecophysiological adaptations. In some cases, the actual mechanisms of stress adaptation may be related, as in some cases of cold and desiccation tolerance (Woodward, 1995). The maintenance of the physiological adaptations must be reconciled with the rest of the life of the organism: there are structural and functional constraints within individual organism (Gould & Lewontin, 1979). The evolutionary potential of the trait will depend on its genetics: how many loci, what kinds of gene action, e.g. dominance or overdominance. Genetic correlations with other traits will also pose limits. The adaptedness of populations will depend not just on selection but on other evolutionary factors as well. For instance, migration can introduce new variation in adaptive traits (e.g. Felsenstein, 1976).

Finally, the environmental conditions may be so severe that populations are less dense and fluctuate more than others in size or in annual seed production. Such influences will have immediate effects on the effective population sizes of the species. This may give rise to increased drift or inbreeding in small populations (e.g. Hedrick, 1983). These processes may in turn influence the genetic composition of the popula-

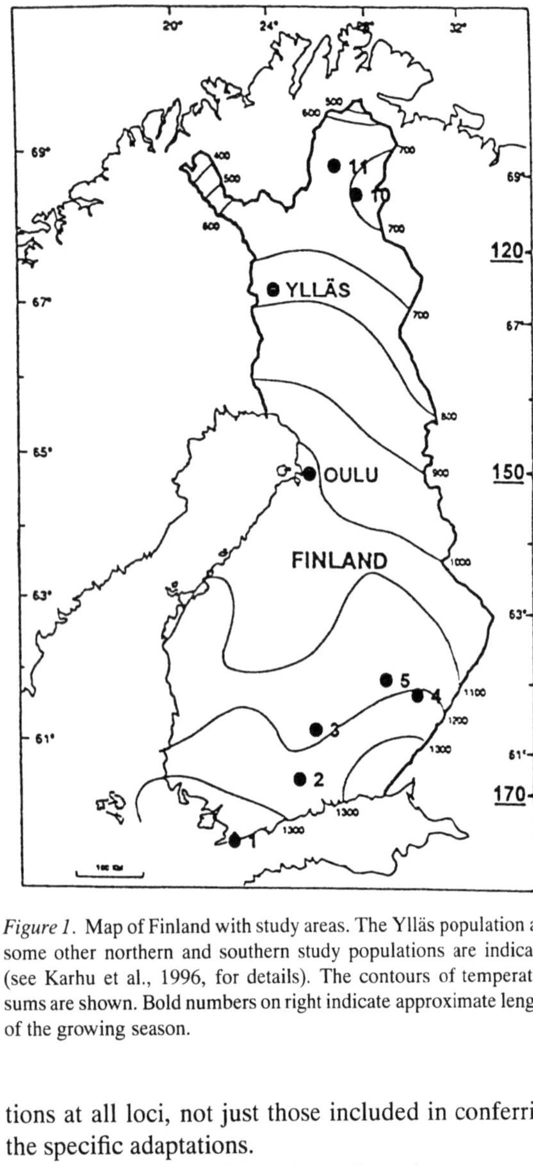

Figure 1. Map of Finland with study areas. The Ylläs population and some other northern and southern study populations are indicated (see Karhu et al., 1996, for details). The contours of temperature sums are shown. Bold numbers on right indicate approximate lengths of the growing season.

tions at all loci, not just those included in conferring the specific adaptations.

In this paper, I will explore adaptation to extreme conditions using *Pinus sylvestris* at the arctic circle as a case study. Comparison of northern and southern populations sheds light on the general problems of stressful environments.

Environment of pines at the polar circle

The distribution area of *Pinus sylvestris* is the widest of all pines, ranging from Scotland to Siberia and from Turkey to northern Scandinavia (Mirov, 1967). The limit of the distribution in Finland is between latitudes 68° N and 69° N. There are, however, scattered populations on the Arctic coast. The studies of the research group at the University of Oulu have mainly been made on different Finnish populations (Figure 1). The environmental gradients are steep in this area. At midsummer, the daylength in Helsinki (60° N) is about 16 h, while our intensively studied population at the Ylliästunturi mountain (67° 30′ N) has a daylength of 24 h. Relevant data on wheather conditions are shown in Table 1, based on Heino & Hellsten (1983). They show that the latitudinal temperature differences translate into a growing season difference of more than 50 days between the north and the south. Furthermore, the variability between years is also much more extreme in the north (see Luomajoki, 1993 for further information).

Ecology of the northern populations

The northern pine populations have colonized these areas fairly recently, after the ice receded some 9000 years ago. Between about 7000 and 5000 years before present, during warmer conditions, pines had a wider distribution in the north (Hyvärinen, 1987). In all this means that there have been no more than some hundred generations of pines in the north. The severe environmental conditions in the north influence the life cycle of the trees in many ways. For many tree species early stages of the life cycle, i.e. germination and seedling establishment are critical (Woodward, 1995; Graumlich & Brubaker, 1995). In northern Finland, years with high enough temperature sums for seed production are rare. The probability of production of a seed crop with 50% germination ability (due to seed maturation) is low, at the extreme range less than 5% (Henttonen et al., 1986). Survival of local populations is low in the north from the seedling to mean height 2.5 m (e.g. Persson, 1994). Survival of local populations in areas with temperature sums of 600 or less was below 25% while survival in areas with temperature sums over 1200 was higher than 75%. Transfers of seedlings to the north or higher up of their origin lowers survival, transfers to the south increase survival (Eiche, 1966; Eriksson et al., 1980). Northern populations grow much slower than southern ones (e.g. Ilvessalo, 1937). This is to a large degree due to shorter growing periods (e.g. Mikola, 1982). For example, the population at Yllästunturi (67° 30′ N, 450 m a.s.l.) has average age 140 years and average height 9.7 m (Savolainen & Hedrick, 1995). Beuker (1994) has studied old pine provenance experiments and found that a transfer to the south (60°) of northern provenances (66°) N could increase their

Table 1. Climatological data of Finland (Latitude, longitude, altitude (m)), January mean temperature (Jan), July mean temperature (July), depth of snow cover on April 15 (Snow), length of thermal growing season, i.e. number of days with mean temperature above 5° C (Grow)

Site	Lat	Long	Alt	Jan	July	Snow	Grow
Helsinki	60 19	24 58	6	- 6.9	16.4	05	174
Oulu	64 56	25 22	12	- 10.4	15.8	22	150
Ivalo	68 37	27 25	145	- 13.9	13.5	55	121

Table 2. Genetic diversity (H_e, [1]), phenotypic variation (Shannon-Weaver index, [2]), or total phenotypic variance in common garden experiment ([3]) and proportion of variation between populations for various characters in the northern (Ylläs) and southern (various) populations

Character	Southern	Northern	Between
Enzymes[1]	0.35	0.35	0.02
RFLP[1]	0.44	0.54	0.02
Microsatellites[1]	0.74	0.80	0.02
rDNA[2]	1.40	1.38	0.14
Bud set[3]	169.0	171.6	0.34

total wood production by up to 100%. This shows that environmental conditions limit tree growth severely. However, local southern provenances still grow better than the northern ones.

The general harshness of the climate is also reflected by timing of reproductive events. Chung (1981) and Luomajoki (1993) have shown that in common garden environments, northern provenances shed pollen and the female strobili become receptive much earlier than southern ones, after accumulation of a lower heat sum. In the short growing season reproduction has to start very early. Male flowering, as measured by anther residues, is also much lower and more variable between years than in the south (Koski & Tallqvist, 1978, see Figure 2). The northern forests also produce less female strobili and seeds than southern ones (Figure 2). Further, variation between years and individuals is more pronounced in the north (Koski & Tallqvist, 1978).

Genetic comparison of southern and northern populations

We have based our study of the genetics of stressed populations on a comparison between south and north. Comparisons of different populations and traits allow population genetic interpretations.

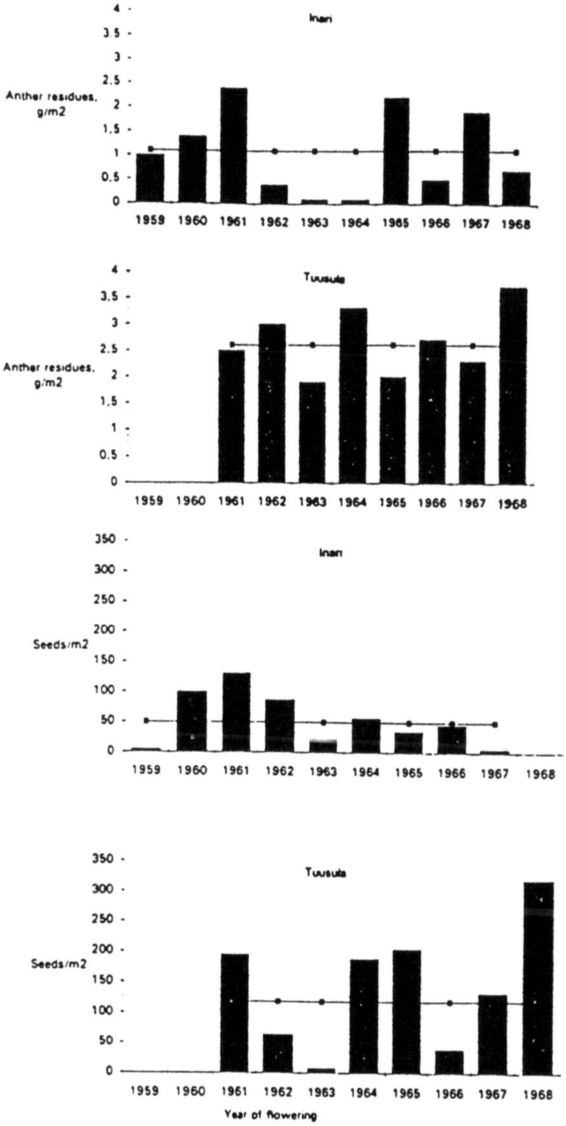

Figure 2. Pollen production (measured as g of anther residues/m^2) and seed production between 1961 and 1968 at Tuusula (lat. 60° N) and Inari (lat. 69° N), based on data given by Koski & Tallqvist (1978).

Marker genes

Several studies have shown that *Pinus sylvestris* is highly variable at isozyme loci (e.g. Gullberg et al., 1985; Szmidt & Muona, 1985; Muona & Harju, 1989). These and later studies have also shown that even the northernmost populations have as high levels of variability as more southern ones (Karhu et al., 1996). At these marker loci, populations are hardly at all differentiated (see Table 2). Earlier studies of Scots pine have shown that differences between such distant populations as Scandinavia and China account for only 12% of total variation (Wang & Szmidt, 1991). This is in accordance with data on other conifers, which also have ample genetic variation, but the differentiation between populations is minor (e.g. Hamrick & Godt, 1992). Also, Tigerstedt (1973) showed that there is equal variation of isozymes in marginal and central populations of *Picea abies*. Our recent work on molecular markers at the DNA level has provided similar results. Northern and southern populations are equally variable and there is little differentiation at RFLPs or microsatellite markers (Karhu et al., 1996, Table 2). Karvonen & Savolainen (1993) found some variation between populations in rDNA phenotype frequencies (14% between populations). This general pattern seems to hold for all of the nuclear DNA markers we have studied. Natural selection does not seem to distinguish between the different alleles, at least not differentially between the north and the south. Thus, the markers behave as neutral genes.

Quantitative variation

Common garden experiments show that northern and southern populations are differentiated with respect to those traits that confer adaptation to the northern harsh conditions (e.g. Mikola, 1982). Our own studies have focused on bud set dates and frost tolerance. As in earlier studies by Mikola (1982), we have found extensive differences between populations in the date of budset of first year seedlings. Planted under natural daylength in a glasshouse at the beginning of June, the seedlings set a terminal bud between August and October. In 1994, 36% of the total variation in an experiment containing samples from latitudes 60° N to 67° N was between populations (Karhu et al., 1996; Hurme et al., in prep.). Corresponding to the differences in bud set, the populations also differed in their development of frost tolerance (Hurme et al., in prep.). Northern populations develop better frost tolerance and more rapidly than southern populations (Aho, 1994; Hurme et al., in prep.).

Outcrossing and genetic load

In Scots pine, reproductive ecology and the genetic system are closely related. Conifers have no incompatibility systems, and self pollen can presumably cause fertilization just as efficiently as foreign pollen (Koski, 1971). The proportion of outcrossed seed at the mature seed stage can depend on the proportion of self pollen grains in the fertilizing pollen cloud and on the viability of selfed seed (Kärkkäinen & Savolainen, 1993). It is well known that most selfed embryos (on average about 85%) die before seed maturation (e.g. Koski, 1971). This very early severe inbreeding depression eliminates most selfed seeds in many conifers (e.g. Sorensen, 1982; Williams & Savolainen, 1996). There is, however considerable variation between trees in the proportion of empty (dead) seed upon selfing. This variation is probably due to variation in the number of deleterious and lethal recessives. Estimates of lethals range from a few per tree to up to 20 (Koski, 1971). We have also found that there is variation between populations in the number of lethals. While southern populations have on average 6.9 lethals, northern ones have 4.5 lethals (Kärkkäinen et al., 1996). Earlier studies have shown that seeds from northern populations have higher homozygosity than southern ones, maybe due to less mortality following self- fertilization (Szmidt & Muona, 1985). However, in later studies, the differences in outcrossing rates turned out to be small, and only one northern population had a significantly lower outcrossing rate than other populations (Kärkkäinen et al., 1995). In all, it seems that northern and southern populations have very similar genetic structures. The reasons for the differences in lethal numbers remain unknown, perhaps one possibility is the recent colonization and historical bottlenecks associated with early stages of colonization, or that the same alleles could be differentially selected in the north than in the south (Kärkkäinen et al., 1995). At present, it seems that large effective population sizes are maintained, and current populations do not have evidence of increased inbreeding or influence of drift in the north.

Genetic basis of the variation

It would be important to know the genetic basis of the variation for understanding its evolution (e.g. Barton & Turelli, 1989). For the quantitative traits, we would like to know the underlying loci and study their frequencies. Studies of this kind have become possible only recently, as dense molecular maps allow the detection of quantitative trait loci (see Tanksley, 1993). Such maps have already been constructed for some conifer traits (Groover et al., 1994). We are developing molecular maps for Scots pine (Hurme et al., in prep.) and are also searching for the QTL associated with bud set date.

The maintenance of the enzyme variation can be due to a mutation-drift balance in large populations. Others have suggested that overdominance of the loci may be responsible (Bush & Smouse, 1992). Our recent studies showed that heterozygosity and fitness were not at all related, which is more in accordance with the mutation-drift-migration balance (Savolainen & Hedrick, 1995). The results on random DNA markers are also in agreement with this hypothesis.

Why is there so much variation?

In the harsh conditions of the north, we assume that there is strong directional selection for the traits conferring adaption. Such selection should eliminate additive genetic variation at these traits. However, in all traits we have studied, there is a high level of genetic variation. This was found for the presumably neutral marker genes, but also for quantitative traits. Table 2 shows that in common garden experiments there is much phenotypic variation for bud set date. Measurement on additive genetic variation are not yet available. Cone and pollen production of northern populations, measured in common garden experiments, have as much total genetic variation as in the south (Savolainen et al., 1993). Seed maturation probabilities vary between individuals despite intense selection (Ryynänen, 1982; Harju et al., in prep.). Studies on pollen dispersal (e.g. Koski, 1970) also support the role of extensive migration and large effective population sizes in accounting for the high levels of variation. In each generation, natural selection may eliminate part of the variation. The influence of selection should be demonstrable as genetic changes over the life time of a cohort. However, such data are not available, and are difficult to obtain (e.g. Ruotsalainen, 1989). The role of migration in lowering adaptation could be studied by comparing seeds produced through open pollination and controlled crosses between local trees. On the other hand, Tigerstedt (1994) has suggested that variable selection maintains all the variation.

Implications for tree breeding and conservation of genetic resources

All these data together suggest that the population structure here can be accounted for by extensive migration between areas in continuous populations, and by strong differential selection for some traits. These results are of evolutionary interest, but also contain many implications for tree breeding. The clear genetic differentiation between the northern and southern populations shows that each area needs its own seed sources, as has been recognized by Finnish tree breeders for long. Seed orchards have been established in southern areas also for northern seed production. This program was initiated by Sarvas (e.g. 1970), after carefully considering the associated biology. This program has some unanticipated associated problems, such as background pollination by southern sources (e.g. Harju & Muona, 1989).

In the last two decades there have been many efforts to base monitoring of genetic resources on easily scored marker genes. In general, data on conifers show that marker genes have very different patterns of variation from adaptive variation, as the latter may be subject to strong differential selection (e.g. Muona, 1990; Yang et al., 1996). There are rare exceptions where the distribution of variation is similar in most traits, e.g. *Pinus monticola* (Rehfeldt, 1979). In most cases, studies of random isozyme or DNA markers cannot be a substitute for studying the polygenic variation of adaptive traits itself. Similar findings have been made in angiosperms species as well (e.g. Podolsky & Holtsford, 1995). There are now efforts underway to locate the genes underlying variation at the quantitative traits (Groover et al., 1994). It remains to be seen how many loci are in general responsible. If the number is low, it may be possible to find corresponding markers. If there are many loci with equal effects, finding single markers that will provide useful information on quantitative traits in populations will be difficult. Still, it may be possible to use these markers when individual crosses are studied (Grattapaglia & Sederoff, 1994).

Selection by tree breeders differs from natural selection. Silviculturists are not interested in total fitness, if seedlings are to be planted: reproductive fitness is then not important. Further, the importance of survival depends on the costs. While natural selection can waste seeds and seedlings to high mortality, silviculturists can not, and have to manipulate survival e.g. by transfer of expensive seedlings. A good understanding of the basic demography and genetics of natural populations in stressed conditions is a prerequisite for tree breeding for such areas.

Acknowledgements

This work reviews research that has been done in collaboration with students. I thank Anni Harju, Päivi Hurme, Matti Karjalainen, Auli Karhu, Päivi Karvonen, Helmi Kuittinen and Katri Kärkkäinen for their contributions. This work would not have been possible without the good collaboration of the Finnish Forest Research Institute and its staff, especially Veikko Koski. The research has been supported by the former Research Council for Agriculture and Forestry.

References

Abrams, M.D., 1994. Genotypic and phenotypic variation as stress adaptations in temperate tree species: a review of several case studies. Tree Physiology 14: 833–842.

Aho, M.-L., 1994. Autumn frost hardening of one-year old *Pinus sylvestris* (L.) seedlings: effect of origin and parent trees. Scand J For Res 9: 17–24.

Barton, N.H. & M. Turelli, 1989. Evolutionary quantitative genetics: how little do we know? Ann Rev Genet 23: 337–370.

Beuker, E., 1994. Long-term effects of temperature on the wood production of *Pinus sylvestris* L. and of *Picea abies* (L.) Karst in old provenance experiments. Scand J For Res 9: 34–45.

Bradshaw, A.D., 1972. Some evolutionary consequences of being a plant. Evolutionary Biology 5: 25–47.

Bush, R.M. & P.E. Smouse, 1992. Evidence for the adaptive significance of allozymes in forest trees. New Forests 6: 179–196.

Chung, M.S., 1981. Flowering characteristics of *Pinus sylvestris* L. with special emphasis on the reproductive adaptation to local temperature factor. Acta Forestalia Fennica 169: 1–69.

Eiche, V., 1966. Cold damage and plant mortality in experimental provenance plantations with Scots pine in Northern Sweden. Studia Forestalia Suecica 36: 1–218.

Eriksson, G., S. Andersson, V. Eiche, J. Ifver & A. Persson, 1980. Severity index and transfer effects on survival and volume production of *Pinus sylvestris* in northern Sweden. Stud For Suecica 156: 1–32.

Felsenstein, J., 1976. The theoretical population genetics of variable selection and migration. Ann Rev Genet 10: 253–280.

Gould, S.J. & R.C. Lewontin, 1979. The spandrels of San Marco and the Panglossian paradigm: a critique of the adaptationist programme. Proc R Soc Lond B 205: 581–598.

Grattapaglia, D. & R. Sederoff, 1994. Genetic linkage maps of *Eucalyptus grandis* and *Eucalyptus urophylla* using a pseudo-testcross: mapping strategy and RAPD markers. Genetics 137: 1121–1137.

Graumlich, L.J. & L.B. Brubaker, 1995. Long-term records of growth and distribution of conifers: integration of paleoecology and physiological ecology.

Groover, A., M. Devey, T. Fiddler, J. Lee, R. Megraw, T. Mitchell-Olds, B. Sherman, S. Vujcic, C. Williams & D. Neale, 1994. Identification of quantitative trait loci influencing wood specific gravity in an outbred pedigree of loblolly pine. Genetics 138: 1293–1300.

Gullberg, U., R. Yazdani, D. Rudin & N. Ryman, 1985. Allozyme variation in Scots pine (*Pinus sylvestris* L.) in Sweden. Silvae Genetica 34: 193–200.

Hamrick, J.L., M.J. Godt & S.L. Sherman-Broyles, 1992. Factors influencing levels of genetic diversity in woody plant species. New Forests 6: 95–124.

Harju, A. & O. Muona, 1989. Background pollination in *Pinus sylvestris* seed orchards. Scand J For Res 4: 513–520.

Heino, R. & E. Helsten, 1983. Climatological statistics of Finland 1961–1980. Meteorological yearbook of Finland. Vol 80 (Supplement, part 1a) pp 1–494.

Henttonen, H., M. Kanninen, M. Nygren & R. Ojansuu, 1986. The maturation of *Pinus sylvestris* seeds in relation to temperature climate in Northern Finland. Scand J For Res 1: 243–249.

Hedrick, P.W., 1983. Genetics of populations. Science Books International. Boston.

Hyvärinen, H., 1987. History of forests in northern Europe since the last glaciation. Ann Acad Scient Fenn Series A III. Geologica – Geographica 145: 7–18.

Ilvessalo, Y., 1937. Perä-Pohjola luonnon normaalien metsiköiden kasvu ja kehitys (Growth of natural normal stands in central North-Suomi (Finland)). Commun Inst For Fenn 24.2: 1–168.

Karhu, A., P. Hurme, M. Karjalainen, P. Karuonen, K. Kärkkäinen, D.B. Neale & O. Savolainen, 1996. Do molecular markers predict patterns of adaptive variation? Theor Appl Genet (in press).

Karvonen, P. & O. Savolainen, 1993. Variation and inheritance of ribosomal DNA in *Pinus sylvestris* L. (Scots pine). Heredity 71: 614–622.

Koski, V., 1970. A study of pollen dispersal as a means of gene flow. Commun Inst For Fenn 70: 1–78.

Koski, V., 1971. Embryonic lethals of *Picea abies* and *Pinus sylvestris*. Commun Inst For Fenn 75.3: 1–30.

Koski, V. & R. Tallqvist, 1978. Results on long-time measurements of the quantity of flowering and seed crop of forest trees. Folia Forestalia 364: 1–60.

Kärkkäinen, K. & O. Savolainen, 1993. The degree of early inbreeding depression determines the selfing rate at the seed stage: model and results from *Pinus sylvestris* (Scots pine). Heredity 71: 160–166.

Kärkkäinen, K., V. Koski & O. Savolainen, 1996. Geographical variation in the inbreeding depression of Scots pine. Evolution 50: 111–119.

Luomajoki, A., 1993. Climatic adaptation of Scots pine (*Pinus sylvestris* L.) in Finland based on male flowering phenology. Acta Forestalia Fennica 237: 1–27.

Mikola, J., 1982. Bud-set phenology as an indicator of climatic adaptation in Scots pine in Finland. Silva Fennica 16: 178–184.

Mirov, N.T., 1967. The genus *Pinus*. The Ronald Press Company, New York.

Muona, O., 1990. Population genetics in forest tree improvement. p. 282–298. In: A.H.D. Brown, M.T. Clegg, A.L. Kahler & B.S.

Weir (Eds). Plant Population Genetics, Breeding, and Genetic Resources. Sinauer, Sunderland, Mass.

Muona, O. & A. Harju, 1989. Effective population sizes, genetic variability, and mating system in natural stands and seed orchards of *Pinus sylvestris*. Silvae Genetica 38: 221–228.

Persson, B., 1994. Effects on provenance transfer on survival in nine experimental series with *Pinus sylvestris* (L.) in northern Sweden. Scand J For Res 9: 275–287.

Podolsky, R.H. & T.P. Holtsford, 1995. Population structure of morphological traits in *Clarkia dudleyana*. I. Comparison of F_{ST} between allozymes and morphological traits. Genetics 140: 733–744.

Rehfeldt, G.E., 1979. Ecotypic differentiation in populations of *Pinus monticola* in north Idaho – myth or reality. Amer Natur 114: 627–636.

Ruotsalainen, S., 1989. Variation of growth rhythm within pine stands. M.Sc. Thesis. Department of Genetics, University of Helsinki.

Ryynänen, M., 1982. Individual variation in seed maturation in marginal populations of Scots pine. Silva Fennica 16: 185–187.

Sarvas, R., 1962. Investigations on the flowering and seed crop of *Pinus sylvestris*. Commun Inst For Fenn 53.4: 1–198.

Sarvas, R., 1970. Establishment and registration of seed orchards. Folia Forestalia Fennica 89: 1–24.

Savolainen, O. & P.W. Hedrick, 1995. Heterozygosity and fitness: no association in Scots pine. Genetics 140: 755–766.

Savolainen, O., K. Kärkkäinen, A. Harju, T. Nikkanen & M. Rusanen, 1993. Fertility variation in *Pinus sylvestris*: a test of sexual allocation theory. Amer J Botany 80: 1016–1020.

Sorensen, F.C., 1982. The roles of polyembryony and embryo viability in the genetic system of conifers. Evolution 36: 725–733.

Szmidt, A.E. & O. Muona, 1985. Genetic effects of Scots pine (*Pinus sylvestris* L.) domestication. p. 241–252. In: H.-R. Gregorius (Ed). Population Genetics in Forestry. Lecture notes in Biomathematics 60. Springer Verlag, Berlin.

Tanksley, S.D., 1993. Mapping polygenes. Ann Rev Genet 27: 205–233.

Tigerstedt, P.M.A., 1973. Studies of isozyme variation in marginal and central populations of *Picea abies*. Hereditas 75: 47–60.

Tigerstedt, P.M.A., 1994. Adaptation, variation and selection in marginal areas. Euphytica 77: 171–174.

Yang, R.-C., F.C. & A.D. Yanchuk, 1996. Genetic structure of *Pinus contorta* spp. latiforlia as revealed by isozymes and quantitative traits. Genetics (in press).

Wang, X.-R., A.E. Szmidt & D. Lindgren, 1991. Allozyme differentiation among populations of *Pinus sylvestris*. Hereditas 114: 219–226.

Williams, C.G. & O. Savolainen, 1996. Inbreeding depression in conifers: implications for breeding strategy. Forest Science (in press).

Woodward, F.I., 1995. Ecophysiological controls of conifer distributions. In: W.K. Smith & T.M. Hinckley (Eds). Ecophysiology of Coniferous Forests. pp. Academic Press.

P.M.A. Tigerstedt (Ed.), *Adaptation in Plant Breeding*, 161–167, 1997.

CIMMYT's approach to breed for drought tolerance

Sanjaya Rajaram[1], Hans-Joachim Braun[2] & Maarten van Ginkel[1]
[1] *International Maize and Wheat Improvement Center (CIMMYT), Lisboa 27, Mexico D.F. 06600, Mexico;*
[2] *CIMMYT, P.K. 39 Emek, 06511, Ankara, Turkey*

Key words: breeding methodology, drought patterns, input-efficiency, wheat, *Triticum aestivum*

Summary

About 32% of the 99 million ha wheat grown in developing countries experiences varying levels of drought stress. Three major drought types have been identified: Late drought (LD) is common in the Mediterranean region, early drought (ED) is found in Latin America and wheat is produced on residual soil moisture (RM) in the Indian subcontinent and part of Australia. Until 1983, CIMMYT selected all germplasm under near optimum conditions for its yield potential and tested only advanced lines under drought. In spite of many critics, this approach proved to be successful, since in the mid 80's CIMMYT germplasm was grown on 45% of the wheat area in LC with annual rainfall from 300–500 mm and on 21% in areas with less than 300 mm. Since 1983, CIMMYT's drought breeding methodology is to alternate segregating populations between drought stressed and fully irrigated conditions (FI) and to test advanced lines under a line source irrigation system. To compare the efficiency of these approach, yield of four, mostly leading varieties, from each of the regions with LD, ED, RM, and FI and twelve recent CIMMYT cultivars selected for high yield under FI and RM conditions (ALT) were compared under four different moisture regimes (FI, LD, ED, and RM) in 89–90 and 90–91 in Yaqui Valley, Mexico. Genotypic correlation between yield and days to flowering, days to maturity, height, grains m^{-2}, TKW, test weight and grain fill period were calculated.

Mean grain yield of the four best lines in the ALT group was highest under all moisture stress regimes, followed by the FI-group. However, the highest yielding cultivar within each moisture regime was from the FI-group under FI, from the LD-group under LD, and from the ALT-group under ED and RM conditions. Estimates for genetic advance suggest that FI is the best environment for increasing grain yield even in all three drought environments. This indicates that yield potential per se is beneficial also in drought environments. The highest yield in drought environments was realized by the CIM cultivars selected under FI and RM. Simultaneous evaluation of the germplasm under near optimum conditions, to utilize high heritabilities and identify lines with high yield potential, and under stress conditions to preserve alleles for drought tolerance seem at present the best strategy.

Introduction

There has been a large transformation in the productivity of wheat due to the application of Green Revolution technology. This has resulted in a doubling and tripling of wheat production in many environments, but especially in irrigated areas. The high yielding varieties of semi-dwarf statured wheats have continuously replaced the older tall types at a rate of 2 million hectares per year since 1977. The overall value of this new germplasm has been estimated at an additional 3 billion dollars in 1990, which have most benefited the agricultural community in developing countries (Byerlee & Moya, 1993; Table 1; Figure 1).

There is a growing recognition that the dissemination, application and adoption of this technology has, however, been slower in marginal environments, especially in the semiarid environments affected by poor distribution of water and drought. The annual gain in genetic yield potential in drought environments is only about half that obtained in irrigated, optimum conditions. Many investigators have attempted to produce wheat varieties adapted to these semi-arid environments, however, with limited success. Others have

Table 1. Estimated effects of spring wheat breeding research on production by region in the post-Green Revolution period, 1977–90 (Byerlee & Moya, 1993)

	Sub-Saharan Africa	West Asia/ North Africa[a]	South Asia	Latin America	All
Total production increase in 1990 (million t)	0.15	2.45	9.34	3.40	15.34
Average wheat price (US$ 1990/t)	210	210	195	195	198
Total value of production increase in 1990 (US$ 1990 millions)[a]	31	515	1822	662	3030
Percent germplasm of CIMMYT origin[b]	39	52	44	60	49
Value of production increase attributed to CIMMYT (US$ 1990 millions)[b]	12	268	802	397	1485

[a] Excludes winter/facultative wheats.
[b] Varieties released since 1972 are weighted as follows: CIMMYT cross, 0.85, NARSs cross with CIMMYT parent, 0.50.

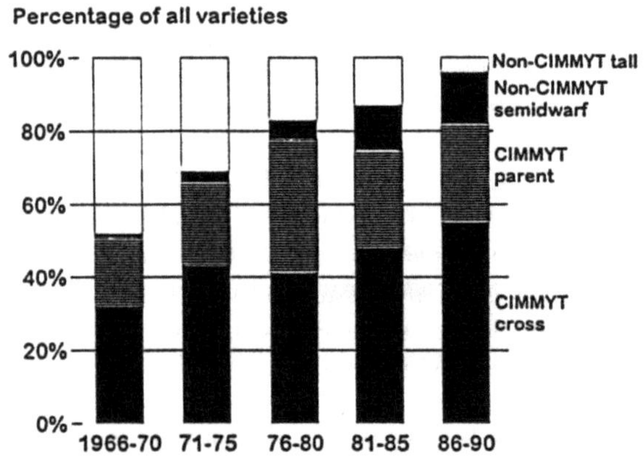

Figure 1. Trends in the origin of spring bread wheat varieties in developing countries, 1966–90.

criticized the Green Revolution technology for failing to adequately address productivity constraints in semi-arid environments, although their own recommended technology has had limited impact, in particular in farmer's fields (Ceccarelli et al., 1987). This criticism is in clear contrast to the actual acceptance of semi dwarf wheat cultivars in rainfed areas, since most of the 16 million ha increase in the area sown to Mexican semi dwarf wheats in the mid 80's occurred in rainfed areas and in 1990, more than 60% of the dryland area in developing countries are planted with semi dwarfs (Byerlee & Moya, 1993).

In this paper, we wish to give a presentation of why CIMMYT wheat germplasm has had considerable adaptive success in semi-arid environments. We also wish to draw conclusions regarding an effective methodology for a breeding program addressing drought prone areas. While doing so, we do not intend to belittle any other methodology or approach followed elsewhere, but do wish to forward the adoption by farmers as the decisive criteria of success for any methodology.

Definition of semiarid environments and description of distinct drought patterns

In Table 2, the major global drought patterns observed in wheat production is presented (Rajaram et al., 1993; Edmeades et al., 1989). Through respectively dealing with spring (ME4A), facultative (ME9), and winter wheat (ME12), these three mega-environments are characterized by sufficient rainfall prior to anthesis, followed by drought during the grain-filling period. The South America, the Southern Cone type of drought (ME4B) is characterized by moisture stress early in the crop season, with rainfall occurring during the post anthesis phase. In the Indian Subcontinent type of drought stress (ME4C), the wheat crop utilizes water reserves left from the monsoon rains during the previous summer season. In the Subcontinent also irrigated wheat crop (ME1) may suffer drought due to a reduced or less than optimum number of irrigations.

Table 2. Megaenvironments (ME) involving drought conditions ($\leq$ 500 mm rainfall)

ME	Main rainfall	Temperate regime	Wheat types	Sown	Example locations
ME_4A	Winter dominant	Temperate	Spring	Autumn	Morocco Settat
ME_4B	Summer dominant	Temperate	Spring	Autumn	Argentina M. Juarez
ME_4C	Monsoon residual	Hot	Spring	Autumn	India Dharward
ME_6	Summer dominant	Temperate	Spring	Spring	Russia Saratov
ME_9	Winter dominant	Moderate cold	Facult.	Autumn	Algeria Sidi Bel Abbes
ME_{12}	Winter dominant	Severe cold	Winter	Autumn	Turkey Ankara

Source : Rajaram et al., 1995. In: Abstracts 8th IWGS, Beijing, China, p. 262.

Traditional methodology of breeding for drought stress

The traditional methodology, which has been practiced for many years in varying forms, is typified by handling of all segregating populations under target conditions of drought, and recommends the use of local landraces in the breeding process (Ceccarelli et al., 1987). What is not particularly evidenced by this methodology is any impact on yield, farmers' adoption or final national production. This traditional methodology is based on the assumption that the agro-ecological situation facing the farmer does not vary in its expression over time. It assumes that responsiveness of varieties to improved growing conditions will not be needed. Also it presumes that there always occurs a crossover below a certain yield level under dry conditions, where modern high yielding varieties of a responsive nature would always yield less than traditional land race based genotypes. Such crossovers may occur for selected genotypes and one should always be open to the possibility that there are real 'drought tolerance' traits operating at the 1 t/ha and below yield level, that adversely affect high yield potential at the 4 t/ha and higher yield levels. So far at CIMMYT such traits were not identified. In any case, cross over would be restricted to such harsh conditions, where in fact farmers choose – rightfully so – not to grow wheat at all, but rather other known more drought tolerant crops such as barley or sorghum, or resort to grazing practices.

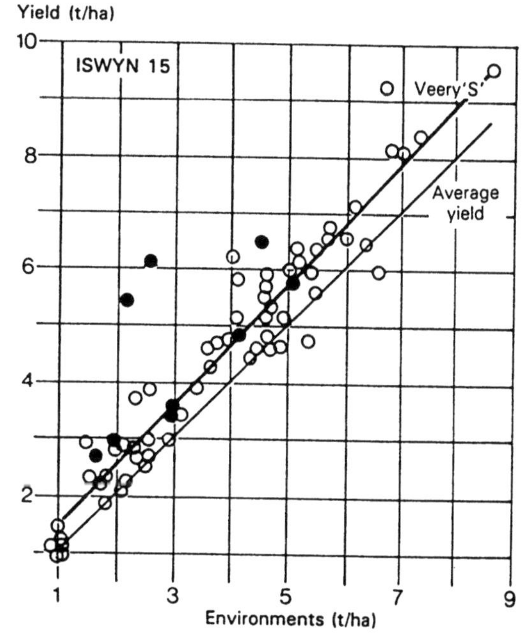

Figure 2. Yield of Veery'S' in the 73 environments of the 15th ISWYN (Pfeiffer, 1984).

Alternative methodology of combining yield responsiveness and adaptation to drought

At CIMMYT we advocate an 'open-ended system' of breeding in which yield responsiveness is combined with adaptation to drought conditions. Most semi-arid environments differ significantly across years in their

Table 3. The effects of the 1BL/1RS translocation on yield characteristics of 28 random F2-derived F6 lines (14 1BL/1RS and 14 1B) from the cross of *Triticum aestivum* L. cvs Nacosari 76/Seri 82, under reduced irrigated conditions during the 1991–92 and 1992–93 crop cycles. Cd. Obregon, Sonora, Mexico

Plant characteristics	1BL/1RS	1B	Mean dif.
Grain yield (kg/ha)	4945	4743	202*
Above-ground biomass			
at maturity (t/ha)	12600	12100	500*
Grains/m^2	14074	13922	152NS
Grains/spike	43.5	40.6	2.9*
1000-grain weight (g)	37.05	36.53	0.52*

Source : Villareal et al., 1994; NS : Not significant,
*: Significant at 0.05 level.

water availability and distribution pattern. Hence it is prudent to construct a genetic system in which plant responsiveness provides a bonus wherever environmental situations improve due to higher rainfall. With such a system, improved moisture conditions immediately translate into greater gain to the farmer. Why do we believe this can be done?

The tale of the VEERY's

In the early 1980's when the advanced lines derived from the spring × winter cross Kavkaz/Buho//Kal/BB (CM33027) was tested in 73 global environments of the 15th International Wheat Yield Nursery (15th ISWYN) (Figure 2), their performance was quite untypical compared to any previously known high yielding varieties. In later tests, we found that these lines, called VEERY's, carry the 1B/1R translocation from rye, and that general performance of such germplasm was superior not only in high yielding environments but particularly under drought conditions (Villareal et al., 1995; Table 3). From the Veery cross 43 varieties were released, excluding those released in Europe.

However, in addition to the creation of a new class of superior germplasm, there is an important lesson in breeding to be learned here. The VEERY's represent a genetic system in which high yield performance in favorable environments and adaptation to drought could be combined in one genotype. The two genetic systems are apparently not always incompatible, although others have claimed that their combination would not be possible. Based on this revelation, it is possible to hypothesize a plant system in which efficient input use and responsiveness to improved levels of external inputs (in this case available water) can be combined to produce germplasm for marginal (in this case semi-arid) environments, that at least maintain minimum traditional yields and express dramatic increases whenever the environment improves.

Evidence supporting promotion of this methodology

1. Contrary to the condemnation by critics (Ceccarelli et al., 1987), by the mid 1980's CIMMYT bred germplasm, occupied 45% of the semiarid wheat areas with rainfall between 300–500 mm, and 21% of the area less than 300 mm (Morris & Belaid, 1991), including large tracks in West Asia/North Africa (WANA). By 1990 63% of the dryland areas, in especially ME4A and ME4B, was planted with semi-dwarf wheats (Byerlee & Moya, 1993), many carrying the 1B/1R translocation. This represents clear acceptance by farmers, who widely adopted the new responsive germplasm over their traditional varieties. The positive trend among the final users of our products can not be ignored. Indirectly, it supports our view that the modern genotypes have adaptation to ME4A and ME4B drought areas while expressing high yields in improved conditions.
2. To support above assumptions, an experiment was conducted (Calhoun et al., 1994; Tables 4, 5, 6) to determine how the most modern and widely (spatially) adapted germplasm compared to commercial germplasm from countries representing the Mediterranean region (ME4A), the Southern Cone of South America (ME4B) and the Indian Subcontinent (ME4C), under conditions artificially simulating those three Mega-environments. The most widely (spatially) adapted CIMMYT lines outyielded the commercial varieties in all artificially simulated environments. The recent adoption trend of CIMMYT germplasm in these difficult marginal environments supports the model of input efficiency/input responsiveness.
3. The story of Nesser. Nesser is an advanced line with superior performance in drought conditions bred at CIMMYT/Mexico and identified at ICARDA/Syria. The cross combines a high yielding CIMMYT variety Jupateco and a drought tolerant Australian variety W3918A. The performance of Nesser in WANA's ME4A environments has been widely publicized (ICARDA, 1993), and the line

Table 4. Wheat genotypes representing adaptation to different moisture environments

ME_1	Irrigation	Super Kauz, Pavon 76, Genaro 81, Opata 85
ME_4A	(Mediterranean)	Almansor, Nesser, Sitta, Siete Cerros
ME_4B	(Southern Cone)	Cruz Alta, Prointa Don Alberto, LAP1376, PSN/BOW CM69560
ME_4	(Subcontinent)	C306, Sonalika, Punjab 81, Barani

Source : Calhoun et al., 1994.

Table 5. Grain yields of selected wheat genotypes grouped by adaptation (Table 4) and tested under 4 moisture regimes in the Yaqui Valley, Mexico, 1989–90 and 1990–91

Adaptation	Group	Full irrigation[1]	Late drought[2]	Early drought[3]	Residual moisture[4]
ME_1	Irrigation	6636 a	4198 a	4576 a	3032 a
ME_4C	Mediterranean	6342 b	3990 ab	4390 b	2883 b
ME_4B	Southern Cone	5028 c	3148 bc	4224 b	2359 c
ME_4C	Subcontinent	4778 c	3245 bc	3657 c	2704 b

Source : Calhoun et al., 1994.
[1] , received 5 irrigations; [2] , received 2 irrigations early before heading; [3] , received one irrigation for germination and two in post heading; [4] , received one irrigation for germination only.
: Means in the same column followed by the same letter are not significantly different at 0.05 probability level.

is considered by ICARDA to represent a uniquely drought tolerant genotype. However, it was selected at CIMMYT/Mexico under favorable environments, and carries a combination of input efficiency and high yield responsiveness. It performs similarly to the VEERY lines in the absence of rust (Figure 2).

Based on the above evidence, our proposed operational methodology is to actively combine input efficiency and input responsiveness.

Application

A breeding scheme is described below we use to achieve the combination of the two genetic systems. Two contrasting selection environments are alternated, allowing alternate selection for input efficiency and input responsiveness.

F1. Crosses involving spatially widely adapted germplasm representing yield stability and yield, with lines with proven drought tolerance in the specific setting of either ME4A, ME4B or ME4C. Winter wheats and synthetic germplasm are emphasized.

F2. The individual plants are raised under irrigated and optimally fertilized conditions, and inoculated with a wide spectrum of rust virulence. Only robust and (horizontally) resistant plants are. These may represent adaptation to favorable environments.

F3, F4. The selected F2 plants are evaluated in modified pedigree/bulk breeding system (Rajaram & van Ginkel, 1995) under rainfed conditions or very low water availability. The selection is based on individual lines rather than on individual plants. The progenies are selected based on such criteria as spike density, biomass/vigor, grains/m^2, and others (Van Ginkel et al., 1995; Table 5). This index helps identify lines which may adapt to low water situations.

F5, F6. The selected lines from F4 are further evaluated under optimum conditions.

F7, F8. Simultaneous evaluations under optimum and low water environments. Selection of those lines showing outstanding performance under both conditions. Further evaluation in international environments is carried out for purposes of verification.

Conclusions

The proposed breeding methodology is supported in research published in recent years by others, not only on wheat (Bramel-Cox et al., 1991; Cooper et al.,

Table 6. Genotypic correlation (rg) between agronomic traits and final grain yield, for optimum environment (full irrigations) and reduced water regime (late drought, Mediterranean type) in wheat

Trait	Moisture regime	
	Full irrigation	Late drought
Days to heading	0.40	0.19
Days to maturity	0.29	0.27
Grain fill period	- 0.32	0.36
Height	- 0.39	0.05
Peduncle length	- 0.46	0.22
Relative peduncle extrusion	- 0.51*	0.25
Spike length	- 0.28	- 0.50*
Spike M^{-2}	- 0.12	0.64**
Grains/spike	0.62*	- 0.42
Grains M^{-2}	0.74**	0.68**
Yield/spike	0.55*	- 0.64**
1000 grain weight	0.08	- 0.45
Test weight	0.13	0.05
Harvest index (HI)	0.83**	- 0.39
Biomass	0.90**	0.94**
Straw yield	0.52*	0.86**
Yield/day (planting)	0.99**	0.57*
Yield/day (heading)	0.94**	0.44
Biomass/day (planting)	0.86**	0.69**
Biomass/day (heading)	0.74**	0.63**
Vegetative growth rate	0.32	0.63**
Spike growth rate	0.62**	- 0.58*
Grain growth rate	0.17	- 0.44

*, ** indicate significance at the 0.05 and 0.01 probability level, respectively.
Source : van Ginkel, M. CIMMYT, 1995.

1994; Duvick, 1990, 1992; Edhaie et al., 1988; Uddin et al., 1992; Zavala-Garcia et al., 1992), where the importance of testing and selecting in a range of environments including well-irrigated ones, has shown to identify superior genotypes for stressed conditions. The methodology aims at combining input efficiency with input responsiveness, by alternating selection environments during the breeding process. This approach results in germplasm that is accepted by farmers because it translates improved environmental conditions into yield grains. The traditional methodology of only selecting under drought conditions, and narrowly relying on the landrace genotypes, does not move yield levels significantly beyond those traditionally obtained, and does not provide the farmer with a bonus yield in the 'fat years'.

References

Bramel-Cox, P.J., T. Barker, F. Zavala-Garcia & J.D. Eastin, 1991. Selection and testing environments for improved performance under reduced-input conditions. p. 29–56. In: Plant Breeding and Sustainable Agriculture: Considerations for objectives and methods. CSSA Special Publication No. 18. CSSA and ASA, Madison, USA.

Byerlee, D. & P. Moya, 1993. Impacts of international wheat breeding research in the developing world, 1966–1990. CIMMYT, Mexico, D.F. 135 pp.

Calhoun, D.S., G. Gebeyehu, A. Miranda, S. Rajaram & M. van Ginkel, 1994. Choosing evaluation environments to increase wheat grain yield under drought conditions. Crop Sci 34: 673–678.

Ceccarelli, S., M. M. Nachit, G.O. Ferrara, M.S. Mekni, M. Tahir, J. Van Leur & J.P. Srivastava, 1987. Breeding strategies for improving cereal yield and stability under drought. p. 101–114. In: J.P. Srivastava, E. Porceddu, E. Acevedo & S. Varma (Eds). Drought Tolerance in Winter Cereals. John Wiley & Sons, New York. 387 pp.

Cooper, M., D.E. Byth & D.R. Woodruff, 1994. An investigation of the grain yield adaptation of CIMMYT wheat lines to water stress environments in Queensland. II. Classification analysis. Agric Res 45: 985–1002.

Duvick, D.N., 1990. Ideotype evolution of hybrid maize in the USA, 1930–1990. p. 557–570. In: ATTI Proceedings, Vol. II, II National Maize Conference: research, economy, environment. Grado (GO) – Italy September 19–20–21, 1990. centro regionale per la sperimentazione agraria, pozzuolo del friuli edagrocole s.p.a., Bologna, Italy.

Duvick, D.N. 1992. Genetic contributions to advances in yield of U.S. maize. Maydica 37: 69–79.

Ethaie, B., J.G. Waines & A.E. Hall, 1988. Differential responses of landrace and improved spring wheat genotypes to stress environments. Crop Sci 28: 838–842.

Edmeades, G.O., J. Bolanos, H.R. Lafitte, S. Rajaram, W. Pfeiffer & R.A. Fischer, 1989. Traditional approaches to breeding for drought resistance in cereals. p. 27–52. In: F.W.G. Baker (Ed). Drought Resistance in Cereals. ICSU, Paris and C.A.B. International, Wallingford, England.

ICARDA, 1993. Cereal Program. Annual Report for 1992. ICARDA.

Morris, M.L., A. Belaid & D. Byrelee, 1991. Wheat and barley production in rainfed marginal environments of developing world. Part I of 1990–91 CIMMYT world wheat facts and trends. CIMMYT, Mexico, D.F. 51 pp.

Rajaram, S., M. van Ginkel & R.A. Fischer, 1994. CIMMYT's wheat breeding mega-environments (ME). In: Proceedings of the 8th International Wheat Genetic Symposium, July 19–24, 1993. Beijing, China (in press).

Rajaram, S. & M. van Ginkel, 1995. Wheat breeding methodology: international perspectives. Proceedings 20th Hard Red Winter Wheat Workers Workshop. Oklahoma City, USA (in press).

Uddin, N., B.F. Carver & A.C. Clutter, 1992. Genetic analysis and selection for wheat yield in drought-stressed and irrigated environments. Euphytica 62: 89–96.

Van Ginkel, M., D.S. Calhoun, G. Gebeyehou, A. Miranda, C. Tianyou, R. Pargas Lara, R.M. Trethowan, K. Sayre, J. Crossa & S. Rajaram, 1995. Plant traits related to yield of wheat in early, late, or continuous drought conditions (in press).

Villareal, R.L., E. Del Toro, A. Mujeeb-Kazi & S. Rajaram. The 1BL/1RS chromosome translocation effect on yield characterization in a *Triticum aestivum* L. cross. Plant Breeding (in press).

Zavala-Garcia, F., P.J. Bramel-Cox, J.D. Eastin, M.D. Witt & D.J. Andrews, 1992. Increasing the efficiency of crop selection for unpredictable environments. Crop Sci 32: 51–57.

P.M.A. Tigerstedt (Ed.), Adaptation in Plant Breeding, 169–176, 1997.

Aluminium uptake by roots of rye seedlings of differing tolerance to aluminium toxicity

Andrzej Anioł
Department of Plant Biochemistry and Physiology, Plant Breeding and Acclimatization Institute, Radzików, P.O. Box 1019, 00-950 Warszawa, Poland

Key words: aluminium toxicity, aluminium accumulation, rye, roots, *Secale cereale*

Summary

The aluminium accumulation pattern was studied in roots and root tips of rye seedlings from two inbred lines, differing in Al-tolerance and their F_1 hybrid.

This investigations were undertaken in order to find whether the mechanism preventing root tips from to high Al-accumulation and damage to apical meristem observed in Al-tolerant isoline of wheat is also operating in rye. Al-uptake into the root tissue was studied in nutrient solution experiments during 24 h of incubation in solution containing aluminium. After Al-incubation roots were harvested, separated into root tips and basal parts, and aluminium content in root tissue was determined after wet digestion using catechol violet colorimetric method. It was found, that at low external Al-concentration in rye roots the mechanism preventing root tips from Al-accumulation and Al-damage operates, similarly as in Al-tolerant wheat isoline. But contrary to wheat in rye seedlings functioning of this mechanism was observed in both Al-tolerant and Al-sensitive lines. It is concluded that observed mechanism of the inhibition of Al-accumulation in apical part of roots plays much more important role in Al-tolerance mechanism in wheat than in rye.

Abbreviations: EDTA – ethylenediaminetetraacetic acid; NIL – near-isogenic line

Introduction

Aluminium toxicity is regarded as the most important yield-limiting factor for cereals grown on acid soils and cannot be always corrected by liming, especially when subsoil acidity occurs (Alva et al., 1986). Breeding cultivars more tolerant to Al-toxicity, using the existing biodiversity of crops and their wild relatives may be a solution to the Al-toxicity problem (Foy, 1983).

Crop tolerance to acidic soils has been extensively investigated and genetic variation between cultivars in reaction to aluminium toxicity has been found in most cultivated species (Duke, 1982). However, several conditions must be fulfilled before the genes controlling Al-tolerance dispersed among cultivars and wild relatives of crops can be exploited in breeding programs. First, a suitable, cheap, and rapid technique for identifying tolerance must be available for screening large numbers of plants. Second, appropriate genetic variation among cultivars and wild relatives must exist and sources of tolerance must be identified. Third, the mode of inheritance of tolerance should be established. For genetic manipulations and utilization of genes from alien species at the chromosome or molecular level, the localization, identification and isolation of genes concerned is necessary (Aniol, 1991).

Based on known genetic variation in plant reaction to Al-stress, conventional breeding programs, oriented towards developing tolerant cultivars, were designed in different crops. Breeding for tolerance is relatively advanced in some crops and significant achievements are reported in releasing germplasm tolerant to aluminium (Fisher & Scott, 1983; Carver et al., 1993). Despite an abundance of information on genetic variability for plant response to Al-toxicity in acid soils, information on genetic systems controlling these responses is limited and fragmentary. For comprehensive reviews summarizing data on genetics of Al-tolerance in cultivated plants see Aniol & Gustafson (1990), Foy et al. (1988), Blum (1988). From the available data it can be concluded that aluminium tolerance in plants is a complex character. It is assumed that several major genes are controlling the expression of this character during the different stages of plant development. Analogously, there are many physiological and biochemical processes associated with plant's reaction to Al-ions. There are many possible mechanisms of metal avoidance or internal tolerance, but our knowledge of these mechanisms are far from complete (Taylor, 1991), and it would be unjustified at this moment to base any selection program on any simple biochemical parameter (Graham, 1988).

Because of the lack of sufficient knowledge about both, genetical control and physiological mechanisms of aluminium tolerance in plants bioassay methods are used for germplasm evaluation and selection work. There are many variants of bioassay tests but all are based on the assessment of the overall effect of aluminium ions on root growth. As a byproduct in selection programs several sets of near-isogenic-lines (NIL's) were developed in wheat and rye (Aniol, 1995; Carver et al., 1993; Falfus et al. 1991). These NILs' pairs should allow for the description of individual genes determining Al-tolerance and possible differences in physiological processes between these NIL's can be ascribed to the gene by which this lines differ.

The present paper reports on studies involving aluminium uptake pattern in seedling roots of rye lines differing in Al-tolerance.

Material and methods

Plant material

Winter rye inbred lines of different Al-tolerance were selected from the breeding program of Dr. L. Madej from IHAR Radzikow.

Two rye inbred lines of different tolerance to aluminium were selected from over 160 lines tested in screening experiments: No. 4475 Al-tolerant and No. 1993, Al-sensitive (Falfus et al., 1991). The Al-uptake pattern into the root segments of these lines and their F1 hybrid was investigated in order to find whether the mechanism preventing root apical parts from Al-damage is also operating in rye.

Rye seeds used for the experiments were from 1993 harvest.

Methods

Al-uptake experiments

Seeds were sterilized with 0.1% $HgCl_2$ aqueous solution for 10 min, rinsed excessively with water, and germinated overnight on filter paper in Petri dishes. Sprouted seeds were sown the next day on polyethylene net fixed in Lucite frames. Styrofoam blocks were attached to the frames with rubber bands and floated on the surface of vigorously aerated nutrient solution. Containers with nutrient solution were placed in water bath at 25 °C under 16 h per day illumination with incandescent low intensity light from Polam incandescent tubes.

For screening tests, a nutrient solution was used as described earlier (Anioł, 1984). Al-uptake experiments were performed in simplified nutrient solutions, e.g. Al^{+3} added in different concentrations in aqueous solution and with Ca^{+2} or Mg^{+2} added at 100 micromolar concentration. Four day old seedlings were used for incubation with aluminum. After each exposure to Al, (as indicated in Figures) seedlings were removed, thoroughly washed for 2–3 min in running tap water, and then placed for 30 min on 0.5 mM aqueous EDTA solution at 4 °C (approximately 130 seedlings per replication). After the desorbtion with EDTA each sample was split into two parts; one, of approximately 30 seedlings was washed again for 2–3 min in running tap water and transferred to Al-free medium for 48 h. Any additional root growth after Al-shock was easily assessed by staining roots with 0.1% aqueous solution of Eriochrome cyanine R for 10 min. After staining the excess dye was removed by washing under the tap water. If the aluminum treatment did not destroy the root apical meristem, part of the root which grew after Al-treatment was white (unstained) and contrasted with heavily stained root part exposed to aluminum. When apical meristem was irreversibly damaged, then after 48 h of growth in Al-free medium, the root did not show any regrowth and root tips remained intensively stained. During all stages of growth, and particularly during the Al-treatment the nutrient solution was maintained at pH 4.5^{+} 0.02. When a ratio of approximately 20 ml of nutrient solution per seedling was maintained pH changes in the medium did not exceed 0.02 during 24 h of aluminium treatment. In the second

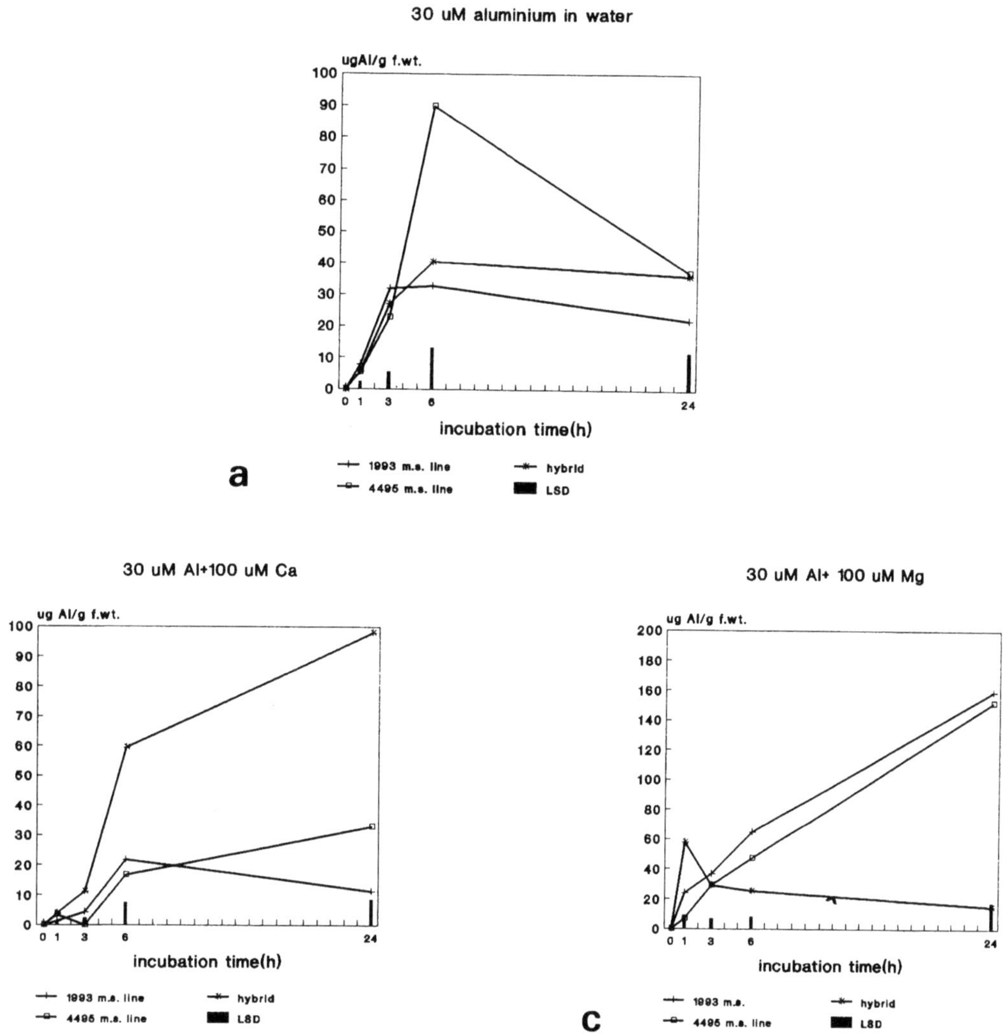

Figure 1. Aluminium accumulation in rye root tips at low external Al-conc.

part of the sample, after EDTA treatment, roots (from approximately 100 seedlings) were separated into root tips (apical 5 mm) and rest of roots, then fresh weight of tips and roots was estimated. Each experiment was replicated at least three times.

Aluminum determination

The roots and root tips from each replication were mineralized in the acid mixture as described by Jones & Thurman (1957). The acid mixture consisted of sulfuric, perchloric and nitric acids in proportion by volume 1 : 2 : 7. Fresh root samples were transferred to digestion tubes and the acid mixture was added (5 ml/g fresh weight). Samples were mineralized in Tecator heating block. Aluminum content was estimated using the catechol violet method (Wilson, 1984).

Analysis of data

Statistical analyses of data were performed using multifactor analysis of variance available on statistical package, Statgraphics Version 2.6. Comparisons among means were performed using Tukey's procedure. Statistics significance is on a confidence level of at least 95%.

Table 1. The root regrowth of tested rye genotypes after aluminium treatment. Data represent percent of seedlings with root regrowth after Al-treatment; mean length of root regrowth in mm is shown in parantheses

Line/treatment	1993 m.s. line Al-sensitive	F_1 hybrid (1993 × 4495)	4495 m.s. line Al-tolerant
Aluminium in water			
30 μM Al	100 (200)	100 (350)	100 (300)
60 μM Al	100 (45)	100 (175)	100 (175)
100 μM Ca^{+2}			
+ 30 μM Al	100 (250)	100 (250)	100 (250)
+ 300 μM Al	100 (40)	100 (150)	100 (150)
100 μM Mg^{+2}			
+ 30 μM Al	100 (250)	100 (180)	100 (300)
+ 300 μM Al	100 (45)	100 (165)	100 (175)

Results and discussion

The detailed results of Al-uptake pattern into root segments of wheat NIL's and their parental genotypes were presented in recent paper (Anioł, 1995). The obtained data indicate that gene or genes located on the long arm of chromosome 2D from Al-tolerant BH 1146 wheat cultivar control the mechanism preventing root apical meristem from high Al-accumulation and by this from Al-damage. This mechanism, however does not operate at higher external Al-concentrations.

Aluminium uptake into rye root segments was analyzed at two levels of Al-concentration in the nutrient medium; non-lethal – where root growth was little affected and sub-lethal when root elongation rate was reduced by 50% in Al-tolerant line and more than 50% in Al-sensitive line (Table 1). But at low and high external Al-concentration root apical meristems were not irreversibly damaged, even in Al-sensitive line, therefore the observed differences in Al-accumulation pattern in tested rye lines represent real differences in Al-uptake mechanism between these genotypes.

There are no differences in Al-accumulation in root tips from tested rye lines exposed to 30 micromoles of Al in water (Figure 1A); after the initial uptake during the first 3 hours the aluminium concentration in root tip tissue stabilizes at a similar level in both rye inbred lines and their hybrid. Only in root tips from Al-tolerant line 4475 the initial uptake phase lasts up to 6 h and results in higher Al-accumulation but this effect disappears during the next 18 h of incubation and after 24 h the level of Al in root tip tissue of all tested rye genotypes is similar.

It can be concluded that in both rye lines and their hybrid Al-accumulation in root tips is controlled by a mechanism preventing root apical meristems from Al-damage at low external Al-concentration. This mechanism is similar to that observed at low external Al-concentration in root tips of Al-tolerant wheat isoline.

In the presence of calcium ions in the medium Al-accumulation pattern in root tips of both rye inbred lines was similar, only in root tips of F1 hybrid Al-accumulation was stabilized at significantly higher level (Figure 1B). When Mg ions were present in the external medium the effect of the mechanism responsible for the stabilization of Al-concentration in root tips was not observed in both tested rye inbred lines; however, its operation was observed in root tips from F1 seedlings (Figure 1C).

It can be concluded that similarly as in root tips of Al-tolerant wheat isoline also in rye lines a mechanism preventing root tips from to high Al-accumulation is operating at low external Al-concentration. But contrary to wheat, in rye this mechanism is present in both lines, irrespective of their Al-tolerance and the presence of divalent cations in the external medium caused disturbances in the functioning of this mechanism.

After incubation in the medium containing 10-times higher Al-concentration root growth rate was reduced, most significantly in seedlings of Al-sensitive 1993 m.s. line (Table 1). The Al-accumulation in root tips continued during the whole 24 h period (Figure

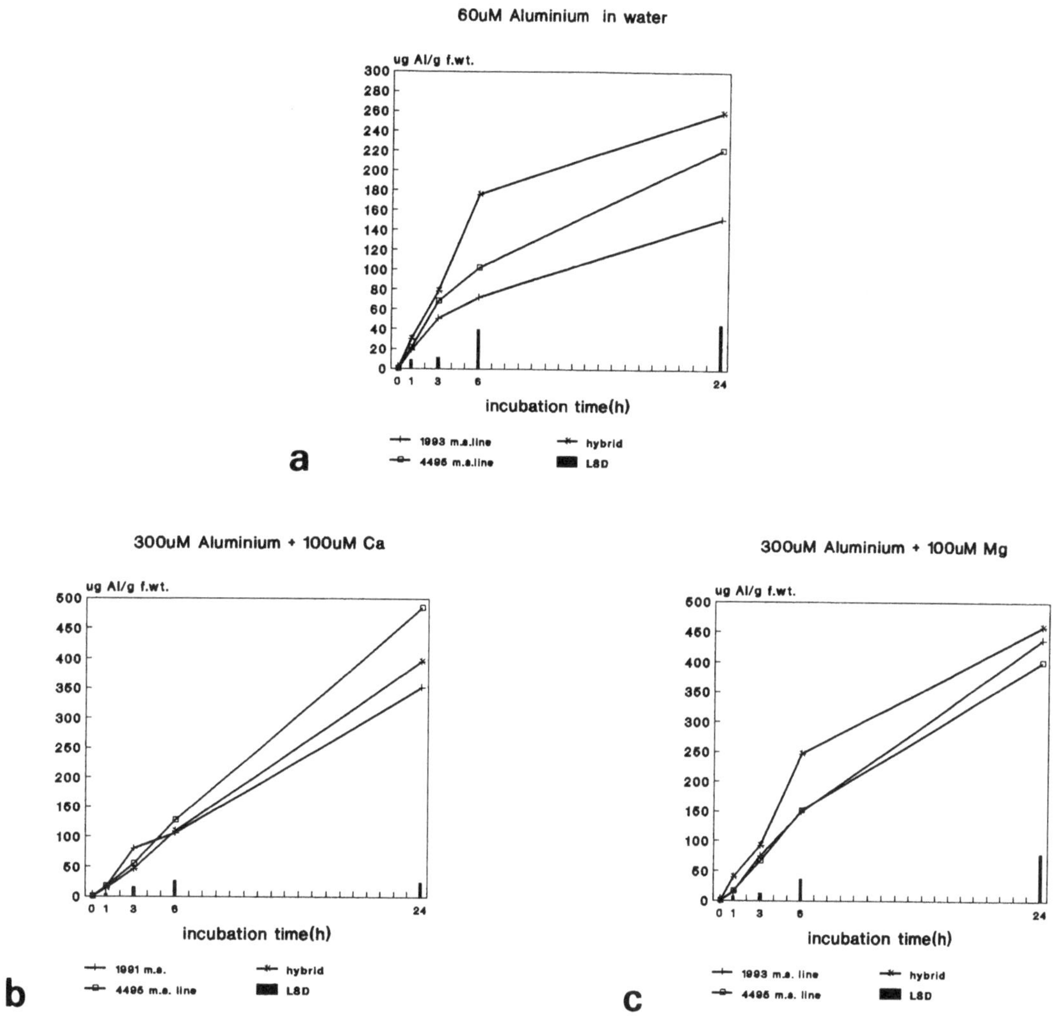

Figure 2. Aluminium accumulation in rye root tips at high external Al-conc.

2A). The presence of divalent cations in the medium caused increased Al-accumulation in root tips; this increased Al-accumulation was similar in roots of all tested rye genotypes (Figures 2B, C).

Al-accumulation in the basal root parts was very similar in tested rye genotypes, both at low and high level of Al in the external medium (Figure 3); only in roots of F1 hybrid exposed to low external Al conc. some tendency to higher Al accumulation was observed, specially when Ca ions were present in the growth medium (Figure 3B); Mg ions stimulated the initial Al-accumulation in F_1 seedling roots, but this effect disappeared after 6 h of incubation (Figure 3C). At higher external Al-concentration Al-accumulation was similar in roots of all tested rye genotypes (Figures 3D–F).

A similar mechanism of Al-tolerance based on the inhibition of Al accumulation in root tips has been described in Al-tolerant wheat cv Atlas 66 by Rincon & Gonzales (1992), and these authors suggest that this mechanism operates as a metabolism-dependent exclusion of Al from the root apical meristem.

Studies on Al uptake in root apices of wheat NIL's differing at single locus, Alt-1, revealed that this gene from Al-tolerant Carazinho cultivar encodes a mechanism responsible for malic acid excretion in root tips under Al stress (Delhaize et al., 1993a, b).

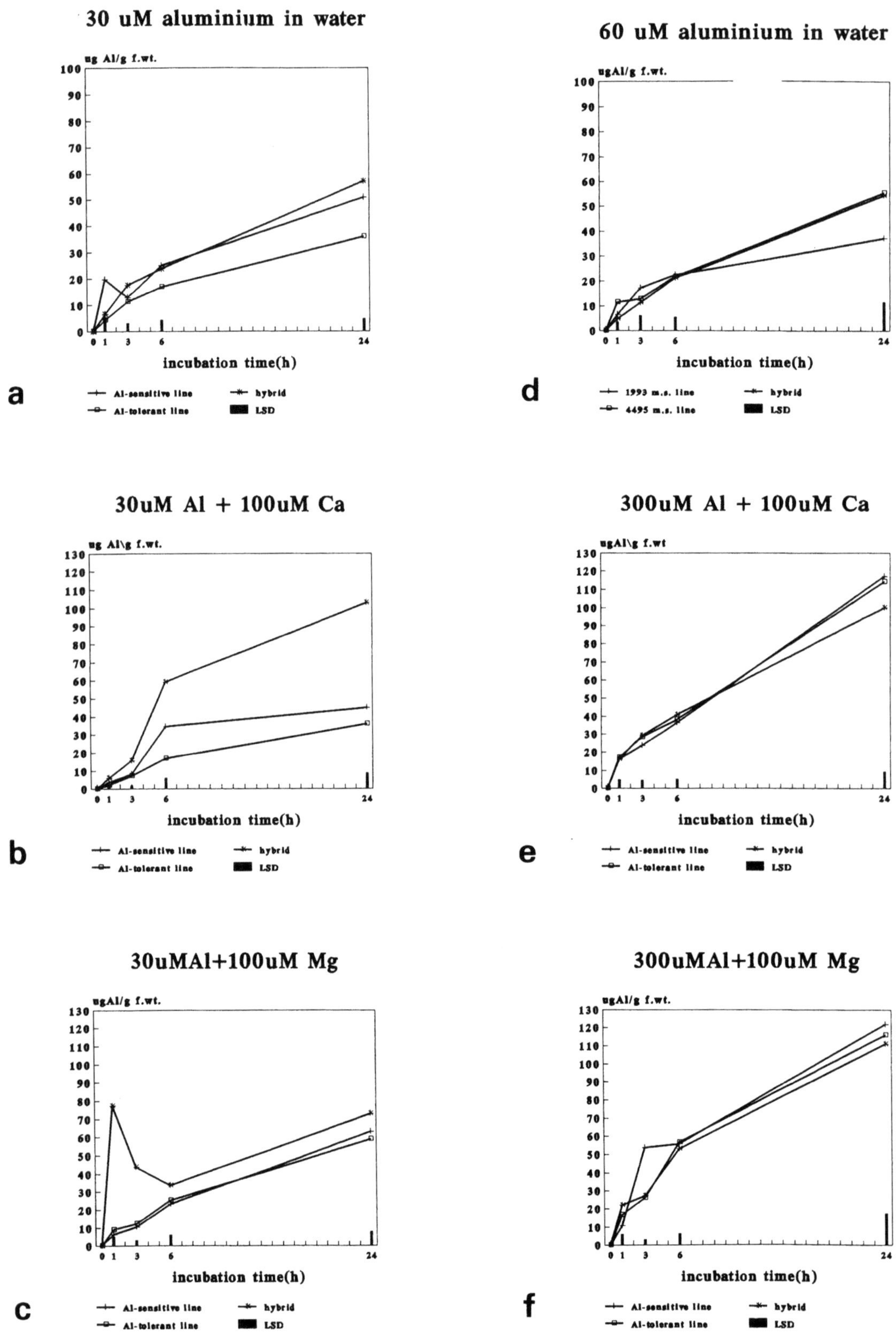

Figure 3. Al-accumulation in rye roots.

Data reported in this paper indicate that the described Al-accumulation inhibition mechanism is efficient only at low external Al concentrations and that some other mechanism must operate at higher Al-concentrations in the medium as well as in root tissue.

The results presented in this paper confirm the ameliorative effect of calcium and magnesium ions on Al rhizotoxicity reported by many other authors (Kinraide & Parker, 1987; Rengel, 1992, just to mention the recent ones). But the role of these cations, particularly Ca^{+2}, in Al-toxicity/tolerance mechanisms is still disputed. The effect of divalent cations described in this paper is puzzling, their ameliorative effect on Al-toxicity as measured by root growth inhibition is evident but on the other hand these cations stimulate Al accumulation in root tissue. These results suggest that the divalent cations, particularly Ca^{+2}, might play a important role not only in the processes occurring on the membrane surfaces but also in the mechanisms preventing cell damage, even at high Al accumulation inside the root tissue. Comparison of the Al-accumulation pattern in wheat and rye roots suggests that in both cereals a mechanism preventing root apical meristems from to high Al-concentration and Al-damage operates at low external Al-concentration. This mechanism was observed only in roots of Al-tolerant wheat isoline and is controlled by gene or genes located on the long arm of 2D chromosome (Anioll, 1995) while in rye roots this mechanism was expressed in all genotypes, irrespective of their Al-tolerance. It must be remembered however, that rye line classified as Al-sensitive under testing regime for rye behives as Al-tolerant under testing regime for wheat. One may speculate therefore, that observed mechanism of maintaining Al-concentration in root tips is a common mechanism operating in both cereals. While it might play a significant role in Al-tolerance in wheat where this character is expressed at low level it seems to be of little importance in Al-tolerance mechanism in rye.

Acknowledgements

The financial support of this work by the Polish-American Joint Commission under the Maria Skłodowska-Curie Foundation, project MR/USDA-92-93, is highly appreciated.

References

Alva, A.K., C.J. Asher & D.G. Edwards, 1986. The role of calcium in alleviating aluminium toxicity. Aust J Agric Res 37: 375–382.

Aniol, A.M., 1995. Physiological aspects of aluminium tolerance associated with the long arm of chromosome 2D of the wheat (*Triticum aestivum* L.) genome. Theor Appl Genet 91: 510–516.

Anioł, A.M., 1991. Genetics of acid tolerant plant. p. 1007–1017. In: R.J. Wright et al. (Ed) Plant-Soil Interactions at Low pH. Kluwer Academic Publishers, The Netherlands.

Aniol, A.M. & J.P. Gustafson, 1990. Genetics of tolerance in agronomic plants. p. 255–267. In: A.J. Shaw (Ed). Heavy Metal Tolerance in Plants: Evolutionary aspects. CRC Press Inc, Boca Raton, USA.

Anioł, A., 1984. Induction of aluminum tolerance in wheat by low doses of aluminum in nutrient solution. PLant Physiol 75: 551–555.

Blum, A., 1988. Plant breeding for stress environments. CRC Press Inc, Boca Raton, FL.

Carver, B.F., W.E. Whitmore, E.L. Smith & L. Bona, 1993. Registration of four aluminum-tolerant winter germplasms and two susceptible near-isolines. Crop Sci 33: 1113–1114.

Delhaize, E., S. Craig, C.D. Beaton, R.J. Bennet, V.C. Jagadish & P.J. Randall, 1993a. Aluminum tolerance in wheat (*Triticum aestivum* L.) I. Uptaken and distribution of aluminum in root apices. Plant Physiol 103: 685–693.

Delhaize, E., P.R. Ryan & P.J. Randall, 1993b. Aluminum tolerance in wheat (*Triticum aestivum* L.) II. Aluminum stimulated excretion of malic acid from root apices. Plant Physiol 103: 695–702.

Duke, J.A., 1982. Plant germplasm resources for breeding crops adapted to marginal environments. In: N.M.N. Chistiansen & C.F. Lewis (Eds). Breeding Plants for Less Favorable Environments. Chapter 12, John Wiley & Sons, New York.

Falfus, J., L. Madej & A. Anioł, 1992. Tolerancyjnośċ linii wsobnych i mieszaśċów żyta na toksyczne działanie jonów glinu. Biul IHAR 183: 135–140.

Fisher, J.A. & B.J. Scott, 1983. Breeding wheats for tolerance to acid soils. Proc Austr Breeding Conf, Adelaide. 333p.

Foy, C.D., 1988. Plant adaptation to acid, aluminum toxic soils. Comm Soil Sci Plant Anal 19: 959–987.

Foy, C.D., 1983. The physiology of plant adaptation to mineral stress. Iowa St J Res 57: 335–391.

Graham, R.D., 1988. Genotypic differences in tolerance to manganese deficiency. p. 261–276. In: R.D. Graham et al. (Ed). Manganese in Soils and Plants. Kluwer Academic Publ, Dordrecht, The Netherlands.

Jones, L.D. & D.A. Thurman, 1957. The determination of aluminium in soil, ash and plant material using Eriochrome cyanine R. Plant Soil 9: 131–141.

Kinraide, T.B. & D.A. Parker, 1987. Cation amelioration of aluminum toxicity in wheat. Plant Physiol 83: 546–551.

Rengel, Z., 1992. Disturbance of cell Ca^{2+} homeostasis as a primary trigger of Al toxicity syndrome. Plant Cell and Environ 15: 931–938.

Rincon, M. & R. Gonzales, 1992. Aluminium partitioning in intact roots of aluminium-tolerant and aluminium-sensitive wheat (*Triticum aestivum* L.) cultivars. Plant Physiol 99: 1021–1028.

Snape, J.W., B.B. Parker & M.D. Gale, 1983. Use of backcross reciprocal monosomic method for evaluating chromosomal variation for quantitative characters. p. 367–373. In: S. Sakamoto (Ed). Proc 6th Int Wheat Genet Symp. Maruzen, Tokyo, Japan.

Taylor, G.J., 1991. Current views of the aluminium stress response: the physiological basis of tolerance. Current Topics Plant Biochem Physiol 10: 57–93.

Wilson, D.O., 1984. Determination of aluminium in plant tissue digest using a catechol violet colorimetric method. Comm Soil Sci Plant Anal 15: 1269–1279.

P.M.A. Tigerstedt (Ed.), *Adaptation in Plant Breeding*, 177–183, 1997.

Structural adaptation of the leaf chlorenchyma to stress condition in the Kola peninsula plants

Irina M. Kravkina
Komarov Botanical Institute, St. Petersburg, 197376, Russia

Key words: mitochondria, chloroplast, pollutant, adaptation

Summary

Compared were the number of mitochondria and chloroplasts in the chlorenchyma cells of pine needles (*Pinus sylvestris*), birch leaves (*Betula pubescens*), and shrub leaves (*Arctostaphylos uva-ursi, Vaccinium vitis-idaea, V. myrtillus, V. uliginosum*). Included also were herbaceous plant leaves *Campanula rotundifolia, Cornus suecica, Potentilla erecta* and *Solidago lapponica*. These plants were growing 10 km (severely polluted area) and 65 km (background area) south (SSW) of the 'Severonickel' smelter complex (Monchegorsk, Kola peninsula), where the dominating gaseous atmospheric pollutant is SO_2. Comparison between plants of polluted and unpolluted areas was made using electron microscopy on intact chlorenchyma cells. Morphometric analysis showed that the chloroplast numbers in the mesophyll cells did not depend on the pollution level. The mitochondria numbers did not differ significantly in leaf cells of the deciduous and herbaceous plants. However, it is considerably higher in evergreen species of the damaged area. The conclusion is that the increase of the mitochondria numbers is the adaptative response of the evergreen plants to atmospheric pollution.

Introduction

The principal part of the investigation associated with a study of the effect of industrial atmospheric pollution on plants is determined by the necessity of their protection and conservation. In order to elaborate the scientific basis for a rational utilization and an effective conservation of the vegetation, it is essential to study the reactions of plants under different anthropogenic effects. The leaf responds to the changes in the growth conditions first. The visible symptoms of the leaf damage due to atmospheric pollution are distinct in the forms of chloroses and necroses on the leaf surface. It is obvious that the leaf vitality must be provided by the green parts of the leaves. However, under the electron microscope both the damaged cells in different degrees and the undamaged cells are shown in the green parts of the leaves. Of course, determining the degree of cell damage is of great interest in leaves before visible symptoms of the effect become evident. And early diagnosis of the leaf damage may be carried out on the ultrastructural level of the cell.

During the last two decades much attention was given to investigations of the ultrastructural alterations which took place in the cell protoplasts of leaves and needles of those plants that had been exposed to atmospheric pollution, and to SO_2 effect in particular. So-called invisible damages in the structure of the cell organells, preceded by a visible injuries of needles and leaves were studied (Soikkeli & Tuovinen, 1979; Soikkeli, 1981; Kravkina et al., 1988). According to the stage of the degradation in cytoplasm organells, it is possible to determine if the damage is reparable or not. But the main role in the vital condition of the leaf is played by cells whose structure has not been disrupted under the impact of atmospheric pollution. It is safe to say that these cells allow the leaves and the plants as a whole to survive under the anthropogenic stress. In the undamaged cells on a structural level, there probably exist adaptive mechanisms which allow the leaf with visible disease symptoms to function normally under stress conditions.

In order to elucidate the cell tolerance to atmospheric pollution, I studied the quantitative dynamics of the

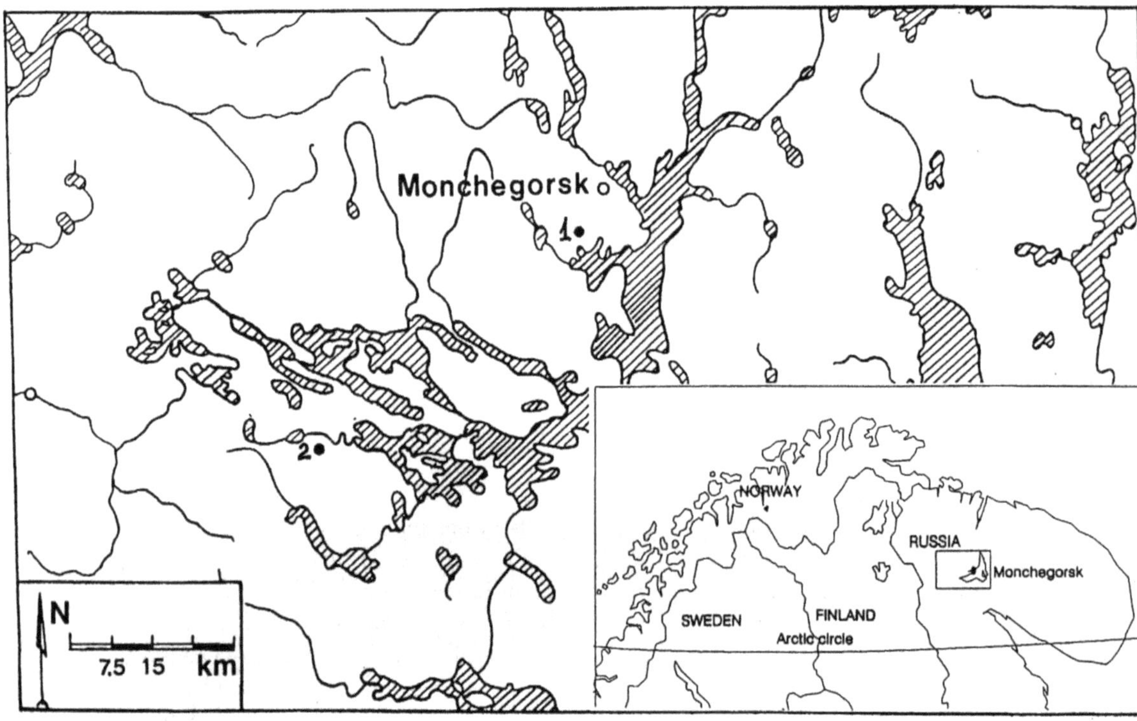

Figure 1. Point 1 – severely polluted area, 8 to 10 km south (SSW) of 'Severonickel' smelter complex, Monchegorsk. Point 2 – background area, 65 to 70 km south (SSW) of 'Severonickel' smelter complex, Monchegorsk.

main cell organells (chloroplasts and mitochondria) in the leaf chlorenchyma of the trees, shrubs and herbaceous plants growing in the background area as well as in the severely polluted area of the pine forests of the Kola peninsula.

Materials and methods

Materials for investigation were collected at the end of July – beginning of August from 1988 to 1992 in the Kola peninsula in an area polluted by the 'Severonickel' smelter complex. Forests in this region are affected by SO_2 and heavy metals: Ni, Cu, Co. The emission of sulphur dioxide reaches a level of 250,000 to 270,000 metric tons/year, and a polymetallic dust level of 15,000 to 20,000 tons/year (Kryuchkov & Makarova, 1989; Pozniakov, 1993). The average concentration of SO_2 in the air was 10 mkg/m^3 (during vegetation period) in the background area, and 35 to 40 mkg/m^3 in the severely polluted area (Boltneva et al., 1982).

The needles of the middle and lower layers of the pine crowns of 30-years old trees (*Pinus sylvestris*) and the leaves of the middle layer of the birch crowns of 50-years old trees (*Betula pubescens*) were investigated. Included also were the leaves of partial shrubs (*Arctostaphylos uva-ursi, Vaccinium vitis-idaea, V. uliginosum, V. myrtillus*), and the leaves of herbaceous plants (*Campanula rotundifolia, Cornus suecica, Potentilla erecta* and *Solidago lapponica*). These plants were growing 65–70 km south (SSW) of the smelter complex in the area of undamaged forest ecosystems, (background area) and 8–10 km south (SSW) of 'Severonickel' in severely polluted area (Figure 1).

The leaf parts without visible damages from the middle third of the leaves were fixed in glutaraldehyde, then post-fixed in buffered osmium tetraoxide, dehydrated and infiltrated by conventional procedures using Epon-Araldite mixture as an embedding medium. The intact chlorenchyma cells were investigated under the electron microscope Hitachi-600. Morphometry was performed on micrographs of longitudinal sections of the middle portions of leaves from each species. The area of cell sections represented by each micrograph was determined, and the number of chloroplast and mitochondrial profiles contained per unit area were calculated.

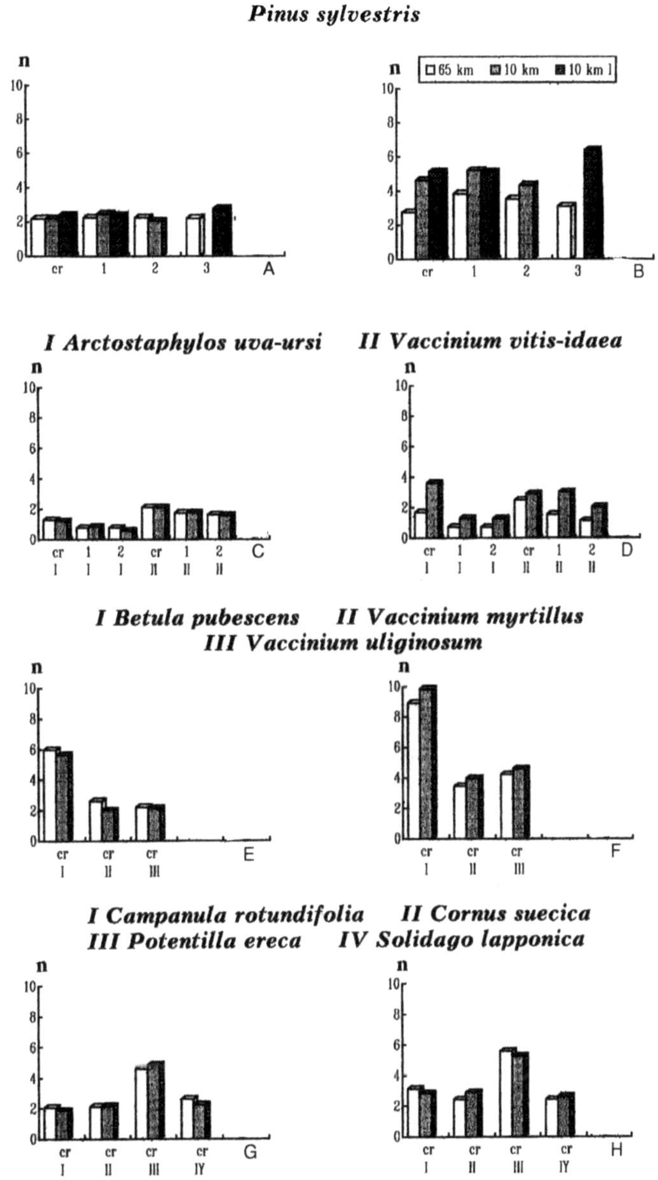

Figure 2. Number (n) of chloroplasts (A, C, E, G) and mitochondria (B, D, F, H) per 100 μm^2 of median sections in the chlorenchyma cells of the current- (cr), one- (1), two- (2), three-year-old (3) needles and leaves. 65 km (background area), 10 km and 10 km l (severely polluted area) – distances from the 'Severonickel' smelter complex and parts of the crowns (65 km and 10 km – middle, 10 km l – lower parts of the crowns) (replicates: number of mitochondria and chloroplasts n = 30; parameters significantly different among the severely polluted and background areas at $P < 0.05$).

Results

The data is given in Table 1, and Table 2 and summarized on histograms in Figure 2.

I have not noticed injuries of the chloroplasts and mitochondria structure in the investigated plant cells in severely polluted area.

The comparative morphometric analysis of the chloroplast numbers per 100 μm^2 of median cell sections in undamaged chlorenchyma cells of the plants from background region, and severely polluted area yielded the following results. In the mesophyll cells of *Pinus sylvestris* needles, the number of chloroplasts had not changed during the 3 years. The number of

Table 1. Number of chloroplasts and mitochondria per 100 μm^2 of median sections in the chlorenchyma cells of the current-, one-, two-, three-year-old needles. 65 km (background area), 10 km and 10 km l (severely polluted area) – distances from the 'Severonickel' smelter complex and parts of the crowns (65 km and 10 km – middle, 10 km l – lower parts of the crowns). (replicates: number of mitochondria and chloroplasts n = 30; parameters significantly different among the severely polluted and background areas at P < 0.05)

Species	Needle age	65 km	10 km	10 km l	65 km	10 km	10 km l
Pinus		mitochondria	mitochondria	mitochondria	chloroplasts	chloroplasts	chloroplasts
sylvestris	Current	2.73 ± 0.13	4.61 ± 0.34	5.11 ± 0.36	2.18 ± 0.13	2.17 ± 0.13	2.38 ± 0.17
	1 year	3.82 ± 0.23	5.18 ± 0.31	4.87 ± 0.34	2.24 ± 0.14	2.48 ± 0.15	2.36 ± 0.16
	2 year	3.50 ± 0.22	4.34 ± 0.30		2.24 ± 0.12	2.03 ± 0.15	
	3 year	3.08 ± 0.21		6.38 ± 0.41	2.21 ± 0.14		2.77 ± 0.16

Table 2. Number of chloroplasts and mitochondria per 100 μm^2 of median sections in the chlorenchyma cells of the current-, one-, two-, three-year-old leaves. 65 km (background area), 10 km and 10 km l (severely polluted area) – distances from the 'Severonickel' smelter complex and parts of the crowns (65 km and 10 km – middle, 10 kml lower parts of the crowns). (replicates: number of mitochondria and chloroplasts n = 30; parameters significantly different among the severely polluted and background areas at P < 0.05)

Species	Leaf age	65 km	10 km	65 km	10 km
		mitochondria	mitochondria	chloroplasts	chloroplasts
Arctostaphylos	Current	1.69 ± 0.09	3.61 ± 0.19	1.29 ± 0.06	1.21 ± 0.06
uva-ursi	1 year	0.73 ± 0.02	1.31 ± 0.07	0.77 ± 0.04	0.84 ± 0.04
	2 year	0.72 ± 0.02	1.30 ± 0.07	0.77 ± 0.04	0.57 ± 0.03
Vaccinium	Current	2.50 ± 0.52	2.91 ± 0.56	2.12 ± 0.19	2.11 ± 0.25
vitis-idaea	1 year	1.56 ± 0.24	3.02 ± 0.53	1.75 ± 0.15	1.76 ± 0.21
	2 year	1.13 ± 0.11	2.05 ± 0.19	1.63 ± 0.15	1.58 ± 0.08
Betula pubescens	Current	8.90 ± 0.45	9.85 ± 0.49	5.96 ± 0.29	5.61 ± 0.32
Vaccinium myrtillus	Current	3.45 ± 0.42	3.95 ± 0.49	2.61 ± 0.18	2.00 ± 0.14
Vaccinium uliginosum	Current	4.21 ± 0.25	4.60 ± 0.44	2.21 ± 0.09	2.12 ± 0.17
Campanula rotundifolia	Current	3.14 ± 0.14	2.84 ± 0.14	2.04 ± 0.10	1.83 ± 0.10
Cornus suecica	Current	2.45 ± 0.12	2.91 ± 0.14	2.10 ± 0.10	2.13 ± 0.10
Potentilla erecta	Current	5.59 ± 0.26	5.29 ± 0.26	4.59 ± 0.18	4.83 ± 0.20
Solidago lapponica	Current	2.45 ± 0.12	2.66 ± 0.13	2.59 ± 0.13	2.21 ± 0.11

chloroplasts increased by 25% only in the four-year needles from the polluted area (Figure 2A). This was the last year of needle life in severely polluted area, whereas the duration of the needle live averages 7 to 8 years and often 9 to 10 years. The changes of the chloroplast number, depending on the pollution effect during 3 years, were not found in *Vaccinium vitis-idaea* (Figure 2C). In *Arctostaphylos uva-ursi* the chloroplast numbers decreased by 26% in severely polluted area (Figure 2C) in the third year of vegetation. In regards to the deciduous plants, the chloroplast numbers decreased by 24% in *V. myrtillus*, and practically did not vary in *V. uliginosum* and *Betula pubescens* (Figure 2E). The chloroplast numbers in the herbaceous plants: *Campanulla rotundifolia, Cornus suecica, Potentilla erecta* and *Solidago lapponica* in severely polluted area either did not vary (*C. suecica* and *P. erecta*) or decreased insignificantly: by 11% in *C. rotundifolia* and by 15% in *S. lapponica*, respectively (Figure 2G).

In undamaged chlorenchyma cells of evergreen plants (*P. sylvestris, A. uva-ursi* and *V. vitis-idae*), the mitochondria number increased, two and more times: in the three-year-old needles in the lower part of the crown of *P. sylvestris* by 107% (Figure 2B), in the current-year leaves of *A. uva-ursi* by 114%, and in the one-year old leaves of *V. vitis-idaea* by 94% (Figure 2D). As for deciduous plants: *B. pubescens, V. myrtillus* and *V. uliginosum*, the mitochondria numbers in their chlorenchyma cells increased only insignificantly (by 10%, 14% and 9%, respectively), Figure 2F. In the herbaceous plants, the variation of the mitochondria numbers is also not big: in *C. suecica* and *S. lapponica* the increase of the mitochondria numbers becomes slightly more visible (19% and 9% respectively), than in the mitochondria numbers in severely polluted area (Figure 2H). In *C. rotundifolia* and *P. erecta* the increase of the mitochondria numbers, in contrast, becomes slightly less visible (10% and 5% respectively), than in the mitochondria numbers in severely polluted area (Figure 2H).

Discussion

Yarmishko et al. (1995) investigated the impact of industrial emissions on the assimilative apparatus of the plants in the 'Severonickel' smelter region. They found that the vital condition of the pine trees is significantly reduced as pollution increases. So, in the background societies of *P. sylvestris* the healthy trees constitute more than 70%, and in severely atmospheric polluted area healthy trees are entirely absent. The central stem and lateral shoots in the upper third or half part of the crowns of the most preserved trees are dry. This explains why the needles, both of the middle part and of the lower part of the crown, were also investigated. Through a severe pollution, almost 100% of the preserved needles of trees of almost all ages are damaged (except for 10 to 30% in one-year needles). In this case the intensity of the injuries (change in colour) exceeds 50% from the needle square.

As it is known, injuries of the leaves and needles become visible after injuries of its cell structure. It is generally recognized that invisible changes in the plant cells take place first on the level of the biochemical and physiological process and then proceed into the structural organization of the plant cells (Malhotra & Khan, 1988). It is indicated that the chloroplast is the first target of the SO_2 effect (Furukawa et al., 1980). It is obvious that any stage or type of degradation in the thylakoid system of the chloroplasts is harmful to photosynthesis. The reaction of photosynthesis under the effect of SO_2 depends on the duration of exposure and concentration. Prolonged effect of low SO_2 concentration causes strong inhibition to photosynthesis (Malhotra & Khan, 1988).

The study of the impact of industrial atmospheric pollution on the photosynthesis of the pine in the same areas where I carried out investigations has revealed that the intensity of photosynthesis with increased degree of pollution effects on the pine trees may be reduced on 28 to 68%, depending on the state of the pine trees (Kaibiyainen et al., 1994). The damaging effect of the pollutants on the needle structure and the disruption of the stomatal functions are, according to Kaibiyainen et al. (1994), the causes of photosynthetic inhibition.

So, only the chloroplasts of the undamaged cells take part in the process of photosynthesis and the main decline of photosynthesis is accounted for by impaired or disintegrated cells.

According to my data, the chloroplast number in undamaged cells of most investigated plants in severely polluted area had not been changing essentially. The chloroplast number has shown marked decrease in *V. myrtillus* leaves by 24% (Figure 2E) and in *A. uva-urvi* three-year old leaves by 26% (Figure 2C), and has been increasing in four-year-old needles of *P. sylvestris* by 25% (Figure 2A).

Consequently, according to the data of Yarmishko (1995) and Kaibiyainen (1994) and according to my results, it may be supposed that photosynthesis of the plants exposed to prolonged affect correlates with the number of preserved undamaged chlorenchyma cells in plants from the severely polluted area.

A less important role in the vital condition of the cells is not played by mitochondria. In undamaged chlorenchyma cells of evergreen plants (*P. sylvestris, A. uva-ursi* and *V. vitis-idaea*), the mitochondria number increases significantly (Figure 2B, D). As for deciduous plants, the mitochondria numbers in their chlorenchyma cells increased insignificantly (Figure 2F). In the herbaceous plants, the variation of the mitochondria number is also not great in severely polluted area (Figure 2H). It is difficult to analyze the variation of the mitochondria numbers in the leaves of the evergreen plants according to ages, but I will try to explain the tendency of increasing numbers.

The variation of metabolism in plants during the plant adaptation to stress conditions is shown by the variation of the energy consumption in the process of adaptation. The energy consumption for the adaptation process in its turn is spent on the adaptation to the constant ecological stress conditions: for example, (1) low temperature, and (2) for the adaptation to changing ecological conditions: temperature, water and salt stresses (Ivanova et al., 1989; Ivanova et al., 1992; Semikhatova, 1990; Semikhatova, 1995).

Atmospheric pollutants also relate to varying ecological conditions. In the investigations on the leaves of *Vicia faba*, it was shown (Kropff, 1990), that the leaves fumigated by SO_2 considerably (30%) increased the leaf respiratory capacity independently from the leaf age.

The comparison of the respiratory activity in leaves of *V. myrtillus* and in needles of *P. sylvestris* in moderately polluted area and severely polluted area (polluted by 'Severonickel' smelter complex), revealed that the respiratory activity was higher in moderately polluted area than in background area, while in severely polluted area the respiratory capacity droped below its level in background area (Yarmishko et al., 1995). It seems that the decrease of the respiratory activity in severely polluted area is due to the decrease in number of undamaged cells in the leaves and needles of the plants. This means that measured values of the respiratory capacity were average values. Then, obviously, the respiratory capacity of the undamaged cells increases abruptly, since even the average values of the respiratory capacity in severely polluted area are comparable to the values of the respiratory capacity in background area.

It should be noted that in severely polluted area plants are under the double pressure of the extreme growth conditions: low temperature and anthropogenic effects. But adaptative variation doesn't touch the main principles of respiration that are connected with the gas-exchange intensity and heat resistance of the respiratory system of plants. In other words, between the tolerance of the respiratory functions and the growth conditions of the plants, the exact link lies in the process of evolution and its mechanism need not to be looked on at an ultrastructural level, but on a molecular level.

Previous investigations of high-mountain and northern plants indicated that the mitochondria numbers increases in the chlorenchyma cells of plants grown at low temperature (Bubolo, 1984; Miroslavov et al., 1990; Miroslavov & Kravkina, 1991). Explanations have been provided that energy consumption in relation to the adaptation to the environment is determined substantially by increasing concentration of the enzyme proteins in the cell (Alexandrov, 1985; Hochachka & Somero, 1977). This is connected with increasing the mitochondria number where, as known, the enzymes of the Krep's cycle are localized. This appears to be the case during the process of adaptation both in the constant and variable ecological stresses. That is the reason why in the undamaged cells of the plants that grow in severely polluted area, I noticed a considerable increase of mitochondria number in evergreen plants, and a small increase in deciduous plants, and in two herbaceous plants (Figure 2B, D, F, H). Then it is believed that the energy consumption during the adaptation process to the anthropogenic effect in the deciduous and herbaceous plants is lower; this is clear since their vital leaf cycle of these plants has one year of vegetation. Mitochondria numbers in the *C. rotundifolia* and the *P. erecta* become lower since their vital conditions are evidently adversely affected, energy reserves are partially or completely exhausted and soon the degradation of the cell organells will probably start.

I think that increases by more than two times of the mitochondria numbers and by 25% of the chloroplast numbers in the needles (Figure 2B, A), which lived last year in severely polluted area, show the way of mobilization of all structural reserves by cells for the maintenance of the main processes of metabolism.

Thus, a quantitative strategy of the cell on the level of the main cell organell – mitochondria – is working inside an adaptative mechanism that regulates the main processes of metabolism and allows the plants to survive under severe atmospheric pollution. The increase in the mitochondria numbers is considered the adaptative response of the plant cell to environmental pollution.

Acknowledgements

I am very grateful to Prof. E.L. Zodrow for his useful advice and corrections of the manuscript.

References

Alexandrov, V.J., 1985. Cell Reactivity and Proteins, 317 pp. Nauka, Leningrad (in Russian).

Boltneva, L.I., A.A. Ignatyev, R.T. Karaban, I.M. Nazarov & T.I. Sisigina, 1982. Forecast model of vegetation injury by atmospheric industrial emissions. Interaction between forest ecosystems and pollutants. Tallinn. Part II: 163–173 (in Russian).

Bubolo, L.S., 1984. A comparative ultrastructural analysis of the leaf chlorenchyma cells in some representatives of the Wrangel Island and Leningrad region floras. Russ J Bot J 69: 1482–1491.

Furukawa, A., T. Natory & T. Totsuka, 1980. The effect of SO_2 on net photosynthesis in sunflower leaf. Studies on the effect of air pollutants on plants and mechanisms of phytotoxicity. Res Rep Natl Inst Environ Stud Japan 11: 1–8.

Hochachka, P.W. & G.N. Somero, 1977. Biochemical Adaptation. 568 pp. Princeton University Press, Princeton, New Jersey.

Ivanova, T.I., O.S. Judina & O.A. Semikhatova, 1992. About the adaptive component of respiration. Gas exchange in agricultural and natural plant phytocenoses. Syktyvkar. p. 28 (in Russian).

Ivanova, T.I., O.A. Semikhatova, O.S. Judina & G.D. Leina, 1989. The effect temperature on the respiration of plants from different zones. In: O.A. Semikhatova (Ed.), Ecophysiological Investi-

gations of Plant Photosynthesis and Respiration, pp. 140–167. Nauka, Leningrad (in Russian).
Kaibiyainen, L.K., P. Hari, G.I. Sofronova & V.K. Bolondinskii, 1994. CO_2 exchange *in vivo*: test for plant conditions as affected by persistent toxic pollutants. Plant Physiol 41 (5): 788–793.
Kravkina, I.M., E.A. Miroslavov & R.E. Crang, 1989. The impact of sulphur dioxide fumigation on photosynthetic and ultrastructural responses of mesophyll cells from developing Pinus strobus needles. I. Mesophyll cells. Proceedings of the second US-USSR symposium on: Air pollution effects on vegetation including forest ecosystems, pp. 71–81. North Forest Exp Station.
Kropff, M.J., 1990. The effect of long-term open-air fumigation with SO_2 on a field crop of broad-bean (*Vicia faba* L.). III. Quantitative analysis of damage components. New Phytol 115 (2): 357–365.
Kryuchkov, V.V. & T.D. Makarova, 1989. Aerotechnogenic impact on the ecosystems of Kola North. Apatity. 96 pp. (in Russian).
Malhotra, S.S. & A.A. Khan, 1988. Biochemical and physiological effect of the priority pollution substances. In: M. Treshow (Ed.), Ear Pollution and Plant Life, pp. 144–189. Leningrad. 535 pp. (in Russian).
Miroslavov, E.A. & L.S. Bubolo, 1980. The ultrastructure of the cells of the leaf chlorenchyma in some arctic plants. Russ J Bot J 65 (11): 1523–1530.
Miroslavov, E.A. & I.M. Kravkina, 1991. Comparative analysis of chloroplasts and mitochondria in leaf chlorenchyma from mountain plants grown at different altitudes. Ann Bot 68: 195–200.
Miroslavov, E.A., I.M. Kravkina & L.S. Bubolo, 1990. The structural adaptation of the plastidom and chondriom to the conditions of high-mountain and far North. Russ J Ecology 4: 36–42.
Pozniakov, V.Ya., 1993. The 'Severonickel' smelter complex: history of development. Aerial pollution in Kola Peninsula (April 14–16, 1992, St.-Petersburg): Proc of the Intern Workshop. St.-Petersburg; Apatity. pp. 16–19.
Semikhatova, O.A., 1990. Energy aspects of plant respiration under normal conditions and ecological stress. The 48th Timiryazev Lecture. Nauka, Leningrad. 73 pp.
Semikhatova, O.A., 1995. Maintenance respiration and the cost of plant adaptation. Russ J Plant Physiol 42 (2): 312–319.
Soikkeli, S., 1981. Comparison of cytological injuries in conifer needles from several polluted industrial environments in Finland. Ann Bot Fenn 18 (1): 47–61.
Soikkeli, S. & T. Tuovinen, 1979. Damage in mesophyll ultrastructure of needles of Norway spruce. Ann Bot Fenn 16 (1): 50–64.
Yarmishko, V.T., N.M. Deyeva, E.A. Maznaya & G.D. Leina, 1995. Impact of industrial emissions on assimilation apparatus of *Pinus sylvestris* L. and *Vaccinium myrtillus* L. of the European north of Russia. Russ J Plant Res 3: 36–51.

P.M.A. Tigerstedt (Ed.), *Adaptation in Plant Breeding*, 185–196, 1997.

Breeding widely adapted, popular maize hybrids

A. Forrest Troyer
611 Joanne Lane, DeKalb, Illinois 60115, U.S.A.

Key words: adaptation, maize breeding, genetic diversity, hybrid-by-environment interaction, information systems, performance testing, *Zea mays*

Summary

World trend is to more maize hectarage grown to fewer, more widely-adapted hbrids. My purpose is to document research events leading to the most popular corn hybrids, Pioneer Brand 3780 and 3732. I provide background on maize adaptedness, on the seed maize business, on commercial maize hybrid development process, and on cultural practice change. I provide historical information on the cultivars: Leaming corn, Reid Yellow Dent, and Minn.13 were widely-adapted open-pollinated varieties. Shift from open-pollinated to hybrid maize caused a shift in emphasis from local to wider adaptation. The first widely-adapted hybrids were Iowa 939 and U.S. 13. DEKALB 404A was the first popular and most popular double-cross hybrid. Pioneer 349 was the second-most popular double-cross hybrid. The first popular single cross hybrids were DEKALB 805 in medium and DEKALB XL45 in early U.S. Corn Belt in the early-1960's. Cultural practices (higher planting density, narrower rows, more nitrogen, better pesticides, and larger equipment allowing more timely operations) changed greatly in the 1960's and 1970's; allowing fewer, more widely-adapted genotypes to prevail. First hybrid 3780, then 3732 became the most popular hybrid; each exceeded one billion (1×10^9) dollars sales. They were the result of wide-area testing, of utilizing higher plant density stress for selecting inbred lines and for screening hybrids, and of modern information management.

Introduction

My charge is to document factors important to developing the two most popular corn hybrids. My paper provides some adaptedness and some seed-business background, provides records of cultural practice change and of widely adapted maize cultivars, describes the commercial hybrid development process, relates a chronological narrative on research events leading to hybrids 3780 and 3732, and offers my views on why they became so popular. I hope this paper provides an approach to breeding for changing times and serves to encourage plant breeders in discouraging circumstances.

Background

Darwin (1868) noted: 'Plants are more strictly adapted to climate than are animals. With almost every plant long cultivated, varieties exist for different climates. The adaptation to climate is often very close. In North America in proceeding northward, the maize varieties cultivated in each zone produce their flowers and ripen their seed within shorter and shorter periods. So, the tall, slowly maturing southern varieties do not succeed in New England, and the New England varieties do not succeed in Canada. Plants may become adapted to climates different from their own from flowering and fruiting either earlier or later in the season.' Darwin cited Alphonse de Candolle in Geographic Botany that the culture of maize extended northward 30 leagues (ca 170 km) in Europe during the first half of the 19th century. In his concluding remarks section Darwin (1868)

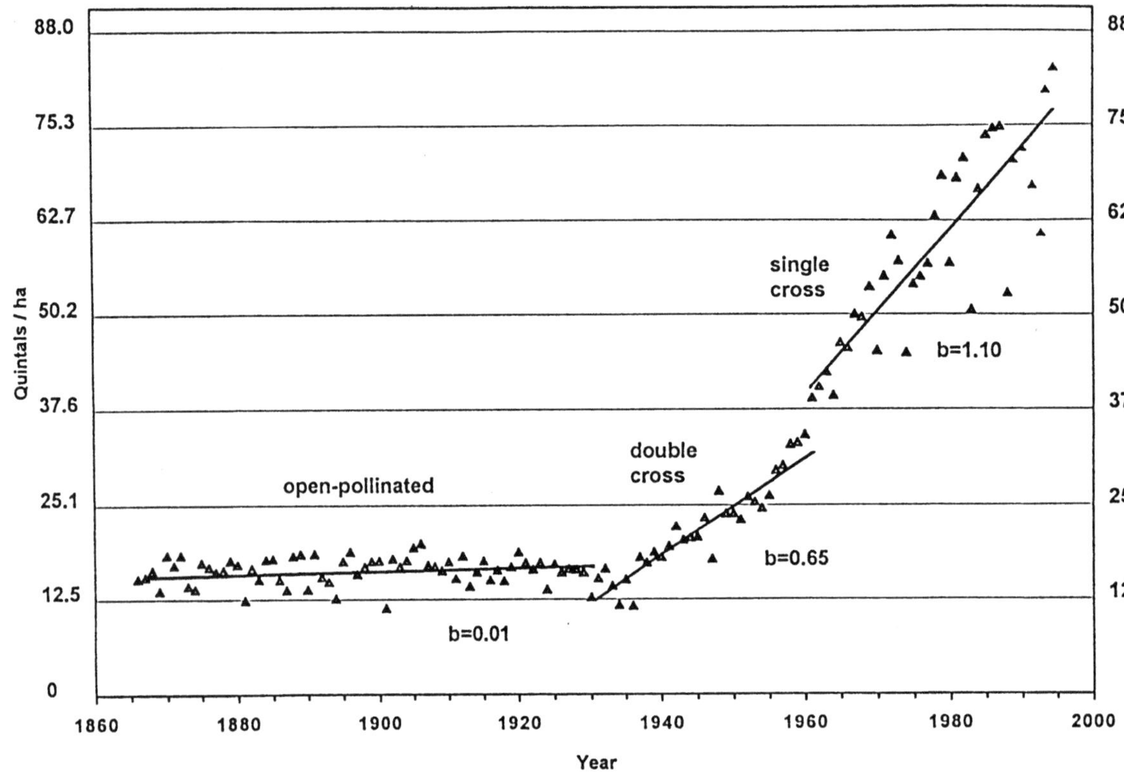

Figure 1. U.S. maize yields and kinds of maize, Civil War to 1994; b values (regressions) indicate gain in quintals per hectare per year.

stated: 'With plants, adaptation to certain soils, the power of resisting frost, the period of flowering and fruiting, the duration of life, the period of shedding leaves or retaining them through winter, the proportion and nature of certain chemical compounds in the tissues of seeds, all vary.'

Maize is grown successfully from 58° N to 40° S including tropical, subtropical, and intermediate climates. In the tropics, corn is grown from near sea level to 4,000 meters elevation. The size of the plant varies greatly from 9 to 44 leaves (Kuleshov, 1933). In temperate regions, the growing seasons are limited by colder temperatures and shorter days. Limitations are more severe in continental than in oceanic areas.

The research, development, marketing, and sales for seed corn are quite different than for other crops. Many hybrids are sold. Minnesota requires annual registration exceeding 1,500 hybrids; perhaps 6,000 hybrids are sold nationwide. Some of these hybrids have the same pedigree; but some may consistently perform either better or worse because of seed production, handling, and storage practices. Relatively few inbred lines (ca 20) are used widely, but hundreds of inbred lines are used in commercial hybrids; some are closely related. Throughout the industry, newer inbred lines are more closely related to recent, elite, widely-used inbred lines providing an unstructured, but very effective, form of recurrent selection (Figure 1). The United States is divided into 10 to 12 horizontal maturity zones. Later maize varieties are not grown in shorter-season zones because they may not mature – as Darwin pointed out. The annual U.S. seed corn market is about 24 million bags; a bag contains 80 thousand kernels.

Commercial companies have a chain of command and a business plan designed for profit. Research expenditures are an investment that either pays a return or is eliminated. Successful companies do not change methods arbitrarily; academic freedom is limited to academia. The entire episode described herein occurred while I was an employee of Pioneer Hi-Bred International, a major progressive, research-oriented seed company. Unless otherwise noted, individuals referred to were fellow employees.

Open-pollinated varieties

Earlier and more drought resistant open-pollinated varieties evolved as agriculture developed from Virginia, Tennessee, and Kentucky north and west to Ohio, Illinois, Missouri, and Iowa during the 1800's. Both natural selection and selection by humans occurred. Leaming corn evolved near Cincinnati OH in the 1850's; Reid Yellow Dent evolved near Peoria IL in the 1890's (Troyer, 1994). Minnesota 13 was developed near St. Paul MN about 1900 by ear-to-row selection for maturity and for yield; it pushed maize production 31 km north in a decade (Hays, 1904). Leaming corn, Reid Yellow Dent, and Minn. 13 were widely adapted; they received an extraordinary amount of selection for earliness by humans.

Farmers realized good seed was their best investment and many became seedsmen. More than 2,000 small seed maize companies evolved. Minnesota recommended a maize breeder for every township (91.6 square km) (Bull 1908) to develop varieties for local adaptation. Farmers bought locally grown seed for proper maturity and for tolerance to chronic, local, disease and insect problems. From 1866 to 1910, U.S. maize production quadrupled to 162 million quintals due to wider adaptation (more hectarage).

Double-cross hybrids

A significant amount (0.1%) of hybrid corn was grown in America in 1933. By 1950 80% and by 1960 virtually all of U.S. maize hectarage was planted to hybrids. Unlike open-pollinated varieties, hybrid corn performance is independent of where the seed is produced. This allowed aggressive, research-oriented seed corn companies to grow at the expense of smaller, local companies. Shifts from local to wider adaptation necessarily occurred: Natural selection aided by mass selection and by ear-to-row selection of varieties shifted to remote performance testing of hybrids. Varietal selection by local seedsmen shifted to hybrid selection by remote committees. In six decades the number of seed corn companies diminished from 2,000 to about 300.

The desire for local adaptation of maize has diminished slowly – it is not dead yet. Iowa 939 and U.S.13 (first widely-adapted maize hybrids) were developed by M.T. Jenkins in 1933 and 1935 respectively at Iowa State. They performed well over the years in several Corn Belt states; however, it was often claimed by seed dealers and generally believed by farmers, that better hybrids were available for any given smaller area. Regional tests by USDA-ARS were used primarily to introduce new inbred lines (*not* hybrids) to additional states. As recently as 1971 the purpose of USDA North Central Region uniform evaluation tests was 'to determine agronomic performance and disease and insect reactions for elite and new inbred lines when grown under several environments.' Open-pedigree hybrids were sponsored by state seed certification organizations who serviced smaller seed companies. The trend to differentiate larger from smaller companies by wider versus narrower adaptation slowly strengthened.

DEKALB 404A (first popular and most popular double-cross hybrid) was developed by Charlie Gunn of Dekalb IL in 1940. It introduced higher yields for shorter growing seasons moving the Corn Belt 125 km north (Crabb, 1948). It was the highest yielding hybrid over a large, short-season area. Within two years it sold more than 300,000 bags annually, which it did seven times during its 25-year period. DEKALB 404A twice exceeded 500,000 bags annually; its total sales were near 5 million bags. DEKALB became known as the early-corn company.

Pioneer 349 (second-most popular double-cross hybrid) was developed by Perry Collins of Algona IA in 1947. It was widely sold in northern Iowa and southern Minnesota where sales exceeded 200,000 bags annually several times in its 20-year period. Its total sales exceeded 4 million bags.

Change is my friend?

Virtually all facets of corn growing underwent large changes during the 1960's and 1970's. These changes in cultural practices affected farmers' choice of hybrids and led to the development of new approaches for phenotypic selection of inbred lines and for identifying superior maize hybrids.

Average maize plant densities increased from 36,000 to 46,000 plants/ha from 1964 to 1974 in Minnesota and Iowa. Typically, higher plant densities cause more barrenness, more stalk breaking, more root lodging, and more dropped ears; all of which decrease harvestable yield. Dropped ears are frustrating to farmers – their hybrid choice may become emotional (noneconomic); farmers value peace of mind. Increased planting densities increased seed use; more seed is used now for 32 million ha than for 52 million ha of maize 60 years ago.

Nitrogen applications on maize more than doubled (50 to 120 kg/ha) from 1965 to 1976 in the United

States. As nitrogen became cheaper and cheaper, farmers used more and more leading to higher and higher corn yields. Before cheaper nitrogen became available, many farmers grew maize only on their better fields. Nitrogen made virtually all land suitable for maize production.

The maize combine made it easier for the farmer to estimate relative yield of hybrids. With ear-maize harvest, grain yield was usually not known until the following spring. Combines moved into the U.S. Corn Belt over several years from east to west. Our company expected combines and grain driers would allow farmers to grow later maize for higher yields. In fact, magnified kernel soundness and test weight differences due to combining caused a change to earlier maize. We should have realized this sooner.

Row widths for corn were reduced about 25% from 101 cm in 1965 to 76 cm in 1980 in the U.S. Corn Belt. New planters, cultivators and combine-heads were required for narrower rows. Tax deductibility helped farmers buy larger, more powerful equipment causing more timely operations. Narrower rows and higher planting densities helped shorter, earlier maize make a more effective canopy. Narrower rows caused more uniform spacing thus complementing increasing planting densities. However, narrower rows and higher planting densities increased border effects (artifacts) in research, yield-test plots.

Single-cross hybrids

Crow's Seed Co. of Milford IL vigorously marketed single-cross hybrids from the early 1940's (R.W. Briggs personal communication). DEKALB sold some foundation seed single-crosses (parents of double-crosses) for dealers to grow from a listing called excess lots (XL). Better inbred lines and improved cultural practices made single-cross hybrids practical on a large scale. Single-cross hybrids affected hybrid choice because they differed enough in appearance to be recognizable. Single-cross breeding and production each require one less generation than double-cross hybrids; much faster feedback to the breeder occurs. Corn breeders had long grown single-cross hybrids to predict the best double-cross hybrids and to evaluate inbred lines. Single-cross hybrids interact more with the environment than double-cross hybrids.

An excellent study of single-cross, three-way, and double-cross hybrids was conducted by S.D. Jensen of York NE (unpublished). The hybrids were all-possible combinations of four unrelated, important, proprietary inbred lines; hence type of cross was the main genetic effect. The study had multiple planting densities, locations, years, and amounts of irrigation. This elegant experiment clearly showed single-cross hybrids to be most superior. Concern about interaction of single-cross hybrids with environments still existed. W.L. Brown, vice president, was among the first to clearly point out that the increased variation of single-cross hybrids was at a higher yield level and would rarely result in lower yields than a comparable double-cross hybrid. Though we analyzed the situation correctly, DEKALB first marketed popular single-cross hybrids.

Role of a maize breeder

My research station, near Mankato in southern Minnesota, was established in 1958 to help protect a 30% market share built by Collins at Algona IA and William Landgren at Willmar MN. Our hybrids (all proprietary) flowered later, were taller, yielded well, and dried faster partly due to less hard starch in the kernels and to premature plant death. Collins' breeding strategy replaced or improved the poorest of four inbred lines in a successful double cross formula; so, we had many commercial hybrids that were 75% and 88% related to each other.

When full size, my station had four full-time employees plus seasonal help. We made 100,000 hand pollinations and serviced 15,000 yield test plots at 15 locations in southern Minnesota, northern Iowa, and central Wisconsin. We tested 250 new stock number hybrids annually plus several hundred new single-cross hybrids primarily to evaluate inbred lines. We developed 100 to 200 new inbred lines annually that were evaluated in testcrosses. In today's dollars, the annual budget would be about $ 350,000. W.B. Ambrose became the assistant manager in 1959. He transferred to Middletown DE as manager in 1960 and G.R. Herrick replaced him.

I left the University of Minnesota with much enthusiasm for the Rinke backcross method to make later flowering inbred lines earlier (Rinke & Sentz, 1961). Raymond Baker, research director, suggested adding F2 populations for selection before backcrossing and helped choose faster-drying, late inbred lines to make earlier (Troyer 1994). (My publications are usually delayed several years for competitive reasons.) He also encouraged studying field drying rate of corn. We studied both artificial treatments on hybrids and also natural differences among 150 open-pollinated varieties

(Troyer & Ambrose, 1971). We learned, among other things, that looser husk cover helps maize dry.

I spent much time with sales people the first few years while the research program was growing; I helped answer service calls – sales agronomists had not yet been invented. Our hybrids' stalks broke in longer- or in warmer-than-average seasons due to premature death and to stalk rot. Sometimes northern leaf blight (*Helminthosporium turcicum*) predisposed the plants to stalk rot. We got approval to establish a remote leaf disease nursery where we routinely did early-generation, phenotypic selection and rated hybrids under heavy leaf blight infection.

DEKALB 805 (first popular, single-cross hybrid) was introduced in 1958 in the central U.S. Corn Belt; it was a tall, large-eared hybrid that yielded very well under good conditions. It was more popular where growing-season rainfall is generally higher. Sales grew slowly to exceed 100,000 bags in 1962 and 250,000 bags in 1964. Its sales declined rapidly after the 1965 drought year. Total sales were about 600,000 bags over a decade.

Our first, early, single-cross hybrid was 3755 in 1962. It was the female parent of some of our best double-cross hybrids. The following year 3775 was launched; it was a non-parental single cross (one inbred line in one parent the other inbred line in the other parent) in some of our best double-cross hybrids. These were Collins' inbred lines and hybrids. They sold 5,000 to 10,000 bags annually. Their uniform appearance was new, but they were uniformly old fashioned in performance traits.

DEKALB XL45 (first early, popular, single-cross hybrid) was introduced in the northern U.S. Corn Belt in 1963; it was developed by G.H. Stringfield, research coordinator at DEKALB, formerly of Ohio State University. XL45 was the right hybrid at the right time. It was widely adapted – grown from Minnesota to Texas and from Colorado to Delaware. It was shorter and earlier flowering; it was perceived to be adapted to higher planting densities and to narrower row width. Its kernels had more hard starch, which coupled with earlier flowering reduced grain damage when combined at higher grain moistures. Crop modelers praised XL45's earlier flowering that provided more grain-filling period (sink). Smaller plant size to reduce lodging with higher nitrogen application was in accord with the Green Revolution philosophy (Chang, 1967). Sales grew rapidly to exceed one million bags in each of three different years in the 1960's. Total sales were about 8.5 million bags over an 18-year period.

XL45/XL45A (reciprocal crosses) changed my outlook on adaptation. Why was it so popular? If one hybrid could be widely adapted, why not all? I literally went to sleep every night and awakened every morning thinking about adaptation. The Green Revolution was happening. Borlaug's shuttle breeding (disruptive selection), using nurseries in central (winter) and in northern (summer) Mexico (Borlaug, 1965), was the subject of much plant breeding debate. Disruptive selection had been criticized by H.K. Hayes of Minnesota; was he finally wrong about something? Something important? The Rice Institute claimed erect leaves were essential to higher yields – are they really?

In 1964, Successful Farming magazine estimated 10% of U.S. maize hectarage was planted to single-cross hybrids. We added test weight as an additional performance trait. I made cups from plastic pipe to take a uniform volume of grain from the moisture sample to weigh on a gram diet scale. We used regression to adjust for moisture content. We were soon determining 200,000 test weights annually on performance test hybrids among our research stations. Our company's largest-selling hybrid, 368A (double cross), dropped ears badly. Of course its closely related sister hybrids 368 and 3658 did too. These hybrids totaled 230,000 bags annually for a three- state area. We had experienced a hot, dry, windy fall after heavy European corn borer (*Ostrinia nubilalis*) infestation in ear shanks. I spoke at 8 or 10 dealer meetings in the worst areas assuring them it was unlikely to occur again, but it did – the following fall. Meanwhile XL45 had a great 1964 and also did well in 1965, a year that had an unusually cool fall causing premature death. I saw tropical origin as the cause of cool temperature intolerance and started selecting routinely for stay-green as an antidote for premature death (Troyer, 1978).

Our company's headquarters in central Iowa was surrounded by corn fields frequently suffering drought stress. Everyone there readily agreed that corn planting densities had increased virtually every year; yet, they expected densities to level off or to decrease next year. The XL45 phenomenon convinced me that adaptation to crowding was a trait desired by farmers. We had been using higher plant densities to save space where large populations were grown to select for earliness. We recognized genotypic differences for delayed silking (Troyer, 1967; Troyer & Brown, 1976). I decided twice the average planting density of local, progressive farmers would be advantageous for early-generation, phenotypic selection. Strong silking and good ear development were most important, but we also selected for erect leaves to reduce self-shading and for stay-green.

Our use of much higher planting densities for phenotypic selection was generally known. I sometimes received admonitions on selecting smaller ears. (Actually we kept larger-than-average ears of survivors.) As it turned out, no one (including me) predicted the most frequent noticeable effect, which was smaller tassels probably due to changed resource partitioning. Most importantly, the resulting inbred lines were tougher as inbred lines and in hybrids. They were more widely adapted.

Jim Weatherspoon of Johnston IA was in charge of yield testing. He expected higher plant densities to decrease accuracy of yield tests – like drought. Baker and I talked with him two or three times in 1964 and 1965 about substantially increasing yield-test plant densities and we finally agreed that I would do experimentation to determine the idea's merit. We grew 84 different paired-density trials over a 9-year period totaling 34,410 replicate-entry plots (Troyer & Rosenbrook, 1983). We handed out current-year and year-to-date summaries of plant density effect on experiments each year at our annual research meeting. Testing at higher than optimum (yield decreasing) plant densities increased barrenness, stalk breakage, root lodging and ears dropped; it also increased differences among hybrids thereby improving effectiveness of selection.

The four northern research stations I worked with applied 300 kgN/ha to nurseries and to yield test locations. We wanted to get ahead of the increasing nitrogen trend. More nitrogen also provided better stalk breakage data. Eight years later the wisdom of our higher nitrogen rates was seriously questioned.

In 1965, 3911 was launched, the first single-cross hybrid from Mankato. It had excellent test weight and attractive field appearance. It helped improve our test-weight image. Hybrid 3773 was launched. Its parents were developed by Collins and by Landgren; it was recommended by an eastern station. When it was good it was very good and when it was bad it was awful. Its nemesis was higher plant density stress. Stress caused the ears to be replaced with smut balls (*Ustilago zea*) – like black flags. It had most of the things we wanted except what we wanted most, which was tolerance to higher plant density stress. It lasted 11 years exceeding 100 thousand bags annually some years. I helped kill it.

Because things were going so poorly, it was a good time to instigate change. Management was looking for solutions, too. I got approval to exchange breeding material inventories among stations by arguing quality was more important than quantity of breeding starts. I encouraged more plants and pollinations in F2 and BC1 populations where greatest segregation occurs and fewer rows and pollinations in advanced-self (S5+) rows where little segregation is expected; and encouraged earlier testing (S4 then S3) to increase the number of families tested, to decrease advanced-self effort, and to decrease cycle time.

Recipe for revelation

I kept thinking about adaptation each night before going to sleep (Moore, 1969; Kauffman & Kauffman, 1994). Obviously a widely-adapted hybrid existed (XL45); the proper question was: How to identify them? One morning in the spring of 1966 I awakened and it was as clear to me as if written on my bedroom wall; TEST WIDELY. I understood this was the answer, but it was contrary to traditional, ripple-in-a-pond progression over time of testing sites.

With some persuasion, six research stations provided number of extra kernels of each advanced hybrid along with number of kernels required for each advanced hybrid experiment; these hybrids were assigned to additional stations for more testing. This approach made more efficient use of seed and also provided more useful hybrid comparisons than would separate, additional, wide-area testing experiments. It still will. An advanced hybrid was tested at 15 to 40 locations instead of 5 or 10; we used existing locations more effectively for wider adaptation. Results for environmentally-affected traits were summarized on index cards (Figure 2). Later, the number of locations was increased at the expense of replications (Sprague & Federer, 1951) and all data were summarized by computer. Now, an advanced hybrid is commonly tested at more than 100 one-replication locations. More locations identify more adaptedness genes.

First scent

Minhybrid 4201 was first grown commercially in 1966. It was developed by E.H. Rinke at the University of Minnesota. It was higher yielding and earlier than XL45 (Table 1A). Nearly all companies in the northern U.S. Corn Belt marketed Minhybrid 4201 under their own brand name and number. The aggregate approached one million bags annually. Our company did not sell it and our market share continued to erode. Baker and I talked about further exploiting the same

Table 1. A. Four-row small plot (1/598 ha) at moderate and at high planting densities with center two rows machine harvested results comparing XL45/XL45A and Minhybrid 4201 (TXS102) at 147 locations in 7 years (Iowa Corn Yield Test)

Hybrid	Yield q/ha		H_2O	Stalk broken	Root lodged	Ears dropped
	Mod.	High	%	%	%	%
XL45/XL45A	46.5 **	47.6 **	23.1 **	5.1	2.2 **	0.3 *
4201	51.7	52.5	21.6	5.4	3.6	0.4

*, ** Performance differs significantly at P 0.10, 0.05, 0.01 respectively.

3956 Stock No.   Pedigree   95 RM   1966 Year

RATE: 1(poor)-5(good)

Exp. No.	Sol. Indx.	% Ck Y - M	Root. Ldgo	Brk. Stlk	Drpd Ears	Retest	Reason
1100	106.7	-2	2	5	-	R	Stalks
1101	00.0	+24	2	5	3	R	Stalks, Y/M
1102	00.0	+23	2	4	4	R	Stalks, Y/M
2111	153.6	+23	4	4	3	R	Y/M
4102	00.0	+25	-	3	-	R	Y/M
5101	119.1	+15	-	5	-	R	Y/M, Stalks

Figure 2. Wide-area testing data summary card.

heterotic pattern; I made additional hunch hybrids. L.J. Hoffbeck joined the company to expand work on early corn at Tipton IN.

Hybrid 3956 (Figure 2) performed well again in 1967. Wide-area testing at higher planting densities was helping, but the right hybrids were not yet in sight. I seriously considered a job offer to work with M.T. Jenkins back home in Indiana near my elderly parents. I could not do it; it was not my nature to leave while we were losing.

A phone call from Brown about Minhybrid 4201 resulted in additional modified, single-cross hybrids. The surviving X7049 became 3538 which was sold in Illinois for several years. J.T. Watson joined the company; he interned a year at Mankato then managed the Woodstock ONT research station. He developed an excellent testing program of 21 locations in southern Ontario.

A very serious 1968 letter about declining market share and research budgets from Brown asked for specific strategies to improve product performance. We had talked at length about the things I was doing. The crucial part of my answer was that Baker and I were impressed with a new faster-drying (looser husks) version of a widely-adapted, elite inbred line developed by Stan Bettoney at Johnston IA. Baker was making several medium maturity hybrids and I was making 50 early hybrids with Bettoney's new inbred line. Baker's best became 3518. The best early hybrid turned out to be X8094. I asked Ambrose to transfer to Algona IA on Collins' retirement. R.W. Rosenbrook joined the company as station manager at Mankato MN.

Hybrids 3582 and 3579, a single cross and modified single cross, were launched in 1968 and 1969. They had harder kernels and earlier flowering like XL45. The earlier inbred line was developed by Landgren. Baker was their chief advocate. Together they exceeded 100 thousand bags annually a few years in their six or seven-year life span; they helped. In 1969, X8094 performed very well.

In the late fall of 1969, I returned from a 10 day business trip to find Herrick and Rosenbrook bursting with information. The recent parent test results indicated X8094 female was low yielding and required a planting delay; X8094 was virtually dead. They had found testcross seed of Bettoney's inbred line with one very closely related to the female of X8094. The related inbred line was later flowering and a good female parent; it was doing well in X7049 as a male. Ambrose at Algona IA, and Walt Vandeventer at Tipton IN had agreed to send us their seed of this cross. Either could have requested a stock number and claimed it; they kindly honored our pursuit on this trail. (The odds of an untested hybrid seedlot becoming a commercial were about 1/500.) Not enough seed existed to maximize testing, but Herrick and Rosenbrook had found where the needed inbred lines were growing. Rosenbrook would help pollinate that winter and carry pollen to make the needed seed. I recruited three of these breeders, and I was proud of what they had done. A

U.S. Army axiom states; 'Leadership is best measured during absence of the leader'. This hybrid of borrowed and of unplanned seed became X9027.

Landgren's 3784 was launched in 1970. It was in normal cytoplasm and unaffected by *Helminthosporium maydis T* leaf blight. It exceeded 100 thousand bags in 1971 and again in 1972, then faded rapidly – not enough stress tolerance. Hybrid 3711 from Mankato found a home in Illinois. It was short and attractive like XL45, but only average overall performance. Herrick resigned after a family tragedy; he was replaced by A.R. Grunst. X9027 performed very well, but we learned that Bettoney's inbred line would not shed enough pollen for western seed production. What next? Baker had modified it to make 3517. Grunst made enough seed of X9027 and X9027A (modified male) side-by-side in winter nursery for advanced wide area testing. We hoped to slipstream X9027's previous testing.

Brown became president in 1971. D.N. Duvick became research director; he improved the testing program by including strip test results for hybrid advancement. Strip tests for marketing corn hybrids were first widely used by Jacques Corn Co. of Prescott WI in the mid 1960's. They are planted and cared for by the farmer using his own cultural practices. Strip test plots are much larger (150–200X) than research plots with no alley effect and with 60 to 80% less side-border effect. I was asked to also coordinate eastern corn research (four more research stations), and to chair commercial hybrid advancement meetings for four of six sales areas. We changed the meetings from small, smoke-filled rooms to more open, more comprehensive, data-driven discussions; more interaction with sales and marketing helped. Weatherspoon did most of the work (data processing). Hoffbeck became manager at Tipton IN.

Weatherspoon had visited me at the University of Minnesota in 1957 to observe computer use for yield trial data. Along with computers, I recommended unique hybrid reference numbers to form a data base; this finally happened in 1971. They had been delayed 14 years because of limited space on an 80 column punch card. Wide-area testing and unique hybrid reference numbers fit together well. Weatherspoon brought together data for hybrid sets. Two hybrids at a time (head-to-head comparisons) provided the most locations (Bradley et al. 1988).

X9027A was twice removed from the list of potential new commercials. I thought it fell through a crack the first time, and put it back on. The second time I looked for and found two senior executives overly concerned about test weight. I asked Grunst to make test weight comparisons with Minhybrid 4201 at different moistures from our research data, then from our maturity-trial data that used official size and shape test weight cups. X9027A stayed on the list.

Ambrose and I made the product launch presentation to sales managers on the Perservence pays off past performance and future expectations of X9027A/3780; it went well. Hybrid 3780 was a paragon of adaptability to higher yield conditions; the higher the yield level the greater its winning margin. Yet, it had good stress tolerance. It showed marked superiority over Minhybrid 4201 for yield, barrenness, stay-green, stalks, and roots (Table 1B). Furthermore the superiority widened with higher plant densities (Troyer & Rosenbrook, 1983).

Pioneer 3780 was first commercially grown in 1972. The introductory sample was 10,000, then 60,000, then more than one million bags every year for more than a decade with multiple near two million bag years. Its total sales were more than 20 million bags. Farm Industry News named 3780 the landmark product maize hybrid having the most impact on farming operations for the past 25 years based on farmers' polls (Zenk, 1992). They noted consistently higher yields and faster dry down. Hybrid 3780 (second-most popular hybrid) generated sales in excess of \$ 1 billion ($1X10^9$). X9027/3780A became the most popular hybrid in Canada.

Another trail

Hybrid 3785 from Tipton IN was an excellent companion for 3780. The hybrids were virtually unrelated with 3785 flowering a day earlier. It was a new heterotic pattern for early maize. In a few years, it was selling 150,000 bags.

In 1973, the energy crisis caused nitrogen price to double. Rumors started that 3780 would lose its competitive advantage at lower levels of nitrogen because it had been developed at higher levels. Duvick asked each research station to find a field with a history of low nitrogen application to compare 10 to 12 appropriate hybrids to 3780 at 0, 60, 120 and 240 kgN/ha. No genotype-by-nitrogen interaction occurred. The net effect of the energy crisis was to increase 3780 sales because of its faster field drying that reduced drying costs.

More research stations and more sales areas meant more travel, more people, and more maize. Baker developed an excellent medium maturity inbred line.

Table 1. B. Small plot (1/1235 ha) results comparing Minhybrid 4201 and 3780 at 299 location in 6 years (Troyer & Rosenbrook, 1983)

Hybrid	Yield q/ha	H_2O %	Barren %	Stay-green %	Stalk broken %	Root lodged %	Ears dropped %
4201	46.7	20.4	2.7	84	11.7	13.2	0.6
	**	**	**	**	**	**	**
3780	49.0	19.1	1.5	101	10.1	3.2	0.9

*, ** Performance differs significantly at P 0.10, 0.05, 0.01 respectively.

It yielded and stood well in several hybrids grown in many environments. Tassel size was limiting, so it was a female only. Then it failed as a female (heat and drought) in 3549 seed production in western U.S. Corn Belt. Who better could make it earlier and more stress tolerant?

I noticed the related S1 pedigree (Baker's new inbred line by female of 3785) in Rosenbrook's section of our remote disease nursery. I requested seed to send to the winter program for the above breeding objective. I used it in a newer method that coordinated inbred line phenotypic selection, hybrid testing, parent testing, and seed increases using two generations per year for minimum time. I selected for earliness and for stress tolerance. A resulting testcross became X4772. Rosenbrook continued with the pedigree at a slower pace.

In 1974, Owen Newlin, president central sales area, produced 600,000 bags of 3780. Hybrid 3710 was launched. It was another Hoffbeck hybrid – a newer-model 3785 that was soon up to 150,000 bags a year. Hybrid 3780B was launched; it was Ambrose's hybrid. It was a modified female hybrid that was predicted from X8094 and from X9027A. It sold 70,000 bags annually. Hybrid 3596 was launched; it was Jensen's hybrid. It was also a modified female hybrid half predicted from X8094. It sold 100,000 bags annually.

In 1975, 3709 was launched, still another very good Hoffbeck hybrid related to 3785. The data from 1974 and 1975 showed it could beat 3780 at below average yield levels. It was genetically diverse from 3780. We produced over 400,000 bags for 1976.

Pioneer 3732 (X4772) was advanced to commercial in Canada in 1976. It was developed by the FAST method (shuttle breeding) in only four years (Troyer, 1990, 1994). Related parents of the breeding population were chosen based on specific combining ability with the potential male parent developed by Hoffbeck. Both parents and three of four grandparents of 3732

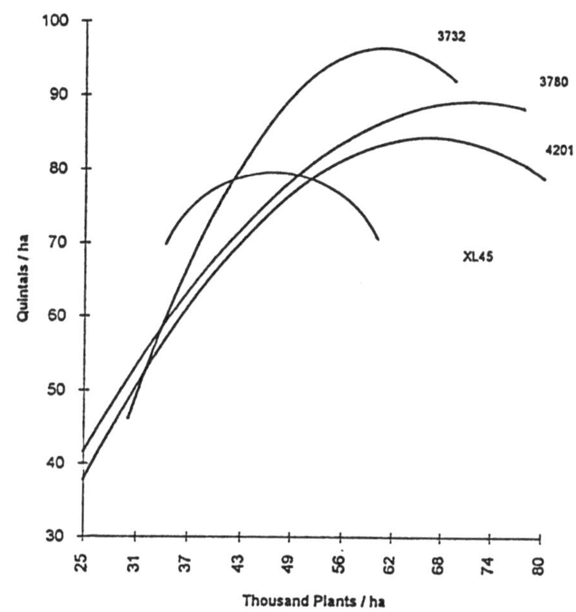

Figure 3. Yield response to plant density for XL45, M4201, 3780, and 3732 (see Table 1).

were developed by phenotypic selection under higher plant density stress. It had a higher level of stress tolerance for the northern U.S. Corn Belt (Figure 3, Tables 1C, D). It won all critical comparisons in 1976. In 1977, its U.S. competition was Ambrose's X5711, which appropriately dropped ears. (Our saga begins and ends with dropped ears.) Hybrid 3732 was 75% genetically diverse from 3780 (Troyer et al., 1988). It had several near two million bag years. Its total sales are more than 25 million bags. It is still sold. Hybrid 3732 (most popular hybrid) has generated sales in excess of $ 1 billion (1X10⁹).

Table 1. C. Strip test (1/8 ha) results comparing 3780 and 3732 at 51 locations in 4 years (Troyer & Rosenbrook, 1983)

Hybrid	Yield q/ha	H_2O %	Stalk broken %	Root lodged %	Ears dropped %
3780	55.7	19.5	5.7	1.6	1.5
	**	**	**		+
3732	57.4	20.8	2.1	1.6	1.0

+, *, ** Performance differs significantly at P 0.10, 0.05, 0.01 respectively.

Table 1. D. Small plot (1/1235 ha) results comparing 3780 and 3732 at 362 locations in 4 years (Troyer & Rosenbrook, 1983)

Hybrid	Yield q/ha	H_2O %	Barren %	Stay-green %	Stalk broken %	Root lodged %	Ears dropped %
3780	51.5	19.9	1.9	100	10.6	2.8	1.0
	**	**		**	**	*	
3732	52.7	21.1	1.1	124	4.8	3.8	0.9

*, ** Performance differs significantly at P 0.10, 0.05, 0.01 respectively.

Conclusions

Good hybrids do not drop from the heavens; they are made – not born. Pioneer 3780 and 3732 were made of necessity. They were needed to meet the challenge of changes occurring in the 1960's and 1970's. To me they are symbols of overcoming adversity and of planning for change – the ultimate triumph of a few human spirits (team efforts). In plant breeding terms, they were the result of wide area testing, of utilizing higher plant density stress for selecting inbred lines and for screening hybrids, and of modern information management.

Hybrid maize and the Green Revolution benefitted mutually from each other's success. Improved cultural practices for hybrid maize influenced the Green Revolution – the farmer is willing to change the way he grows his crop. Disruptive selection, wider adaptation, crop geometry, and plant morphology for the Green Revolution influenced hybrid maize; they certainly influenced me.

What are the adaptation consequences of genotype-by-environment interaction? From the genotype standpoint, genetic diversity (multiple genotypes) is an adaptive response to different environments – different genotypes for different niches. From the environment standpoint, a more homogenous environment allows fewer genotypes to prevail. A completely homogenous environment allows a single genotype to always be superior. Cultural practices are ways of adapting environments to genotypes. Increasing planting densities (more factories), narrowing rows (more energy), and increasing nitrogen rates (more fuel) provided additional resources to make the environment more homogenous. Fewer genotypes, particularly those having survived natural selection and methodical selection for these additional resources, prevailed.

The general rules of adaptation apply. XL45 was grown virtually everywhere in the United States except in the extreme northern parts where it was too late; 3780 and 3732 were too. Earlier maize always has the potential to be more widely adapted than later corn, but hybrids that use all the season, where and when it is limiting, have more time to produce yield. Stay-green helps earlier corn use more season.

Hybrid-by-year interaction is largest. Growing seasons may be long, short, warm, cool, wet, or dry. Earlier flowering increases grain filling time in shorter or in cooler seasons and usually avoids stress in warmer or in drier seasons. Stay-green and deeper kernels allow earlier hybrids to increase yield in longer, in warmer, or in wetter seasons.

Hybrid-by-location interaction is most frequently caused by differing moisture availability at flowering. Synchronization of maize imperfect flower signals better water use (Troyer & Brown, 1976). Selecting for stronger silk emergence under higher planting densities (fresh silks are 90% water) allows drought tolerance selection under non-drought conditions (Troyer, 1983). Disease and insects can cause interaction, but moderate tolerance is usually enough; evolution couples higher pest tolerance with lower yield to preserve the species in varying conditions (Allard, 1990).

Hybrid-by-soil-type interaction is rare. However, in the early 1970's, popular hybrid MO17 × N28 performed very well on older, unglaciated, hardpan soils (located beyond the Wisconsin glacier) because of inbred line N28's exceptional root system that could penetrate hardpans. Soil type affects water holding capacity thus it can affect moisture availability at flowering. Previous crop also affects subsoil moisture; soybeans remove less, sorghum and sunflowers remove more water than maize. Water is the key.

Ceccarelli (1989) generates variety crossover interaction with widely different (0.80 vs. 1.17) stability regressions extended to very low (< 2.0 t/ha) yield levels. His logical approach does not apply to hybrid maize as I know it for two reasons: Stability regressions for popular maize hybrids usually range from 0.95 to 1.05, and maize growers expect yields higher than 2.0 t/ha.

Maturity and selection for earlier maturity are important to adaptedness in corn. Mixing maize maturities increases adaptedness; yield from late southern dents and earliness from northern flints resulted in U.S. Corn Belt maize. Selection for earlier flowering has increased yield-for-maturity (efficiency) of Reid Yellow Dent: L.C. Burnett at Iowa State developed Iodent variety, Baker developed Modern Iodent, and Troyer and Hoffbeck developed Early Iodent (Troyer, 1994). Selection for earlier flowering in the Rinke method develops elite, earlier inbred lines (Rinke & Sentz, 1961); it also provides genetically diverse materials for their new, shorter-season area of adaptation.

More locations identify more adaptedness genes by sampling more environments. More environments better evaluate the genotype's daylength, temperature, and efficient-water-use adaptedness. Temperature-effect and water-use interact with growth stage; daylength response interacts with latitude and with planting date; so, more environments help. Also, more locations increase chances of collecting rare, discerning, agronomic data.

I sometimes hear people who have never developed a popular hybrid profess that corn breeding is a numbers game – number of test crosses or number of inbred lines. Perhaps this is an alibi or a self-fulfilling prophecy; games are what we make them. Over a 4-year period, between 100 and 200 hybrids (depending on classification) were made and tested between good inbred lines in one heterotic pattern to find 3780. Hybrid 3732 resulted from developing a new inbred line from an elite, related-line, S2 population to cross with Hoffbeck's very elite inbred line. I developed only two inbred lines and discarded one based on per se barrenness. One of three testcrosses became 3732. You can probably guess my next hybrid: My inbred line by Baker's as female parent with Hoffbeck's inbred line as male became 3713. No new inbred lines, no new testcrosses – only maize breeding logic. The numbers that seem important to me are number of hearts and minds, and number of eyes and hands.

The quickest and greatest benefit from higher plant density performance testing was in central and northern Minnesota and in Canada where market shares went from an all-time high of 7 or 8% in the mid-1960's, when yield test plant densities were increased, to more than 30% in the mid-1970's. Early Iodent material was very helpful in these areas. Selection at higher plant density, as practiced in maize breeding, is survival of the fittest in its purest form resulting in tougher inbred lines and hybrids (Troyer, 1994). Tougher hybrids are more needed in areas farther removed from the tropics where domesticated maize originated.

What happened to market share? It went up. Our station met its obligation. Our company's annual seed maize sales in North America increased more then 2.5 million bags from 1972 to 1977. This increase was approximately incremental to sales of 3780 and 3732. We became No. 1.

Acknowledgement

I appreciated the support of sales and marketing personnel during the lean years – I never doubted increased market share would follow superior products. I appreciated R.W. Rosenbrook and A.R. Grunst adjusting their busy schedules to accommodate my many business trips and my corn breeding project; thank you Bob and Al. I thank R.W. Allard and A.R. Hallauer for reviewing the manuscript and offering many helpful suggestions.

References

Allard, R.W., 1990. The genetics of host-pathogen coevolution: implications for genetic resource conservation. J Hered 81: 1–6.

Bull, C.P., 1908. Corn breeding in Minnesota. Minn AES Bul 107: 170–256.

Borlaug, N.E., 1965. Wheat, rust, and people. Phytopathology 55: 1088–1098.

Bradley, J.P., K.H. Knittle & A.F. Troyer, 1988. Statistical methods in seed corn production. J Prod Agric 1: 34–38.

Ceccarelli, S., 1989. Wide adaptation: How wide? Euphytica 40: 197–205.

Chang, T.T., 1967. Growth characteristics, lodging and grain development. International Rice Commission Newsletter Special Issue: 54–60.

Crabb, A.R., 1948. The hybrid corn makers. Rutgers University Press, New Brunswick.

Darwin, C.R., 1868. The variation of animals and plants under domestication. John Murray, London.

Hays, W.M., 1904. Breeding animals and plants. The University Press, Minneapolis.

Kauffman, G.B. & L.M. Kauffman, 1994. Linus Pauling: Reflections. Amer Sci 82: 522–524.

Kuleshov, N.N., 1933. World's diversity of phenotypes of maize. J Amer Soc Agron 25: 688–700.

Moore, A.D., 1969. Invention, discovery, and creativity. Anchor Books, Doubleday & Co., Inc Garden City, NY.

Rinke, E.H. & J.C. Sentz, 1961. Moving Corn Belt germplasm northward. Proc ASTA Hybrid Corn Res Conf 16: 53–56.

Sprague, G.F. & W.T. Federer, 1951. A comparison of variance components in yield trials. II. Agron J 43: 535–541.

Troyer, A.F., 1967. Yield as influenced by maturity and population. Proc ASTA Hybrid Corn Res Conf 22: 91–98.

Troyer, A.F., 1978. Corn yield as influenced by flowering date and season. Proc ASTA Corn Sorghum Res Conf 33: 1–15.

Troyer, A.F., 1983. Breeding corn for heat and drought tolerance. Proc ASTA Corn Sorghum Res Conf 38: 128–143.

Troyer, A.F., 1990. A retrospective view of corn genetic resources. J Hered 81: 17–24.

Troyer, A.F., 1994. Breeding early corn. In: A.R. Hallauer (Ed). Specialty Corns. CRC Press, Boca Raton.

Troyer, A.F. & W.B. Ambrose, 1971. Plant characteristics affecting field drying rate of ear corn. Crop Sci 11: 529–531.

Troyer, A.F. & W.L. Brown, 1976. Selection for early flowering in corn: seven late synthetics. Crop Sci 16: 767–772.

Troyer, A.F. & R.W. Rosenbrook, 1983. Utility of higher plant densities for corn performance testing. Crop Sci 23: 863–867.

Troyer, A.F., S.J. Openshaw & K.H. Knittle, 1988. Measurement of genetic diversity among popular commercial corn hybrids. Crop Sci 28: 481–485.

Zenk, P., 1992. Readers choose landmark products. Farm Industry News vol 25 no 11. Webb Publishing, Minneapolis.

P.M.A. Tigerstedt (Ed.), *Adaptation in Plant Breeding*, 197–205, 1997.

CIMMYT's approach to breeding for wide adaptation

Hans-Joachim Braun[1], Sanjaya Rajaram[2] & Maarten van Ginkel[1]
[1] *International Maize and Wheat Improvement Center (CIMMYT), P.K. 39 Emek, 06511 Ankara, Turkey;*
[2] *CIMMYT, Apdo Postal 6-641, 06600 Mexico D.F., Mexico*

Key words: adaptation, durable resistance, N-use efficiency, stability, wheat, yield potential, *Triticum aestivum*

Summary

The wheat area in developing countries, including China, is around 100 million ha. To address the needs of these very diverse wheat growing areas, CIMMYT has defined 12 wheat mega-environments (ME). A ME is defined as broad, not necessarily continuous often transcontinental area with similar biotic and abiotic stresses, cropping systems and consumer preferences. The factors describing each ME are presented.

CIMMYT's breeding methodology is centered around the development of widely adapted germplasm with high and stable yield across a wide range of environments. Segregating populations are alternating screened in two diverse environments in Mexico. One key requirement is that all germplasm is tested under near optimum conditions for its yield potential. The second one is multi-locational testing of advanced lines at sites that represent a given ME (key locations) and careful screening of germplasm for tolerance to abiotic and biotic stresses specific to that environment. This methodology has permitted the pyramiding of a large number of multiple resistance genes for use against a wide spectrum of diseases and tolerance to abiotic stresses within each ME. In addition, the widespread testing of lines allows the identification of traits which are beneficial in several environments. Data from international nurseries are used to further delineate environments within an ME. This approach has proven to be successful since around 70% of the spring wheat area in developing countries (excluding China) is planted to varieties derived directly or indirectly from CIMMYT germplasm. The performance of the bread wheat cultivar Pastor in international trials is given as an example for a wide adaptation.

Introduction

CIMMYT's breeding methodology is tailored to develop widely adapted, disease resistant germplasm with high and stable yield across a wide range of environments. The impact of this approach has been significant. The total spring bread wheat (*Triticum aestivum* L.) area in developing countries, excluding China, is around 63 million ha of which 36 million ha or 58% are planted to varieties derived from CIMMYT germplasm (Byerlee & Moya, 1993; Rajaram, 1995). During the period of 1966 to 1990 1317 bread wheat cultivars were released by developing countries, of which 70% were either direct releases from CIMMYT advanced lines or had at least one CIMMYT parent (Byerlee & Moya, 1993). For the period from 1986 to 1990 84% of all bread wheat cultivars released in developing countries had CIMMYT germplasm in the pedigree. Simultaneously the use of dwarfing genes has continued to increase over time and today, regardless of the type of wheat, more than 90% of all wheat varieties released in developing countries are semidwarfs, which covered by the end of 1990 70% of the total area in developing countries (Byerlee & Moya, 1993). The continuous adoption of semi dwarf spring wheat cultivars in the post-Green Revolution period from 1977–1990 resulted in about 15.5 million tons of additional wheat production in 1990, valued at about 3 billion US$, of which 50% or 1.5 billion US$ are attributed to the adoption of new Mexican semi dwarf wheat cultivars (Byerlee & Moya, 1993). In 1990, an estimated 93% of the total spring bread wheat production in developing countries, excluding China, comes from semi dwarf spring wheats, which cover about 83% of the

Table 1. Classification of megaenvironments (MEs) used by the CIMMYT wheat program

ME	Latitude (degrees)	Area (mill./ha)	Moisture regime	Temperature regime	Growth habit	Sown	Major breeding objectives	Representative locations/regions	Year breeding began at CIMMYT
Spring Wheat									
1	Low	32.0	Low rainfall irrigated	Temperate	Spring	A	Resistance to lodging, SR, LR, YR	Yaqui Valley, Mexico Indus Valley, Pakistan Gangetic Valley, India Nile Valley, Egypt	1945
2	Low	10.0	High rainfall	Temperate	Spring	A	As for ME1 + resistance to YR, *Septoria* spp., sprouting	North African Coast, Highlands of East Africa, Andes, and Mexico	1972
3	Low	1.7	High rainfall	Temperate	Spring	A	As for ME2 + acid soil tolerance	Passo Fundo, Brazil	1974
4A	Low	10.0	Low rainfall, winter dominant	Temperate	Spring	A	Resistance to drought, *Septoria* spp., YR	Aleppo, Syria; Settat, Morocco	1974
4B	Low	5.8	Low rainfall, summer dominant	Temperate	Spring	A	Resistance to drought, *Septoria* spp. *Fusarium* spp., LR, SR	Marcos Juarez, Argentina	1974
4C	Low	5.8	Mostly residual moisture	Hot	Spring	A	Resistance to drought, and heat in seedling stage	Indore, India	1974
5A	Low	3.9	High rainfall/ irrigated, humid	Hot	Spring	A	Resistance to heat, *Helminthosporium* spp., *Fusarium* spp., sprouting	Joydepur, Bangladesh Londrina, Brazil	1981
5B	Low	3.2	Irrigated, low humidity	Hot	Spring	A	Resistance to heat and SR	Gezira, Sudan; Kano, Nigeria	1975
6	High	5.4	Moderate rainfall/ summer dominant	Temperate	Spring	S	Resistance to SR, LR, *Helminthosporium* spp., *Fusarium* spp., sprouting photoperiod sensitivity	Harbin, China	1989
Winter/Facultative wheat									
7	High		Irrigated	Moderate cold	Facultative	A	Rapid grain fill, resistance to cold, YR, PM, BYD	Zhenzhou, China	1986
8A	High		High rainfall/ irrigated, long season	Moderate cold	Facultative	A	Resistance to cold, YR, *Septoria* spp.	Chillan, Chile	1986
8B	High		High rainfall/ irrigated, short season	Moderate cold	Facultative	A	Resistance to *Septoria* spp., YR, PM, *Fusarium* spp., sprouting	Edirne, Turkey	1986
9	High		Low rainfall	Moderate cold	Facultative	A	Resistance to cold, drought	Diyarbakir, Turkey	1986
10	High		Irrigated	Severe cold	Winter	A	Resistance to winterkill, YR, LR, PM, BYD	Beijing, China	1986
11A	High		High rainfall/ irrigated, long season	Moderate Cold	Winter	A	Resistance to *Septoria* spp., *Fusarium* spp., YR, LR, PM	Temuco, Chile	1986

Table 2. Continued

ME	Latitude (degrees)	Area (mill./ha)	Moisture regime	Temperature regime	Growth habit	Sown	Major breeding objectives	Representative locations/regions	Year breeding began at CIMMYT
11B	High		High rainfall/ irrigated, short season	Severe cold	Winter	A	Resistance to LR, SR, PM, winterkill, sprouting	Lovrin, Romania	1986
12	High		Low rainfall	Severe cold	Winter	A	Resistance to winterkill, drought, YR, bunts	Ankara, Turkey	1986

Source: Adapted from Rajaram et al. (1995)
a Low = less than about 35–40 degrees.
b Refers to rainfall just before and during the crop cycle. High = > 500 mm; low = < 500 mm.
c Hot = mean temperature of the cooloes month > 17.5 degrees; cold = < 5.0 degrees.
d A = autunm, S = spring.
e Factors additional to yield and industrial quality. SR = stem rust, LR = leaf rust, YR = yellow (stripe) rust, PM = powdery mildew, and BYD = barley yellow dwarf virus.
f Further subdivided into (1) optimum growing conditions, (2) presence of Karnal bunt, (3) late planted, and (4) problems of salinity.

total spring bread wheat area in developing countries (Byerlee & Moya, 1993).

The cornerstone of CIMMYT's breeding methodology are targeted breeding for ME, the use of a diverse gene pool for crossing, shuttle breeding, selection for yield under optimum conditions, and multi-locational testing to identify superior germplasm with good disease resistance. In this paper we would like to present some of the recent developments at CIMMYT's wheat program.

Targeted breeding – the mega-environment concept

To address the needs of diverse wheat growing areas, CIMMYT has introduced in 1988 the concepts of mega-environments (ME) (Rajaram et al., 1995). An ME is defined as a broad, not necessarily contiguous area, occurring in more than one country and frequently transcontinental, defined by similar biotic and abiotic stresses, cropping system requirements, consumer preferences, and, for convenience, by a volume of production. Germplasm generated for a given ME is useful throughout it, accommodating major stresses, but perhaps not all the significant secondary stresses. Within an ME, millions of ha are addressed with a certain degree of homogeneity as it relates to wheat. By 1993, 12 ME have been defined, 6 for spring wheats (ME1–ME6), 3 for facultative wheats (ME7–ME9) and 3 for winter wheats (ME9–ME12). Details for each ME are given in Table 1.

Use of a diverse genepool for crossing to maintain genetic diversity

Broad based plant germplasm resources are imperative for a sound and successful breeding program. Utmost attention is given to the genetic diversity within the CIMMYT germplasm to minimize the risk of genetic vulnerability, since it is grown on large areas and is widely used by NARS. We also believe that the use of genetically diverse material is mandatory for future increase of yield potential and yield stability. Parental group of lines considered for crossing in any year consist of 500–800 lines. Twice a year around 30% of the parental stocks are replaced with outstanding introductions. About 2000 out of 8000 crosses/year are made to these introductions. In addition, commercial varieties from NARS, and non conventional sources such as durum wheat, and alien species are used to incorporate desired traits by recombination or translocation. The introductions are mostly used as female to preserve cytoplasmatic diversity.

The most recent example for the potential impact of generating new diversity is the reconstitution of bread wheat by the CIMMYT wide crossing program by crossing durum wheat (*Triticum durum*) with the D-genome donor *Triticum tauschii*. Lines derived from backcrosses to bread wheat showed substantial morpho-agronomic variation, resistance to Karnal Bunt (*Tilletia indica*) and scab (*Fusarium graminearum*) and a TKW of up to 53 g (Villareal 1995). Yield potential is close to that of bread wheat and grain yield of the best synthetic wheat reached 7.7 t/ha (Table 2).

Other sources exploited for new variation are *Triticum dicoccoides* (emmer wheat) as source for resistance to stripe rust, leaf rust, powdery mildew, *Septoria* spp. and Wheat Streak Mosaic Virus, tolerance to drought, high protein content and higher yield potential.

Bread wheat is crossed with durum wheat to increase grain size. The 6 highest yielding lines derived from this program outyielded their breadwheat parent by 5 to 20% in yield trials in Cd. Obregon, Mexico.

Table 3. Grain yield and TKW of two crosses of bread wheat with synthetic wheats in yield trials at Cd. Obregon, Mexico in 1993

Entry	Grain yield (kg/ha)	1000 KW (g)
Chen/*T. tauschii*/BCN	7740a[1]	53a
Cndo/R143/Ente/Mexi/3/		
T. tauschii/4/Weaver	6830b	52a
Bacanora 88 (BW check)	6770b	40b

[1] Means within columns followed by different letters are significantly different at the 0.05 level of probability.

Shuttle breeding between Cd. Obregon and Toluca

Young & Frey (1994) provide two factors which influence the success of a shuttle program: 'a) the use of a germplasm pool encompassing genotypes with broad adaptation and b) the use of selection environments eliciting different responses from plant types' and state further 'The wheat breeding program of Borlaug met these conditions'. When N.E. Borlaug started the shuttle breeding approach in 1944 the only objective was to speed up breeding for stem rust resistance. Since then, segregating populations have been shuttled 100 times between the two environmentally contrasting sites in Mexico, Cd. Obregon and Toluca (for a detailed description of the two locations see Braun et al., 1992). We would like to stress this point since the discussion of breeding for wide versus specific adaptation has not come to an end (Ceccarelli, 1989). We believe, comparisons between breeding methodologies for different species should not only consider selection parameters like heritabilities but also the breeding investments into the respective crop. Alone at CIMMYT more than 200<DS>000 crosses have been made since 1944 and today, more than 4000 advanced spring wheat lines are annually screened worldwide. We are of the opinion that such a large scale testing and crossing program using a diverse genepool will most likely have higher chances of identifying widely adapted germplasm, break genetic linkages and pyramid desired genes than programs which test the germplasm in a narrow environmental range and restrict the genetic diversity in the crossing program. The wide acceptance of CIMMYT germplasm by NARS and by farmers supports this approach. This is said without disparaging other approaches.

One of the important results of this shuttle was the selection of photo insensitive wheat genotypes. Initially, selection for photoperiodic insensitivity was unconscious, but only this trait permitted the wide spread of the Mexican semi-dwarfs (Borlaug, 1995). Today, this trait has been incorporated into basically all spring wheat cultivars grown below 48° latitude and is now also spreading to wheat areas above 48° N (Worland, 1994).

Selection under optimum conditions and breeding for yield potential

Selection of segregating populations and consequent yield testing of advanced lines is paramount for identification of high yielding and input responsive wheat genotypes. The increase in yield potential of CIMMYT cultivars developed since the 60's is shown in Figure 1 (Rees et al. 1993). The average increase per year was 0.9% and there is no evidence that a yield plateau is reached. This genetic progress in increasing the yield potential is closely associated with an increase in the photosynthetic activity (Rees et al., 1993). Both, photosynthetic activity and yield potential increased over the 30 year period by some 25%. These findings may have major implications on CIMMYT's future selection strategy since there is evidence that wheat genotypes with a higher photosynthesis rate have a lower canopy temperature, which can be easily, fast and cheaply measured using a hand-held thermometer. If verified in future trials, breeders may be able to use this trait to increase selection efficiency for yield potential. This technique may be in particular useful to select

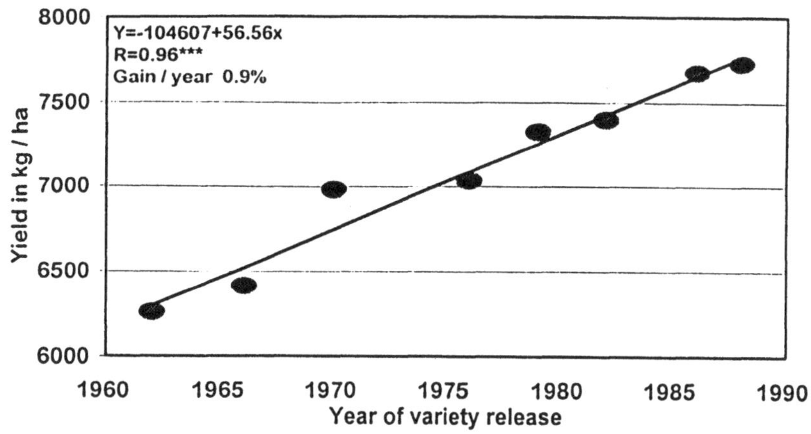

Figure 1. Mean grain yields for the historical series of bread wheat varieties for the years 1990–93 at Cd. Obregon, Mexico. (Data from Rees et al., 1993.)

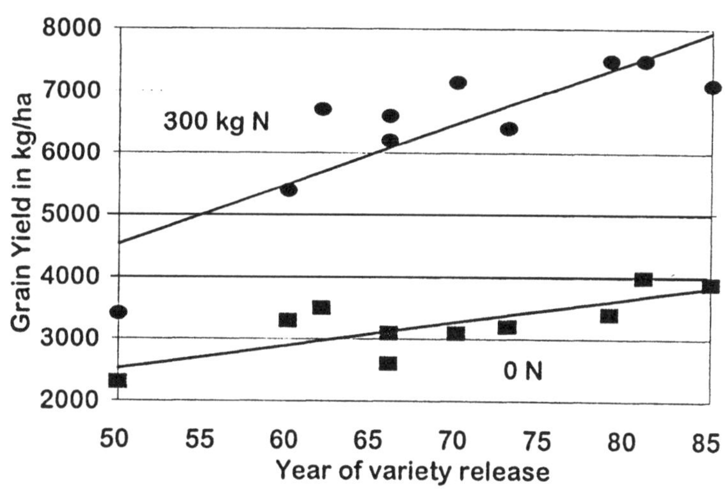

Figure 2. Grain yield of the historical series of bread wheats at Cd. Obregon, Mexico at 0 and 300 kg/ha N application. (Data from J.I. Ortiz-Monasterio et al., 1995.)

wheat genotypes adapted to environments where heat is a production constraint.

Yield per se is closely associated with input responsiveness. Increasing the input efficiency at low production levels can shift cross over points, provided they exist, and enhance residual effects of high genetic yield potential. Furthermore, combining input efficiency with high yield potential will allow a farmer to benefit from such cultivars over a wide range of input levels. The increase in N-use-efficiency is shown in Figure 2 (Ortiz-Monestario et al., 1995).

Multi-locational testing

Around 1500 sets of yield trials and screening nurseries consisting of around 4000 advanced bread wheat lines are annually sent to more than 200 locations. Multi-locational testing plays a key-role in identifying best performing entries for crossing. Since the shuttle program (see above) permits two full breeding cycles/year, it takes around five to six years from crossing to international distribution of advanced lines to cooperators. This 'recurrent selection program' ensures a continuous and fast pyramiding of desirable genes.

Ceccarelli (1989) pointed out that the wide spread cultivation of some wheat cultivars should not be taken as a demonstration of wide adaptation, since a large fraction of these area is similar or made similar by use of irrigation and/or fertilizer and therefore, the term wide adaptation has been used mainly to describe geographical rather than environmental differences. If this is true, the genotypic variation should be considerably higher than the G × E-interaction in ANOVAs of CIMMYT trials. Braun et al. (1992) showed that this is not the case. When subsets of locations were grouped on geographical and/or environmental similarities, the G × E interaction was mostly greater than the genotypic variance. The environmental diversity of sites where CIMMYT's 21st International Bread Wheat Screening Nursery was grown and the diversity amongst genotypes in this nursery was demonstrated by Bull et al. (1994). They classified similarities among environments by forming subsets of genotypes from the total dataset and compared it with the classification based on the remaining genotypes. Using this procedure they concluded that it was not possible to come to a stable grouping of environments, because little or no relationship existed among them.

Conclusions drawn from trials carried out on research stations are always open to critics who argue, that these results do not necessarily reflect farmers field conditions. However, the wide acceptance of CIMMYT germplasm by farmers in ME 1 to ME 5 (see above) does not support the view, that the wide adaptation of CIMMYT germplasm is based on geographical rather than environmental differences.

CIMMYT's strategy for durability of resistance

From its beginning, incorporation of durable, non specific disease resistance into CIMMYT's germplasm was a high priority since breeding of widely adapted germplasm with stable yields without adequate resistance against the major diseases would be impossible. The concept goes back to Niederhauser et al. (1954), Borlaug (1966) and Caldwell (1968) who proposed the application of general resistance in the CIMMYT program versus the specific or hypersensitive type. Intentionally very diverse sources of resistance for rusts and other diseases are used in the crossing program. The major sources are germplasm from national programs, advanced CIMMYT lines, germplasm received from the CIMMYT or other genebanks and CIMMYT's wide crossing program.

CIMMYT's strategy in the case of cereal rusts is to breed for general resistance (slow rusting) based on historically proven stable genes. This non specific resistance can be further diversified by accumulating several minor genes and combine them with different specific genes to provide a certain degree of additional genetic diversity. This concept is also applied to other diseases like septoria leaf blotch, helminthosporium spot blotch, fusarium head scab etc. Following is the present situation of the CIMMYT germplasm regarding resistance to major diseases.

*Stem rust (*Puccinia graminis *f.sp.* tritici*) resistance.* This resistance has been stable after 40 years of utilization of the genes derived from the variety Hope and losses due to stem rust have been negligible since the late 60's. The resistance is based on the gene complex Sr2, which actually consists of Sr2 plus 8–10 minor genes pyramided into three to four gene combinations (Rajaram et al., 1988). Sr 2 alone behaves as a slow rusting gene. Since there has been no major stem rust epidemic in areas where CIMMYT germplasm is grown, the resistance seems to be durable.

*Leaf rust (*Puccinia recondita *f.sp.* tritici*) resistance.* This resistance has been stabilized by using genes derived from many sources, in particular the Brazilian cultivar Frontana (Singh & Rajaram, 1992). No major epidemic has been observed for almost twenty years. Four partial resistance genes including Lr 34 give a slow rusting response and have been the reason for the containment of leaf rust epidemics in the developing world during the last 15 years wherever the varieties carry this minor genes. About 60% of the CIMMYT germplasm carry one to four of these partial resistance genes. Lr 34 is linked to Yr 18 as well as to a morphological marker leaf tip necrosis which makes the gene particular attractive for breeders (Singh, 1992a, b). CIMMYT continues to look for new sources of partial resistance.

*Stripe rust (*Puccinia striiformis*).* Slow rusting genes like Yr 18 have been identified (Singh, 1992b); however their interaction is less additive than for leaf and stem rust. More basic research is needed to understand the status of durable resistance in high yielding germplasm. The breakdown of Yr 9 in West Asia and North Africa and the present yellow rust epidemics underline the need for the release of cultivars with accumulated durable resistance.

Septoria tritici. Initially all semi dwarf cultivars developed for irrigated conditions were susceptible.

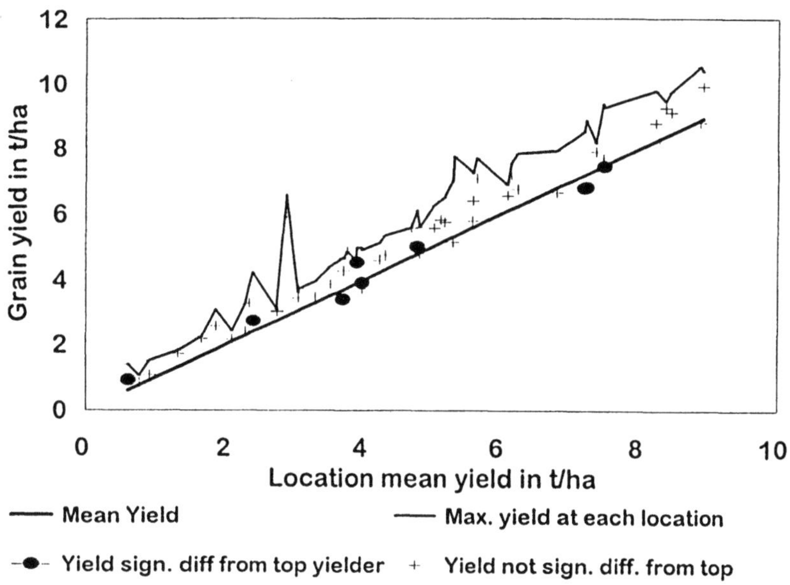

Figure 3. Yield of Pastor at 50 locations of 13th ESWYT.

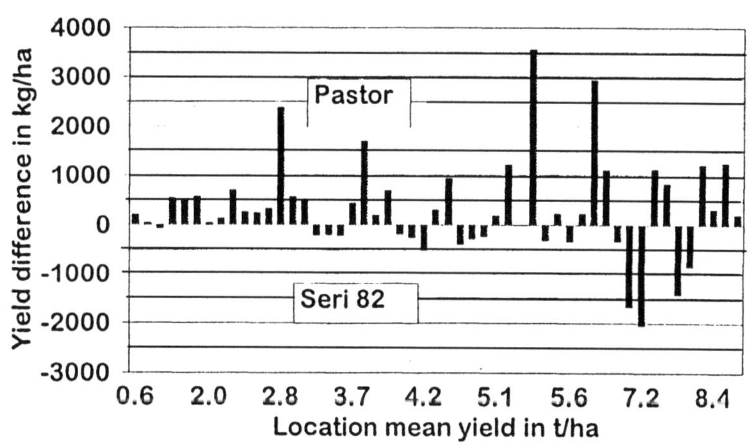

Figure 4. Grain yield difference between spring bread wheat cultivars Pastor and Seri 82 at 50 locations of the 13th ESWYT.

Today more than eight genes have been identified in CIMMYT germplasm and two to three genes in combination provide acceptable resistance. Future activities will concentrate on pyramiding these genes and spread them more widely in the CIMMYT germplasm (Jlibene, 1992; Matus-Tejos, 1993).

*Karnal Bunt (*Tilletia indica*).* More than five genes have been identified and most of them are partially dominant. Genes providing resistance to Karnal Bunt have been incorporated into high yielding lines (Singh et al., 1995).

*Powdery mildew (*Erysiphe graminis *f.sp.* tritici*).* CIMMYT's germplasm is considered to be vulnerable to this disease. The disease is absent in Mexico and the responsibility to transfer resistance genes has been delegated to CIMMYT's regional breeder in South America.

Adaptation of recent CIMMYT cultivars

CIMMYT's breeding strategy has resulted in the development of widely grown varieties, such as Siete Cerros, Anza, Sonalika, Seri 82 which at their peak were grown on several million ha's. Seri 82 was released for irrigated as well as rainfed environments. Reynolds et al. (1994) reported that Seri 82 was the highest yielding entry in the 1st and 2nd International Heat Stress Genotype Experiment. Seri 82 can be considered as the first wheat genotype truly adapted to several ME, particularly to ME1, ME2, ME4 and ME5. A comparison between Seri 82 and a Pastor, a recently developed CIMMYT cultivar, demonstrates the progress made in widening adaptation during the last ten years. Figure 3 shows the performance of Pastor (Pfau/Seri//Bow), in the CIMMYT's 13th Elite Spring Wheat Yield Nursery. In 50 trials grown in al 6 ME, Pastor yielded only in 8 trials significantly ($P = 0.01$) lower than the highest yielding entry. This figure also demonstrates that Pastor has no tendency for a cross over at any yield level. While we do not reject, that such a cross over may exist for some cultivars, Pastor and Seri 82 are clear examples that it is possible to combine abiotic stress tolerance with high yield potential. Figure 4 shows the yield difference between Seri 82 and Pastor. In only 16 out of 50 trials Seri had a higher yield than Pastor. The latter cultivar proves that breeding for wide adaptation has not yet reached its limit.

Conclusion

The selection scheme applied at CIMMYT has resulted in the development of high-yielding cultivars with stable disease resistance to stem- and leaf rust which are now grown on more than 60% of the wheat area in the developing world. Yield potential has increased over the last 25 years at an annual rate of 0.9% and there is no evidence that it has reached a plateau. Simultaneously the N-use-efficiency increased. Wheat cultivars with improved non specific, durable resistance to other diseases than leaf and stem rust will enable farmers to increase wheat production. This is a must since the final target is sustainable production. This requires that wheat cultivars have higher yield potential, better yield stability, improved disease resistance and high input efficiency. Only with such cultivars it will be possible to meet the food needs of the 21st century without harming the environment.

References

Antony, G. & J.P. Brennan, 1987. Progress in yield potential and bread-making characteristics in wheat in New South Wales, 1925–26 to 1984–85. Agric Economics Bull, Div of Marketing and Economic Services. NSW, Australia, Dep of Agric.

Borlaug, N.E., 1966. Basic concepts which influence the choice of methods for use in breeding for diverse resistance in cross pollinated and self pollinated crop plants. In: H.D. Gerold et al. (Ed) Breeding Pest Resistant Trees. Pergamon Press, Oxford.

Borlaug, N.E., 1995. Wheat breeding at CIMMYT. Commemorating 50 years of research in Mexico for global wheat improvement. Wheat Special Report No 29. p. IV–VI.

Braun, H.-J., W.H. Pfeiffer & W.G. Pollmer, 1992. Environments for selecting widely adapted spring wheat. Crop Sci 32 (6): 1420–1427.

Bull, J.K., M. Cooper & K.E. Basford, 1994. A procedure for investigating the number of genotypes required to provide a stable classification of environments. Field Crop Res 38: 47–56.

Byerlee, D. & P. Moya, 1993. Impacts of international wheat breeding research in the developing world, 1969–90. Mexico, D.F.: CIMMYT. 135pp.

Caldwell, R.M., 1968. Breeding for general and/or specific plant disease resistance. In: Proc 3rd Int Wheat Genetics Symp Canberra, Australia.

Ceccarelli, S., 1989. Wide adaptation: How wide? Euphytica 40: 197–205.

Jlibene, M., J.P. Gustafson & S. Rajaram, 1992. A field disease evaluation method for selecting wheat resistance to *Mycosphaerella graminicola*. Plant Breed 108: 26–32.

Matus-Tejos, I.A., 1993. Genetica de la resistencia a Septoria tritici en trigos harineros. Tesis de Maestria en Ciencia, Montecillo, Mexico.

Niederhauser, J.S., J. Servantes & L. Servin, 1954. Late blight in Mexico and its implications. Phytopath 44: 406–408.

Ortiz-Monasterio, J.I., K.D. Sayre, S. Rajaram & M. McMahon, 1995. Genetic progress of CIMMYT's bread wheat germplasm under different levels of nitrogen. I. Grain yield and nitrogen use efficiency.

Rajaram, S., R.P. Singh & E. Torres, 1988. Current CIMMYT approaches in breeding wheat for rust resistance. p. 101–118. In: S. Rajaram & N.W. Simmonds (Eds). Breeding Strategies for Resistance to the Rusts of Wheat. CIMMYT, Mexico D.F.

Rajaram, S., 1995. Yield stability and avoiding genetic vulnerability in bread wheat. p. 11–15. In: S. Rajaram & G. Hettel (Eds). Wheat Breeding at CIMMYT. Commemorating 50 years of research in Mexico for global wheat improvement. Wheat Special Report No 29. Mexico, D.F. CIMMYT.

Rajaram, S., M. van Ginkel & R.A. Fischer, 1995. CIMMYT's wheat breeding mega-environments (ME). In: Proceedings of the 8th International Wheat Genetic Symposium, July 19–24, 1993. Beijing, China (in press).

Rees, D., K. Sayre, E. Acevedo, T.N. Sanchez, Z. Lu, E. Zeiger & L. Limon, 1993. Canopy temperatures of wheat: Relationship with yield and potential as a technique for early generation selection. Wheat Special Report No 10. Mexico D.F.: CIMMYT.

Reynolds, M.P., M. Balota, M.I.B. Delgado, I. Amani & R.A. Fischer, 1994. Physiological and morphological traits associated with spring wheat yield under hot, irrigated conditions. Aust J Plant Physiol 21: 717–730.

Singh, G., S. Rajaram, J. Montaya & G. Fuentes-Davila, 1995. Genetic analysis of resistance to Karnal bunt (*Tilletia indica, mitra*) in bread wheat. Euphytica 81: 117–120.

Singh, R.P. & S. Rajaram, 1992. Genetics of adult plant resistance to leaf rust in 'Frontana' and three CIMMYT wheats. Genome 35: 24–31.

Singh, R.P., 1992a. Association between gene *Lr34* for leaf rust resistance and leaf tip necrosis in wheat. Crop Sci 32: 874–878.

Singh, R.P., 1992b. Genetic association of leaf rust resistance gene *Lr34* with adult plant resistance to stripe rust in bread wheat. Phytopathology 82: 835–838.

Villareal, R.L., 1995. Expanding the genetic base of CIMMYT bread wheat germplasm. p. 16–21. In: S. Rajaram & G. Hettel (Eds). Wheat Breeding at CIMMYT. Commemorating 50 years of research in Mexico for global wheat improvement. Wheat Special Report No 29. Mexico, D.F. CIMMYT.

Worland, A.J., M.L. Appendino & E.J. Sayers, 1994. The distribution, in European winter wheats, of genes that influence ecoclimatic adaptability whilst determining photoperiodic intensivity and plant height. Euphytica (in press).

Young, C. & J. Frey, 1994. Grain-yield characteristics of oat lines surviving uniform and shuttle selection strategies. Euphytica 76: 63–71.

P.M.A. Tigerstedt (Ed.), *Adaptation in Plant Breeding*, 207–212, 1997.

Breeding for wide adaptation in faba bean

Wolfgang Link[1], Bruno Schill[2] & Ernst von Kittlitz[3]
[1] *Institute of Agronomy and Plant Breeding, Georg-August-University, 37075 Göttingen, Germany;* [2] *Institute of Plant Breeding, Seed Science and Population Genetics, University of Hohenheim, 70593 Stuttgart, Germany;*
[3] *State Plant Breeding Institute, University of Hohenheim, 70593 Stuttgart, Germany*

Key words: AMMI, genetic resources, germplasm pools, heterosis, wide adaptation, yield stability, *Vicia faba*, faba bean

Summary

Broadening agronomic adaptation will improve yield stability in the grain legume *Vicia faba* L. We gathered information on the adaptation of European and Mediterranean material to European and Mediterranean environments. The material comprised 20 inbred lines (12 European and 8 Mediterranean lines) and 99 intra- and interpool-crosses in generation F_1. These were evaluated in 9 environments: two spring-sown Southern German environments (SGermE), and seven autumn-sown Mediterranean environments (MedE) in Sicily, Puglia, Andalucia and South Africa. Standard ANOVA, stability analyses and AMMI analysis were performed. Mean yield in F_1 was 257 g/row, the overall parental mean was 144 g/row. The range of environmental means was from 94 g/row to 411 g/row. The average regression coefficient in F_1 was $b_i = 1.07$, being significantly greater than for the parents ($b_i = 0.68$). The opposite was true for the relative magnitude of the deviations from the regressions, which were highly correlated to the AMMI-PC1-results. The AMMI analysis clearly separated the SGermE from the MedE, as well as the germplasm pools. Though the superiority of the F_1-hybrids over their parents was striking, their pattern of interactions with the environments strictly reflected that of their parents. A number of promising crosses was identified as a nucleus of a widely adapted faba bean genepool.

Introduction

Faba bean (*Vicia faba* L.) is widely grown in the Mediterranean region, and to a lesser extent in Central and Northern Europe. Its small-seeded (minor) cultivars are mainly used as animal feed, whereas large seeded faba beans (major) are used for human consumption. The main agronomic problem is instability of yield. Several lines from Mediterranean accessions were found to show relatively stable yield across a range of climatic conditions and sowing dates. Nevertheless, their lodging resistance was rather poor (Kittlitz, 1993a, b).

The present project was started to investigate intra- and interpool-crosses of the type Mediterranean × European (Schill et al., 1995). With the analyses presented here, we focus on wide adaptation of this material referring to a number of very diverse agro-ecological conditions. The question was raised whether widely adapted genotypes can be bred to create a widely adapted germplasm pool, that later on may be used as source of inbred lines for crossing and for creating synthetic or hybrid cultivars. The term wide adaptation is used in the sense of small genotype-by-environment interactions for grain yield, combined with high genotypic yield level, hence, in the sense of unspecific, general agronomic adaptation across diverse environments.

Materials and methods

Twenty *Vicia faba* inbred lines, comprising six European minor lines, six European major lines, and eight Mediterranean lines, were selected to represent the corresponding faba bean germplasm pools (Schill et al. 1995). The minor and major lines are spring beans

Table 1. Characterization of the nine agro-ecological macro-environments

Environments	Sowing (month/ year)	Plant height (cm)	Maturity (month)	Yield (g/row)	Precipitation during growth period (mm)
Mediterranean environments					
1992:					
San Biagio, Sicily	Jan. 92	126	June	142	467
Catania, Sicily	Dec. 91	115	June	189	320
Cordoba, South Spain	Nov. 91	120	May	340	397
Gouda, South Africa	May 92	108	Nov.	94	367
1993:					
Bari, South Italy	Dec. 92	100	June	154	338
San Biagio, Sicily	Dec. 92	102	June	259	377
Catania, Sicily	Dec. 92	111	June	352	367
Southern German environments					
1993:					
Hohenheim, Germany	March 93	107	Aug.	411	501
Lindenhof, Germany	April 93	155	Sept.	364	560

with a thousand kernel weight of approx. 450 g and 1000 g, respectively, the Mediterranean progenitor varieties are sown in late fall or winter, their average thousand kernel weight being approx. 470 g, except one large-seeded parental line (1200 g). These lines were crossed according to a factorial pattern. Eleven of the parents (three minor, three major, and five Mediterranean lines) were crossed with the remaining nine parents (three minor, three major, and three Mediterranean lines), resulting in 99 crosses (F_1-hybrids). Six groups of entries resulted: three intrapool groups (minor × minor, major × major, and Mediterranean × Mediterranean), and three interpool groups (minor × major, minor × Mediterranean, and major × Mediterranean).

The 20 parental lines and their 99 F_1-progenies were tested for their agronomic value in two series of field trials, (1) at seven Mediterranean environments (MedE) and (2) at two southern German environments (SGermE, Table 1). The trials were conducted in lattice designs with two replicates. A plot consisted of 20 plants, sown in two rows of 1 m length. The trials were sown between November 6 and January 8 in the MedE and at March 17 and April 13 in the SGermE. The month of May at Gouda, South Africa, i.e., in the southern hemisphere, corresponds to November in the northern hemisphere. The geographical location of Gouda at 33 degree southern longitude corresponds approximately to that of Casablanca and Damascus. No lodging occured in the MedE. In the SGermE, lodging was prevented technically using a mash net, stretched across the plots.

In a first step, the results were analyzed as lattice design for each environment. In a second step, line and F_1-hybrid means were used in a combined analysis of variance across environments. The genotype × environment interactions were analysed following two paths. First, the regression of genotype i on the environmental means, (b_i), and the mean square of deviations from the regression, $(s^2{}_d)_i$, were computed (Eberhart & Russell, 1966). Additionally, ecovalence, W_i, was computed according to Weber & Wricke (1990). Second, an AMMI analysis was performed following Crossa et al. (1990). This analysis fits first additive effects of genotypes and environments in the same way as a standard ANOVA, and then fits multiplicative effects for genotype-environment interaction by principal components analysis. The biplot display was used showing the main effects of genotypes and environments on the abscissa and the values of the first principal component axis on the ordinate. The AMMI-expected interaction term among a genotype and an environment is the genotype score on the ordinate of the biplot multiplied by the environment score.

Results and discussion

In the MedE, maturity was reached in May/June, whereas in the SGermE, the trials were harvest-

Table 2. Partition of genotype × environment interactions for yield (g/row) of 20 faba bean inbred lines and their 99 factorial F_1-hybrids, evaluated at seven Mediterranean and two southern German environments

Source of variation	df	Component of variance
Environments	8	11 258**
Genotypes	118	4 424**
G × E interaction	944	3 479**
Heterogeneity due to		
regression of genotypes (b_i)	118	43
Deviation from regression $(s^2{}_d)_i$	826	3 431***

* Bartlett's test for heterogeneity of deviation from regression variances showed significance at 0.01 level of probability.

ed in August and September (Table 1). Broomrape (*Orobanche crenata* L.) attack occured in three of the MedE, in 1992 at Cordoba and San Biagio, and in 1993 at San Biagio, but did not reach a damaging level. Overall yield level of the lines in the MedE was 57% of that in the SGermE, for the F_1s the relation was 62%.

The minor lines flowered later than the other lines, both in the MedE (about 10 days) and SGermE (about 6 days), whereas the major lines were the earliest to flower. When compared with the local checks in the MedE, end of flowering appeared too late in the minor lines, causing severe stress from terminal drought. The Mediterranean lines were the earliest that stopped flowering. Correspondingly, the correlation between mean yield and end of flowering for the lines at the MedE was r = - 0.90**. The minor lines were the tallest, whereas the major and Mediterranean lines consisted of shorter genotypes (except one line, named Peleponnes), both in the MedE and SGermE. In the MedE, the minor lines gave the lowest average yield (57 g/row) and the Mediterranean lines the greatest (177 g/row), whereas in the SGermE, the contrary was true (265 vs. 186 g/row). The inferiority of the Mediterranean lines at the SGermE is partly explained by their earlier maturity (12 days). The correlation between mean yield and maturity at the two SGermE was r = 0.62**. In the SGermE, the differences in thousand kernel weight gave rise to strictly corresponding differences in the number of pods and seeds per row for both, lines and F_1-hybrids (- 0.85** < r < - 0.70**), whereas in the MedE, these traits were only loosely correlated. There, in spite of their small seed size, the minor lines and minor × minor-hybrids produced very low pod and seed numbers. These results are in good accordance with those of von Kittlitz et al. (1993b).

In the MedE, the minor × minor hybrids reached only 31%, whereas the major × major hybrids reached 54% of the Mediterranean intrapool hybrids' yield: in the SGermE, the Mediterranean intrapool-hybrids performed relatively better, showing 93% of the European (minor and major) intrapool hybrid yield. At Gouda, unlike to the other MedE, the major × major hybrids did not outyield the minor × minor hybrids, and the minor × Mediterranean interpool-crosses performed better than the major × Mediterranean crosses.

The individual Mediterranean environments differed markedly for the mean expression of the traits. The range in yield among the MedE for the generation F_1 was from 94 to 352 g/row. The correlation coefficients among the MedE (except Gouda) for yield of the 20 lines were 0.77** < r < 0.95**, the range for the F_1-hybrids was 0.67** < r < 0.88**. The correlation of the F_1-hybrids at Gouda with their mean yield at the other MedE was only r = 0.45** in spite of r = 0.85** for the parental lines. The correlation among the two SGermE environments for yield was r = 0.82** for the 20 lines and r = 0.66** for their 99 F_1s. The correlations between the single MedE and SGermE were - 0.30 > r > - 0.63** for the lines and 0.00 > r > - 0.29** for the hybrids, reflecting great interactions between genotypes and these two geographical regions.

The ANOVA showed (Table 2) that the environments were the most important source of variation, explaining 59% of the variance in yield, followed by the genotypes (23%) and the genotypes × environment interactions (18%). Splitting the genotype × environment term demonstrated that the regressions of the genotypes on the environmental means explained only 1.2% of these interactions, in spite of their wide range ($0.52 < b_i < 1.52$). Hence, the correlation coefficient

Table 3. Stability parameters for yield of 20 faba bean parental inbred lines and their 99 F_1-hybrids evaluated at nine diverse agro-ecological macro-environments

Entry	Stability parameter				
	Mean (g/row)	Regression b_i	Deviations from regression		AMMI-PC1 values (cf. Figure 1) $(g/row)^{0.5}$
			Mean $(g/row)^2$	Min.–Max. $(g/row)^2$	
Inbred lines:					
General mean	144	0.68	3554	435– 8899	- 0.32
Minor	103	0.76	4399	1945– 8899	3.32
Major	139	0.75	1358	435– 2341	0.54
Mediterranean	179	0.57	4567	2170– 6352	- 3.70
F_1-hybrids:					
General mean	257	1.07	5084	641–18303	0.07
Minor × Minor	164	1.04	7574	4740–12347	4.92
Major × Major	221	1.10	2607	1325– 5849	1.65
Medit. × Medit.	331	0.95	9148	3386–18303	- 4.63
Minor × Major	202	1.14	4896	878–15447	3.68
Minor × Medit.	261	1.04	2917	641– 7227	0.10
Major × Medit.	295	1.10	4849	1965– 8538	- 2.16

among the genotypes' deviations from their regressions $(s^2{}_d)_i$ and the genotypes' W_i-values was very high (r = 0.98** for the hybrids and r = 0.89** for the lines). The regression coefficients of the lines were lower than those of the hybrids (Table 3). Low $(s^2{}_d)_i$-values were obviously exhibited by the major lines and the major × major hybrids. Similar wide adaptation was found for some minor × Mediterranean hybrids and some other interpool-crosses, caused by the intermediate reaction of these hybrids with regard to that of their parental lines. Whether this can be fixed in inbred lines will be studied in the more inbred generations of these crosses. The average $(s^2{}_d)_i$ of the hybrids was 43% greater than that of the lines. Taking the level of performance of the hybrids into consideration, which surpassed that of their parental lines by 78%, it becomes obvious that the interactions of the hybrids with environments were relatively smaller than those of their parents.

The first principal component axis of the AMMI analysis accounted for 57% of the genotype × environment interaction variance. The correlation coefficient of the F_1-scores (ignoring their sign) on this axis with their b_i-values were r = - 0.31**, whereas with their $(s^2{}_d)_i$-values it was r = 88.**. The same correlations for the parental lines were r = - 0.43 and r = 0.91**, respectively. Hence, at one hand, the AMMI results contained nearly all information that were displayed by the deviations from regressions. At the other hand, AMMI allowed the presentation of the adaptation of groups of genotypes to specific environments, which could not be seen from plotting mere $(s^2{}_d)_i$-values. A clear pattern of specific adaptations of the 20 parental lines was identified by the AMMI-based biplot (Figure 1). It resulted from great positive interaction terms of the minor lines and the SGermE at one side and the Mediterranean lines and the MedE (except Gouda) at the other side. This pattern was strictly reflected by the respective minor and Mediterranean intrapool-hybrids. Hence, heterozygosity, in spite of generally increasing performance, did not alter the general pattern of adaptation and mis-adaptation.

The highest average yields were obtained in two of the MedE (Cordoba and Catania in 1992) and in both SGermE (cf. Figure 1). Obviously, the Mediterranean group of lines and crosses caused these high average yields at the two MedE, wheras the European material caused the high average at the two SGermE. Due to this constellation, large deviations form regression occurred at these environments. The outcome was that the regression approach could explain only very few of this pattern (Table 2), and that the correlation between the $(s^2{}_d)_i$-values and the AMMI-scores was very high.

Many interpool-hybrids and some major intrapool-hybrids displayed scores near to zero along the ordinate of the AMMI-biplot; two major × Mediterranean and four Mediterranean intrapool-crosses combined

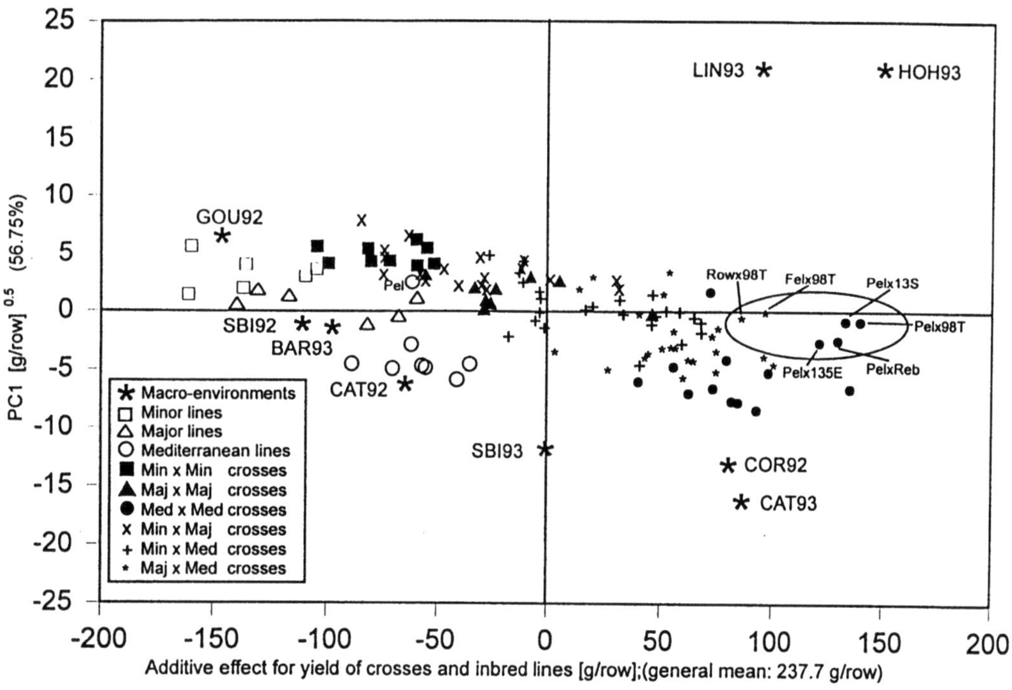

Figure 1. AMMI-biplot of 99 *Vicia faba* F_1-hybrids and their 20 parental inbred lines evaluated at two Southern German (LIN93, HOH93) and seven Mediterranean environments for grain yield.

this with outstanding overall performance, hence, exhibited an agronomically useful wide adaptation. These crosses included several times the Mediterranean inbred line 98Tunesia (98T) and the line Peleponnes, together with other Mediterranean and Major lines (Figure 1). Peleponnes, though entering the project as Mediterranean line, behaved to some extent like an European genotype (Link et al., 1995; Schill et al., 1995). Its yield level was 70% of the Mediterranean lines at the MedE, but 137% of the European lines at the SGermE. The overall impression is that the encircled crosses in Figure 1 may be good candidates to establish a widely adapted germplasm pool in faba bean. This germplasm pool could serve as new opposite pool for breeding synthetic or hybrid faba bean cultivars in Europe (Ederer & Link, 1995). Ongoing experiments with more inbred generations (F_2–F_5) will further increase our knowledge on the usefulness of these germplasm pools and their intra- and interpool progenies with regard to wide agronomic adaptation.

Acknowledgements

We thankfully acknowledge the fruitful collaboration with Profs. V. Abbate, A. Filippetti, and L. Stringi, Italy, and J.I. Cubero, Spain, as well as with Dr. A. van Jaarsvelt, South Africa. The financial contributions of the EISELEN-foundation, Ulm, Germany, is thankfully acknowledged.

References

Crossa, J., H.G. Gauch Jr & R.W. Zobel, 1990. Additive main effects and multiplicative interaction analysis of two international maize cultivar trials. Crop Sci 30: 493–500.

Eberhart, S.A. & W.A. Russell, 1966. Stability parameters for comparing varieties. Crop Sci 6: 36–40.

Ederer, W. & W. Link, 1995. Two new sources of CMS in faba bean (*Vicia faba* L.) segregated from wide crosses. Proc 2nd European Conference on Grain Legumes, Copenhagen, Denmark, 225.

Schill, B., E. von Kittlitz, A.E. Melchinger & W. Link, 1995. Heterosis within and between *Vicia faba* L. germplasm pools. Proc 2nd European Conference on Grain Legumes, Copenhagen, Denmark, 220–221.

Kittlitz, E. von, 1993. Enwicklung neuer Genpools bei der Fababohne (*Vicia faba* L. zur Verbesserung der Klimatoleranz. Bericht Arbeitstagung 1992 der Arbeitsgem. Österr Saatzuchtleiter, Gumpenstein, Austria.

Kittlitz, E. von, K.I.M. Ibrahim, P. Ruckenbauer & L.D. Robertson, 1993. Analysis and use of interpool-crosses (Mediterranean × Central European) in faba beans (*Vicia faba* L.). I. Performance of Mediterranean and Central European faba beans in Syria and Germany. Plant Breeding 110: 307–314.

Link, W., C. Dixkens, M. Singh, M. Schwall & A.E. Melchinger, 1995. Genetic diversity in European and Mediterranean faba bean germplasm revealed by RAPD markers. Theor Appl Genet 90: 27–32.

Weber, W.E. & G. Wricke, 1990. Genotype × environment interaction and its implication in plant breeding. p. 1–19. In: M.S. Kang (Ed). Genotype-by-environment interaction and plant breeding. Louisiana State University Agricultural Center, USA.

P.M.A. Tigerstedt (Ed.), *Adaptation in Plant Breeding*, 213–224, 1997.

Yield stability and adaptation of Nordic barleys

M. Nurminiemi[1], Å. Bjørnstad[2] & O.A. Rognli[1]
[1] *Agric. University of Norway, Department of Biotechnological Sciences/Genetics, POB 5040, N-1432 Ås, Norway;* [2] *Agric. University of Norway, Department of Horticulture and Crop Sciences, POB 5022, N-1432 Ås, Norway*

Key words: genotype × environment (GE) interaction, *Hordeum vulgare* L., stability variance

Summary

Grain yield was studied in a collection of 220 Nordic barley lines at diverse locations in the Nordic countries. Two-row (2r) and six-row (6r) lines differed very significantly in reaction to the growing conditions within and between the two locations, Svalöv (in southern Sweden) and Højbakkegård (in Denmark). This difference was also highly significant at Viikki (in Finland), but not at Ås (in Norway) or between Viikki and Ås. Genotype × location (GL) and genotype × year (GY) variance components were used to estimate phenotypic yield stability by Shukla's stability variance ($\sigma^2{}_i$). Only 7 lines did not contribute significantly to GL- and GY-interactions, and their yield levels were 7–27% lower than that of the highest yielding line (5057 kg/ha). Estimates of GL- and GY-stability parameters were not significantly correlated. Neither responsiveness, measured by the regression coefficient (b_i), nor phenotypic yield stability, measured by the deviations from regression (Tai's λ_i) were correlated with yield. Pedigree studies showed that both b_i and $\sigma^2{}_i$ can be changed by recombination and/or induced mutations. Mixing of near isogenic lines with different resistance genes, and selection within a landrace, also resulted in changes in responsiveness. Recently released 2r-cultivars were more unstable than older 2r-cultivars revealed by positive correlation between the year of release and λ_i. Cultivars originating from southern Scandinavia were higher yielding than cultivars originating from the central or the northern regions of Scandinavia.

Introduction

Despite the large variation in climate, soil and growing conditions, barley breeding has had rather similar objectives in Scandinavia. In general, yield and quality have had the highest priority. In addition, yield and other characters must be stable over years and larger cultivation areas. Due to considerable variation in the growing conditions within rather small geographic regions we need more information on the extent of genotype × environment (GE) interaction and stability over larger cultivation areas in order to improve selection for wide adaptation (SNP, 1982).

Predictable variation in weather conditions, together with variations in edaphic and biotic factors and management create genotype × location (GL) interaction, while unpredictable climatic factors contribute more to GY- and GLY-interactions (Allard & Bradshaw, 1964). Yield in low and high yielding environments can be considered as separate traits, which are not necessarily maximized by identical sets of alleles (Falconer, 1990; Ceccarelli et al., 1992). Plant breeding strategies should therefore be different in stress and non-stress environments (Rosielle & Hamblin, 1981). The most widely used selection criteria for high and stable performance are mean yield, regression response on site mean yield (b_i), and deviations from regressions ($S^2{}_{di}$)) (Finlay & Wilkinson, 1963; Eberhart & Russell, 1966; Becker & Léon, 1988). Both positive and zero correlations between mean yield and regression coefficient have been reported (Perkins & Jinks, 1968a; Langer et al., 1979; Becker, 1983; Helms, 1993; Reysack et al., 1993). Any relation will depend upon the test environment and history of selection (Ceccarelli & Grando, 1991). Several studies have shown that $S^2{}_{di}$ is not consistently associated with b_i or the mean

Table 1. ACC = accession no., C = country of origin, Identification = variety name or line number, ROW = row type and e = early, and REL = year of release or known since

ACC	C	Identification	Pedigree	Breeder[a]	ROW	REL
1	F	63945	Manker/2 × Pokko	HJA	6	
2	F	71384	Hankkija-673/Pomo	HJA	6e	
3	F	77028	Eero NaN_3 M4	HJA	6e	
4	I	Sigur-F	LV from the Faroe Islands	RALA	6e	
5	I	Tampar	LV from the Faroe Islands	RALA	6e	
6	I	046	Mari/Tampar//Akka/Sigur-F	RALA	2e	
7	I	051	Q.B.59.28/Birgitta//Mari	RALA	2	
8	I	054	Q.B.59.28/Birgitta//Mari	RALA	2	
9	N	Vo-H-10660	Otra/Vigdis//Agneta	SFK	6e	88?
10	N	Bode	Pirkka/Nordlys	SFV Å	6e	78
11	N	Fræg	Asplund/Maskin	SFV	6e	48
12	N	Fg-672-2-10-1	Fræg/mutant	IG	6e	64
13	N	M-268	Fg-672-2-10-1/Dcγ-69b	IG	6e	68
14	N	Lise	Asplund/DS 295//Varde	IPF	6	63
15	N	Varde	Asplund/Maskin	VIDA	6e	41
16	N	Yrjar	Jarle/Varde	SFK	6e	75
17	S	Å 80152	Mari//Åsa/Edda II/3/Birgitta/2 × Arla	SV	2	
18	S	Å 80139	Birgitta/Polar//2 × Å 6518	SV	2	
19	S	Å 73109	Birgitta/Å 56888/3/Mari//Åsa/Edda II	SV	2	
20	S	Å 72112 (Kymppi)	Ingrid/3/Mari//Åsa/Edda II	SV	2	86
21	S	Å 74172	Birgitta/Mari//Gunilla	SV	2	
22	F	78045	Eero/Pomo//Potra	HJA	6e	
23	F	77110	Aapo/Kajsa//Eero	HJA	6	
24	D	46925 (Catola)	Alva/Ortolan	CA	2	
25	F	62403	Aapo/Berenice	HJA	2	
26	F	72244	Birgitta/Pirkka	HJA	2	
27	F	78175	Risø 9265//Aapo/Adorra-mutant	HJA	2	
28	S	Å-8293	Pernilla//Mari/Pullserien	SV	2	
29	S	Å-8379	Pernilla//Mari/Pullserien	SV	2	
30	N	Tunga	Fræg//Juli/Rigel	SFK	6	75
31	N	Agneta	Åsa/Frisia//Monte Cristo/4 × Edda II	SV	6e	78
32	N	Bamse	2 × //Åsa/Frisia/3/Monte Cristo/4 × Edda II	SV	6e	81
33	F	Silja	Seger/Vega (6r)//Suvi	JO	6e	79
34	F	Pomo	Seger/Vega (6r//Bonus	JO	6	68
35	F	Vega	Kristina/Lofa	ABED	2	76
36	F	Gunhild	Algerian/Lone//MGH 63199	PF	2	80
37	N	H-3051 (Tyra)	Sold/3/Birgitta/Mari//Gunilla	IPF	2	88
38	S	Arla	Maja/3/Hanna/Svanhals//Opal/4/Tammi	WW	2e	62
39	S	Akka	Monte Cristo/6 × Arla	WW	2e	70
40	N	H-2207	Lise/Paavo//Agneta	IPF	6e	
41	S	Pernilla	Birgitta/Mari//Gunilla	SV	2	84
42	S	Roland	Lud/Tellus M1D	WW	2	82
43	N	P-693	Domen/Fræg	IG	6	68
44	S	Tellus	Strengs Franken III/3/Balder//Binder/Opal	WW	2	70
45	S	Flavina	Emir/4/Herta/3/Balder//Binder/Opal	WW	2	74
46	D	Duks	Carlsberg II/Lyallpur 2	SJ	2	74
47	S	Harry	Arla M1/Tellus	WW	2	78

Table 1. Continued

48	S	80294	Salka/Alva	SV	2	
49	S	Ida	Arla M1/Tellus	WW	2	79
50	S	Annika	WW 5925 M5/WW 6060//Zephyr	WW	2	83
51	N	H-87-130	Bomi/Møyjar	IG	2	
52	D	Salka	Elbo/Vada	PF	2	73
53	S	82248	Tellus/Alva	SV	2	
54	N	Vo-H-2845	Bamse/4/Otra/3/Anita//Bonus/Varde	SFK	6e	
55	S	Welam	Monte Cristo/Clara//2 × WW 5793/3/3 × WW 5853	WW	2	76
56	D	807909	Odin/Aramir	CA	2	
57	D	Keti	Rupal//Ingrid/Proctor	CA	2	82
58	S	81222	Alva/Mazurka	SV	2	
59	F	Jenny	Rupee/5 × Pallas//2 × Hellas/3/Kristina	SV	2	80
60	N	Vo-1789-73	Osiris/Nordlys	SFK	6	91?
61	S	Agneta	Ȧsa/Frisia//Monte Cristo/4 × Edda II	SV	6e	78
62	D	Zita	Herta/KVL 191//Ymer/3/Vada	PF	2	73
63	D	Amdi	Aramir/Nordal	CA	2	
64	F	1309	Ingrid/3/Herta//Seger/Vega (6r)	JO	2	
65	D	Mikkel	Visir/4/KVL 191/8 × Herta//Ingrid/3/Minerva	SJ	2	84
66	N	H-3036	Audley/Jessica	IPF	2	
67	D	Triumph	Alsa/S 3170//11719-59/3/Union/4/Diamant	VEB	2	73
68	N	D-10-27	Domen-mutant	IG	2	
69	D	8456	Loti/3/Kristina//Cambrinus/Mala	ABED	2	
70	F	78023	Olli/Hiproly//Kajsa	HJA	6e	
71	N	P-58	Domen/Fræg	IG	6e	68
72	N	H-349-220	Bode/Agneta	IG	6e	
73	D	Nery	KVL 191/8 × Herta//Ingrid/3/Minerva	SJ	2	75
74	F	78133	Hankkija-673/Aapo	HJA	2	
75	F	1382	Herta//Seger/Vega (6r)/3/Goliat//Seger/Vega (6r)	JO	2	
76	N	Dcγ/69b	Domen-mutant	IG	2	64
77	D	Pallas	Bonus-mutant	SV	2	58
78	D	Pallas-M10	near isogenic lines of Pallas	KVL	2	
79	S	Ansgar	KVL 191/8 × Herta//5 × Ingrid/3/Wing	WW	2	73
80	D	Lami	Anla/Minerva	SJ	2	73
81	N	Vo-H-2825	Bamse/Yrjar	SFK	6e	
82	S	80515	Salka/Drossel	SV	2	
83	S	Birka	Baladi16/2 × Rika//2 × Tellus/3/2 × //Monte Cristo/4 × Tellus	WW	2	79
84	F	1346	Fort/Etu	JO	6	
85	F	1352	Bonus/Varde	JO	2	
86	F	78012	Tammi//Seger/Vega (6r)/3/Paavo/4/Hankkija-673/Hiproly	HJA	6	
87	F	78003 (Pohto)	Hiproly/2 × Hankkija-673	HJA	6	87
88	S	Ȧ 80190	Salka//Cilla/Aramir	SV	2	
89	S	Lina	Multan/5 × Mari/3/Mari//Ȧsa/EddaII/4/Lofa	SV	2	85
90	D	Jonna	Salka/Lauda	PF	2	81
91	S	82212	Aramir/Alva	SV	2	
92	S	Jenny	Rupee/5 × Pallas//2 × Hellas/3/Kristina	SV	2	80
93	S	Golf	Armelle/Lud//Luke	NRPB	2	83

Table 1. Continued

94	F	1315	Tammi-mutant	JO	6e	
95	D	Susan	Salka/Sultan	PF	2	80
96	F	1344	Jessica/Otra	JO	6e	
97	D	4.61146	Georgie/Tyra	PF	2	
98	D	0267	Kristina/Dina//Anna	ABED	2	
99	F	1220	Herta//Seger/Vega (6r)	JO	2	
100	S	77188 (Frida)	Cilla/Aramir	SV	2	84
101	D	01050	Triumpf/Dina	ABED	2	
102	D	Birger	Mona/Tron	SJ	2	
103	F	1279	Otra/Etu	JO	6e	
104	N	M-138	Fg-672-2-10-1/Dcγ/69b	IG	6	68
105	D	0391	Kristina/Dina//Mala/Gula	ABED	2	
106	D	Romi	Kristina//Denso/Minerva/3/Rupal	ABED	2	83
107	D	01065	Torkel/Dina	ABED	2	
108	S	Marlies	Lud/Tellus M1D	WW	2	83
109	N	H-349-10	Bode/Agneta	IG	6e	
110	F	62951	Aapo/2 × Adorra-mutant	HJA	2	
111	D	Nordal	Heine 4808/Dana	CA	2	71
112	F	77061 (Eero-80)	Olli/Eero	HJA	6e	85
113	D	Odin	Sv 665433//Ingrid/Proctor	CA	2	81
114	F	Triumph	Alsa/S/3170//11719-59/3/Union/4/Diamant	VEB	2	73
115	N	H-161-212	Møyjar/M-65//Møyjar/Lise	IG	2	
116	D	Gula	Impala/Emir	ABED	2	77
117	N	D-50-89	Domen-mutant	IG	2	
118	S	81353	Tellus/Alva	SV	2	
119	N	D-25-78	Domen-mutant	IG	2	
120	N	H-349-348	Bode/Agneta	IG	6	
121	D	37906 (Charlie)	Aramir/Seru	CA	2	84
122	S	Cilla	Ingrid/Hadostreng	WW	2	70
123	S	81324	Tellus/Carina	SV	2	
124	D	802183	MC 20/3/51-1722/MR Iowa//Mari/4/Lofa/5/2 × Duks	SJ	2	
125	F	1252	Varde/Otra	JO	6e	
126	D	Vega	Kristina/Lofa	ABED	2	76
127	F	1263	Herta//Seger/Vega (6r)	JO	2	
128	D	Carlsberg II	Carlsberg	CA	2	53
129	N	H-25-23	Mona//Morocco/4 × Pallas	IG	2	
130	S	Prisca	KVL 191/7 × Herta//Clara/3/4 × WW 5793/4/Monte Cristo /Clara//5 × WW 5793	WW	2	76
131	N	H-3208	Lise/Paavo//Agneta	IPF	6	
132	F	1328	Suvi/Otra	JO	6e	
133	N	H-25-195	Mona//Morocco/4 × Pallas	IG	2	
134	S	Herta	Kenia/Isaria	WW	2	49
135	F	1259	Pomo/Vega (2r)	JO	2	
136	F	77082	Eero/Pomo//Aapo	HJA	6	
137	N	H-161-30	Møyjar/M-65//Møyjar/Lise	IG	2	
138	D	Robert	Nery/Duks	SJ	2	85
139	N	H-3003	Vada/203-7489//Jessica	IPF	2	

Table 1. Continued

140	S	Ingrid	Balder//Binder/Opal	WW	2	58
141	S	Benedicte	Ingrid M9/2 × Ansgar//2 × Tellus/3/2 × WW 6484	WW	2	83
142	F	1330 (Niina)	Varde/Otra	JO	6	86
143	N	H-25-112	Mona//Morocco/4 × Pallas	IG	2	
144	F	77065	Etu/Pirkka	HJA	6e	
145	S	80512 (Semira)	Salka/Drossel	SV	2	
146	N	H-351-6	M-268/Agneta	IG	6	
147	D	Gunhild	Algerian/Lone//MGH 63199	PF	2	80
148	S	Vanja	Tradak/5 × Cilla/Cilla M1	WW	2	
149	F	72446	Aapo/Birgitta	HJA	2	
150	N	Vo-H-2859	Forus/Edda II//Bode	SFK	6	
151	N	Vo-H-1844	Forus/Edda II//Forus/Fræg	SFK	6	
152	N	H-2202	Bode/Clermont	IPF	6	
153	N	H-351-90	M-268/Agneta	IG	6	
154	F	1103	Varde//Opal/Perttu	JO	6e	
155	N	H-87-168	Bomi/Møyjar	IG	2	
156	F	1310	Jet/Ingrid	JO	6e	
157	F	1359	Herta//Seger/Vega (6r)/3/Otra	JO	2	
158	D	Jarl	Anla/Minerva/4/KVL 191/8 × Herta//Ingrid/3/Minerva	SJ	2	81
159	S	Liberty	Midas//Tellus M1D/3 × Zephyr	WW	2	82
160	N	Møyjar	Domen/Herta	Mø	2	69
161	D	4.61844	Salka/Lauda//Tyra	PF	2	
162	D	Tron	Impala/unknown	SJ	2	78
163	N	H-3202	Paavo/Vigdis//CI 8158/7 × Vigdis	IPF	6	
164	D	4.61836	Susan/Gunhild	PF	2	
165	F	1183	Fort/3/Tammi//Gull/OAC 21	JO	6	
166	N	H-1014 (Trine)	Lise/Clermont	IPF	6	86
167	N	M-5	Fg-672-2-10-1/Dcγ-69b	IG	6	68
168	D	Bingo	Lami/Rupal	CA	2	84
169	F	1374	Opal/Perttu/3/Balder//Seger/Vega (6r)	JO	6	
170	F	77059	Gunilla/Aapo	HJA	2	
171	D	Pallas-M36	near isogenic lines of Pallas	KVL	2	
172	S	Balder	Gull/LV from Scania//Maja	WW	2	42
173	S	81323	Tellus/Carina	SV	2	
174	F	1297	Suvi/Pirkka	JO	6e	
175	F	62485	Pirkka-933/Lola//Salve	HJA	2	
176	S	Therese	Patrik/3/2 × Engeldow India//Proctor/WW 5414	WW	2	
177	S	Patrik	Emir/4/Herta/3/Binder/Opal//Balder	WW	2	80
178	D	Caja	Herta/KVL 191//Ymer/3/Rika/4/PF 62/6/6-4	PF	2	80
179	F	63914	Manker/Pokko	HJA	6	
180	D	4.6386	Salka/Sultan/Gunhild	PF	2	
181	S	82358	Aramir/Salka	SV	2	
182	S	Särla	Laschkego Tibetanski/3/Hanna/Svanhals//Kenia/4/Domen	WW	2	65
183	F	1184 (Arra)	Varde/Otra	JO	6e	82
184	D	813197	Mirjam/Tyra	SJ	2	
185	S	Brita	Fon Tien/5 × Tellus	WW	2	
186	F	Hankkija-673	Otra/Paavo	HJA	6e	73
187	D	813234	Nery/3/Goldfield//Hiproly/MILN 155-39	SJ	2	

Table 1. Continued

188	S	82364	Salka/Menuet	SV	2	
189	N	H-349-347	Bode/Agneta	IG	6	
190	D	602216 (Catrin)	Nordal/Triumph	CA	2	86
191	D	011 (Alis)	Triumph/Rosie	ABED	2	
192	D	4.4682	Caja/Aramir	PF	2	
193	S	Stallar II	Kenia/Isaria	WW	2	52
194	S	Rika	Kenia/Isaria	WW	2	51
195	S	82908	Aramir/Triumph	SV	2	
196	D	0879	Harry/Gula	ABED	2	
197	S	Wing	Lyallpur 2/6 × Ingrid	WW	2	70
198	D	55106	Salka/Pirouette	CA	2	
199	S	80809	Alva/3/Birgitta/Mari//Gunilla	SV	2	
200	D	361671 (Calasc)	Lauda/Seru	CA	2	
201	N	Vo-H-1807	Yrjar//Forus/Edda II	SFK	6	
202	S	Agneta	Åsa/Frisia//Monte Cristo/4 × Edda II	SV	6e	78
203	S	Caroline	Lud/Tellus M1D	WW	2	83
204	F	78104	Eero/Pomo//Aapo/3/Kajsa	HJA	6	
205	N	Vo-H-2813	Agneta/Yrjar	SFK	6	
206	D	Inga	Vatonga//Goldfield/Mala	ABED	2	82
207	F	70185	Pomo/Pirkka	HJA	6	
208	N	H-1027 (Tore)	Lise/Clermont	IPF	6	86
209	N	M-65	Fg-672-2-10-1/Dcγ-69b	IG	6	68
210	D	Tyra	Algerian/8 × Herta//Rika/Drost	PF	2	75
211	N	H-10066	Gunilla/Bomi	IPF	2	
212	S	82204	Aramir/Alva	SV	2	
213	S	Torkel	Monte Cristo/Clara//3 × WW 5853/3/WW 5926	WW	2	80
214	S	Bente	Asplund//Laschkego Tibetanski/WW 4088/3/Denso /Weihenstephan II/4/Pauline	WW	2	82
215	F	1343	Varde/Otra	JO	6e	
216	D	Lofa	Proctor/Minverva	ABED	2	68
217	D	Gorm	Kristina/4/KVL 191/8 × Herta//Ingrid/3/Minerva	SJ	2	81
218	N	Vo-H-2912	Sv 71558/unknown 2r	SFK	6	
219	S	82368	Salka/Menuet	SV	2	
220	F	62407	Berenice//Karri/4 × Ingrid	HJA	2	

[a] Abbreviations for plant breeding stations:
ABED = Abed Plant Breeding Station, Denmark; CD = Carlsberg AS, Denmark; HJA = Hankkija Plant Breeding Institute (joined with JO in 1994; Boreal Plant Breeding, Finland), JO = Agric. Res. Centre, Inst. of Plant Breeding, Finland, IG = Agric. Univ. of Norway, Dept. of Biotechnological Sciences/Genetics, Norway, IPF = Agric. Univ. of Norway, Dept. of Plant Science, Norway, KVL = Royal Vet. and Agric. Univ., Dept. of Crop Science, Denmark, Mø; = former State Exp. Station Møystad, Norway, NRPB = Nickerson Seeds Ltd, England, PF = Pajbjerg Foundation, Denmark, RALA = Agric. Res. Institute, Iceland, SFK = Kvithamar Agr. Research Station, Norway, SFV = former State Exp. Station Voll, Norway, SFVÅ = Vågønes Agr. Research Station, Norway, SJ = Sejet Plantbreeding, Denmark, SV = Svalöf AB (joined with Weibullsholm Plant Breeding Institute: Svalöf Weibull AB), Sweden, VEB = VEB Saat- and Pflanzgut, Germany, VIDA = former Felleskjøpet Agri. Exp. Station at Vidarshov, Norway, WW = Weibullsholm Plant Breeding Institute, Sweden.

yield (Perkins & Jinks, 1968b; Langer et al., 1979; Jalaluddin & Harrison, 1993).

The aim of the present study was to investigate yield stability of Nordic barley lines grown in diverse regions of the Nordic countries.

Material and methods

Plant material

The 220 barley breeding lines or cultivars used in this study originated from Nordic plant breeding companies and research stations. A detailed description of origin and pedigrees of the barley lines is presented in the Table 1. Adapted landraces occurred much more recently in pedigrees of 6r-lines than 2r-lines, while the pedigrees of 2r-lines showed a much wider geographic origin than 6r-lines. The material was divided into subsets on the basis of earliness or row type. Subset EARLY consisted of 37 six-row and 3 two-row early maturing feed barleys. Subset 6R included all 71 6r-barleys, while subset 2R comprised 149 2r-barleys.

Description of experimental locations

Field trials were conducted from 1987 to 1989 at the following experimental locations: Ås (59° 40′ N 10° 48′ E) in South-East Norway, Højbakkegård (55° 40′ N 12° 20′ E) in Denmark, Korpa (63° 46′ N 20° 15′ W) in southern Iceland, Röbäcksdalen (63° 51′ N 20° 16′ E) in northern Sweden, Svalöv (55° 57′ N 13° 8′ E) in southern Sweden and Viikki (60° 10′ N 25° 3′ E) in southern Finland. Subset EARLY was grown at all locations, while subsets 6R and 2R were grown only at the four southern locations. Detailed information about climatic conditions, soil characteristics and management practise for the field sites has been presented by Nurminiemi (1995).

Experimental design and statistical analyses

The experimental design was a special type of incomplete blocks with two complete replications at each location. Plot size varied from 5 m^2 to 10 m^2, depending on location, and the sowing rate was 440 seeds/m^2. Analyses of variance were performed using PROC ANOVA of SAS/STAT (SAS, 1987) according to the following model:

$$y_{ijkr} = \mu + G_i + L_j + Y_k + B_{jkr} + (GL)_{ij} + (GY)_{ik} + (GLY)_{ijk} + e_{ijkr}$$

where y_{ijkr} is the 15% moisture corrected grain yield observation (kg/ha) of the i-th genotype, at the j-th location, and the k-th year, in the r-th block. e_{ijkr} stands for the unexplained residuals. $(GL)_{ij}$, $(GY)_{ik}$, and $(GLY)_{ijk}$ designate genotype × environment (GE) interactions. GL-interactions were partitioned by Shukla's (1972a) stability variance, $\sigma^2{}_i$, as follows:

$$\sigma_i^2 = [1/(L-1)(G-1)(G-2)] [G(G-1)\Sigma_j(y_{ij} - y_{i.} - y_{.j} + y_{..})^2 - \Sigma_i\Sigma_j(y_{ij} - y_{i.} - y_{.j} + y_{..})^2]$$

where G = no. of genotypes, L = no. of locations, y_{ij} is the grain yield of the i-th genotype in the j-th location (over years), $y_{i.}$ is the mean yield of the i-th genotype over j locations, $y_{.j}$ is the mean yield of the j-th location over years, and $y_{..}$ is the grand mean. F-tests for significance of GL-stability variances (Shukla, 1972a; Kang, 1989) can be obtained as: r $\sigma^2{}_i/MS_e = F_{\alpha,L-1,G(r-1)}$, where MS_e is the error mean square. Difference in GL-stability between lines can be estimated for any pairs of genotypes (Shukla, 1972b). If $\sigma^2{}_i/\sigma^2{}_{i'} > F_{\alpha/2,dfl,df2}$, then lines i and i' have different stabilities. $\sigma^2{}_i$-values were obtained using a SAS/IML-programme kindly provided by Dr Kang (Kang, 1989). A similar procedure was used to estimate the contribution of different lines to genotype × year (GY) interactions.

In order to reveal the nature of GE-interactions, regression analysis was also performed according to the model of Eberhardt & Russell (1966):

$$Y_{ij} = \mu_i + b_i I_j + \sigma_{ij},$$

where μ_i is the mean of the i-th genotype over all environments, I_j the environmental index ($I_j = y_{.j} - y_{..}$), b_i the linear regression coefficient of the i-th genotype, and σ_{ij} the deviation from regression. Tai's (1971) lambda-value, $\lambda_i = [\Sigma_j \sigma^2{}_{ij}/(n-2)] \times r/MS_e$, was used to estimate the yield stability of genotypes. If $\lambda_i \leq 1$, then deviations from regression are small. Hypothesis that $b_i = 1$, or b_i of the genotype i is not different from $b_{i'}$ of the genotype i' was tested by t-tests.

Results and discussion

Yield level and adaptation of row types

Temperature is the most important climatic factor restricting plant growth in northern Scandinavia (SNP 1992). However, the heat sum requirements are lower at higher latitudes, due to a compensating effect of longer photoperiods (Bergthórsson et al., 1987). Early autumn frost and strong winds decrease grain yield potential at locations like Korpa (Bragason, 1985). Low temperatures occurred more frequently at Korpa than at Röbäcksdalen, and mean yields at Korpa are therefore permanently below 2500 kg/ha. Barley can only be cultivated on the southern coastal region of Iceland, and Korpa represents the marginal growing conditions in the biological sense. The growth season is short and intensive at Röbäcksdalen. Night frost may occur in late summer/early autumn (Wiberg,

Table 2. Means of grain yields (kg/ha) of 2r- and 6r-lines of the whole and early maturing material

Location	2r	6r	Difference (2r–6r)	Difference bw. locations	Difference bw. regions[a]
all 220 lines					
Ås = Å	4500	4415	85 ± 68.00 ns	Å vs. V ns	
Viikki = V	4254	4150	104 ± 43.83 **		Middle vs. South
Svalöv = S	5198	4588	61 ± 45.48 ***	S vs. H **	***
Højbakkegård = H	3733	2634	1099 ± 62.78 ***		
Average	4421	3947	474		
40 early lines					
Korpa = K	1914	1467	447 ± 91.77 ***	K vs. R **	North vs.
Röbäcksd. = R	2348	2465	- 117 ± 178.77 ns		(Middle + South)
Ås	3207	3579	- 372 ± 173.19 *	Å vs. V ns	**
Viikki	3420	3907	- 487 ± 171.63 **		
Svalöv	4521	4821	- 300 ± 131.19 *	S vs. H **	Middle vs. South
Højbakkegård	3227	3065	162 ± 90.36 *		**
Average	3106	3217	111		

[a] Regions are: 'North' = Korpa and Röbäcksdalen, 'Middle' = Viikki and Ås, and 'South' = Højbakkegård and Svalöv. p > 0.05 = ns, p < 0.05 = *, p < 0.01 = **, and p < 0.001 = ***.

1993). The type of stress, its severity and timing, and combinations of different types of stress vary more at Röbäcksdalen than at Korpa. The most favourable growth season in 1989 resulted in a grain yield level above 4000 kg/ha for the best lines, while this level was 2300 kg/ha at Korpa. Denmark, Finland, areas in northern Sweden, and in South-East Norway can be described as sub-arid environments.

Grain yield differences between 2r- and 6r-lines within the four southern locations are presented in Table 2. Orthogonal linear contrasts were used to test differences between row types within regions. The differences were 19 kg/ha (not significant) within region 'Middle' (Ås and Viikki), 489 kg/ha ($p < 0.01$) within region 'South' (Højbakkegård and Svalöv), and 760 kg/ha ($p < 0.001$) between regions 'Middle' and 'South'. Table 2 shows that 2r-barleys yielded more and were better adapted to southern Scandinavia than 6r-barleys. This difference was levelled out when one approaches 60° N latitude, and increased again towards the northern limit of barley cultivation in favour of 6r-barleys. This can partly be explained by the stronger daylength response of 6r-lines (Wiberg, 1993), and disease and lodging resistance of 2r-lines (data not shown).

The difference between row types among the 40 early maturing lines was highly significant ($p < 0.001$) at Korpa where 2r-lines outyielded 6r-lines by 447 kg/ha (Table 2). This was due to the better shedding and lodging resistance of 2r-barleys in the Icelandic climate. Also at Højbakkegård 2r-lines outyielded 6r-lines significantly ($p < 0.05$). 6r-lines gave higher yields than 2r-lines at Ås ($p < 0.05$), Svalöv ($p < 0.05$) and Viikki ($p < 0.01$). The difference between row types was not significant at Röbäcksdalen (117 kg/ha) (Table 2). Two of the three early maturing 2r-barleys, 'Arla' and 'Akka', are specifically bred for the short growth season of northern Sweden. Orthogonal linear constrasts were used to split the variation in difference between row types. In the 'North'-region the differences between row types in reaction to growing conditions between Korpa and Röbäcksdalen were highly significant ($p < 0.01$). This was also the case between locations in 'South', but not in region 'Middle' (Table 2). The row types reacted differently ($p < 0.01$) between regions 'Middle' and 'South', and between 'North' and 'Middle + South'. Wiberg (1993) stated that the superiority of 6r-lines in northern Sweden may partly be caused by a closer origin to locally adapted landraces than that of 2r-barleys.

The long day conditions and the maritime climate makes Iceland unique in barley breeding. High yielding plants combined with sufficient earliness, and with shedding and lodging resistance are desirable (Bragason, 1985). When the stress environment has a low yield potential (under 2 500 kg/ha), direct selection

Table 3. Results from the analyses of variance over L locations, Y years and r blocks of different subsets

Source	df	Early		6R		2R	
		df	MS[a]	df	MS	df	MS
Genotypes	G-1	39	1316	70	2739	148	2697
Genotypes × Locations	(G-1)(L-1)	195	400	210	516	444	403
Genotypes × Years	(G-1)(Y-1)	39	263[b]	140	460	296	335
G × L × Y	(G-1)(L-1)(Y-1)	195	176[c]	420	293	888	296
Error: G × $r/_{L\times Y}$	(r-1)(G- 1)LY	468	176	840	189	1776	173

[a] All MS-values should be multiplied by 10^3.
[b] Significant at $p < 0.05$, while other mean squares are significant at $p < 0.001$.
[c] Not significant.

in the target environment is the most efficient strategy (Ceccarelli, 1989; Ceccarelli & Grando, 1989). Between 62° N and 65° N in Scandinavia the long day combined with a more continental climate than in Iceland create a region where 6r-varieties dominate barley cultivation. In this region different types of stress and non-stress environments occur randomly, but equally frequent. Co-operation between the countries in this northern region should be more intensive in selection and testing of breeding lines.

At Viikki and Ås 2r- and 6r-lines were more or less equal both within and between locations. Similar growing conditions are found in Sweden (SNP, 1986). This area around the 60° N latitude makes another 'central' region where the co-operation of breeding for geographically wide adaptation could be improved. The most southern region differed highly significantly from the 'central' region. However, the great difference in mean grain yields between Svalöv and Højbakkegård is very difficult to explain, keeping in mind that this region is considered to be relatively uniform in growing conditions (SNP, 1986).

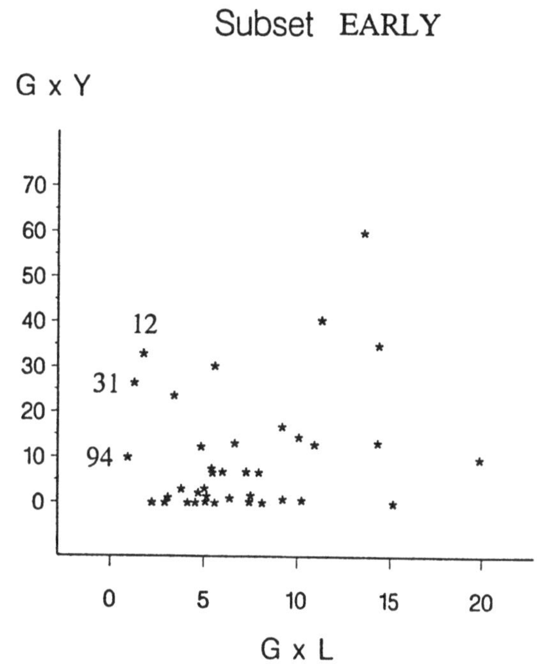

Figure 1. Relation between stability parameters (F_α) for GL- and GY-interactions of 39 early maturing lines. 'Sigur-F' had its values out of range. Nos. 12, 31, and 94 had GL-stability.

Analysis of variance and phenotypic stability

All GE-interactions were highly significant ($p < 0.001$) in both subset 6R and 2R (Table 3). GL-interactions of grain yield were always greater than GY- or GLY-interactions. In subset EARLY GL- and GY-interactions were significant at the $p < 0.001$ and $p < 0.05$ levels, respectively, while the GLY-interaction was not significant. If the year 1987 is omitted from the analyses the significance level of GY-interactions decreased remarkably (data not shown). This means that the years 1988 and 1989 had a similar climate, and that the cool and rainy growth season of 1987 caused large GY-interactions when compared to 1988 and/or 1989.

GY-interactions were not correlated with GL-interactions in subsets 6R and 2R. In subset EARLY the correlation between GY- and GL-interactions was highly significant, $r = 0.82$, ($p < 0.001$). This was mainly caused by the low number of years, and similarity in climate of these years. Taken together these results indicate that stability for GL- and GY-interactions are separate characters.

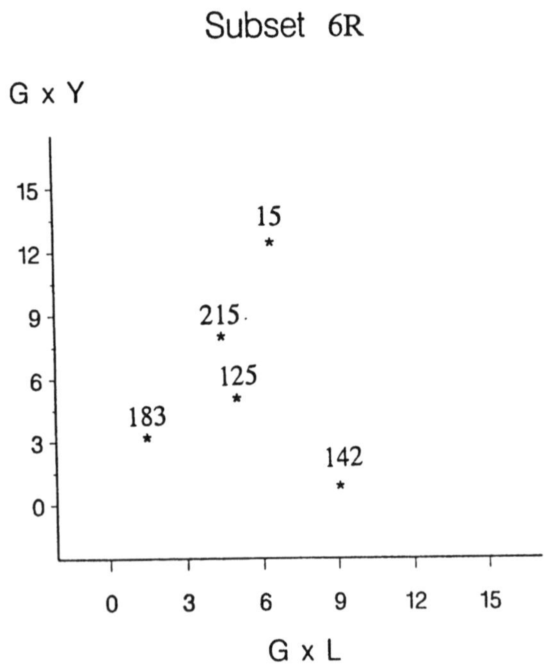

Figure 2. Relation between stability parameters (F_α) for GL- and GY-interactions of 'Varde' (15) and the progeny of the cross 'Varde × Otra' (125, 142, 183 and 215) in subset 6R.

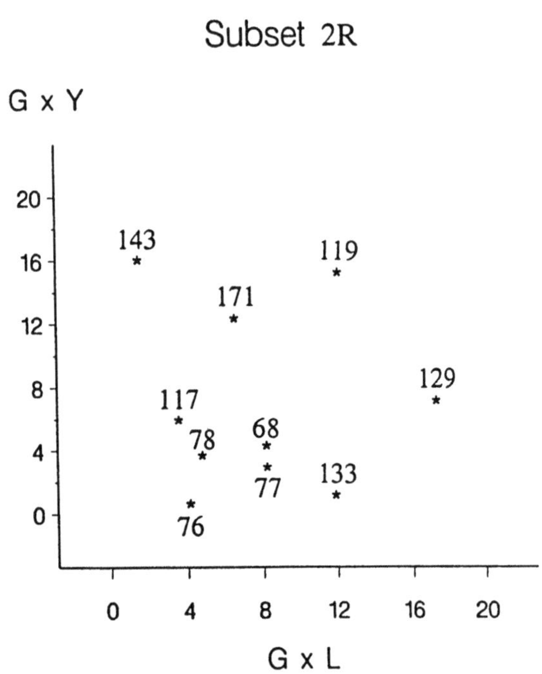

Figure 3. Relation between stability parameters (F_α) for GL- and GY-interactions of 'Domen-mutants' (68, 76, 117, 119), progeny from the cross 'Mona × (Morocco × Pallas^4)' (129, 133, 143), multilines of 'Pallas' (78, 171), and 'Pallas' (77) in subset 2R.

F_{alpha}-values from testing of stability variances over years and locations are shown in Figures 1, 2 and 3 for some of the lines in different subsets. In subset EARLY none of the genotypes was widely adapted over locations and years (Figure 1). The 'Tammi'-mutant (94), 'Fræg'-mutant (12) and 'Agneta' (31) had low GL-interactions, but high GY-interactions. The most unstable genotype, 'Sigur-F', had stability variances out of the range of Figure 1 ($F_{\alpha G\times L}$ = 67.7, and $F_{\alpha G\times Y}$ = 154.1). On the other hand, 'Tampar', which has been selected from the same landrace as 'Sigur-F', showed very low GY-interactions ($F_{\alpha G\times L}$ = 2.2, and $F_{\alpha G\times Y}$ = 0). In subset 6R, progeny from the cross 'Varde (no. 15) × Otra' gave a line with low GY-interactions (no. 142, 'Niina'), a line with low GL-interactions (no. 183, 'Arra'), and two lines with an average stability (nos. 125 and 215) (Figure 2). Arra was widely cultivated in Finland in the end of the, 1980'ies, while 'Niina' has been an export variety to Russia and Estonia.

In subset 2R, F_α-values of two multilines of 'Pallas' (nos. 78 and 171) and the original 'Pallas' (no. 77) are plotted in Figure 3. From a cross of 'Mona × (Morocco × Pallas^4)', line no. 143 had low GL-interactions, line no. 133 had low GY-interactions, and line no. 129 was unstable. Of the four induced mutants of 'Domen' (Nos. 68, 76, 117 and 119) only no. 76 had low GY-interactions. Only one 6r-line from Norway and six 2r-lines did not contribute significantly to both GL- and GY-interactions, and thus could be stated to have wide adaptation. The mean grain yields of those 2r-lines were 7–27% lower than the mean yield of the highest yielding 2r-accession, 'Golf'.

Regression analysis of GE-interactions

Testing of the regression mean squares against the deviation mean squares demonstrated that the linear regression explained the GE-interactions well in subsets EARLY and 6R, but not in subset 2R. The suitability of the regression method, and repeatability of regression coefficient and yield stability between years will be dealt with a separate paper. Here we would like to focus on the conditions between yield and responsiveness (b_i) on the one hand, and important characteristics as origin, row type and year of release on the other hand.

The regression coefficients were negatively correlated with λ_i-values in subset EARLY (r = - 0.34), i.e. genotypes with high b_i-value (responsive) were more stable than less responsive genotypes (Table 4).

Table 4. Correlations between responsiveness (b_i), phenotypic stability (λ_i), mean yield, origin, and year of release within different subsets

Subset		λ_i	Yield	Origin	Release
EARLY	b_i	- 0.340*	0.463**	0.179	- 0.121
	λ_i		- 0.265	- 0.064	0.147
	yield			0.462**	0.425
	origin				- 0.095
6R	b_i	- 0.065	0.129	- 0.004	0.060
	λ_i		0.124	0.143	0.229
	yield			0.390***	0.542***
	origin				- 0.142
2R	b_i	0.156	0.007	- 0.012	- 0.191
	λ_i		- 0.131	- 0.121	0.260*
	yield			0.569***	0.607***
	origin				0.195

Significance of correlation at $p < 0.05$ = *, at $p < 0.01$ = **, and at $p < 0.001$ = ***.

This was not observed in other subsets which must be due to changes in the number of environments, years and background material among subsets. Correlation between b_i and grain yield was not significant in subsets 6R and 2R. However, this correlation was positive and significant in subset EARLY which is most probably due to selection environment of this material. Origin was assigned in north-south direction so that Iceland, Finland, Norway, Sweden and Denmark were given values from 1 to 5, respectively. Most of the 6r-material came from Finland and Norway, while the majority of high yielding 2r barleys originated from Denmark and Sweden. This resulted in a positive association between grain yield and origin. Cultivars bred recently yielded more than older varieties, and especially 2r-cultivars released recently seemed to have a greater instability than older cultivars revealed by positive correlation between λ_i and the year of release. This correlation was not significant in other subsets. Due to variation in the correlations between b_i and grain yield, responsiveness and yield are assumed to be at least partly independent of each other, and it should be possible to combine high yielding ability with desired level of responsiveness. Also association between λ_i and grain yield was not significant, indicating that high yield can be combined at least with average stability.

6r-barleys had larger regression coefficients than 2r-barleys when both row types were analyzed together (data not shown). Responsiveness of genotypes in separate tests of 6r- and 2r-lines was only slightly changed if all the lines were analyzed together. However, yield stability was remarkably changed when the row types were analyzed together. Therefore, 2r- and 6r-materials should be tested separately due to their different intensity of response to growth conditions in the Nordic countries. Grouping of material by earliness would indentify stable genotypes within different maturity groups, and reasons for instability, e.g. unfavourable weather conditions at a certain physiological growth stage, would be easier to detect.

Changes in the growth response, b_i

'Fræg' (no. 11) and its mutant line (no. 12) were significantly ($p < 0.05$) different in response to the growing conditions in subset EARLY ($b_{11} = 1.17 \pm 0.08$ and $b_{12} = 0.96 \pm 0.06$). An opposite response pattern has previously been observed for these barleys in studying GY-interactions at Ås (Aastveit & Aastveit, 1984). 'Sigur-F' and 'Tampar' (nos. 4 and 5, respectively) are selections from a landrace, and originate from the same farm on the Faroe Islands. They were highly significantly different in responsiveness ($b_4 = 0.36 \pm 0.19$ and $b_5 = 0.95 \pm 0.07$, $p < 0.01$).

Sister lines from the cross 'Varde × Otra' (nos. 125 and 183) showed a different response to environments in subset 6R ($b_{125} = 0.796 \pm 0.03$ and $b_{183} = 0.989 \pm 0.08$, $p < 0.05$), indicating that they exhibit recombination of different alleles for responsiveness. The two 'Pallas' multilines (nos. 78 and 171) differed significantly in responsiveness from each other ($b_{78} = 0.66 \pm 0.05$ and $b_{171} = 0.89 \pm 0.09$, $p < 0.05$), but not from the original 'Pallas'. Thus, mixing of near isogenic lines, with different powdery mildew resistance genes, evidently resulted in changes in responsiveness.

More attention should be paid to positive and large deviations from regression at some location(s) which reveals that the genotype has a better, specific adaptation to these growing conditions than the average of the whole material. This is often the case with resistance to biotic or edaphic factors that limit plant growth. Emphasis should be devoted to GL-stability which is more predictable than GY-stability.

Acknowledgements

We thank Nordic plant breeding companies, research stations and agricultural universities for providing plant materials, experimental sites and technical assistance. Weather data have been provided by Nordic climate recording stations. This work was supported by grants from the Nordic Plant Breeding (SNP-project 87:1) under the Nordic Council of Ministers, The Norwegian Research Council (NFR), and the Department of Horticulture and Crop Sciences, Agricultural University of Norway. We thank Prof. K. Aastveit for critical reading of earlier drafts of the manuscript, and for helpful advice and fruitful discussions.

References

Aastveit, A.H. & K. Aastveit, 1984. Genetic variation of developmental stability in barley. Hereditas 101: 155–170.

Allard, R.W. & A.D. Bradshaw, 1964. Implications of genotype-environmental interactions in applied plant breeding. Crop Sci 4:503–508.

Becker, H.C., 1983. Züchterische Möglichkeiten zur Verbesserung der Ertragssicherheit. Vortr Pflanzenzüchtg 3: 203–225.

Becker, H.C. & J. Léon, 1988. Stability analysis in plant breeding. Plant Breed 101: 1–23.

Bergthórsson, P., H. Bjornsson, O. Dýrmundsson, B. Gudmundsson, A. Helgadóttir & J.V. Jónmundsson, 1987. The effect of climatical variations on agriculture in Iceland. p. 383–509. In: M.L. Parry, T.R. Carter & N.T. Konijn (Eds). The Impact of Climatic Variations on Agriculture, Vol I: Assessments in cool temperature and cold regions. Kluwer Academic Publishers Group, Dordrecht.

Bragason, A., 1985. Sammenligning af vårbygpopulationer i Danmark og Island. Den kgl. Veterinær- og Landbohøjskole, Afdelingen for Landbrugets Plantekultur, København, 90 pp.

Ceccarelli, S., 1989. Wide adaptation: How wide? Euphytica 40: 197–205.

Ceccarelli, S. & S. Grando, 1989. Efficiency of empirical selection under stress conditions in barley. J Genet & Breed 43: 25–31.

Ceccarelli, S. & S. Grando, 1991. Selection environment and environmental sensitivity in barley. Euphytica 57: 157–167.

Ceccarelli, S., S. Grando & J. Hamblin, 1992. Relationship between barley grain yield measured in low- and high-yielding environments. Euphytica 64: 49–58.

Eberhart, S.A. & W.A. Russell, 1966. Stability parameters for comparing varieties. Crop Sci 6: 36–40.

Falconer, D.S., 1990. Selection in different environments: effects on environmental sensitivity (reaction norm) and on mean performance. Genet Res Camb 56: 57–70.

Finlay, K.W. & G.N. Wilkinson, 1963. The analysis of adaptation in a plant-breeding programme. Austr J Agric Res 14: 742–754.

Helms, T.C., 1993. Selection for yield and stability among oat lines. Crop Sci 33: 423–426.

Jalaluddin, Md. & S.A. Harrison, 1993. Repeatability of stability estimators for grain yield in wheat. Crop Sci 33: 720–724.

Kang, M.S., 1989. A new SAS program for calculating stability-variance parameters. J Hered 80: 415.

Langer, I., K.J. Frey & T. Bailey, 1979. Associations among productivity, production response, and stability indexes in oat varieties. Euphytica 28: 17–24.

Nurminiemi, M., 1995. Environmental stability of Nordic barley materials. Agric Univ of Norway. Doctor Scientiarum Thesis 1995: 7. 157 pp.

Perkins, J.M. & J.L. Jinks, 1968a. Environmental and genotype-environmental components of variability. III Multiple lines and crosses. Heredity 23: 339–356.

Perkins, J.M. & J.L. Jinks, 1968b. Environmental and genotype-environmental components of variability. IV Non-linear interactions for multiple inbred lines. Heredity 23: 525–535.

Reysack, J.J., D.D. Stuthman & R.E. Stucker, 1993. Recurrent selection in oat: Stability of yield and changes in unselected traits. Crop Sci 33: 919–924.

Rosielle, A.A. & J. Hamblin, 1981. Theoretical aspects of selection for yield in stress and non-stress environments. Crop Sci 21: 943–946.

SAS, 1987. SAS/STAT Guide for Personal Computers, Version 6 Edition, 1028 pp. Cary, NC, SAS Institute Inc.

Shukla, G.K., 1972a. Some statistical aspects of partitioning genotype-environmental components of variability. Heredity 29: 237–245.

Shukla, G.K., 1972b. An invariant test for the homogeneity of variances in a two-way classification. Biometrics 28: 1063–1072.

SNP, 1982. Planteforedlingsteori. Samnordisk Planteforedling, publikation nr. 3, 33 pp. Nordisk Ministerråd, Oslo.

SNP, 1986. Samnordnad odlingsvärdesprovning i Norden. SNP-publikation nr. 14, 115 pp. Nordisk Ministerråd, Oslo.

SNP, 1992. Agroklimatisk kartlegging av Norden. Samnordisk planteforedling. Skrifter och rapporter nr 5, 97 pp. BTJ Tryck AB, Lund.

Tai, G.C.C., 1971. Genotypic stability analysis and its application to potato regional trials. Crop Sci 11: 184–190.

Wiberg, A., 1993. Norrländskt jordbruk – finns det behov av speciell sortframställning? Sveriges Utsädesf Tidskr 103: 17–24.

P.M.A. Tigerstedt (Ed.), Adaptation in Plant Breeding, 225–236, 1997.

Adaptation to low/high input cultivation

Salvatore Ceccarelli
The International Center for Agricultural Research in the Dry Areas (ICARDA), P.O. Box 5466 Aleppo, Syria

Key words: *Hordeum vulgare*, barley, genotype by environment interaction, landraces, low-input agriculture, specific adaptation

Summary

Many areas of world, particular those where agriculture is largely practiced by resource-poor farmers with little or no use of external inputs, have not benefitted from the spectacular yield increases achieved by the combination of modern breeding technologies and use of inputs. The paper argues that because breeding is mostly conducted in presence of high inputs, it has systematically missed the opportunity to exploit genetic differences at low levels of inputs. Many studies show that these differences do exist, particularly in the case of fertilizers, and that these differences can only be identified is selection is conducted under the target level of inputs. Although this was predicted by theory more than 40 years ago, and has been supported by a large body of experimental data, very few breeders select in sub-optimal or stress conditions. The most common justification is the high environmental variation, and hence the lower heritability expected in low input conditions. While this is not supported by experimental evidence, the paper shows that in the case of a typical crop grown in low-input and climatically marginal conditions such as barley, genetic gains are possible by using locally adapted germplasm and by selecting in the target environment. Similar conclusions, in relation to the use of a low-input selection environment, have been reached recently in maize. It is concluded that the best avenue to a sustainable increase of agricultural production in low-input agricultural systems is through locally based breeding programs.

Introduction

Adaptation to low/high input cultivation will be discussed throughout this paper in an agronomic, rather than in a biological context. Hence, a genotype, or breeding line, or cultivar will be considered adapted to a given type of conditions (climatic, agronomic, edaphic) when it is able to give an economic production, and not necessarily only survives in that set of conditions. This recognizes that there could be adaptation without economic productivity, but not economic productivity without adaptation.

Adaptation of agricultural crops to their environment has been a key factor to the yield increases which have occurred since the spread of crops to new environments (Evans, 1980). From the early stages of domestication most of the selection was by farmers for better adaptation to very specific environments (Bramel-Cox et al., 1991). Beginning towards the end of the last century, the agricultural environment began to be modified, first by the use of fertilizers, later by the use of pesticides, fungicides and herbicides. The process of adapting crops to gradually changing agronomic conditions has become more efficient through the use of systematic breeding programs, and selection by breeders has been for high yield potential and broad adaptation. Therefore, the consequent yield increases have been achieved as a combined effect of an agronomic environment made more favorable for plant growth and of cultivars better adapted to that environment.

It has been estimated that the spectacular yield increases of crops during the second part of this century have to be attributed in almost equal measure to breeding and to the use of inputs. These yield increases have occurred in most of the crops in most of the developed countries but only in some crops in some developing countries. Large areas of Africa, Central and South America, and Asia have been by-passed by the high-

Table 1. Average grain yield (kg ha^{-1}) of the ICARDA Barley Regional Yield Trials for low- and high-rainfall areas (1985–1991) conducted on research stations and national average yields in a number of countries

Country	Average yield in exp. stations		National average yield
	Low rainfall trials	High rainfall trials	
Algeria	2993	4168	862
Afghanistan	1565	3462	1070
China	5349	3890	2836
Cyprus	3160	4804	1948
Egypt	2798	3855	2698
Ethiopia	1690	3042	1131
Iran	2782	3597	1143
Iraq	945	819	741
Jordan	2294	2820	684
Lebanon	2319	2828	1884
Libya	1465	2886	596
Morocco	2926	3715	1140
Pakistan	1897	2275	772
S. arabia	6610	4280	3887
Syria	3417	3367	716
Tunisia	3261	3883	735
Turkey	2077	3635	2043

input type of agricultural development (Maurya et al., 1988; Haugerud & Collinson, 1990; Pimbert, 1994). Despite the large benefits obtained, the use (or perhaps the abuse) of agricultural inputs – not only chemicals but also water – has caused an increasing concern for the environmental consequences of a strategy based on modifying the environments to fit crops and varieties with an increasing yield potential – a strategy that may be not sustainable in the long term (Poutala et al., 1994). Environmental degradation, abuse of physical resources (mostly water and soil) and of biological resources have become an international concern during the last ten years, and terms such as sustainability, friendly or ecological or biological agriculture, and similar, have become part of the every-day jargon of agricultural scientists and administrators.

The debate is often polarized between scientists advocating a 'clean agriculture' and scientists concerned with feeding a growing population, and it is best summarized in the following joint statement by the U.S. National Academy of Sciences and the U.K. Royal Society: 'If current predictions of population growth prove accurate and patterns of human activity on the planet remain unchanged, science and technology may not be able to prevent either irreversible degradation of the environment or continued poverty for much of the world' (Norse, 1992). Although this issue provides a general framework for discussing adaptation to low/high inputs, this paper will address the more specific question of why the adaptation to increasing levels of inputs, which has been a constant objective in modern plant breeding, did not have beneficial effects in those traditional agricultural systems which are characterized by harsh environments and limited use of external inputs. It has been estimated that some 1.4 billion people are dependent on this type of agriculture (Pimbert, 1994) and that resource-poor farmers practice approximately 60% of global agriculture, and produce 15–20% of the world food (Francis, 1986). These farmers have never known the Green Revolution.

Institutional breeding and low-input agriculture

Although there are as many different ways of running a breeding program as there are breeders, most breeding programs, both in developed and in developing countries, share some general concepts and, consequently, some common ways of handling breeding materials.

Some examples of what most breeders commonly do are: (a) selection is conducted in research stations with optimum amount of fertilizers, weed control and water supply, often in crop rotations that little have in common with reality (Rathjen & Pederson, 1986; Atlin & Frey, 1989a, 1989b; Simmonds, 1991). This is done because under these conditions environmental noises can be kept under control, error variances are small, and response to selection high; (b) breed cultivars that are genetically homogenous (pure lines, hybrids, clones) to fit the rules for variety registration which in many developing countries are as strict as in developed countries; (c) new varieties are released based on their average performance in trials conducted for a number of years and in a number of locations, often at near-optimum levels of crop management (some breeding programs in Australia are an exception); (d) in developing countries locally adapted landraces are seen as one of the reasons for low yield and therefore they must be replaced; (e) seed of improved varieties is made available to farmers through mechanisms and institutions such as variety release committees, seed certification schemes and seed production organizations; (f) the end users of new varieties are not involved in selection and testing; they are only involved at the end of the consolidated routine (breeding, researcher managed trials, verification trials), to verify if the choices made for them by others are appropriate or not.

Breeding programs based on this set of concepts have been very efficient in developed countries, and in general in environments which are either favorable or which can be made favorable by the use of external inputs. This is because the agronomic and climatic conditions in farmers' fields and in the experiment station are not very different from each other. Therefore genotype × environment interaction is unlikely to pose major problems. Uniformity of varieties is needed to meet the requirements of a market oriented agriculture, broadly adapted varieties fit the requirements of the seed industry, seed distribution mechanisms are efficient and new varieties can quickly reach the farmers.

Agriculture in many developing countries, and particularly in unfavorable conditions, offers a very different picture.

Yield levels in farmers' fields are often several fold lower than those obtained in research stations. An example is given in Table 1 for a crop such as barley which is traditionally grown in unfavorable conditions. Low yields in farmers' fields are often due to low levels of inputs (particularly fertilizers, herbicides and pesticides), but also to what researchers consider inappropriate time of sowing, poor soil preparation, in general poor agronomic techniques. These 'primitive' agronomic practices are only partly due to the ignorance of farmers as too often it is said. Fertilizers, herbicides and pesticides are often not unknown to farmers, but they are either not available or too expensive, or their use is perceived by farmers as risky in an environment with a high probability of crop losses due to climatic event such as drought (McDonald, 1989). Herbicides may not be used because what an agricultural scientist considers a weed is often seen by a farmer as a source of animal feed, and what the same agricultural scientist defines as "hand weeding" is often "harvesting a forage crop". The yield advantage of changing sowing date or the phenology of the crop can be easily demonstrated on research stations and that particular technology ends in an Annual Report as a 'recommended practice'. Farmers, particularly small and resource-poor farmers, often have to seek off-farm sources of income which can prevent them from changing a number of operations on the farms and which can be much greater than those derived from the 'recommended' agricultural practices. Demonstrating the benefit of an appropriate and timely seed-bed preparation is too often purely academic because obtained in the experiment station with machinery acquired through development aids and available nowhere else in the country.

Resource-poor farmers in many regions of the word practicing agriculture in these situations have adopted a strategy based on both intraspecific and interspecific diversity (Martin & Adams, 1987). A number of different crops are usually grown in the same field at the same time (intraspecific diversity), and the cultivars of the different crops are usually genetically heterogeneous (interspecific diversity). A second level of interspecific diversity is obtained by growing, at the same time and in the same field, different cultivars of the same crops (Haugerud & Collinson, 1990). The type of diversity which prevails in different areas depends on both climatic and socioeconomic conditions. In central Africa and central America both intra- and interspecific diversity are exploited at the same time. In other areas, interspecific diversity represented by one heterogeneous cultivar, is predominant. In the dry areas of West Asia and North Africa, for example, barley is often the only possible rainfed crop, and the cultivars which are grown at present, and which have been grown for centuries, are genetically heterogeneous (Ceccarelli et al., 1987; Weltzien & Fischbeck, 1990).

While the typical concepts of institutional breeding listed earlier seem to fit well advanced agricultural systems, they seem to be in a sharp contrast with the realities of low-input agriculture practiced by resource-poor farmers. The rest of the paper will discuss in details two of these concepts, namely the environment where selection is conducted and the type of germplasm, in relation to adaptation to high/low input conditions. Before doing that it may be useful to clarify that low/high input cultivation means different things in different countries. In high-input agricultural systems low-input cultivation is seen as one way of responding to environmental concerns. In low-input agricultural systems low-input agriculture, as practiced by resource-poor farmers, is not an option but in the majority of cases is dictated by the physical and economic environment. In this case adapting crops to low-input cultivation is the only avenue to increase agricultural production in the short term.

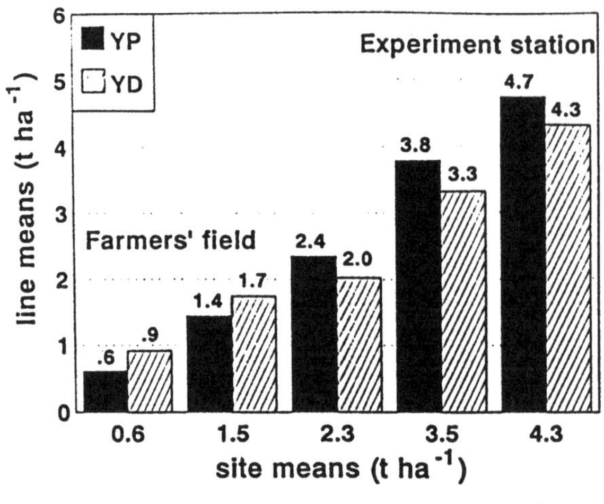

Figure 1. An example of cross-over type of G $\times$ E interaction in barley: YP and YD are lines selected in high and low yielding environments, respectively and then tested in 21 location-year combinations. On the X-axis the location-year combinations are combined in five groups according to the average grain yield of 64 barley lines. The two lowest and the two highest yielding groups are representative of framers' fields and of research stations, respectively.

Table 2. Grain yield (t ha^{-1}) of two barley cultivars in Sardinia with standard agronomic practices (weed control and 80 kg ha^{-1} of N) and with reduced inputs (no weed control and 40 kg ha^{-1} of N) (from: Attene et al., in press)

Cultivar	Grain yield[a]	
	weed control and 80 kg ha^{-1} of N	no weed control and 40 kg ha^{-1} of N
Formula	6.3a	3.6b
Local landrace	5.4b	4.3a

[a] Values followed by different letters are significantly different at $P < 0.05$.

Selection environment and adaptation

The theoretical framework of this issue, which essentially is a case of Genotype $\times$ Environment (G $\times$ E) Interaction and which continues to be hotly debated among plant breeders, has been developed by Falconer more than 40 years ago (Falconer, 1952). A recent review of experiments, mostly on animals (Falconer, 1990), showed that 'If a breeder wants to improve performance in environment A he should select in environment A' (see also Falconer, 1993). The conclusions of Falconer have been confirmed by theoretical studies such as those of Rosielle & Hamblin (1981) and Simmonds (1991), and are supported by experimental data on both self-pollinated and cross-pollinated crops (reviewed by Ceccarelli, 1994) and on small ruminants (reviewed by Bradford & Berger, 1988).

An example of this type of experimental data is shown in Figure 1 with data from barley in Syria. Barley lines selected for high grain yield either in high yielding sites (YP) or in low yielding sites (YD) between 1988 and 1990 were compared in 21 location-year combinations between 1991 and 1994. The YP lines yielded more than the YD lines in the medium to high yielding location-year combinations and the YD lines yielded more than the YP lines in the low and very low yielding location-year combinations. Between the medium and low-yielding location-year combinations the YP and YD lines cross-over. These data, like others published earlier (Ceccarelli & Grando, 1991a, 1991b; van Oosterom et al., 1993), indicate that the top yielding lines a breeder selects in a high yielding environment (high inputs and favorable climatic conditions) would not be selected in a low yielding environment (low inputs and unfavorable climatic conditions, and viceversa. The implications of the data in Figure 1 are clearer by considering the cumulative probability distribution of grain yield of barley in two agroecological zones of Syria (Figure 2). In the drier areas (where the use of inputs is limited by the high frequency of abiotic stresses) the probability of yields less than 1.5 t ha^{-1} is about six times higher than the probability of yields exceeding 3 t ha^{-1}. Therefore, the lower yield potential of the YD lines, specifically selected for the most frequent type of conditions is not a serious problem.

The majority of experimental data indicating a G $\times$ E interaction of cross-over type are from studies where the environment is characterized by the mean yield of the lines under testing, and low- and high-yielding environments are not related to specific climatic or edaphic causes. However, most of the studies which

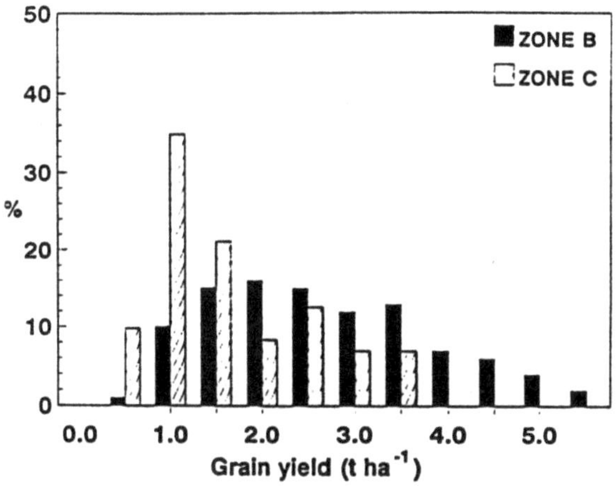

Figure 2. Frequency distribution of barley yields in two climatic zones of Syria (zone C receiving less than 250–300 mm annual rainfall, and zone B receiving 300–350 mm annual rainfall) between 1983 and 1994. The distributions are based on 72 and 101 location-year combinations for zone C and B, respectively.

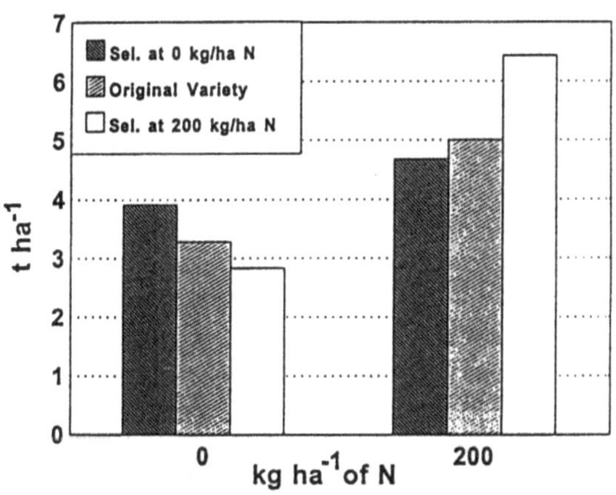

Figure 3. Grain yield, at 0 and 200 kg/ha of N, of an open pollinated maize population, and of two synthetics developed from the ten highest or the ten lowest yielding half-sib progenies at 0 and 200 kg/ha of N (from Muruli & Paulsen, 1981).

deal with genotypic responses to specific inputs show a similar pattern of responses. Balko & Russell (1980) found a strong genotype × nitrogen level interaction in maize, and Muruli & Paulsen (1981) showed that it is possible to obtain response to selection conducted in a low-nitrogen environment (Figure 3). This suggests that the alleles controlling yield in the low- and high-nitrogen level are not identical. Similarly Atlin & Frey (1989a) found genotype × phosphate interaction in oats, predicting that direct selection at low-P lev-

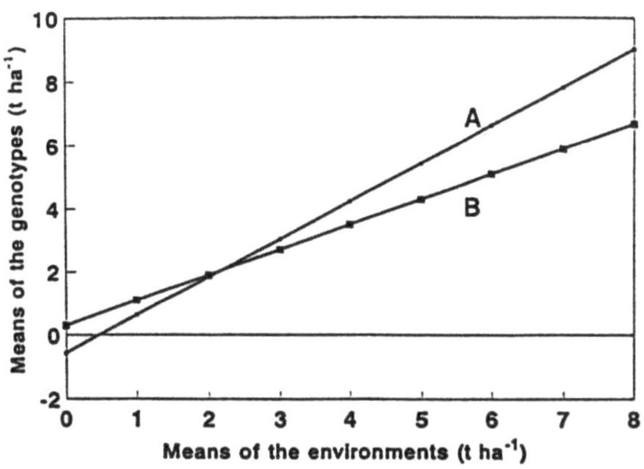

Figure 4. Cross-over type of G × E interaction: A and B are typical genotypes selected in high and low yielding environments, respectively (redrawn from Simmonds, 1984).

els was superior to indirect selection, i.e. selection at high P-levels. However, no interaction was found with nitrogen levels. Other examples of interaction between genotypes and fertilizer levels (nitrogen, phosphate and potash) in oats, barley, white clover, sorghum, common bean and tomato are given by Atlin & Frey (1989b). Recently Attene et al. (1995) showed an example of genotype × agronomic practices (nitrogen fertilizer and weed control) interactions in barley (Table 2).

The frequency with which genotype × input interactions occur indicates that alleles that confer specific adaptation to low-input conditions may exist in many crop species (Atlin & Frey, 1989b).

These results are not surprising in the light of a review of plant productivity in relation to nutrients, types of management and stresses published in 1982 (Boyer, 1982). The review concludes 1) that there are genetic differences in the ability of plants to accumulate nutrients such as iron, nitrogen, phosphorus and a number of micronutrient, and 2) that 'considerable success can be expected if plant improvement includes selection under conditions that are often unfavorable for growth. . . so that genotypes capable of exploiting limited resources can be identified. With these genotypes, large scale modifications of the environment is less necessary'.

If we broadly define as stress environments those environments where plant productivity is limited by either climatic or nutritional factors, these results also agree with the circumstantial evidence that plant breeding has not been as effective in stress environments as it has been in optimum or near-optimum environments

(Maurya et al., 1988; Haugerud & Collinson, 1990; Jansen et al., 1990; Grisley & Shamambo, 1993; Sperling et al., 1993; Hardon & de Boef, 1993).

The experimental data discussed so far can be summarized schematically as in Figure 4 which shows the environmental response of hypothetical genotypes selected under high (A) and low (B) input conditions (or in a high- and low-yielding environment). The figure, first published by Simmonds (1984) indicates that, in general, genotypes selected for adaptation and performance in high input conditions, are not well adapted to low input conditions. The question of whether the same genotype can be equally well adapted to high and low yield levels, caused by different levels of climatic or nutritional stresses, is hotly debated among breeders while some animal and plant geneticists (Allard & Hansche, 1964; Falconer, 1990) and plant physiologists (Evans & Wardlaw, 1976; Fischer & Maurer, 1978; Fischer & Wood, 1979; Blum, 1993) have long since recognized that species and varieties adapted to favorable growing conditions are, in general, not well adapted to stress conditions.

A number of studies did not reveal the presence of cross-over interactions (Castleberry et al., 1984; Austin et al., 1989; Khan & Spilde, 1992; Bulman et al., 1993) but indicated that rates of gain in yield from genetic improvement are smaller or even nonexistent in low input or stress environments as compared to high input, highly productive conditions (Smith et al., 1990). These studies do not necessarily contradict those showing the presence of cross-over interactions.

Besides the experimental difficulties of comparing old and modern varieties, the comparison which is missing in these studies is between cultivars specifically selected for high-input cultivation and cultivars specifically selected by modern breeding for low-input cultivation.

Most of the evidence would therefore suggest that 1) in most crops there is genetic variation for adaptation, and hence yield, in low-input conditions, 2) that the most efficient way to improve adaptation and yield in low-input conditions is by direct selection in low-input conditions.

Selection in low input conditions

Despite the evidence given above, very few breeding programs conduct early generation selection in low input, and more in general, in stress conditions. The most common justification given for this widespread reluctance is low heritability. The justification is largely groundless. Firstly, theory of indirect selection shows that response to selection done in a different environment is a function of not only heritability but also of the genetic correlation coefficient between the selected trait measured in the two environments (Falconer, 1989). Secondly, the evidence that heritability in stress environments is lower than in non-stress environments is not as obvious as many breeders claim (Table 3). Furthermore there have been many reports that narrow sense heritability increases in response to increased environmental stresses (Ward, 1994). This agrees with a cross-over type of genotype × environment interactions (Figure 4), where the lowest heritability is expected at the cross-over point and explains some of the disagreement among breeders on the issue of the optimum environment of selection.

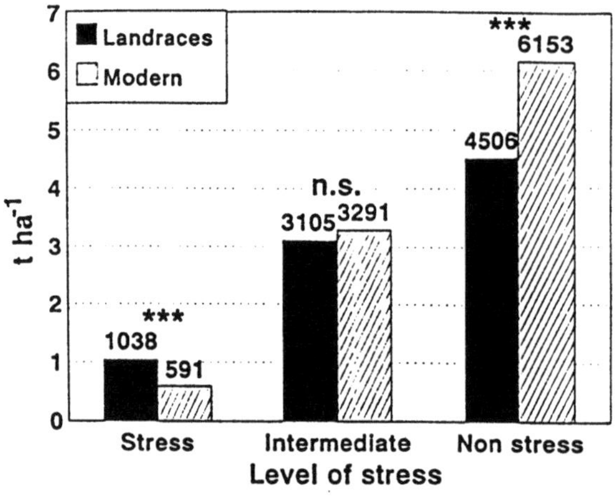

Figure 5. Differences in grain yield between modern barley cultivars and Syrian barley landraces at three different yield levels. (n.s. = non significant, *** significant at P < 0.001 based on *t*-test for groups of unequal size).

When the target environments lie above the cross-over point the best discrimination between breeding lines occurs in the highest yielding environments. In the lowest yielding environments (close to the cross-over point) breeding lines tend to look alike, hence the low heritability which is often found. When the target environments lie below the cross-over point, the largest differences between genotypes are found at the lowest yield levels and the identification of the best genotypes becomes increasingly difficult as yield levels increase. An example is given in Figure 5 where 44 pure lines from Syrian barley landraces and 206 modern barley cultivars are compared at three levels of

Table 3. Heritability estimates of grain yield at low- and high yield levels and ratio between the root squares of the two heritabilities in different crops (modified from Ceccarelli, 1994)

Crop	High (h_y)	Low (h_x)	Ratio h_y/h_x	Reference
Wheat	0.33	0.68	0.70	Pederson & Rathjen, 1981
Barley	0.47	0.68	0.83	Singh & Ceccarelli, 1995
Oats	0.38	0.52	0.85	Johnson & Frey, 1967
Maize	0.52	0.71	0.86	Selmani & Wassom, 1993
Flax	0.44	0.56	0.89	Allen et al., 1978
Maize	0.46	0.58	0.89	Lafitte & Edmeades, 1994
Barley	0.47	0.54	0.93	Allen et al., 1978
Oats	0.56	0.63	0.94	Allen et al., 1978
Barley	0.65	0.66	0.99	Weltzien & Fischbeck, 1990
Wheat	0.89	0.74	1.10	Pfeiffer, 1988
Oats	0.45	0.32	1.18	Frey, 1964
Sorghum	0.63	0.43	1.21	Zavala-Garcia et al., 1992
Cocksfoot	0.89	0.50	1.33	Breese, 1969
Soybeans	0.56	0.31	1.34	Allen et al., 1978
Oats	0.67	0.32	1.45	Atlin & Frey, 1990
Sorghum	0.69	0.32	1.47	Zavala-Garcia et al., 1992
Wheat	0.78	0.32	1.56	Allen et al., 1978
Wheat	0.25	0.03	2.89	Roy & Murty, 1970

Table 4. Grain yield (mean, standard error and range in kg ha^{-1}) of Syrian barley landraces and of modern cultivars in low and high yielding locations. The number of landraces was 44 and 76 and the number of modern cultivars was 206 and 165 in the first and second set, respectively

Location/year[a]	Rainfall (mm)	Landraces		Non landraces	
		Mean ± s.e.	Range	Mean ± s.e.	Range
First set					
Bouider, 1989	186	1038 ± 25.4	790–1479	591 ± 16.7	0–1364
Athalassa, 1989	344	4506 ± 114.6	3003–6316	6153 ± 60.3	3509–8820
Days to heading		120 ± 0.3		118 ± 0.2	
Second set					
Breda, 1990	185	610 ± 4.9	302– 918	388 ± 10.9	0–747
Tel Hadya, 1988	504	3611 ± 16.2	1326–6190	5126 ± 97.1	1580–7852
Days to heading		115 ± 0.8		113 ± 0.3	

[a] Bouider, Breda and Tel Hadya are three selection and testing sites in northern Syria: in Bouider and Breda no inputs were used. In Tel Hadya and Athalassa (Cyprus) both nitrogen (60 kg ha^{-1}), phosphate (50 kg ha^{-1}) and weed control were applied.

stress resulting from different amounts of inputs and different rainfall. At the lowest and highest yield levels there were significant differences between the two groups of germplasm, while the difference was not significant at the intermediate level, near to the cross-over point.

The data of Figure 5 raise the issue of whether the type of cultivars developed by breeding in high-inputs and climatically favorable conditions, is the best genetic material to test the potential of low-input conditions.

In many developing countries, and for crops grown in low-input conditions and in stress environments, landraces are still the backbone of agricultural production. The reasons why farmers prefer to grow only landraces or continue to grow landraces even after partial adoption of modern cultivars are not well documented. Food and feed quality and seed storability are often quoted as common reasons for this preference, but also important is the belief that landraces are able to produce some yield even in difficult conditions where modern vari-

Table 5. Average increase in grain yield (kg ha^{-1}) over local landraces of three barley lines obtained through pure-line selection from the two landraces commonly grown in Syria. The data are from on-farm trials

Line	Nr. of environments[a]	Grain yield	% increase
Arta	51	2433	22.3
Tadmor	11	780	23.6
Zanbaka	8	945	22.4

[a] Location-year combinations.

Table 6. Grain yield (t ha^{-1}) at low (N-) and high (N+) nitrogen level of experimental maize cultivars selected for high grain yield at either low or high nitrogen level (modified from: Lafitte & Edmeades, 1994)

Testing conditions	Selected for	
	Grain yield at N+	Grain yield at N-
N-	1.89	2.22
N+	6.20	5.70

eties are less reliable. This is in fact the case in those areas where farmers have adopted modern cultivars but have kept the landraces in the most unfavorable areas of the farm.

To illustrate the value of this type of germplasm I will use the example of barley in Syria. Barley landraces are still the predominant cultivars in countries such as Ethiopia, Eritrea, Nepal, Yemen, and in all the Middle East and large areas of North Africa. In Syria, barley production is entirely based on two landraces, Arabi Abiad (white seed) and Arabi Aswad (black seed). The first is common in slightly better environments (between 250 and 400 mm rainfall) and the second in harsher environments (less than 250 mm rainfall) – a clear example that farmers recognize the value of specific adaptation.

About 10 years ago pure lines extracted from Syrian barley landraces began to be routinely evaluated in the breeding program at ICARDA. Therefore it has been possible to compare landraces and modern cultivars in a range of conditions ranging from severe stress (low-input and low rainfall) to moderately favorable conditions (use of inputs and high rainfall). These comparisons (such as the example shown in Table 4) have consistently indicated that 1) landraces, as a group, yield more than modern cultivars under low-input and stress conditions, 2) the superiority of landraces is not associated with escape mechanisms as shown by heading date; 3) within landraces there is considerable variation for grain yield under low-input and stress conditions, but all the landraces-lines yield something whereas some modern cultivars fail; 4) landraces are responsive to both inputs and rainfall and the yield potential of some lines is high, though not as high as modern cultivars, and 5) it is possible to find modern cultivars which under low-input and stress conditions yield almost as well as landraces, but their frequency is very low. The data of Table 4 indicate that selection conducted only in high-input conditions is likely to miss most of the lines that would have performed well under low-input conditions.

The adaptation of landraces to the physical environment in which they evolved, coupled with the genetic diversity they harbor, suggests that pure-line selection within landraces conducted in the target environment could be a promising avenue for short terms gains. Indeed the combined use of the target environment for selection and of locally adapted germplasm has generated lines which outyield the local landraces by more than 20% in farmers' fields (Table 5). A longer term approach is the use of selected lines from landraces to reconstitute mixtures with the objective of using population buffering to stabilize production over time.

A recent example of breeding progress with direct selection in low-input conditions is in maize where a divergent selection experiment (Table 6) showed that the largest grain yields at low soil nitrogen levels were obtained from direct selection at low soil nitrogen levels (Lafitte & Edmeades, 1994), thus confirming the role of the selection environment.

Breeding in variable, unpredictable, low yielding conditions is with no doubts much more difficult than breeding in the uniform, predictable, fully controlled, highly productive conditions of the experiment station. Plot techniques and experimental designs which work well in the experiment station are usually unsuited to control environmental variability in low-input conditions. It has been shown elsewhere that a sufficient degree of precision (in experimental terms) can be obtained in conditions which are often considered, *a priori*, unsuitable to breeding work when innovative experimental techniques are used (Ceccarelli & Grando, 1996).

Not by grain alone

Adaptation to low-input cultivation, particularly in third word countries, requires a full understanding of

Table 7. Grain yield (kg/ha^{-1}), plant height (cm) and harvest index in stress (low rainfall and no fertilizer) and no stress (high rainfall and addition of phosphate) of barley lines with the highest grain yield in stress (HYS) and non stress (HYP) conditions

Trait	HYP	HYS	Difference
	Mean ± s.e.	Mean ± s.e.	
Grain yield – no stress	3577 ± 62	2822 ± 64	755***
Grain yield – stress	745 ± 49	1096 ± 74	- 351***
Plant height – no stress	62 ± 2	58 ± 1	4 n.s.
Plant height – stress	29 ± 0.6	30 ± 0.5	- 1 n.s.
Harvest index – no stress	0.46 ± 0.02	0.37 ± 0.02	0.09***
Harvest index - Stress	0.27 ± 0.02	0.40 ± 0.01	- 0.13***

n.s. = non significant; *** significant at P < 0.001.

Table 8. Grain yield (kg/ha(cm) and harvest index in stress (low rainfall and no fertilizer) and no stress (high rainfall and addition of phosphate) of Syrian barley landraces and of modern cultivars with similar phenology

Trait	Landraces	Modern	Difference
	Mean ± s.e.	Mean ± s.e.	
Grain yield – no stress	3596 ± 114	3749 ± 147	- 153 n.s.
Grain yield – stress	1198 ± 78	489 ± 66	709***
Plant height – no stress	69 ± 1	67 ± 3	2 n.s.
Plant height – stress	29 ± 1	27 ± 1	2 n.s.
Harvest index – no stress	0.48 ± 0.03	0.51 ± 0.03	-0.03 n.s.
Harvest index – stress	0.42 ± 0.01	0.22 ± 0.03	0.20***
Lodging % – no stress	100	0	100***
Lodging % – stress	0	0	0 n.s.
Days to heading	105 ± 0.4	106 ± 2.0	1 n.s.

n.s. = non significant; *** significant at P < 0.001.

the role and utilization of different crops in a given system. The example of plant height in cereals will be used to illustrate this general principle. One of the few issues in plant breeding where the consensus is almost unanimous is that the impressive yield increases obtained in some of the major cereal crops is associated with a higher harvest index. The introduction of dwarfing genes and stiff straw has made crops able to respond to higher fertilizer doses without lodging, thus expressing fully the higher yield potential associated with the higher harvest index. In many countries not only the grain but also the straw may have considerable value to the farmer mostly as source of animal feed. This is certainly the case of barley in West Asia, North Africa, Ethiopia, Nepal, only to mention some countries, but also the case of pearl millet, sorghum, maize, and to some an extent wheat. In many low-input agricultural systems not only the amount of straw but also its quality, particularly softness, is important both when the straw is utilized on the farm and when it is sold on the market.

In the case of barley, genotypes with high harvest index and high grain yield in high-input conditions seem unable to maintain a high harvest index when exposed to moisture and nutritional stress (Table 7). By contrast, barley genotypes specifically selected for high grain yield in low-input and moisture stress conditions have a low harvest index in high-input conditions (and consequently a low grain yield), but are able to maintain the harvest index at the same level when exposed to low-input conditions. Therefore, although a high harvest index seems to be an advantage also in low-input conditions, it can only be selected for by exposing breeding material to those conditions.

Syrian barley landraces offer an additional insight into the relationships between objectives of modern breeding and specific adaptation to the farming systems of low-input agriculture. Surprisingly, their harvest index is not extremely low in high-input conditions (Table 8), probably because of their extremely thin stems which makes them very susceptible to lodging (Ceccarelli, 1993). In low-input conditions their harvest index is only marginally reduced, they become very short and they do not lodge. Although it may sound as a paradox, it looks as if landraces express the typical traits of a high yielding cultivar in high input conditions, but only when grown under low-input and stress conditions.

Both examples show that low-input and moisture stress conditions cause invariably a reduction in plant height. The crop may become too short to be combine-harvested and farmers do not have alternatives to the costly hand-harvesting. Therefore, one of the main breeding objectives in this situation is to increase plant height under drought while keeping the straw sufficiently palatable. The consequences are inevitable: a tall crop with soft straw is precisely the opposite of what a modern breeder would be looking for.

Conclusions

The topic of this paper has been the object of a number of reviews in the last few years (for example Atlin & Frey, 1989b; Smith et al., 1990; Bramel-Cox et al., 1991). These reviews recognize that there is a general agreement on a number of points. Firstly, that the occurrence of genotype by environment interactions of cross-over type is a widespread fenomenon. Second, that in case of genotype by environment interactions, the magnitude of heritability is not a useful criterion to decide on the optimum environment where to do selection. Third, that the assessment of genetic correlation between performance under contrasting situations is essential to predict the likely inpact of genotype by environment interactions. Fourth, that the magnitude and the type of genotype by environment interactions depends on how wide is the range of conditions (input levels, rainfall, soil fertility) under which a given crop is grown.

The major point of disagreement in the literature is on what to do when genotype by environment interactions of cross-over type occur. The most common view is that breeding is ineffective below the cross-over point and therefore the use of inputs is a prerequisite before embarking on plant improvement programs. By improving the agronomic environment, yields are raised above the cross-over point, and therefore selection can be conveniently done within the comfortable world of the research stations. This view is extremely popular in developing countries and produces brilliant results on research stations. It serves, but not always, the minority of farmers whose agronomic and climatic environment is not too dissimilar from the research stations.

The main conclusion of this paper is that resource-poor farmers and low-input agriculture can only be served by taking the opposite view which recognizes that breeding can be effective below the cross-over point. This view is much less popular, and consequently there are not very many examples to quote. The few which are available indicate that yield increases with varieties specifically adapted to low-input conditions, though far for being spectacular, are sufficiently high to stimulate not only adoption of the variety but also adoption of better crop husbandry. However, specific adaptation is not a very popular concept. Presumably because it does not bring the same prestige to the breeder as the concept of wide adaptation and is not in the interest of the seed industry.

References

Allard, R.W. & P.E. Hansche, 1964. Some parameters of population variability and their implications in plant breeding. Advances in Agronomy 16: 281–325.

Allen, F.L., R.E. Comstock & D.C. Rasmusson, 1978. Optimal Environments for Yield Testing. Crop Science 18: 747–751.

Atlin, G.N. & K.J. Frey, 1989a. Predicting the relative effectiveness of direct versus indirect selection for oat yield in three types of stress environments. Euphytica 44: 137–142.

Atlin, G.N. & K.J. Frey, 1989b. Breeding crop varieties for low-input agriculture. American Journal of Alternative Agriculture 4: 53–58.

Atlin, G.N. & K.J. Frey, 1990. Selecting oat lines for yield in low-productivity environments. Crop Science 30: 556–561.

Attene, G., G.F. Delogu, M. Lendini, R. Papa, R. Santilocchi & F. Veronesi, 1995. Risposta dell'orzo (*Hordeum vulgare* L.) a diversi livelli di intensificazione culturale. Rivista di Agronomia (in press).

Austin, R.B., M.A. Ford & C.L. Morgan, 1989. Genetic improvement in the yield of winter wheat: a further evaluation. J. Agric. Sci. Camb. 112: 295–301

Balko, L.G. & W.A. Russell, 1980. Effects of rates of nitrogen fertilizer on maize inbred lines and hybrid progeny. I. Prediction of yield responses. Maydica 25: 65–79.

Blum, A., 1988. Plant breeding for stress environments. CRC Press, Boca Raton, Florida.

Blum, A., 1993. Selection for sustained production in water-deficit environments. pp 343–347 in: International Crop Science I, Crop Science Society of America, Madison, WI.

Boyer, J.S., 1982. Plant Productivity and Environment. Science 218: 443–448.

Bradford, G.E. & Y.M. Berger, 1988. Breeding strategies for Small Ruminants in Arid and Semi-Arid Areas. In: E.F. Thomson & F.S. Thomson (Eds) Increasing Small Ruminant Productivity in Semi-Arid Areas pp. 95–109.

Bramel-Cox, P.J., T. Barker, F. Zavala-Garcia & J.D. Eastin, 1991. Selection and testing environments for improved performance under reduced-input conditions. In: D.A. Sleeper, T.C. Barker & P.J. Bramel-Cox (Eds) Plant breeding and sustainable agriculture, Considerations for Objectives and Methods, pp. 29–56. CSSA Special Publication no. 18.

Breese, E.L., 1969. The measurement and significance of genotype-environment interactions in grasses. Heredity 24: 27–44.

Bulman, P., D.E. Mather & D.L. Smith, 1993. Genetic improvement of spring barley cultivars grown in eastern Canada from 1910 to 1988. Euphytica 71: 35–48.

Castleberry, R.M., C.W. Crum & C.F. Krull, 1984. Genetic Yield Improvement of US Cultivars under Varying Fertility and Climatic Environments. Crop Science 24: 33–36.

Ceccarelli, S., 1993. Plant breeding technologies relevant to developing countries. In: M. Gill, E. Owen, G.E. Pollot & T.L.J. Lawrence (Eds) Animal production in developing countries, Occasional Publication No. 16 – British Society of Animal Production, pp. 37–46.

Ceccarelli, S., 1994. Specific adaptation and breeding for marginal conditions. Euphytica 77: 205–219.

Ceccarelli, S., S. Grando & J.A.G. van Leur, 1987. Genetic diversity in barley landraces from Syria and Jordan. Euphytica 36: 389–405.

Ceccarelli, S. & S. Grando, 1991a. Selection environment and environmental sensitivity in barley. Euphytica 57: 157–219.

Ceccarelli, S. & S. Grando, 1991b. Environment of selection and type of germplasm in barley breeding for stress conditions. Euphytica 57: 207–219.

Ceccarelli, S. & S. Grando, 1995. Drought as a challenge for the plant breeder. In: E. Belhassen (Ed) Drought Tolerance in Higher Plants: Genetical, Physiological and Molecular Biological Analysis. Kluwer Academic Publishers (in press).

Evans, L.T., 1980. The Natural History of Crop Yield. American Scientist 68: 388–396.

Evans, L.T. & I.F. Wardlaw, 1976. Aspects of the comparative physiology of grain yield in cereals. Advances in Agronomy 28: 301–359.

Falconer, D.S., 1952. The problem of environment and selection. American Naturalist 86: 293–298.

Falconer, D.S., 1989. Introduction to quantitative genetics. 3rd edn. Longmann, London.

Falconer, D.S., 1990. Selection in different environments, effects on environmental sensitivity (reaction norm) and on mean performance. Genetic Research Cambridge 56: 57–70.

Falconer, D.S., 1993. Quantitative Genetics in Edinburgh: 1947–1980. Genetics 133: 137–142.

Fischer, R.A. & R. Maurer, 1978. Drought Resistance in Spring Wheat Cultivars. I. Grain Yield Responses. Aust J Agric Res 29: 897–912.

Fischer, R.A. & J.T. Wood, 1979. Drought Resistance in Spring Wheat Cultivars. III* Yield Association with Morpho-physiological Traits. Aust J Agric Res 30: 1001–1020.

Francis, C.A., 1986. Multiple croppping systems. New York: Macmillan 383 pp.

Frey, K.J., 1964. Adaptation Reaction of Oat Strains Selected Under Stress and Non-Stress Environmental Conditions. Crop Science 4: 55–58.

Grisley, W. & M. Shamambo, 1993. An analysis of the adoption and diffusion of carioca beans in Zambia resulting from an experimental distribution of seed. Expl Agric 29: 379–386.

Hardon, J.J. & W.S. de Boef, 1993. Linking farmers and breeders in local crop development. In: W. de Boef, K. Amanor, K. Wellard & A. Bebbington (Eds) Cultivating Knowledge. Genetic diversity, farmer experimentation and crop research, pp. 64–71. Int Techn Publ Ltd.

Haugerud, A. & M.P. Collinson, 1990. Plants, genes and people: improving the relevance of plant breeding in Africa. Expl Agric 26: 341–362.

Jansen, H.G.P., T.S. Walker & R. Barker, 1990. Adoption Ceilings and Modern Coarse Cereal Cultivars in India. Amer J Agr Econ 72: 653–663.

Johnson, G.R. & K.J. Frey, 1967. Heritabilities of Quantitative Attributes of Oats (*Avena* sp.) at Varying Levels of Environmental Stress. Crop Science 7: 43–46.

Khan, A. & L.A. Spilde, 1992. Response of hard red spring wheat genotypes to management systems. Crop Science 32: 206–212.

Lafitte, H.R. & G.O. Edmeades, 1994. Improvement for tolerance to low soil nitrogen in tropical maize. I. Selection criteria. Field Crops Research 39: 1–14.

Martin, G.B. & M.W. Adams, 1987. Landraces of *Phaseolus vulgaris* (*Fabaceae*) in Northern Malawi. II. Generation and Maintenance of Variability. Economic Botany 41: 204–215.

Maurya, D.M., A. Bottral & J. Farrington, 1988. Improved livelihoods, Genetic Diversity and Farmer participation: A strategy for breeding in rainfed areas of India. Expl. Agric. 24: 311–320.

McDonald, G.K., 1989. The contribution of nitrogen fertiliser to the nitrogen nutrition of rainfed wheat crops in Australia: a review. Aust J Expl Agric 29: 455–481.

Muruli, B.I. & G.M. Paulsen, 1981. Improvement of Nitrogen Use Efficiency and its relationship to other traits in maize. Maydica 26: 63–73.

Norse, D., 1992. A new strategy for feeding a crowded planet. Environment 34: 6–39.

Pederson, D.G. & A.J. Rathjen, 1981. Choosing trial sites to maximize selection response for grain yield in spring wheat. Aust J Agric Res 32: 411–424.

Pfeiffer, W.H., 1988. Drought Tolerance in Bread Wheat – Analysis of Yield Improvement over the Years in CIMMYT Germplasm. In: A.R. Klatt (Ed.) Wheat production constraints in tropical environments, pp. 274–284. CIMMYT, Mexico DF, Mexico.

Pimbert, M.P., 1994. The need for another research paradigm. Seedling, July 1994: 20–25.

Poutala, R.T., O. Kuoppamäki, J. Korva & E. Varis, 1994. The performance of ecological, integrated and conventional nutrient management in cereal cropping in Finland. Field Crops Research 37: 3–10.

Rathjen, A.J. & D.G. Pederson, 1986. Selecting for improved grain yields in variable environments. In: Plant Breeding Symposium, pp. 104–115. DSIR, Christchurch, N.Z. Agronomy Society Special Publication No. 3.

Rosielle, A.A. & J. Hamblin, 1981. Theoretical aspects of Selection for Yield in Stress and Non-Stress Environments. Crop Science 21: 943–946.

Roy, N.M. & B.R. Murty, 1970. A selection procedure in wheat for stress environments. Euphytica 19: 509–521.

Selmani, A. & C.E. Wassom, 1993. Daytime chlorophyll fluorescence measurement in field grown maize and its genetic variability under well-watered and water-stressed conditions. Field Crops Research 31: 173–184.

Simmonds, N.W., 1984. Principles of crop improvement. Longman, London and New York, pp. 408.

Simmonds, N.W., 1991. Selection for local adaptation in a plant breeding programme. Theoretical Applied Genetics 82: 363–367.

Singh, M. & S. Ceccarelli, 1995. Estimation of heritability using varietal trials data from incomplete blocks. Theoretical Applied Genetics 90: 142–145.

Smith, M.E., W.R. Coffman & T.C. Barker, 1990. Environmental effects on selection under high and low input conditions. In: M.S. Kang (Ed.) Genotype-by-environment interaction and plant breeding, pp. 261–271. Dept of Agron Louisiana State Agric Experiment Stn, Baton Rouge, USA.

Sperling, L., M.E. Loevinsohn & B. Ntabomvura, 1993. Rethinking the farmer's role in plant breeding: local bean experts and on-station selection in Rwanda. Expl Agric 29: 509–519.

van Oosterom, E.J., D. Kleijn, S. Ceccarelli & M.M. Nachit, 1993. Genotype × Environment Interactions of barley in the Mediterranean region. Crop Science 33: 669–674.

Ward, P.J., 1994. Parent-offspring regression and extreme environments. Heredity 72: 574–581.

Weltzien, E. & G. Fischbeck, 1990. Performance and Variability of Local Barley Landraces in near-Eastern Environments. Plant Breeding 104: 58–67.

Zavala-Garcia, F., P.J. Bramel-Cox, J.D. Eastin, M.D. Witt & D.J. Andrews, 1992. Increasing the Efficiency of Crop Selection for Unpredictable Environments. Crop Science 32: 51–57.

P.M.A. Tigerstedt (Ed.), *Adaptation in Plant Breeding*, 237–241, 1997.

Molecular adaptation of barley to cold and drought conditions

Antonio Michele Stanca, Cristina Crosatti, Maria Grossi, Nadia Gloria Lacerenza, Fulvia Rizza & Luigi Cattivelli
Istituto Sperimentale per la Cerealicoltura Sezione di Fiorenzuola d'Arda, Via S. Protaso 302-I-29017 Fiorenzuola d'Arda (PC), Italy

Key words: barley, cold-acclimation, cold-regulated genes, dehydration induced genes, drought stress, desiccation, hardening, *Hordeum vulgare*

Summary

Molecular adaptation to cold and drought involves a series of biochemical and molecular changes leading plants to improve their winter hardiness or drought resistance.

We are interested to study the molecular basis of cold acclimation and drought response of barley to survive under stress. Several genes regulated by low temperatures and sometimes by drought have been isolated from the barley genome. In this review the most significant results of our recent work will be presented and discussed.

The protein encoded by cDNA clone *pt59* and induced in barley by cold was over-expressed in *E. coli* to produce the matching antibody, which *in vivo* recognizes a cold-induced protein of 14 kDa (COR14). The COR14 is stored in amounts only slightly greater in the cold resistant 'Onice' than in the susceptible 'Gitane', although the former has a higher induction-temperature threshold of COR14 than the latter. This fact is an evolutionary advantage enabling the resistant varieties in the field to prepare for the cold well ahead of the susceptible ones.

Two other cDNA clones, *paf93* and *cdr29*, are regulated by low temperature and drought stress but not by exogenous ABA treatment. Indeed during the early stage of dehydration, the mRNAs are expressed before the induction of known ABA regulated genes such as dehydrins and when only a small increase occurs in ABA content. The sequence analysis revealed that *paf93* encodes for a protein homologous to the cold-regulated protein COR47 of *Arabidopsis*, whereas *cdr29* represents a plant gene homologous to yeast and mammalian sequences coding for acyl-Coenzyme A oxidase.

Abbreviations: ABA – abscisic acid; COR – cold regulated

Introduction

Barley (*Hordeum vulgare* L.), despite its proven adaptability to a range of ecological situations throughout its evolution, is nonetheless susceptible to extreme environmental situations which determine crop stress and ensuing yield losses of varying magnitude. Studies of genetic resistance have intensely focused on the selection of those plant genotypes less subject than others to the insults of abiotic nature so as to increase yields and yield stability. Thus, genetic variation plays a primary role in determining positive adaptation to environmental stresses and, hence, in supporting the spread of various barley genotypes to extreme climatic conditions (Stanca et al. 1992).

Frost resistance in barley is important because barley growing has been shifting for a number of years from spring to autumn sowing as winter varieties are higher-yielding than spring cultivars. It is, however, a known fact that the winter barley varieties are less hardy than winter wheat, rye and triticale.

Drought is another important factor limiting barley yield. Yet, compared to many other cereals, barley is well adapted to this condition, as it is to shallow and poorly fertile soils and high temperatures, thanks to its better water use efficiency. Numerous studies have been conducted to gain further insight into the biochemical and molecular changes induced in barley by water stress. The response to this condition is an increased level of ABA, the accumulation of unusual metabolites (such as proline and polyamines), alterations in activity of certain enzymes, and the induction of a specific set of genes. Most of these changes appear to be controlled by ABA, which is also responsible for activating a broad set of genes (Skriver & Mundy, 1990).

Studies of the physiological mechanisms controlling the adaptation to stress conditions have focused on two areas: a) metabolic changes associated with cold adaptation of barley and b) modification of gene expression and cold-regulated (COR) genes.

The review attempts to give an overview by considering in detail several specific issues of the above two key topics.

Metabolic changes associated with cold-acclimation in contrasting cultivars of barley

Barley plants become more resistant to freezing when first exposed to a period of cold-acclimation. Many physiological and molecular changes have been shown to occur at low temperatures, but the role and the contribution of each to frost resistance is still poorly understood. Two cultivars of barley (*H. vulgare*), the winter Onice and the spring Gitane, were acclimated under controlled conditions (8 h photoperiod at + 4 °C (light) and + 2 °C (dark) for 21 days). Changes in free proline, ABA, water-soluble carbohydrates and free fatty acids were measured to assess their involvement in cold-acclimation and to explain the different frost-resistance capacities of the two cultivars. Exposure of barley plants to low temperature resulted in an increase in proline, equal in both cultivars. During the first days of cold acclimation, ABA levels showed a peak in the frost-resistant cultivar, lasting about 24 h, followed by a decrease. The water soluble carbohydrates reached their highest content after 3 days of hardening, although after 14 to 21 days of acclimation the carbohydrate content was similar to that of unhardened plants. The frost-resistant Onice had a much higher free fatty acids content than the frost-sensitive Gitane. Furthermore, in Onice 86% of free fatty acids was represented by unsaturated molecular species, linolenic acid alone being 71%. In contrast, in the frost-sensitive cultivar only 31% of free fatty acids was unsaturated and a large amount of 9-oxo-nonanoic acid, a product present in the linolenic acid cascade, was also detected.

The ABA content after 2 days of hardening and the free fatty acids composition were clearly different between the two cultivars and may explain, at least in part, the different frost-resistance capacities of Onice and Gitane (Murelli et al., 1995).

An increase in membrane stability has been associated with the adaptation of plant tissues to cold. The level of membrane damage after freezing in standard conditions can therefore be used as a test to assess the genetic variation for frost tolerance. Studies for assessing the influence of hardening on cold tolerance of barley have been carried out by evaluating 30 barley genotypes in the field and laboratory conditions. The data from the field experiments indicate that most of the tested cultivars show a degree of cold tolerance in a narrow range. High cold injury adversely affected yield.

The outcoming results of the artificial test for membrane damage run in controlled conditions on first-leaf plantlets of the five cultivars, representing the variation evinced in the field trial, were highly correlated with the field data. As a result, laboratory test appears to be able to collect in a breeding programme meaningful data on the resistance capacity of varying genotypes. Employing an early stage and a relatively short hardening time, makes it possible to achieve rapid results with a reliability comparable to field selection (Rizza et al., 1994).

Modification of gene expression and cold-regulated genes (COR)

Molecular cloning of cold-regulated genes in barley

It is well known that under low temperature conditions plant cells modify several physiological parameters leading to an improved cold resistance (Levitt, 1980) and that cold-treatment leads to a modification in gene expression (Cattivelli & Bartels, 1989). However, it is not understood how the transcription of a new set of genes can lead to an improved cold tolerance. Therefore, the isolation and analysis of cold-induced sequences is an interesting approach to studying the function of these newly induced proteins.

A set of cDNA clones homologous to cold-regulated mRNAs have been isolated in our laboratory from barley plants exposed to low temperatures. The analyses of their expression pattern indicate that the transcripts accumulate to different levels during cold-treatment. Homology has been found between the isolated cDNAs and cold-induced transcripts of related cereals.

The possibility that several pathways are involved in the cold response in barley is supported by the tissue specificity of some cold-induced transcripts. The mRNAs homologous to cDNA clone *pt59* have been found only in leaves whereas the *pao29, pao86*, and *paf93* homologous transcripts are expressed in leaves as well as in roots during cold stress but at different levels. As the plants show different cold-hardening capacities depending on the developmental stage (Blum, 1988), we examined the appearance of cold-induced transcripts in leaves from plants of three different ages. With the exception of the *pv60* homologous mRNAs, all other cold-regulated transcripts were induced after cold treatment also in mature leaves (Cattivelli & Bartels 1990).

The accumulation kinetic of COR14 is different in forst-resistant and frost-sensitive barley cultivars. The cold regulated cDNA clone *pt59* encode for a cold-induced protein of 14 kDa (COR14). The accumulation of COR14 was monitored by using a specific antibody throughout hardening and subsequent dehardening in the cold-resistant Onice and the cold-susceptible Gitane barley cultivars. The protein COR14 was detected in both cultivars after about 7 d exposure to low temperatures and gradually accumulated over a period of up to four weeks. When the plants were subsequently transferred to 20 °C, COR14 persisted for at least 6 d, a finding in contrast with a previous study showing that the messengers corresponding to *pt59* disappeared a few hours after the temperature was raised (Cattivelli & Bartels, 1990).

The resistant cultivar Onice and the susceptible cultivar Gitane were monitored under hardening at 4, 6, 8 and 10 °C. The amount of COR14 accumulated by Onice plants over one week was limited at 10 °C and higher at 8, 6 and 4 °C. By contrast, COR14 was synthesised in Gitane only at the 4 °C hardening temperature. These data indicate that the difference between the resistant and susceptible cultivars has more to do with the regulation of hardening than with the hardening process itself. An early induction of hardening is an evolutionary advantage because under real field conditions of progressively falling temperatures it enables the resistant cultivar to prepare for the cold well before the susceptible cultivar. These results also show that the COR14 antibody might be a potential tool for selecting cultivars with a greater capacity of resistance to low temperatures.

Analysis of the proteins in the subcellular fractions (nucleus, endoplasm reticulum, intact chloroplasts, thylacoids) indicates that COR14 is present in the proteins extracted from intact chloroplasts but not in the thylacoid fraction. This suggests that COR14 is localized in the stroma of the chloroplasts (Crosatti et al., 1995).

Genes regulated by low temperatures and expressed under desiccation link the molecular adaptation of barley to frost and drought conditions

It has been widely demonstrated that drought and desiccation cause changes in gene expression, and many genes induced under these conditions have been isolated and characterized (Skriver & Mundy, 1990; Bray, 1991). Most of these genes are ABA-regulated and can be induced by its exogenous application. It is well known that ABA is involved in plant adaptation to environmental stress. In plant species such as *Arabidopsis* cold acclimation can be triggered by exogenous ABA treatment and also many ABA regulated genes are expressed upon low temperatures and desiccation conditions. Nevertheless, our experience on barley indicates that only few ABA regulated proteins are accumulated under both stress conditions. An example is the protein named COR75. This protein is the predominant COR protein in barley and immunologically related to the high molecular weight COR proteins of wheat. Yet the extent of a given genotype frost resistance was not found to be related to the amount of COR75 in its tissues. When induced by a low temperature, COR75 registered peak accumulation, which was essentially identical in the all genotypes tested, over the brief span of one week of hardening. The expression pattern of COR75 in response to various environmental stresses was monitored in plants at the first-leaf stage; COR75, which is expressed both in foliar and root tissues, can be induced by exogenous treatment with ABA or drought stress, as well as by low temperatures. It has also been postulated that proteins inducible by such environmental factors (low temperatures and drought) may be the plant response to variations in tissue moisture status, conditions brought about by both these stresses (Guy et al., 1992). The data reported in the present study, i.e. protein induction by low temperatures, ABA and drought, suggest a similar response for COR75 (Crosatti et al., 1994).

Although COR75 is the only example of ABA-regulated protein accumulated upon cold and desiccation in barley, other ABA-independent genes are known to be involved in plant adaptation to both these environmental factors. We have recently reported the characterization of cold and drought regulated cDNA clones, *paf93* and *cdr29*, which respond rapidly to dehydration before a great increase in ABA content occurs. In a time-course experiment made with barley leaves, the induction of known ABA-regulated mRNAs, as detected by hybridization with the *Dhn1* probe, occurred after 5 h of desiccation (4.3% loss of fresh weight), when the ABA level had already increased to 466 ng/g fresh weight, whereas the expression of the mRNAs corresponding to *paf93* was detected after 30 min (0.8% fresh weight loss), when the ABA level was only 20.5 ng/g fresh weight (1.5 times the background). An mRNA of 1.8 kb, homologous to clone *cdr29*, was induced in the same conditions. These findings, together with the impossibility of inducing the mRNAs corresponding to *paf93* (Grossi et al., 1992) and the mRNAs of 1.8 kb corresponding to *cdr29* with an exogenous ABA treatment, suggest that two mechanisms are involved in the barley molecular response to drought. Genes such as dehydrins are induced only after a great increase of ABA, whereas the genes rapidly induced by dehydration are either ABA-independent or triggered by very small variation in ABA content.

Sequence analysis of the cDNA clones revealed that *paf93* encodes for a protein with some similarity to Group II LEA proteins. These proteins accumulate in late embryogenesis and in response to water stress and ABA (Skriver & Mundy, 1990). *paf93* represents the only example to date of a gene coding for a protein homologous to a class of ABA-regulated genes whose induction is ABA-independent. Another gene that is cold-induced and ABA-independent in wheat, *wcs120*, codes for a protein containing numerous domains homologous to the sequences of the dehydrin family, although the mRNAs corresponding to this sequence are not induced by drought (Houde et al., 1992). By contrast, the protein sequence most similar to PAF93, COR47 of *Arabidopsis*, is encoded by a gene inducibile either by exogenous ABA treatment or by low temperature and drought (Gilmour et al., 1992). The comparison of CDR29 with other known protein sequences revealed homologies with several peroxisome localized acyl-Coenzyme A oxidases. These results lead to the hypothesis that *cdr29* encodes for a potential acyl-Coenzyme A oxidase (Grossi et al., 1995).

COR genes are expressed under drought field conditions

The relationship of the expression of drought (*paf93* – Grossi et al., 1995) and ABA-regulated genes (*Dhn1* – Close et al., 1989), water potential and ABA content were analyzed in durum wheat at the grain-filling stage in irrigated and non irrigated field conditions, to determine the involvement of two drought-related genes in the response to water stress. Northern Blot analyses were performed with mRNAs isolated from flag leaves of both irrigated and non-irrigated plants harvested throughout the trial, so as to determine the expression of drought induced genes. The gene *paf93* showed an expression level linked to water potential reduction. The non-irrigated plants exhibited a generally higher level of expression than the irrigated ones, although even in the latter the progressive water potential decline was associated with a parallel rise in the level of induction of messengers homologous to *paf93*. The same filter was also hybridized with the *Dhn1* probe, but no signal was detected.

It would therefore seem that the expression of dehydrin genes under field conditions requires a more severe water deficit and/or a greater rise in the ABA content than that recorded here. On the other hand, the fact that there is another gene class represented by *paf93*, whose control appears directly linked to the water potential value, suggests that some ABA independent responses to drought may be more sensitive to variations in water availability than the ABA-mediated response (Lacerenza et al., 1995).

Conclusions

The interrelation of physiological, genetical and molecular evidences suggests that further studies employing a greater integration of physiological and molecular data should provide in the near future a better understanding of the mechanism of drought and cold tolerance in barley as model plant.

Having cold-inducible cDNA clones available it opens the way to study the properties of the corresponding proteins *in vitro* and to the isolation of promoter sequences which would enable agronomically useful genes in transgenic plants to be expressed during exposure to cold.

Acknowledgement

This work was supported by "Piano Nazionale Biotechnologie Vegetali" MIRAAF, Italy.

References

Blum, A., 1988. Plant breeding for stress environments. CRC Press, Boca Raton, FL.

Bray, E.A., 1991. Regulation of gene expression by endogenous ABA during drought stress. p. 81–98. In: W.J. Davies & H.G. Jones (Eds). Abscisic Acid: Physiology and biochemistry. Bios Scientific Publishers, Oxford.

Cattivelli, L. & D. Bartels, 1989. Cold-induced mRNAs accummulate with different kinetics in barley coleoptiles. Planta 178: 184–188.

Cattivelli, L. & D. Bartels, 1990. Molecular cloning and characterization of cold-regulated genes in barley. Plant Physiol. 93: 1504–1510.

Close, T.J., A.A. Kortt & P.M. Chandler, 1989. A cDNA-based comparison of dehydration-induced proteins (dehydrins) in barley and corn. Plant Mol Biol 13: 95–108.

Crosatti, C., F. Rizza & L. Cattivelli, 1994. Accumulation and characterization of the 75 kDa protein induced by low temperature in barley. Plant Sci 97: 39–46.

Crosatti, C., C. Soncini, A.M. Stanca & L. Cattivelli, 1995. The accumulation of a cold-regulated chloroplastic protein is light-dependent. Planta 196: 458–463.

Gilmour, S.J., N.N. Artus & M.F. Thomashow, 1992. cDNA sequence analysis and expression of two cold-regulated genes of *Arabidopsis thaliana*. Plant Mol Biol 18: 13–21.

Grossi, M., L. Cattivelli, V. Terzi & A.M. Stanca, 1992. Modification of gene expression induced by ABA, in relation to drought and cold stress in barley shoots. Plant Physiol Biochem 30(1): 97–103.

Grossi, M., M. Gulli, A.M. Stanca & L. Cattivelli, 1995. Characterization of two barley genes that respond rapidly to dehydration stress. Plant Sci 105: 71–80.

Guy, C., D. Haskell, L. Neven, P. Klein & C. Smelser, 1992. Hydration-state-responsive proteins link cold and drought stress in spinach. Plant 188: 265–270.

Houde, M., J. Danyluk, J.-F. Laliberté, E. Rassart, R.S. Dhindsa & F. Sarhan, 1992. Cloning, characterization, and expression of a cDNA encoding a 50-kilodalton protein specifically induced by cold acclimation in wheat. Plant Physiol 99: 1381–1387.

Lacerenza, N.G., L. Cattivelli & N. Di Fonzo, 1995. Expression of drought induced genes in durum wheat under stress field conditions. J Plant Physiol 146: 169–172.

Levitt, J., 1980. Responses of plants to environmental stresses, Vol I, Chilling, freezing and high temperature stresses. Academic Press, Orlando, FL.

Murelli, C., F. Rizza, F. Marinone Albini, A. Dulio, V. Terzi & L. Cattivelli, 1995. Metabolic changes associated with cold-acclimation in contrasting cultivars of barley. Physiol Plant 94: 87–93.

Rizza, F., C. Crosatti, A.M. Stanca & L. Cattivelli, 1994. Studies for assessing the influence of hardening on cold tolerance of barley genotypes. Euphytica 75: 131–138.

Skriver, K. & J. Mundy, 1990. Gene expression in response to abscisic acid and osmotic stress. Plant Cell 2: 503–512.

Stanca, A.M., V. Terzi & L. Cattivelli, 1992. Biochemical and molecular studies of stress tolerance in barley. p. 277–288. In: P.R. Shewry (Ed). Barley: Genetics, biochemistry, molecular biology and biotechnology. C.A.B. International, Wallingford, UK.

P.M.A. Tigerstedt (Ed.), Adaptation in Plant Breeding, 243–246, 1997.

Genetic variation for nitrogen use efficiency in winter wheat (*Triticum aestivum* L.)

Jacques Le Gouis & Pierre Pluchard
Institut National de la Recherche Agronomique (INRA), Domaine de Brunehaut, 80200 Estrées-Mons, France

Key words: hybrid lines, nitrogen efficiency, variation, wheat, *Triticum aestivum*

Summary

The new European Common Agricultural Policy and environmental considerations are certainly to change agricultural practices toward low input cultivation systems. Nitrogen is one of the main inputs of winter wheat in northern France and it contributes highly to phreatic water pollution. A research programme has then been set up in order to study whether it is possible to breed for winter wheat cultivars using more efficiently N fertilisers. Less nitrogen would be applied, decreasing pollution risks and operational costs. It has been shown that a large variation exists for N related traits and for the resistance against N deficiency. On the one hand the cv Arche is very resistant to N deficiency, its yield on low N conditions (with no N fertiliser) is on average 89% of its yield on high N conditions (with a high N application). On the other hand, cv Récital is very susceptible to N deficiency as this same percentage is only 61%. A study on 10 hybrids showed that heterosis for grain yield was higher at low N level than at high N level. This was due to a higher number of grains per m^2.

Introduction

Since the beginning of the sixties grain yield of winter wheat (*Triticum aestivum* L.) has increased of about 120 $kg.ha^{-1}.year^{-1}$ in France. Along to a better control of weeds, diseases and lodging through chemicals and a longer development cycle through earlier sowing dates, the increased use of nitrogenous fertilisers has been a very important change in agricultural practices. In northern France the amount of nitrogen applied went from 120 to 180 $kg.ha^{-1}$ between 1977 and 1984 (Varlet & Pluchard, 1986). Grain yield went accordingly with an increase from about 4 to 7 $t.ha^{-1}$ showing that breeding has released varieties adapted to these new agricultural conditions.

From the point of view of production high input management was very successful. In recent years this policy has however been contested. The first reason was the appearance of overproduction in the European Union for the most common crops including winter wheat. This was followed by a reform of the European Common Agricultural Policy in 1993 and the application of lower selling prices. For the farmers, higher income is to be expected at a lower level of inputs. The second reason was that ecological risks caused by high input management are more and more taken into account.

Nitrogen is one of the main inputs in France and it is responsible for an important part of phreatic water pollution. It is however not well known whether it would be possible to breed for varieties which could use more efficiently nitrogen.

Pure lines

In the first step of our research programme, we wanted to assess the genetic variation existing for N related traits. Two parameters are of interest for the plant breeder: (1) grain yield at a defined N level and (2) the stability of a line or in this case its resistance against N deficiency.

In order to investigate these two traits we evaluated in 1994 at two N level (0 kg N/ha and 170 kg N/ha) 26 winter wheat varieties or breeding lines. Soil residual

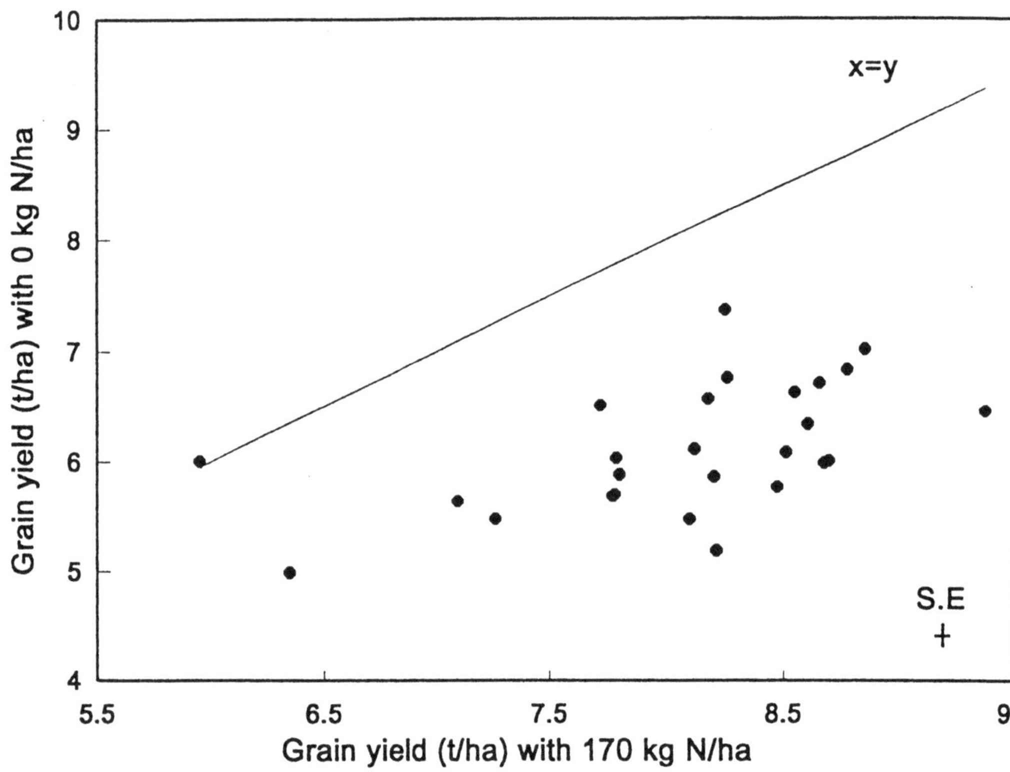

Figure 1. Grain yield of 26 winter wheat varieties measured at two levels of nitrogen supply in northern France.

N in the 0–120 cm layer was found to be 64 kg N/ha on 1 February 1994.

Grain yield ranged from 4.0 to 7.4 t/ha at low N level and from 6.0 to 9.4 t/ha at high N level (Figure 1). N use efficiency is defined as the ration of grain weight (Gw) on N soil supply (Ns) by Moll et al. (1982). Assuming that Ns was the same for all varieties in the trial, differences in grain yield may be related to differences in N use efficiency. Furthermore, a significant Varieties X N levels interaction for yield signifies that ranking differs from one N level to the other. In Figure 1, the resistance of a cultivar to N deficiency may be visualised by the distance to the bisecting line: the nearer the line the more stable the variety. Not taking into account some old and low yielding cultivars (grain yield inferior to 7.5 t/ha at high N level) variation is still large.

The main conclusions which can be drawn from this study and others (Austin, 1977; Loffler et al., 1985; Dhugga & Waines, 1989) are: (1) in defining total plant N either at anthesis or harvest, biomass production is more important than N content, (2) at harvest, grain N quantity explained a large part of total N and (3) N uptake efficiency (Nt/Ns where Nt is total plant N at harvest) is all the more important than N utilisation efficiency (Gw/Nt) in determining N use efficiency as N supply increases. It has also been shown that total N at anthesis accounts for 80–90% of Nt (Loffler et al., 1985; Van Sanford & McKnown, 1987).

In an attempt to better understand N deficiency resistance we chose to study two cultivars. They have similar yields in high N conditions but very different yields in low N conditions. On five trials over three years, cv Arche has always showed a high resistance against N deficiency: on average its grain yield in low N conditions was 89% of its yield in high N conditions. On the opposite, cv Récital has always showed a susceptibility to N deficiency: its grain yield in low N conditions averaged 61% of its grain yield in high N conditions. Cv Soissons, the most cultivated variety in France, was used as a control since its behaviour is intermediate (76%).

We observed the change with time of number of shoots per m^2 for cvs Récital and Arche. At high N level, the two varieties were very similar, either for the maximum number of shoots (about 5.2 per plant) or for the number of shoots at harvest (about 600 per m^2). At low N level, tillering of cv Récital was very affected by nitrogen deficiency. It produced at its maximum 3.2 shoots per plant which gave 320 shoots per m^2 at harvest. In the same conditions cv Arche produced respectively 4.1 shoots per plant and 470 shoots per m^2. The number of shoots per m^2 is the yield component which is the more affected by nitrogen deficiency in our conditions. This is not a surprising result since nitrogen deficiency certainly occurred early due to the absence of any nitrogen fertiliser.

We also investigated the senescence of the leaves. Most interesting results are obtained on the third leaf counting from the ear. The change with time of senescence (weekly visual measures of leaf blades yellowing) showed also striking differences between cv Arche and Récital. Almost no differences appeared between the two nitrogen levels for cv Arche. On the opposite, the senescence of the third leaf is quicker at high N level for cv Récital. This may be related to differences in response to various source-sink ratio as demonstrated with experiments on defoliated or ear-excised plants (Guitman et al., 1991).

Hybrid lines

Three winter wheat hybrid lines have been registered in France in 1994. The use of chemical hybridising agents made possible the production of commercial hybrids and the realisation of research trials on large experimental plots. In agricultural conditions where potential yield is high (8 or 9 t/ha) heterosis is low. Oury et al. (1990b) showed in an experiment carried out in 3 locations with 35 hybrid lines and their parents that on average best-parent heterosis was about 4%. Equivalent results were obtained on one location over 5 years and with 34 hybrid lines (Oury et al., 1990a). In official trials conducted in 1994 for registration, the better hybrid line outyielded the better pure line of only 250 kg/ha (Bernicot, 1995).

Due to their heterozygous genetic structure, it is often supposed that hybrids are more stable than pure lines. This has yet not been observed in high yielding conditions (Oury et al., 1990b). In unfavourable agricultural conditions, like in the case of a nitrogen deficiency, hybrids may be of renewed interest.

Ten hybrids and their parents were evaluated in an experiment with two rates of nitrogen fertiliser (0 and 170 kg N/ha). Average grain yield mid-parent heterosis was 14.6%** at low nitrogen level and 4.0%* at high nitrogen level. Best-parent heterosis was respectively 9.9%* and $1.1\%^{ns}$. These results confirmed the idea of a better behaviour of hybrid lines in adverse conditions. Oury et al. (1990a, b) showed that for the same number of grains per m^2, hybrid lines filled their grain better than pure lines. In our experiment, hybrids and pure lines had equal thousand kernel weight (44.7 and 43.7) although hybrids had a higher number of grain per m^2 (14200 and 15500).

Hybrids have a higher grain yield and a higher total plant weight than their mid-parent. There is however no mid-parent heterosis for N stem or grain concentrations. They did not show either any superiority for N harvest index or N utilisation efficiency. As far as N metabolism is concerned it does not seem that hybrids are different from pure lines.

Conclusions

Cultivar differences exist for N use efficiency or resistance against N deficiency. Up to now we however do not know the genetic determinism or the physiological basis of these differences. This is then difficult to suggest breeding criteria or methods which could be used in a breeding programme. We hope that the comparison of Arche and Récital will help us in understanding these phenomenons. A higher heterosis was found to exist at a low N level than at a high N level. This result will have to be confirmed with other hybrids and in other conditions. The commercial future of hybrid lines will however much depend on the ratio between the cost of the seed and the increase in yield.

References

Austin, R.B., M.A. Ford, J.A. Edrich & R.D. Blackwell, 1977. The economy of winter wheat. J Agric Sci Camb 88: 159–167.

Bernicot, M.H., 1995. Le point sur les blés hybrides. Perspectives Agricoles 202: 30–37.

Dhugga, K.S. & J.G. Waines. Analysis of nitrogen accumulation and use in bread and durum wheat. Crop Sci 29: 1232–1239.

Loffler, C.M., T.L. Rauch & R.H. Busch, 1985. Grain and plant protein relationships in hard red spring wheat. Crop Sci 25: 251–524.

Moll, R.H., E.J. Kamprath & W.A. Jackson, 1982. Analysis of factors which contributes to efficiency of nitrogen utilization. Crop Sci 74: 562–564.

Oury, F.X., J. Koenig, P. Bérard & M. Rousset, 1990a. Une comparaison entre des blés hybrides produits par voie chimique et leur parents: niveaux d'hétérosis et élaboration du rendement. Agronomie 10: 291–304.

Oury, F.X., P. Brabant, P. Pluchard, P. Bérard & M. Rousset, 1990b. Etude multilocale de blés hybrides: niveaux d'hétérosis et élaboration du rendement. Agronomie 10: 735–748.

Van Sanford, D.A. & C.T. McKown, 1987. Cultivar differences in nitrogen remobilization during grain fill in soft red winter wheat. Crop Sci 27: 295–300.

Varlet-Granchet, C. & P. Pluchard, 1986. Blé tendre: jusqu'où améliorer la productivité? Cultivar 195: 31–35.

P.M.A. Tigerstedt (Ed.), *Adaptation in Plant Breeding*, 247–256, 1997.

Selection for adaptation in multipurpose trees and shrubs for production and function in agroforestry systems

F. Owino
International Centre for Research in Agroforestry (ICRAF) P.O. Box 30677, Nairobi, Kenya

Key words: adaptedness, agroforestry systems, phenotypic flexibility, tree-soil relations, tree ideotypes, tree management

Summary

With the greater uniformity in site conditions in cultivated fields and higher levels of inputs (soil amendments and labour) for modern agroforestry systems, selection and breeding for optimal tree adaptation to site conditions and management practices call for strategies radically different from those for plantation forestry. Furthermore, specific requirements for the most efficient growth resource sharing arrangements between the tree and its companion crop call for evaluation both above-ground and below-ground tree growth characteristics in much greater details than has been the case for forestry practice.

Specific adaptations in tree-soil relations are highlighted. Tree ideotype profiles are defined for a valuable fodder species (*Calliandra calothyrsus*) and a valuable timber species (*Grevillea robusta*). Results from half-sib progeny testing experiment on *C. calothyrsus* in alley farming system indicate differential adaptedness to frequent cutting for fodder production among the families after 6 cutting cycles (over a period of 2 years). If this differential response to frequent cutting among genotypes is confirmed, phyllomorphs which are more tolerant to this management practice could be selected for further improvement research.

Excavation of 2 trees per family within *G. robusta* half-sib progeny testing experiments at two locations in Kenya provided the base for studying (a) variations in root growth patterns and (b) correlations between stem and root growth patterns in the background of efficient soil resource sharing strategies and of potentials for indirect selection for root growth characteristics. Analyses based on 7 categories of root growth and 7 categories of stem growth patterns involving 94 families selected in Australia and 60 families selected from landraces in Kenya indicate (a) insignificant variation among families in root growth patterns and (b) non existent or weak correlations between stem and root growth patterns.

Introduction

Trees and shrubs grown as components of modern agroforestry systems such as alley farming will, in general, thrive in more uniform and higher input environments compared to those for natural forests and tree monocultures as farmers apply 'agriculture grade' site preparation and plant management. Under such circumstances, tree adaptation to site and competition with companion crops have to be assessed in ways radically different from those for plantation forestry. For example, species, provenance and family evaluation trials have to be carried beyond broadly defined climatic and soil conditions into such identified constraining soil conditions like low phosphorus, low pH etc. Thus conventional methods for testing tree species and provenances for their adaptation to broad classes of site as practiced for plantation forestry (Burley & Wood, 1976; Eldridge et al., 1994) will not be sufficient. Furthermore, search for and utilization of existing adaptability, in natural populations and in exotic landraces, to such constraining soil conditions is a much needed area for research.

Above and below-ground resource sharing strategies between the tree and its companion crops in agroforestry system is another important aspect in assessing adaptation in species, provenances, and families. Obviously, negative interactions between tree and companion crops such as allelopathic effects reported for eucalypts (Kohli & Singh, 1990) are to be avoided in agroforestry combinations. Beyond this, however, tree characteristics such as reverse deciduous behavior in *Acacia albida* and deep rootedness are beneficial in more efficient resource sharing between trees and companion crops. For example, it has long been established that differences in root structure among tree genotypes are significant determinants of both their competitive ability and competition influence (Stern & Roche, 1974; Burdett et al., 1983). Several workers have reported interesting variation in tree root distribution patterns and root dynamics (Stonecypher et al., 1965; Bilan, 1969; Atkinson, 1983; Van Noordwijk et al., 1991) at the species, family and individual tree levels. Other workers have reported on the relationships between tree root and shoot biomass (Ledig & Perry, 1965; Wareing & Patrick, 1975). While this remains a valuable line of research towards selecting and improving genotypes with optimal root distribution patterns for nutrient sharing with companion crops, the study methods currently in use are prohibitively laborious necessitating recourse to the more recently proposed study methods such as fractal analyses (Van Noordwijk et al., 1994). Results of a study of variation in root distribution of 94 families of *Grevillea robusta* is presented in this paper in the background of determined heritabilities.

Another important aspect of tree adaptation in agroforestry systems is the tree's response to management. Trees are intensively managed in some agroforestry systems like alley cropping which require frequent cutting to low hedges twice to several times a year. Such management is necessary both to reduce light competition with companion crops and for the desired yield of tree products such as mulch and fodder. Significant differences have been reported for tree management response traits such as coppicing ability at species and provenance levels in *Sesbania* spp. and *Erythrina berteroana* (Owino, 1992; Beer, 1987). Results presented in this paper indicate that important variation among half-sib families of *Calliandra calothyrsus* in coppicing ability – which would provide a valuable base for selection.

Genetic bases of tree adaptations

There are two contrasting genetic backgrounds in evaluating and improving tree adaptation in forestry and agroforestry development. One background is founded on adaptability i.e. the ability to become adapted to a range of known and unknown environments while the other is founded on adaptedness i.e. adaptation features of a given population or individual in specified environment. Thoday (1953) discussed adaptability in populations as components of fitness and highlighted the interplay between developmental homeostasis and phenotypic plasticity. Since planting sites in plantation forestry are only slightly manipulated (compared to agricultural fields), it has been the tradition among tree breeders to move with a broad variety of genotypes with optimal adaptation to a variety of microsite conditions (Palmberg-Lerche, 1993). Indeed this consideration has driven forest geneticists focusing on evolutionary adaptability to stress that preservation of adaptability should be the fundamental consideration in gene conservation, provenance evaluation and transfer of forest reproductive material (Namkoong, 1984; Gregorius, 1991; Eriksson, 1991; Muhs, 1993). Given the very long rotation periods for forest plantation species and microsite heterogeneity this emphasis is appropriate.

Some evidence of phenotypic or physiological adaptability being under partial genetic control has been reported in *Fagus sylvatica* (Muller-Starck, 1985). It is also widely recognized that plasticity in quantitative growth traits of trees is under genetic control and that tree populations can be phenotypically plastic as well as genotypically heterogenous. If the above holds, genotypes can be selected and further improved on the basis of their ability to modify their physiology and morphology in different competitive environments. Although there remains important conceptual (adaptation can only be meaningfully assessed at the end of life of the individual) and methodological limitations in quantifying such adaptations, they deserve much higher attention in agroforestry development than in plantation forestry. For example, selection for plasticity in root distribution of trees in agroforestry systems can have significant consequences on nutrient resource sharing with companion crops. In this regard, important differences have been observed in tree root distributions among species and cultivars which could be relied on in selecting the most desirable types (Bilan, 1969; Atkinson, 1983). Results from *G. robusta* root excavation study reported here adds to our knowledge of variation in root morphological characteristics which could used in selection.

In the background of the influence of tree management practices, such as frequent cutting, it is important to recognize that cyclophysis and topophysis are consequences of developmental homeostasis. The adjustments which are possible in growth and architecture of individual trees in response to such stimuli together with adaptive value and inheritance basis of phyllomorphs has been discussed elsewhere (Oldeman, 1992; Oldeman et al., 1994). Variation among families of *C. calothyrsus* in biomass production upon frequent cutting, as reported in this paper would provide a valuable basis for selecting genotypes with optimal adaptation to that specific tree management regime.

Materials and methods

Root distribution variation in Grevillea robusta

Studies on variation in root distribution, inheritance or root traits and correlations between stem and root growth categories were based on 22 months old open-pollinated progeny evaluation experiments with 94 plus trees selected from within the species natural range in Australia and 60 plus trees selected from cultivars introduced into Kenya from the turn of this century. The progeny trial with 94 Australian selections was established at Kakamega forest station while the trial with Kenyan selections was established at Kodera forest station – both sites being in western Kenya at medium altitude (1560 m above sea level) with bimodal rainfall averaging 1800 mm per annum and with deep well-drained fertile lateritic soils which previously had been under natural forest vegetation.

At both sites the trials were designed as balanced incomplete blocks with 3 replications. Family entries were in the form of 8-tree row plots at spacing of 2 m between trees within row plots and 4 m between row plots. It was in the experimental plan to effect systematic thinning within row plots to 50% reduction in density thereby opening up the spacing to 4 m within rows and 4 m between rows. The studies reported here took advantage of this planned thinning. Two trees per plot (trees in positions 2 and 6 in each row plot) were assessed as follows:

(i) Trees were manually excavated to expose distribution of large roots with the extent of excavation varying with observed horizontal and vertical extention of large roots.

(ii) For each excavated tree, both stem and root growth forms were classified into the various categories detailed in Appendix 1.

(iii) Measurements were made on each excavated root to include (a) proximal diameters (in cm) of 4 largest lateral roots, (b) length of taproot (in m), (c) proximal diameter of taproot (in cm), (d) total number of lateral roots, (e) widest spread of lateral roots i.e. diameter of rooting zone (in m) and (f) fresh weight (in kg) of excavated root biomass.

Data and observations from the above exercise formed the basis of three sets of analyses: analysis of variance for root characteristics in (iii) above for each of the two sites; variance components estimation and determination of narrow sense heritabilities for each of the two sites; Pearson correlation analyses on stem versus root growth categories depicted in Appendix 1.

Table 1. Means, components of variance and narrow sense heritability (h^2) estimates at Kakamega forest station

Trait	Mean	σ^2_f	σ^2_w	σ^2_e	h2
SUMDIAN	11.16	0.14	0.0	6.41	0.06
LTR	0.75	0.0	0.0	0.04	0.00
DTR	2.59	0.0	0.0	0.45	0.00
NOPR	19.04	1.97	0.0	47.9	0.01
RWCW	1.87	0.01	0.0	0.28	0.10
ROOTBIO	4.28	0.11	0.0	2.67	0.11

(i) The variance components and heritability estimates may not be very reliable as the number of trees per plot was only.
(ii) Negative variance components are set to zero.

Table 2. Means, components of variance and narrow sense heritability (h^2) estimates at Kakamega forest station

Trait	Mean	σ^2_f	σ^2_w	σ^2_e	h2
SUMDIAM	12.11	0.0	0.58	7.40	0.00
LTR	0.86	0.02	0.0	0.13	0.32
DTR	3.33	0.14	0.0	1.40	0.23
NOPR	15.41	1.43	0.0	16.4	0.07
RWCW	1.43	0.01	0.0	0.11	0.21
ROOTBIO	4.97	0.50	0.0	2.91	0.34

(i) The variance components and heritability estimates may not be very reliable as the number of trees per plot was only 2.
(ii) Negative variance components are set to zero.

Studies on response to frequent cutting in Calliandra calothyrsus

Observations and data reported are based on a longer-term open-pollinated progeny evaluation experiment with 11 plus trees, selected on the basis of their profuse foliage production, from the limited existing woodlots in Kenya as founded on a recent seed introduction from Guatemala. In this particular experiment the half sib families are being evaluated for performance in alley cropping system where trees are cut back to 0.5 m high hedges three times a year to minimize competition with maize crop grown in alleys between the tree hedges. Family entries are in the form of 7-tree row plots planted at close spacing of 0.5 m to form compact hedges. Alleys for maize crop are 4 m wide. The experiment was established in randomized complete block design with three replications. Tree management entailed free growth for one year to be followed three cuttings each year with the first such cutting scheduled for April 1993. At the time of each cutting, fresh biomass weights were determined separately for leaves (much valued for fodder) and woody components as based on total harvest from the hedge of 7-tree plot.

Results and discussion

Inheritance of and correlation between root and stem characteristics in Grevillea robusta

Variance components and heritability estimates for root characteristics of *Grevillea robusta* at Kakamega and Kodera forest sites are presented in Tables 1 and 2. Heritability estimated on a family mean basis from components of variance (Falconer, 1989) for the following parameters:

SUMDIAM	–	Sum of proximal diameters of 4 largest lateral roots (cm)
LTR	–	Length of tap root (m)
DTR	–	Proximaly Diameter of tap root (cm)
NOPR	–	Number of lateral roots
RWCW	–	Widest spread of lateral roots (diameter of rooting zone (m)
ROOTBIO	–	Fresh weight of excavated root biomass (kg)

Heritability is given by:

$$h^2 = \sigma_f^2/[\sigma_f^2 + \sigma_e^2/r + \sigma_w^2/rn]$$

where
$\sigma^2{}_f$ = Family Variance Component
$\sigma^2{}_e$ = Error Variance Component
$\sigma^2{}_w$ = Within plot variance component
r = Number of replicates
n = Number of trees per plot

for our case r = 3, n = 2.

Correlations between root and stem growth categories

Pearson correlation analysis was performed between stem and root growth categories for the experiment at Kakamega at age 22 months. The various stem and root growth categories for scoring are detailed in Appendix 1.

Response to frequent cutting in Calliandra calothyrsus

Figures 1 and 2 show trends in fresh biomass production, leaves and woody components respectively, over a period of two years since first cutting in April 1993. The 12th entry (CONTROL) was derived from a commercial seedlot supplied by the Kenya Tree Seed Centre. Although this experiment will be assessed to the end of 1996 and to include assessment of root development at the terminal excavation assessment, several preliminary trends are pertinent in considering tree adaptation to management. Firstly, the general response to first cutting is increased biomass production for most families. Notable exceptions in this regard are K6CC012 which showed little response in leaf biomass production (but which also showed consistently low responses in subsequent cuttings) and K7CC008 which showed a conspicous decline in leaf biomass production after first cutting (but which subsequently showed the best response to frequent cutting). If such big changes in differential responses to cutting over time are confirmed, this would provide justification to recommend that selection for such responses among families be delayed to later periods in the cutting cycles (in this particular case – after the 6th cutting).

Secondly, it is noteworthy that there appears to be a covergence among families in reduced leaf and woody biomass production from the second to the sixth cutting. This may represent the period when there is more allocation of the trees' growth resources to root relative to shoot. Obviously, during such periods no meaningful selection for shoot biomass production can be made. Thirdly, that big differences among families in both leaf and woody production emerge after the

Table 3. Pearson's correlation coefficients between above-ground and below-ground growth categories of *Grevillea robusta* at the age of 22 months

Stem Categories	Root Categories 1	2	3	4	5	6	7
1	0.06	- 0.01	0.03	- 0.05	- 0.09*	0	0.02
2	- 0.03	- 0.06	0.02	- 0.04	0	- 0.05	0.13**
3	- 0.01	0.02	- 0.01	0.04	0.01	0.03	- 0.06
4	- 0.01	- 0.03	- 0.01	0.10*	0.10*	- 0.03	- 0.06
5	- 0.02	0.13**	- 0.04	- 0.06	0.02	- 0.03	- 0.04
6	- 0.01	- 0.04	- 0.04	0.06	- 0.01	0.28***	- 0.03

* – Correlation significant at 5% level.
* – Correlation significant at 1% level.
** – Correlation significant at 0.1% level.

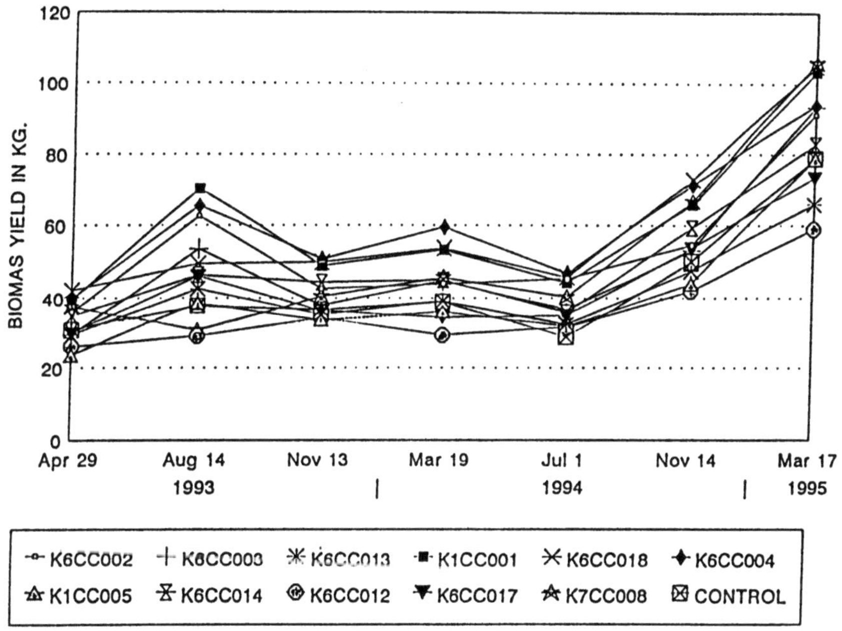

Figure 1. Trends in leaf biomass production among half sib families of *Calliandra calothyrsus* on successive hedging.

sixth cutting offering opportunity for selection at that stage.

Summary and conclusions

Root excavation research with *Grevillea robusta* has revealed a wide variation in root distribution patterns as depicted in artistic representations in Appendix 1. If this observed variation in gross root morphology is associated with nutrient capture strategies, it would form a valuable basis for selecting genotypes of trees which are more efficient in nutrient resource sharing with companion crops in agroforestry systems. However, analysis of variance for the various root characteristics showed no significant differences among families. Moreover, narrow sense heritability determinations for the various root characteristics were non-existent to low. Thus, there is indication of great plasticity in the characteristics. Field observations also confirmed root binding at the nursery growth stage when seedlings were raised in 5 cm diameter, 10 cm long polyethylene tubes for 3 months. This root binding could have a major confounding effect on some of the characteristics reported here. Furthermore, the generally weak correlations between the various root and shoot growth

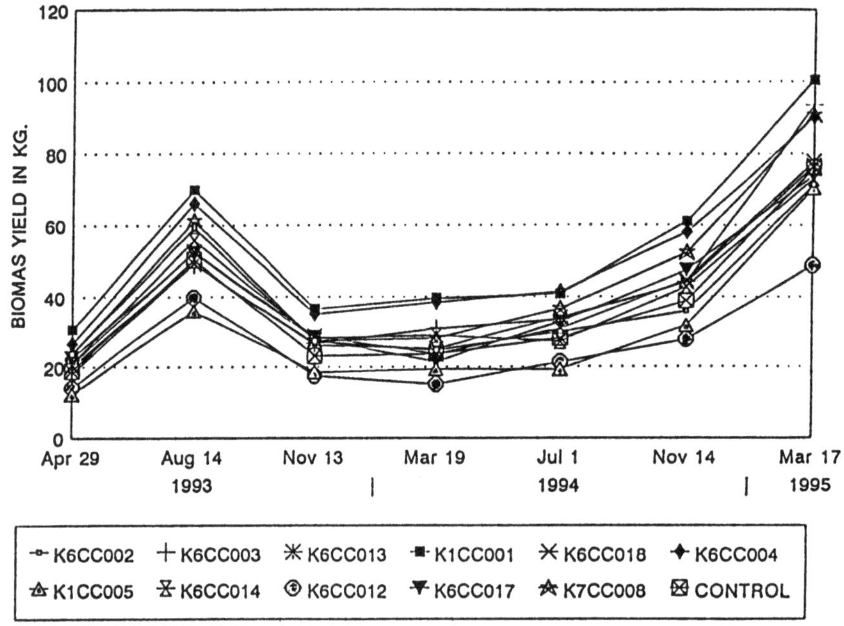

Figure 2. Trends in woody biomass production among half-sib families of *Calliandra calothysus* on successive hedging.

categories indicate limitations for indirect selection for root characteristics on the basis of stem growth characteristics.

Results from the study of *Calliandra calothyrsus* response to frequent cutting, on the other hand, indicate good potential gains from selection at the family level for response to that management regime but only after six cutting cycles. This interesting finding needs further confirmation using to be obtained during the planned assessments in 1996.

Acknowledgements

This research was supported by grant from the German Bundesministerium für wirtschaftliche Zusammenarbeit (BMZ)/Deutsche Gesellschaft für Zusammenarbeit (GTZ). The valuable inputs from J.F. Esegu, J. Aduwo and F. Gone in field assessments and from M. Magadi in data analyses are highly appreciated.

References

Atkinson, D., 1983. The growth, activity and distribution of the fruit tree root system. Soil management affects the balance of short (lateral) to long (extension) roots of apples – with grass competition, there are more lateral roots. Plant and Soil 71: 23–35.

Beer, J., 1987. Experiences with fence line fodder trees in Costa Rica and Nicaragua. p. 215–222. In: J.W. Beer, H.W. Fassbender & J. Heuveldop (Eds). Advances in Agroforestry Research. Proceedings of a Seminar, 1–2 September, 1985. Turrialba, Costa Rica.

Bilan, M.V., 1960. Some aspects of tree root distribution. In: Proc 1st North America Conf on Mycorrhizae. Misc Publ 1189. USDA Forest Service pp. 69–80.

Burdett, A.N., D.G. Simpson & C.F. Thompson, 1983. Root development and plantation establishment success. Plant and Soil 71: 103–110.

Eriksson, G., 1991. Challenges for forest geneticists. Silva Fenn 25: 257–269.

Falconer, D.S., 1989. Introduction to quantitative genetics. 3rd Edition. Longman Scientific & Technical. John Wiley & Sons. 438 pp.

Gregorius, H.-R., 1991. Gene conservation and the preservation of adaptability. p. 31–47. In: A. Seitz & V. Loeschks (Eds). Species Conservation: A population-biological approach. Basel, Birkhauser Verlag.

Hamkoong, G., 1984. A control concept of gene conservation. Silvae Genetica 33: 160–163.

Kohli, R.K. & D. Singh, 1990. Allelopathic potential of Eucalypts in India. In: J. Burley (Ed). Proceedings of the XIX World Congress of IUFRO, Division 2. Montreal, 5–11 August, 1995. The Canadian IUFRO Congress Organizing Committee.

Ledig, F.T. & T.O. Perry, 1965. Physiological genetics of the shoot-root ratio. p. 39–43. In: Proc Soc Amer Foresters, Detroit Michigan.

Muhs, H.-J., 1993. Does the OECD need revision to be better adapted to the trade with tropical and subtropical forest reproductive material? p. 281–293. In: H. Wolf (Ed). Seed Procurement and

Legal Regulations for Forest Reproductive Material in Tropical and Subtropical Countries: Proceedings of GTZ, IUFRO, FAO International Symposium.

Muller-Starck, G., 1985. Genetic differences between 'tolerant' and 'sensitive' beeches (*Fagus sulvatica* L.) in an environmentally stressed adult forest stand. Silvae Genetica 34: 241–247.

Oldeman, R.A.A., 1992. Architectural models, fractals and agroforestry design. Agriculture, Ecosystems and Environment 41: 179–188.

Oldeman, R.A.A. & A.H.M. Sieben-Binnekamp, 1994. Timber Trees: architecture and ecology. p. 25–33. In: R.R.B. Leakey & A.C. Newton (Eds). Tropical Trees: The potential for domestication and the rebuilding of forest resources. HMSO, London.

Owino, F., 1992. Improving multipurpose tree and shrub species for agroforestry systems. Agroforestry Systems 19: 131–137.

Palmberg-Lerche, 1993. The conservation of forest genetic resources as an intergral part of forest management and tree improvement. In: H. Wolf (Ed). Seed Procurement and Legal Regulations for Forest Reproductive Material in Tropical and Subtropical Countries. Proc Int Symp, 4–10 October, 1992. Nairobi, Kenya.

Stern, K. & L. Roche, 1974. Genetics of forest ecosystems. Springer-Verlag. 330 pp.

Stonecypher, R.W., F.C. Cech & B.J. Zobel, 1965. Estimates of components of variance and covariance in root and shoot characteristics of loblolly pine after one growing season. 7th South For Tree Improv Conf Proc pp 86–94.

Thoday, J.M., 1953. Components of fitness. Symp Soc Exp Biol Cambr 7:

Van Noordwijk, M., K. Hairish, M.S. Syekhfani & B. Flach, 1991. Peltophoram pterocarpa: a tree with a root distribution suitable for alley cropping. In: H. Person & B.I. McMichael (Eds). Plant Roots and their Environment.

Van Noordwijk, M., L.Y. Spek & P. de Willigen, 1994. Proximal root diameter as predictor of total root size for fractal branching models. I. Theor Plant and Soil 164: 107–117.

Wareing, P.F. & J. Patrick, 1975. Source-sink relations in the partition of assimilates in plants. In: J.P. Cooper (Ed). Photosynthesis and Productivity in Different Environments. Camb Univ Press, London.

Appendix: Artistic representation of shoot and root growth categories for *Grevillea robusta*

Root categories

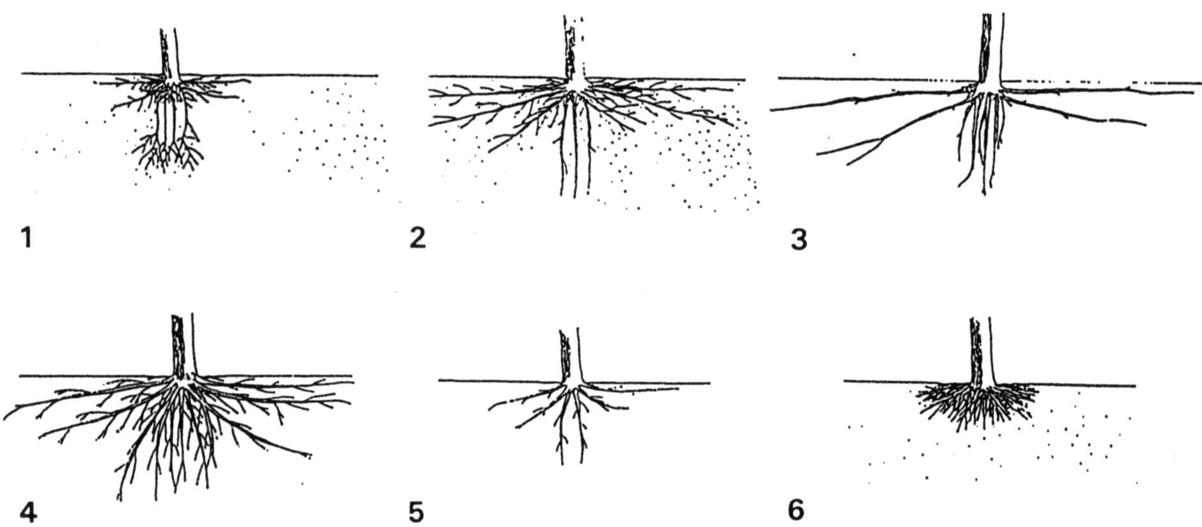

1 – Multiple tap roots with secondary branching to deep soil layers and narrow spread of lateral roots in the top (30 cm) soil layers.
2 – Multiple tap roots with little secondary branching in the deeper soil layers, and narrow spread of laterals.
3 – Multiple tap roots with little secondary branching in the deep soil layers and wide spreading lateral roots in the top soil layers.
4 – Weakly defined tap root system with little secondary branching and spreading lateral roots sinking to deep (i.e. 1 m) layers of the soil.
5 – No definite tap with wide spreading lateral roots in upper soil layers.
6 – Compact tap roots and lateral roots within a narrow diameter (i.e. 1 m diameter and 1 m depth).

Crown growth categories

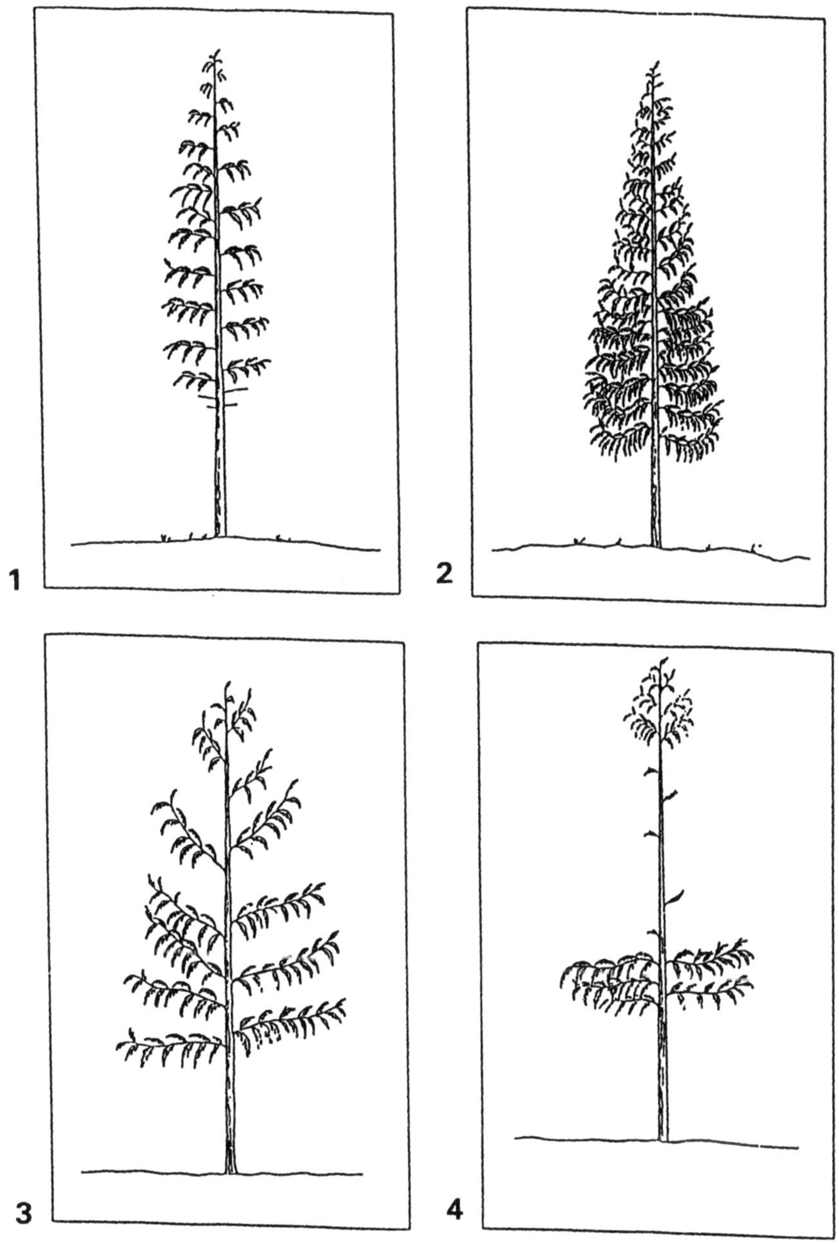

1 – Narrow crown with small evenly spaced branches, sparse foliage and self pruning from below.
2 – Narrow crown with small densely spaced branches, dense foliage and no self pruning.
3 – Wide crown with irregular primary branching and sparse foliage.
4 – Wide crown with conspicuous secondary branching and dense foliage.

Crown growth categories continued

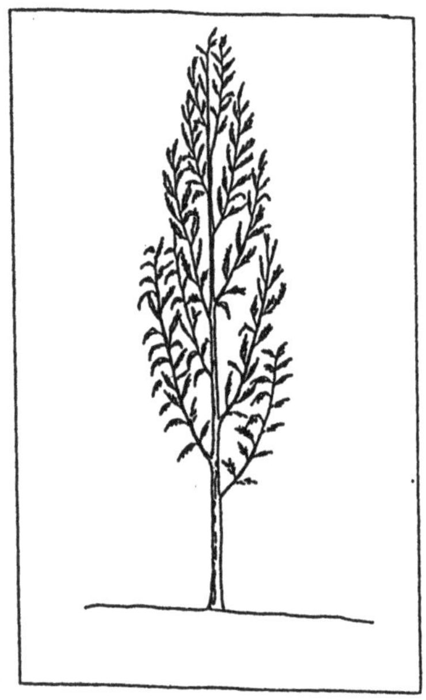

5

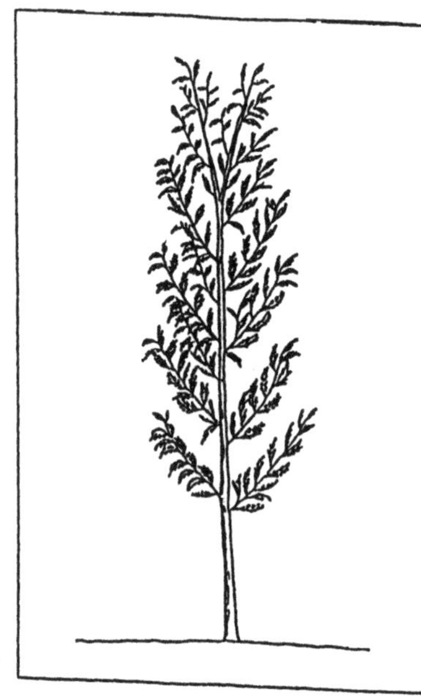

6

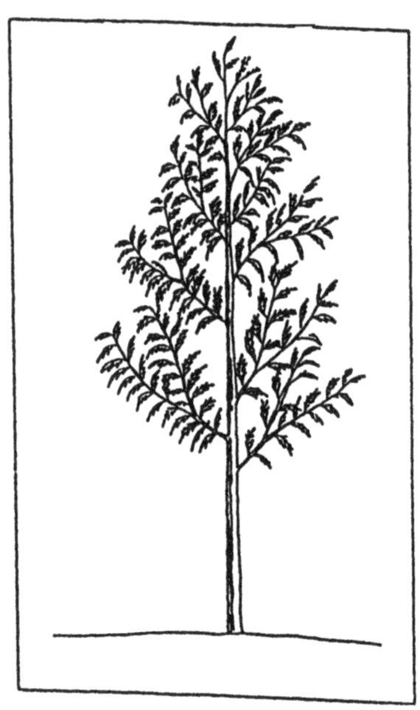

7

5 – Medium width crown with large branches, heavy primary branching at acute angles and dense foliage.
6 – Medium with crown with large branches, heavy primary branching at acute angles, dense foliage and high forking.
7 – Wide crown at the base, followed by irregular and very sparse.

P.M.A. Tigerstedt (Ed.), Adaptation in Plant Breeding, 257–261, 1997.

Breeding plans in case of global warming

Veikko Koski
Finnish Forest Research Institute, P.O. Box 18, FIN-01301 Vantaa, Finland

Key words: acclimatization, adaptation, annual rhythm, diversity, photoperiod, temperature

Summary

Many cultivated crop plants of temperate and cool temperate zones have during their domestication been transferred far from their original habitats and been distributed over a wide range of climates. In general crop plants seem to possess potential to climatic adaptation. Processes in the past have taken hundreds or thousands of years and over large number of generations. The concern on the velocity of the expected global warming and of subsequent changes in other factors justifies planning of breeding for global warming. Perennial plants, forest trees in particular are more at the focal point than annual plants. Most agricultural annuals may be replaced any year with another cultivar or species better adapted to contemporary circumstances. Trees need tens of years to reach commercial dimensions or fructiferous age.

Capriciousness is the main problem of climate change. Large variation of temperature, precipitation among years actually masks the gradual change of average values. The amplitude of variation among years exceeds the predicted average change. Thus, plasticity of individuals besides within population variation in adaptive traits is a corner stone of breeding plans. From the ecological point of view two factors are significant: 1. mild winters may interfere dormancy release and frost tolerance, and also alter host-parasite relationships. 2. The timing of thermal seasons will change in relation to photoperiod. This phenomenon is most prominent in the far north, where the relative warming is largest and the photoperiodic pattern steepest. Even though perennial plants display potential of acclimatization, selection and breeding are eligible measures to improve productivity. As long as the magnitude and velocity of the global warming remain uncertain, it is impossible to set specified aims to breeding. The only reasonable method is to run parallelly several lines with differing patterns of adaptive traits. Multiple population breeding system (MPBS) seems to be tailored to this purpose.

Introduction

The attitudes of people, even those of scientists, to the predicted global warming, or in other words, climate change, vary from a fear of a rapid ecocatastrophe to disputation of the whole phenomenon. Anthropogenic emissions of CO_2 and other greenhouse gases into the atmosphere have been substantial, and in spite of several international efforts to reduce emissions no decline in the combustion of fossil fuels and deforestation can be expected in the near future. In global scale the quantities of carbon (C) are hard to comprehend; the invisible atmospheric CO_2 carries 740×10^9 tons of carbon (McCarthy, 1994). A fact is that the concentration of CO_2 increased since the middle of the 19th century from c. 300 ppm to c. 350 ppm until now (Friendli et al., 1986), and it is expected to increase still. The global system of earth surface, oceans, biosphere, atmosphere, and solar radiation is huge in relation to our normal environment. Interrelationships and feed back mechanisms are complex and poorly understood. Natural fluctuations and stochastic processes cause large deviations from average values. Huge masses have their inertia. Reaching a new equilibrium takes tens of years if energy fluxes change. Due to those factors small gradual changes are difficult to detect. By now an increase of 0.5 °C in the global average surface temperature has been reported in long time series (Folland et al., 1990). The trend is expected to go on so that in 100 years the warming would be 4 ± 2 °C.

A higher temperature means that the energy contents of the atmosphere has increased. Additional energy is released in more frequent cyclones, storms and other disturbances. Faster evapo-transpiration may cause drought but on the other hand more clouds and subsequent rainfall. It is annoying to see that models of climate change are unable to produce accurate predictions for the future climate. The inaccuracy, often called uncertainty, does not eliminate the fundamental phenomenon. The laws of thermodynamics have not yet been falsified. The vast mass of the atmosphere and very long duration of green house gases results in a century long process, even if emissions could be abruptly reduced.

I have used many words in order to describe the global warming issue. One cannot deny the fact that the earth is facing a climate change that will last at least hundred years. It is true that larger climatic fluctuations have occurred during the history of the earth, but for modern man and civilization it will be a new epoch.

Pessimistic scenarios predict substantial changes, such as shifts of vegetation zones, decline of boreal forests, and loss of arable lands. The main argument is that the change will be so rapid that species and ecosystems are unable to adapt parallelly. Therefore, it is appropriate to ask if plant breeding can accelerate adaptation and mitigate the harmful effects of climate change. Transplantation experiments and successful introductions of exotic crop plants show that plant species do have potential to adapt to new environments. One may even ask if simply geographic transfers and new species make specific breeding project needless.

Optional breeding goals

In spite of the fast progress of molecular biology and computer facilities development and multiplication of new cultivars and varieties takes its time. Especially, if we think of long living plants, such as forest trees or fruit trees, tens of years are needed. Thus, breeding programs ought to be initiated quite soon in order to have the products in due time. The goal is, however, obscure, when compared to that of conventional breeding programs.

Breeding plans coupled with global warming must aim at adequate adaptedness to the future circumstances. A dilemma emerges from the earlier mentioned fact that, the uncertainty of the future circumstances is still broader on regional basis than on global basis. Also it must be kept in mind that the variation among years does not level out. On the contrary, extreme weather types may become more frequent. The general conclusion is then that instead of one specified aim several alternative options must be included, and secondly, that in any case wide flexibility (phenotypic plasticity) is a necessary condition of successful cultivars.

Breeding for climate change must be focussed on adaptive traits. It is therefore appropriate to discuss and define the climatological and meteorological characteristics to which adaptedness is needed. Secondly, it is necessary to limit the geographical zone, because world wide perspective is so variable. As the warming is most prominent in cool regions, I will concentrate on the boreal zone. In regions, where the annual mean temperature is now close to freezing point, a rise of 4 degrees causes significant change of growing conditions. Another reason is the light climate of high latitudes.

As such a rise of 4 °C in the annual mean temperature has no direct effect on plants. It is always the ambient temperature with which plants have to come along. The length of the growing season is crucial to all plants, and, in addition, the length and harshness of the winter to perennial plants. Warming of climate turns growing season longer and winters shorter. As regards annual crop plants circumstances obviously become more favourable; varieties with longer growing period and higher yield can replace current ones. Perennial plants are not as simple to deal with. They have been adapted to the sequence of seasons by means of alternation of active growth stage and frost resistant rest stage.

The crucial role of the annual cycle

The synchrony between changing seasons and the physiological stage of perennial plants is based on thermal time and the fluctuation of photoperiod. The climate change has no effect on the photoperiodic pattern. If the regulation of the annual cycle were based on photoperiodic signals only, the existing ecotypes would not be properly adapted to the future climate. The start in spring would be late and the cessation in late summer too early. On the other hand, if dormancy release were only driven by close to zero, chilling temperatures, prolonged autumn and mild winter would trigger bud burst untimely and lead to frost injury.

Empirical studies on forest trees (*Betula pendula, Picea abies, Pinus sylvestris*) have revealed that the regulatory system of the annual cycle is based on more than one factor. Growth cessation is determined by a joint effect of thermal time and night length (Koski & Sievänen, 1985). In warm years growth cessation takes place earlier than in cool ones. In cool years long nights give a progressive signal of the approaching fall. This kind of system can be seen as an adaptation to the variation among years, but it also provides a buffer against climate changes. Provenance trials with forest trees seem to support the flexibility hypothesis. Origins transferred southwards have been exposed to 'climatic warming' for many years, over 60 years in the oldest Finnish experiments. The longest transfers are equal to 4 to 5 degrees' rise in the annual mean temperature. Survival in the south is good, and the total yield is considerably higher than on the original site. In comparison to local southern origins the yield is, however, significantly lower (Beuker, 1994a). Better growth indicates partial acclimatization to warmer climate. Higher total radiation during a longer growing period, together with more fertile soils in the south explain the enhanced growth rate. From the ecological point of view the empirical results show that the natural populations of boreal forest trees are not endangered by direct effects of global warming. As the photoperiodic pattern will not change accordingly, plant breeding is obviously purposeful to obtain optimal combinations of temperature and light climate.

Even though winter is a limiting factor of plant production in the boreal zone, indigenous as well as many introduced species require a cold period in their annual cycle. Low winter temperatures also prevent the spreading of many harmful pests and diseases. Mild winters therefore cause potential risks both directly and indirectly. Winter rest, during which meristems are inactive and all living tissues frost hardy, is divided to two main parts: dormancy and quiescence. During the dormant period growth processes do not start if plants are subjected to above + 10 °C temperature. A longer period of chilling temperatures, i.e. slightly over zero, release dormancy. Quiescence means that plants respond to growth promoting temperatures but with a delay. If dormancy release only depended on the accumulation on chilling days, prolonged autumn and thaw periods in winter would trigger growth initiation and make plants susceptible to cold spells. Experimental studies with young seedlings have supported this hypothesis (e.g. Hänninen, 1990). Experimental studies with older trees, however, indicated that chilling alone may not be the only driving force of dormancy release (Hänninen et al., 1993). Our very recent experiments with 10 years old stecklings of *Picea abies* show that other factors are involved besides chilling. Stecklings were exposed to several combinations of temperature and photoperiod after they had got natural chilling. Temperature regime + 5 . . . + 20 °C, either constant or fluctuating, and photoperiods from natural short day of mid winter, through constant 6 hours to either increasing or declining artificial photoperiod were applied. Most surprising result was that increasing night length, started from 12 hour photoperiod blocked bud burst rather totally. Fluctuating temperature with lower level of + 5 °C accelerated bud burst more than increasing day length. In every case days to bud burst were 70–90 days at mean temperature of + 10 °C and 20–40 days at + 15 °C. In degree day scale 300–600 units (> + 5 °C) were needed. In natural circumstances bud burst takes place with c. 150 d.d. (Beuker, 1994b). The above mentioned forcing experiments are at pilot stage, but it seems to me that even mild winters do not induce untimely dehardening and bud burst.

The main concern of matching the annual rhythm with new proportions of growing season and winter seems to be on summer. In spring growth initiation is controlled by temperature and it can be simulated with thermal time. The regulation of growth cessation is undeniably based on both thermal time and increasing night length. The responses to these two factors are likely inherited separately. Many boreal tree species grow also rather far in the south (Scots pine in Spain and Turkey) on higher elevations, where the temperature regime is rather 'boreal', but the light climate is quite different. So it is possible to produce new combinations of thermal time-photoperiod for warmer climate. Simple transfer of more southern origins, if available, may result in increased risk of frost injury and susceptibility to pests and diseases. If short northern nights delay growth cessation and subsequent inwintering processes, trees become less hardy and less resistant. As mentioned earlier, more pests and diseases may appear. Specific breeding for resistance at long living plants is hardly successful. Tolerance to harmful agents correlates positively with general adaptedness. Genetic diversity of the planting stock is advantageous from this point of view, too. Another justification is the large year to year variation and the uncertainty of the extent and speed of the change. Wide phenotypic plasticity is a necessity on individual level. Out-crossing and high degree of heterozygosity probably are the fundamental elements of good survival and performance.

Breeding plans must be based on diversity

The global warming will have an impact on boreal forests. Experience from provenance trials as well as other kind of empirical knowledge suggest that boreal forests tolerate the warming. In the coldest parts the response is probably accelerated growth. Plant breeding has potential to develop material whose adaptedness is better than that of natural populations in a new environment. At the same time actual economically important traits, quality, growth rate, etc., must be kept in mind, because good survival rate is not the only requirement of genetically improved stock. In general there is no controversy. Poorly adapted trees often suffer from loss of the leader, herbivore attacks, etc., which reduce growth and spoil the stem form. However, the genetic back-ground of adaptive and commercial traits is worth studying in order to establish effective breeding projects.

The basic problem still remains unsolved: to what kind of new environment we should breed? We must not remain waiting until the climatologists and physicists deliver accurate and reliable predictions of the future climate in each region. Maybe they will never be available. The only way out is to establish several optional lines, each for a certain combination of temperature rise and light regime. The number of alternatives need not be superfluous. The densest grid is needed north of latitude 65 degrees, where the photoperiodic pattern changes rapidly. Multiple population breeding strategy has been suggested for forest trees (Namkoong, 1989) and I consider that system ideal in case of optional aims.

> To provide genetic variations to accommodate future needs, however, either very large populations must be carried, or selection intensity must be reduced. Alternatively, intensive selection can be used in multiple independent populations to create a Multiple Population Breeding System (MPBS) with broader differentiation among populations both in their source germplasm and their environmental adaptedness.

It is easy to suggest application of multiple populations as the core idea of breeding plans. In this particular case they ought to be established in environments simulating various options of the future climate. That may be quite difficult to realize without controlled artificial environments where trials can be run over many years. Growth chambers and phytotron are out of question, and even greenhouses and open top systems cause considerable expenses. It was stated earlier that large genetic diversity is a basic requirement; it means large numbers of genotypes. When dealing with trees, we have the problem that the behaviour of young seedlings is different from that of adult trees. It means long duration and plenty of space. My point is that plans of breeding for climate warming need to be realistic. Establishment of multiple populations *per se* is just playing unless there is a certainty of continuity and maintenance of diversifying climatic conditions.

A breeding project for adaptedness to changed climate is a long-term investment. To my mind breeding is worthwhile and efforts need to be started. I do not recommend 'a Trial and error method'. Useful knowledge of the genetic back-ground can be obtained from already existing transplantation experiments (= provenance trials) and short time experiments in manipulated environments. Studies on genetic markers progress so rapidly that we can expect localization of appropriate adaptive genes in a reasonable time. We may even anticipate that gene technology will be able to replace conventional crossing methods in recombination breeding. Furthermore, the hypotheses of wide adaptability due to polyploidy and heterozygosity are worth testing at least in some groups of cultivated plants. Tolerance to large fluctuations and gradual change is crucial at perennial plants, and therefore all means to improve it are desirable.

We have to keep in mind that besides climatic changes biosphere is subject to pollution, increased level of UV radiation and higher concentrations of ozone. Unfortunately, plant breeding is not a solution to those problems, but apparently breeders need to adapt to the new harmful factors, too. Molecular biology and physiology provide powerful tools to breeders, but perhaps we should not totally forget about morphological and anatomical traits either. E.g. xeromorphic structure is without doubt coupled with drought resistance.

There is one more aspect that needs to be taken into account. The range of species moves towards north after climate warming. The natural spreading of species is so slow that from man's viewpoint it has no significance. Cultivation of introduced species probably will be an option, especially if successful cultivars have been developed. In case of forest trees and fruit trees photoperiod again comes into picture. It may be that Central European origins of beech or cultivars of grape are not successful as such, but after being bred for long summer days compete in quality with products from more southern areas.

The expected global warming gives a challenge to plant breeders. From the viewpoint of boreal zone the future is not threatening, but in large densely inhabited areas drought is a real nightmare. The situation is uneasy, because so many things are uncertain. Therefore, I am unable to suggest any concrete plans. The matter is so serious and far fetched that the plans behind actions must be based on knowledge of genetic background of climatic adaptation. More research and experiments are to be carried out in order to design the programs adequately. The pattern, however, ought to be as simple as possible. Carrying through of sophisticated programs over generations may turn out to be a burden. Pilot studies with model plants and even computer simulations would help in finding optimal size and complexity to actual breeding populations.

References

Beuker, E., 1994a. Long-term effects of temperature on the wood production of *Pinus sylvestris* L. and *Picea abies* (L.) Karst. in old provenance experiments. Scand. J. Forest Res. 9: 34–45.

Beuker, E., 1994b. Adaptation to climatic changes of the timing of bud burst in populations of *Pinus sylvestris* L. and *Picea abies* (L.) Karst. Tree Physiol. 14: 961–970.

Folland, C.K., T.R. Karl & K.Ya. Vinnikov, 1990. Observed climate variations and change. In: J.T. Houghton, G.J. Jenkins & J.J. Ephraums (Eds). Climate Change. The IPCC Scientific Assessment. pp. 195–238. WMO, UNEP.

Friendli, H., H. Lötscher, H. Oeschger, U. Siegenthaler & B. Stauffer, 1986. Ice core record of the $^{13}C/^{12}C$ ratio of atmospheric CO^2 in the past two centuries. Nature 324: 237–238.

Hänninen, H., 1991. Does climatic warming increase the risk of frost damage in northern trees? Plant Cell Env. 14: 449–454.

Hänninen, H., S. Kellomäki, K. Laitinen, B. Pajari & T. Repo, 1993. Effect of increased winter temperature on the onset of height growth of Scots pine. Silva Fennica 27: 251–257.

Koski, V. & R. Sievänen, 1985. Timing of growth cessation in relation to the variations in the growing season. In: Tigerstedt, Puttonen & Koski (Eds). Crop Physiology of Forest Trees. pp. 167–193. University Press, Helsinki.

McCarthy, J.J., 1994. Key questions concerning ecosystems and the biosphere. In: Megascience: the OECD Forum, Global Change of Planet Earth, pp. 37–47.

Namkoong, G., 1989. Population genetics and the dynamics of conservation. In: L. Knutson & A.K. Stoner (Eds). Biotic Diversity and Germplasm Preservation, Global Imperatives. pp. 161–181. Kluwer Academic Publishers.

P.M.A. Tigerstedt (Ed.), Adaptation in Plant Breeding, 263–269, 1997.

Six cycles of selection for adaptation in two exotic populations of maize

A. Ordás, I. Santiago, R.A. Malvar & M.I. Vales
Misión Biológica de Galicia, CSIC, Apartado 28, 36080 Pontevedra, Spain

Key words: earliness, selection, adaptation, exotic germplasm, maize, *Zea mays*

Summary

The maturity rating of maize (*Zea mays* L.) grown in Galicia (Northwestern Spain) varies from FAO 200 to 600. Later germplasm has usually too much moisture content in the kernel at normal harvesting time so serious problems with storing could appear. Besides, farmers have tended to use earlier varieties in the last few years. All this imposes limitations on the amount of germplasm available to develop varieties adapted to this area.

To study the possibility of adapting late, exotic material to the environment of the Atlantic coast of Galicia we carried out six cycles of individual selection on two non-adapted populations of maize (Purdue A and Purdue B). The criterion of selection was early silking and the criterion of response was moisture content of kernel at harvesting.

The original populations and the populations obtained after each cycle of selection were crossed to the hybrids CM105 × CM109 (tester Reid) and H99 × H95 (tester Lancaster) and were evaluated in four environments to study the efficiency of the selection scheme. There were significant reductions in days to silking (6.1 and 6.7 days from cycle 0 to cycle 6 for Purdue A and Purdue B, respectively) and grain moisture at harvesting (3.0 and 3.9% from cycle 0 to cycle 6 for Purdue A and Purdue B, respectively). There were also reductions in plant height and yield in both populations.

In general, the crosses 'population × Lancaster' were higher for yield than the crosses 'population × Reid'. Yield of the population crosses by both testers decreased after the six cycles of selection probably because of the earlier maturity of the selected populations. Some inbreeding depression may also have occurred.

Introduction

The maturity rating of the maize (*Zea mays* L.) varieties usually grown in Galicia (Northwestern Spain) varies from FAO 200 in the highlands of the interior part of the region to FAO 600 in the valleys of the South. Hybrids with maturity ratings FAO 300–400 are the most commonly used. Later material normally shows too much moisture at normal harvesting time and then presents problems with storage. This limits severely the scope of the germplasm that can be used in a breeding program located in this area. A way to adapt later material to areas that require earlier maize is through selection by early flowering (Troyer & Brown, 1976).

Too much moisture in the kernel at harvesting time is only one of the aspects of lack of adaptation of a determined population to a particular agroecological area. In Galicia, besides, springs are usually wet and cool and frequently the environmental conditions at sowing time are far from optimum. All that can affect the final yield achieved when nonadapted genotypes are planted.

The genetic basis of most of the maize grown in temperate areas is quite narrow. Most hybrid maize seed now sold in the U.S.A. derives from the Corn Belt Dent race, and only a small portion of the variation available in this race is used. For instance, three inbred lines (B73, Mo17, and A632) were used as parents in 38% of the total hybrid maize grown in 1979 in U.S.A. (Zuber & Darrah, 1980). Although Darrah & Zuber (1986) reported a major reduction in production of hybrids based on these lines in 1984, Smith (1988) shows that replacements of these extensively used public lines represent either identical or closely

Table 1. List of agronomic traits measured and units used to evaluate two populations of maize, the populations obtained after six cycles of selection, the crosses of all the populations by two testers and eight hybrid checks grown in four environments in Northwestern Spain

Trait	Unit
Pollen shedding	number of days from planting to 50% pollen shed[1]
Silking	number of days from planting to 50% silking[1]
Plant height	cm from ground to top of tassel[2]
Yield	weight of grain at 14.5% moisture expressed as t/ha[3]
Moisture	percent of water in the kernel at harvesting time[4]

[1] taken on each plot.
[2] mean of five plants per row (ten plants per plot).
[3] taken from all the ears in a plot.
[4] mean of five random ears per plot.

related germplasm. In Europe, the situation is quite similar. The replacement of traditional germplasm by material from the Corn Belt is observed in commercial cultivars as well as in breeding programs. Furthermore, most hybrids grown in Europe are obtained by crossing U.S. Corn Belt Dent with European flint inbred lines, and only a few flint lines are used extensively in this continent (Moreno-González, 1988).

There is much exotic material that could be used in temperate areas to enhance the genetic basis of the varieties grown by farmers. Exotic germplasm includes all germplasm that does not have immediate usefulness without selection for adaptation for a given area (Hallauer & Miranda, 1988). Such germplasm usually shows a very long growing season for Galicia, where early spring vigor is also a very important trait for maize varieties. Besides, farmers have tended to use earlier varieties in the last few years. All this imposes severe limitations on the amount of germplasm available to develop varieties adapted to this area.

In order to study the possibility of adapting late, nonadapted material, a selection program for female flowering was carried out at the Misión Biológica de Galicia (Pontevedra, Spain) with two populations of maize (Purdue A and Purdue B) that were late for the conditions of Galicia. The objective of this work was to assess the efficiency of six cycles of selection.

Materials and methods

The populations of maize 'Purdue Synthetic A o2' and 'Purdue Synthetic B o2', originated in the maize quality program of the University of Purdue (Indiana, U.S.A.) and called 'Purdue A' and 'Purdue B' for simplicity from here on, were chosen for this study. These populations are well suited for the purposes of this work as they had proven to have a good yield potential but had shown a high kernel moisture in the grain at normal harvesting time under the conditions of Pontevedra.

Each population was managed as an independent experiment. Individual selection for early flowering was carried out in each population as follows. One thousand and two hundred kernels were sown in 600 hills, with two plants per hill. When the plants were in the four or five-leaves state, the less vigorous plant from each hill was removed obtaining, then, a population of 600 plants. The distance between rows was 0.80 m and between plants in a row 0.30 m for a planting density of approximately 42,000 plants/ha. Individual selection was carried out crossing plant to plant the first 60 plants that showed the ear apex for an intensity of selection of ten percent. The selection criterion was, then, early female flowering. Seed was bulked after harvesting from the 30 ears obtained and put in cold storage. The seed for the next cycle of selection was randomly picked from the bulk.

Six cycles of selection were carried out under this scheme. Prior to performance testing the original population and the populations obtained after each cycle of selection were multiplied to have homogeneous seed. All the populations were crossed to the single-cross hybrids CM105 × CM109 (Reid germplasm) and H99 × H95 (Lancaster germplasm) in 1992.

In 1993 and 1994 Purdue A and Purdue B, the populations resulting from the six cycles of selection, the crosses of each population to the single-cross hybrids mentioned above, the single-crosses CM105 × CM109 and H99 × H95, and five hybrid checks covering matu-

Table 2. Analyses of variance of two populations of maize, the populations obtained after six cycles of selection, the crosses of all the populations by two testers and eight hybrid checks grown in four environments in Northwestern Spain

Source	df	Pollen shedding	Silking	Plant height	Yield	Kernel moisture
Entries	49	42.34*	41.22*	1343*	6.868*	31.70*
Populations	13	19.01*	31.76*	776*	2.204*	13.34*
Purdue A	6	15.17*	21.70*	705*	2.365*	10.86*
Linear	1	74.07*	110.72*	4056*	13.096*	53.24*
Quadratic	1	1.42	0.97	5	0.124	0.06
Purdue B	6	19.34*	33.04*	890*	2.393*	17.68*
Linear	1	111.42*	184.83*	5032*	13.038*	85.92*
Quadratic	1	0.23	5.28	8	0.121	0.39
PA vs PB[1]	1	40.09*	84.47*	516	0.100	2.24
Topcrosses	27	35.86*	35.60*	278*	1.229*	28.95*
PA × Reid	6	2.73	7.21	203*	0.862*	4.32
Linear	1	12.95*	20.35	773*	3.189*	8.18*
Quadratic	1	0.84	9.87	105	1.240*	5.36
PA × Lancaster	6	10.69*	11.19*	198*	1.053*	4.32
Linear	1	50.86*	58.10*	703*	2.421*	19.62*
Quadratic	1	1.54	3.20	247*	0.538	2.02
PB × Reid	6	5.42*	6.78*	243*	1.399	4.06
Linear	1	30.58*	36.44*	1241*	7.408*	11.55
Quadratic	1	0.10	0.15	194	0.989	0.47
PB × Lancaster	6	13.92*	15.16*	433*	0.886	6.29
Linear	1	78.73*	85.68*	1175*	2.600*	16.04*
Quadratic	1	0.33	0.87	261	0.707	1.93
Among TC[2]	3	256.89*	239.76*	351	2.660	222.60*
Checks	7	93.73*	67.86*	889*	6.583*	82.44*
Among groups	2	101.92*	85.21*	15991*	114.320*	10.50*
E × Entries[3]	143	2.01	3.27	85	0.498	2.79
E × Populations	39	1.42	2.94	101	0.308	2.54
E × Purdue A	18	1.63	2.11	102	0.313	1.86
E × Purdue B	18	1.04	2.98	97	0.287	3.50
E × PA vs PB	9	2.33	7.70*	122	0.411	0.87
E × Topcrosses	81	2.37	3.24	77	0.374	2.50
E × PA × Reid	18	1.86	3.00	50	0.159	1.84
E × PA × Lanc.	18	1.67	1.85	43	0.385	2.52
E × PB × Reid	18	1.52	1.93	76	0.566	2.75
E × PB × Lanc.	18	2.05	1.50	76	0.400	3.13
E × Among TC	9	7.16*	12.59*	199	0.350	2.03
E × Checks	17	1.60	3.47	84	1.353	2.75
E × Among groups	6	2.33	5.31	84	0.994	8.34*
Error	144	2.61	3.61	152	0.926	2.85

[1] PA = Purdue A; PB = Purdue B.
[2] TC = Topcrosses.
[3] E = Environments.

Table 3. Means of the population Purdue A, the populations obtained after six cycles of selection for early flowering and their crosses to two testers

Population	Pollen shedding	Silking	Plant height	Yield	Kernel moisture
	d	d	cm	t/ha	%
C0	79.7	81.0	191	4.83	35.0
C1	78.8	79.8	190	4.63	35.0
C2	77.1	78.1	187	4.73	35.2
C3	78.9	79.9	177	3.93	33.9
C4	75.9	76.2	167	3.42	31.8
C5	75.9	76.0	164	3.44	31.6
C6	74.5	74.9	159	2.87	32.0
Mean	77.2	78.0	176	3.98	33.5
LSD (5%)	1.9	2.2	15	0.83	2.0
C0 × Reid	76.4	76.7	208	6.26	30.6
C1 × Reid	76.6	76.4	209	6.17	31.3
C2 × Reid	76.2	75.3	202	5.84	29.7
C3 × Reid	76.4	77.1	200	5.32	31.8
C4 × Reid	74.9	76.1	199	5.17	30.7
C5 × Reid	75.1	75.5	187	5.12	29.5
C6 × Reid	74.7	73.1	199	5.61	28.9
Mean	75.7	75.7	201	5.64	30.4
LSD (5%)	2.0	2.6	10	0.59	2.0
C0 × Lancaster	83.1	84.6	223	6.86	36.0
C1 × Lancaster	83.1	84.2	211	6.30	36.5
C2 × Lancaster	81.9	82.1	197	5.90	33.7
C3 × Lancaster	80.4	81.3	200	5.75	34.6
C4 × Lancaster	80.3	81.1	203	6.52	35.8
C5 × Lancaster	79.8	80.6	199	5.72	33.3
C6 × Lancaster	78.0	79.2	192	5.62	33.9
Mean	81.2	80.6	201	6.18	35.4
LSD (5%)	1.9	2.0	10	0.92	2.4

rity ranges from FAO 300 to FAO 700 were evaluated in four environments in the Atlantic coast of Galicia to study the efficiency of the selection scheme. The hybrid checks were Misión 301 (FAO300), Demar (FAO400), Misión 502 (FAO500), Logos (FAO600) and Ivana (FAO700) in 1993. In 1994 Logos was not available and was changed by Luana. Each test was arranged as a 7 × 7 simple lattice. Each experimental plot consisted of two rows. The distance between rows was 0.80 m and between hills in a row 0.25 m. There were 25 hills per row. Two kernels were sown per hill. The hills were thinned to one plant for a planting density of approximately 50,000 plants/ha. The size of the experimental plot was 10 m^2.

From each plot the data shown in Table 1 were taken. Individual analyses of variance were made and then combined following the approach of Cochran and Cox (1957) for lattice designs. Orthogonal partitions of the mean square for entries and of the interaction 'entries × environments' were also made. Each partition of the mean square for entries was tested with the corresponding partition of the mean square of the interaction. The means were compared by the FLSD method. All the statistical analyses were made with the SAS package (SAS, 1989).

Results and discussion

There were significant differences among the original population and the populations resulting from the selection process for both Purdue A and Purdue B for

Table 4. Means of the population Purdue B, the populations obtained after cycles six of selection for early flowering and their crosses to two testers

Population	Pollen shedding	Silking	Plant height	Yield	Kernel moisture
	d	d	cm	t/ha	%
C0	81.3	84.3	198	4.88	34.7
C1	81.3	83.1	198	4.96	35.2
C2	80.5	82.7	196	4.26	34.8
C3	79.1	79.5	179	4.48	33.4
C4	77.5	78.6	174	3.56	32.9
C5	76.7	77.3	169	3.24	29.8
C6	76.1	77.6	162	3.08	30.8
Mean	78.9	80.4	182	4.06	33.1
LSD (5%)	1.5	2.6	15	0.80	2.8
C0 × Reid	77.7	78.6	223	6.56	31.7
C1 × Reid	76.6	77.5	213	6.14	30.6
C2 × Reid	76.4	78.2	209	5.68	30.1
C3 × Reid	76.2	77.2	205	5.41	30.2
C4 × Reid	75.0	76.0	207	5.14	28.6
C5 × Reid	75.0	75.8	200	4.96	30.6
C6 × Reid	74.4	75.2	201	5.13	29.2
Mean	75.9	76.9	208	5.57	30.1
LSD (5%)	1.8	2.1	13	1.12	2.5
C0 × Lancaster	82.2	82.4	214	7.19	36.2
C1 × Lancaster	82.9	81.6	205	6.16	36.5
C2 × Lancaster	83.2	82.1	195	5.86	35.4
C3 × Lancaster	81.2	81.6	200	5.99	35.8
C4 × Lancaster	80.4	79.5	199	6.03	35.7
C5 × Lancaster	79.2	79.2	195	6.41	34.5
C6 × Lancaster	79.2	78.1	196	5.59	33.5
Mean	80.9	81.9	203	6.10	34.8
LSD (5%)	2.1	1.8	13	0.94	2.6

Table 5. Regression lines of the gains for five traits in the population Purdue A and its crosses to two testers after six cycles of selection for early flowering

Trait	Population		
	Purdue A	Purdue A × Reid	Purdue A × Lanc.
Pollen shedding (d)	$y = 79.7–0.8x$	$y = 76.8–0.3x$	$y = 83.2–0.7x$
Silking (d)	$y = 81.0–1.0x$	$y = 77.0–0.4x$	$y = 82.8–0.7x$
Plant height (cm)	$y = 194–6x$	$y = 209–3x$	$y = 212–7x+x^2$
Yield (t/ha)	$y = 5.00–0.34x$	$y = 6.44–0.53x+0.06x^2$	$y = 6.62–0.15x$
Kernel moisture (%)	$y = 35.6–0.7x$	$y = 31.2–0.3x$	$y = 36.6–0.4x$

the five traits. All the responses were linear without detecting any quadratic effect (Table 2). The gain from selection is expecting, then, to continue in successive cycles of selection. The reduction in days to silking after the six cycles of selection was 6.1 days in Purdue A (Table 3) and 6.7 days in Purdue B (Table 4). There was a correlated response on kernel moisture at harvest, with reductions of 3.0% and 3.9% in Purdue A and Purdue B, respectively (Tables 3 and 4).

Table 6. Regression lines of the gains for five traits in the population Purdue B and its crosses to two testers after six cycles of selection for early flowering

Trait	Population		
	Purdue B	Purdue B × Reid	Purdue B × Lanc.
Pollen shedding (d)	y = 81.9–1.0x	y = 77.5–0.5x	y = 83.5–0.8x
Silking (d)	y = 84.3–1.3x	y = 78.6–0.6x	y = 84.5–0.9x
Plant height (cm)	y = 203–7x	y = 218–3x	y = 215–4x
Yield (t/ha)	y = 5.09–0.34x	y = 6.35–0.26x	y = 6.55–0.15x
Kernel moisture (%)	y = 35.7–0.9x	Not significant	y = 36.0–0.4x

The reduction in the number of days to silking per cycle of selection was one day for Purdue A (Table 5) and 1.3 days for Purdue B (Table 6). These results are in line with those reported by Troyer & Brown (1972), Troyer (1986) and Subandi (1985), and somewhat less than the 1.7 days per cycle reported by Troyer & Brown (1976) after five cycles of selection in seven late synthetics.

The corresponding values for kernel moisture were 0.7% and 0.9% for Purdue A and Purdue B, respectively (Tables 5 and 6). Other authors (Troyer & Brown, 1972, 1976; Troyer & Larkins, 1985; Troyer, 1986) have reported gains greater than ours, from 1.0% to 1.6%.

Comparing the kernel moisture content of the populations Purdue A and Purdue B with that of the check hybrids, the maturity rating of Purdue A changed from FAO500–600 in cycle 0 to FAO400 in cycle 6. Purdue A changed from FAO500–600 in cycle 0 to FAO300–400 in cycle 6. These results are in agreement with what was known from previous experiments carried out in the Atlantic coast of Northwestern Spain, in which a reduction of two percent in kernel moisture at harvest would correspond to, approximately, 100 points in the FAO rating. The selection program has been, then, successful as it has lowered the maturity rating of the populations in about 150 to 200 points.

Maize is generally considered protandrous (Russell & Eberhart, 1980). This characteristic is not so clear in the populations grown in the past in Northwestern Spain as in the germplasm from the U.S. Corn Belt. The populations Purdue A and Purdue B belong to the Corn Belt race and showed protandry all along the selection process (Tables 3 and 4). The selection for earlier female flowering has brought about an earlier male flowering too, as it was expected.

In a program of selection it is of paramount importance that there is no undesirable responses on correlated agronomic characters. Undesirable effects could appear as a consequence of inbreeding because the number of individuals used to advance each cycle of selection was not very high. Some of the effects of inbreeding are a decrease in plant size and a reduction of yield. Plant height decreased significantly and linearly in both populations (Table 2), with a total reduction after the sixth cycle of 32 cm in Purdue A (Table 3) and 36 cm in Purdue B (Table 4), while the reductions per cycle were 6 cm for Purdue A (Table 5) and 7 cm for Purdue B (Table 6). Given the small entity of the mean squares for quadratic effects, we can expect that the reduction will continue still for several cycles more of selection. In Purdue B almost half of the reduction occurred in the third cycle of selection with small decreases in the remaining cycles (Table 4). In Purdue A most of the height reduction took place in the third and fourth cycles (Table 3).

Although the primary interest of the populations Purdue A and Purdue B is for using them as sources of inbred lines, a good yielding potential is also desirable. The pattern of yield reduction follows closely that of plant height, with significant linear effects for both populations and very small mean squares for quadratic effects (Table 2). There were reductions of 1.96 and 1.80 t/ha for Purdue A and Purdue B, respectively, after the sixth cycle of selection (Tables 3 and 4). The yield decrease per cycle was 0.34 t/ha for both Purdue A and Purdue B (Tables 5 and 6), a value greater than those reported by Troyer & Brown (1972, 1976) and Troyer (1986). It is interesting to note that Troyer & Larkins (1985), after 11 cycles of selection for early flowering in ten late synthetics, found a yield increase of 0.167 t/ha per cycle which they attribute to maturity adaptation. At any case the populations yielded (Tables 3 and 4) much less than the hybrids of similar maturity. For instance, the average yields of Misión 301 (FAO300), Demar (FAO400) and Misión 502 (FAO500) were 7.40, 7.33 and 7.52 t/ha, respectively.

There must be pointed out that the interactions were mostly not significant. Then, we should expect a consistent performance across environments of both the populations and their topcrosses.

There were significant differences among the topcrosses for the five traits (Table 2). There is no a clear trend in the variation shown by the topcrosses when the crosses of each population to each tester are considered. The quadratic effects were not significant at any case, with the exception of yield for Purdue A × Reid. The crosses of the populations by the Lancaster tester tended to produce more yield than the crosses by the Reid tester (Tables 3 and 4). It must be pointed out that, although we used the earliest Lancaster germplasm available at the moment, the Lancaster tester is later than the Reid tester so the crosses by the former are expected to yield more than the crosses by the latter. The greater reductions for silking, kernel moisture and yield occurred in the crosses by the Lancaster tester.

Earlier plants tend to be shorter and less yielder. It seems, then, that the reduction in the traits under study is a combination of an earlier maturity and, maybe, a consequence of some inbreeding depression. The data obtained do not allow to separate the contribution of each one of the two causes.

Acknowledgements

The authors thanks the financial support of the Diputación Provincial of Pontevedra.

References

Cochran, W.G. & G.M. Cox, 1957. Experimental designs. 2nd ed. John Wiley and Sons, New York, USA.

Darrah, L.L. & M.S. Zuber, 1986. 1985 United States farm maize germplasm base and commercial breeding strategies. Crop Sci. 26: 1109–1113.

Hallauer, A.R. & J.B. Miranda Fo., 1988. Quantitative genetics in maize breeding, 2nd ed. Iowa State Univ. Press, Ames, Iowa, USA.

Moreno-González, J., 1988. Diallel crossing system in sets of flint and dent inbred lines of maize (*Zea mays* L.). Maydica 33: 37–49.

Russell, W.A. & A.R. Hallauer, 1980. Corn. p. 299–312. In: W.F. Fehr & H.H. Hadley (Eds). Hybridization of crop plants, ASA & CSSA, Madison, Wisconsin, USA.

SAS Institute Inc., 1989 SAS/STAT user's guide, Version 6, 4th ed., Vols. 1 and 2. SAS Institute Inc., Cary, North Carolina, USA.

Smith, J.S.C., 1988. Diversity of United States hybrid maize germplasm; isozymic and chromatographic evidence. Crop Sci. 28: 63–69.

Subandi, 1985. Selection for early silking and maturity in three populations of corn (*Zea mays* L.) in Bogor, Indonesia. Indones J. Crop Sci. 1: 65–72.

Troyer, A.F., 1986. Selection for early flowering in corn: 18 adapted F_2 populations. Crop Sci. 26: 283–285.

Troyer, A.F. & W.L. Brown, 1972. Selection for early flowering in corn. Crop Sci. 12: 301–304.

Troyer, A.F. & W.L. Brown, 1976. Selection for early flowering in corn: seven late synthetics. Crop Sci. 16: 767–772.

Troyer, A.F. & J.R. Larkins, 1985. Selection for early flowering in corn: 10 late synthetics. Crop Sci. 25: 695–697.

Zuber, M.S. & L.L. Darrah, 1980. 1979 U.S. corn germplasm base. Proc. Annu. Corn Sorghum Res. Conf. 35: 234–249.

P.M.A. Tigerstedt (Ed.), Adaptation in Plant Breeding, 271–275, 1997.

Overwintering of winter cereals in Hungary in the case of global warming

Ottó Veisz, Noémi Harnos, László Szunics & Tibor Tischner
Agricultural Research Institute of the Hungarian Academy of Sciences, 2462 Martonvásár, Hungary

Key words: cereals, frost resistance, global warming, increased CO_2, overwintering

Summary

Under phytotronic conditions investigations were made on the effect of important environmental factors, such as temperature, water and an increasing concentration of atmospheric CO_2, on the hardening of young cereal plants.

In all the varieties derived from the major wheat growing regions of the world the hardening process was favourably influenced by a doubling of atmospheric CO_2 content, so that a significantly larger number of plants survived the frost test than for plants of the same variety raised under normal conditions.

A reduction in freezing temperature and an increase in soil moisture content caused a slight reduction in survival % for varieties with excellent frost resistance and a great reduction for those with medium or poor frost resistance.

Predictions suggest that in Central Europe, as the result of global climatic changes, there will be a reduction in the quantity of winter precipitation, a considerable rise in winter temperatures and an increase in atmospheric CO_2 concentration. Judging by the experimental results, these changes could improve the overwintering of winter cereals; at the same time, however, a number of factors (mainly the reduction of precipitation) leading to yield losses must be expected during the vegetation period.

Introduction

The frost resistance of winter wheat varieties is of great importance in Hungary, since winter weather conditions are extremely variable, ranging from mild conditions of the Mediterranean type, with plentiful precipitation and a minimum temperature of - 8 or - 10 °C, to long, dry cold spells with no snow cover and temperatures as low as - 25 or - 30 °C. Frost resistance is one component of winter hardiness, but is not identical with it. Wheats which are resistant to frost generally have good winter hardiness, which is why conclusions on winter hardiness can be reliably drawn from a knowledge of frost resistance (Rajki, 1980).

Hardening plays a very important role in the development of winter hardiness. The cold hardening of a given species depends on two factors:

- the genetically determined adaptability of the species,
- the environmental factors influencing the manifestation of the trait.

Temperature is of key importance in the hardening process. According to Olien (1967) the optimum hardening temperature for winter cereals is around 3 °C. In plants kept at higher temperatures the degree of hardiness decreased after a time, even if the nutrient and light conditions were optimum (Andrews, 1958; Gusta & Fowler, 1976; Roberts & Grant, 1968). In experiments carried out using various methods on different varieties, Gusta & Fowler (1976), Chen & Gusta (1978), Tyler et al. (1981) and Nass (1983) concluded that there was a negative correlation between the killed plant rate and the moisture content of the soil, the plant tissues and the tillering node.

Over the last few decades the increasing concentrations of carbon dioxide, methane, nitrogen oxides and other gases in the world have enhanced the greenhouse effect (White, 1992). In all probability, this will be associated with a further rise in temperature and a reduction in precipitation. Among the gases causing the greenhouse effect, the concentration of CO_2, which is present in the greatest quantity, is rising at the rate of 0.4% a year (Houghton & Woodwell, 1989).

Measurements carried out since 1958 indicate that the concentration of CO_2 has increased from 315 ppm to 350 ppm, with an increase of 25% since 1990. It has been predicted that if CO_2 continues to increase at this rate, its concentration may double by the middle of the 21st century.

Plant responses to an increase in atmospheric CO_2 concentration have been studied by a number of authors (Kimball, 1983; Mitchell et al., 1993; Thompson & Woodward, 1994). The majority of cereal species exhibit the 'CO_2 fertilisation' effect (McKee & Woodward, 1994; Smart et al., 1994; Nicolas et al., 1993).

A doubling in the CO_2 level results in a 15–50% rise in biomass production if requirements for other environmental conditions are adequately met. However, changes in other factors may well mean that the positive effect of CO_2 is completely negated (Thompson & Woodward, 1994; Mitchell et al., 1993).

A phytotron offers excellent opportunities for the study of frost resistance under exact conditions. In the Martonvásár phytotron a frost testing method was elaborated (Rajki, 1980). The results achieved up to now provide information on the effect of temperature and of the duration and intensity of illumination on the hardening process in winter cereals (Veisz & Sutka, 1991). The present paper reports on the effect of an increasing atmospheric concentration of CO_2 and on that of temperature and rainfall on hardening processes and overwintering in plants.

Materials and methods

The work was carried out under controlled conditions in the phytotron of the Agricultural Research Institute of the Hungarian Academy of Sciences, Martonvásár (Tischner, 1981). The winter wheat varieties used in the experiments were bred and produced in the major wheat producing zones of the world and have frost resistance values ranging from poor to excellent according to the scores achieved in previous phytotron tests.

For the frost resistance test germinated wheat seeds were planted into a 4 : 1 mixture of earth and sand in wooden boxes. The seeds were planted at a depth of 4–5 cm, so the tillering node was at a depth of 3.5–4 cm. Each box consisted of nine rows with 20 plants to a row. The varieties were planted in four replications, with 20 plants per replication. After planting, the boxes were transferred to an autumn-winter type growth chamber (Conviron PGV-36), where they were kept for 6 weeks. During this period the temperature, light intensity and length of illumination were gradually reduced with a weekly change of programme similar to autumn conditions in the field. Within each weekly programme, the temperature fluctuation, light intensity and daylength were the same each day.

The preliminary growth stage was followed by 2-phase hardening. The plants were exposed to the first phase in the autumn-winter chamber, where the temperature fluctuated daily between + 3 °C and - 3 °C with a 21-hour daylength and a photosynthetic photon flux density (PPFD) of 190 μmolm^{-2}s^{-1}. This phase lasted for a week. The second phase of hardening, which lasted for 4 days, took place in a frost resistance testing chamber (Conviron C-812) immediately prior to freezing. The temperature was a constant - 4 °C with no illumination.

Freezing was carried out in the frost resistance testing chamber set up in the phytotron specifically for this purpose. Freezing took place at - 15 °C for 24 hours. Thawing took place in the freezing chamber at + 0.5 °C for 2 days.

After thawing the boxes were transferred to growth benches (Conviron GB-48). The plants were grown for a further 3 weeks at a day/night temperature of 17 °/16 °C, with a 14-hour daylength and a photon flux density of 125 μmolm^{-2}s^{-1}. At the end of the third week, plants which had survived freezing and had begun to develop were clearly distinguishable from those which had died. The results of freezing are given as a percentage of the plant number prior to freezing.

The testing method described above was used as the starting point and the effect of each ecological component on the hardening process was determined by altering them independently of each other.

Table 1. Effect of soil moisture content on the freezing out of wheat (killed plant rate)

Varieties	Soil moisture		
	Dry	Normal	Moist
Martonvásári 4	0.0	10.7	9.1
Martonvásári 8	7.5	34.1	39.9
Bánkúti 1201	40.7	56.3	92.5
NS Rana 2	79.1	88.1	100.0

Freezing temperature: - 15 °C $LSD_{5\%}$ = 14.9%

Soil water content as a % of natural water capacity.
Dry = 30%.
Normal = 45%.
Moist = 60%.

Table 2. Effect of low temperature on the freezing out of wheat (killed plant rate)

Varieties	Freezing temperature (°C)			
	- 6	- 12	- 15	- 18
Martonvásári 4	0.0	0.0	5.6	25.1
Martonvásári 8	0.0	5.2	13.9	38.3
Bánkúti 1201	2.5	5.3	45.8	73.6
NS Rana 2	2.4	68.5	88.1	100.0

$LSD_{5\%}$ = 17.3%.

Results and discussion

Effect of soil moisture content and freezing temperature on the death rate

In the course of both field overwintering and testing in the phytotron the temperature and duration of freezing and the moisture content of the soil, and thus of the plant tissues, play an important role. It is well known from practical observations, that more plants are frozen out on wet, water-logged areas than on drier ground. In the present experiments an exact determination was made of the extent of damage to different varieties in the case of various soil moisture contents (Table 1).

There were significant differences in the mean killed plant rate as the result of different soil water contents, with less damage at lower soil moisture values and greater damage as the water content rose. In the case of varieties with medium or poorer frost resistance (e.g. Bánkúti 1201) the extent of freezing out was sometimes as much as 100% on wet areas. The results indicate, however, that the survival of varieties with excellent frost resistance is not influenced by the soil moisture content.

From the point of view of freezing out, one of the most important factors is low temperature and its duration. The freezing temperature applied for 24 hours in the phytotron is the value measured at the depth at which the tillering node is found. In nature, when the air temperature is - 15 or - 18 °C and there is no snow cover, the temperature 3–5 cm below the soil surface, where the tillering node is to be found in the case of optimum sowing depth, is not lower than - 5 or - 7 °C. It is thus extremely important to make a correct choice of sowing depth. Table 2 illustrates the effect of freezing temperature on a number of winter wheat varieties.

Effect of atmospheric CO_2 concentration on the hardening of cereals

Under phytotronic conditions investigations were also made on the effect of an increasing concentration of atmospheric CO_2 on the hardening process (cold adaptation) of young wheat plants. The experiment was carried out simultaneously in two plant growth chambers (Conviron PGV-36). The growing conditions were the same in both chambers except for the CO_2 content of the air. In one unit the CO_2 level was that normal for the environment (normally 375 μmolmol^{-1}), while in the other the atmospheric CO_2 concentration was adjusted to 750 μmolmol^{-1}.

The plant number determined in this way is presented in Table 3 as a percentage of the number of plants before freezing. With the exception of a single variety, an increase in the atmospheric CO_2 content had a favourable effect on the hardening process, so that more plants survived than for those raised under normal conditions (Figure 1). In the course of the preliminary growth period, measurements were also made on the leaf area, number of shoots, plant height and dry weight of the plants. For most of the varieties a rise in atmospheric CO_2 concentration led to an increase in leaf area, dry weight and plant height. This difference was significant in some varieties.

A close correlation (0.86) was found between the phytotronic frost resistance values and the field winter hardiness of winter wheat varieties, averaged over 15 years of field experimentation. Consequently, the frost resistance values obtained with the model used in the experiments led to the following conclusions.

Field overwintering values are substantially influenced by low temperature and its duration. The freezing temperature applied for 24 hours in the phytotron is the value measured at the depth at which the tillering node is found. In nature, when the air temperature is - 15 or - 18 °C and there is no snow cover, the temperature 5 cm below the surface, where the tillering node is to be found in the case of optimum sowing depth, is not lower than - 5 or - 7 °C (Veisz & Sutka, 1991). If the tillering node is closer to the soil surface, the temperature at the critical depth may well be as low as - 10 or - 15 °C, depending on how long the cold lasts (Andrews et al., 1974; Olien, 1967; Siminovitch et al., 1967; Veisz & Tischner, 1989).

The environmental factors discussed here are those which have an influence on the overwintering of winter cereals and which are predicted to change in Hungary, as elsewhere. There is likely to be a drop in the quan-

Table 3. Effect of atmospheric CO_2 concentration on the initial development and frost resistance of cereals

Varieties	Number of shoots/plant		Leaf area (cm^2)		Plant height (cm)		Dry weight/plant (g)		Number of plants surviving freezing %	
	375 ppm CO_2	750 ppm CO_2	375 ppm CO_2	750 ppm CO_2	375 ppm CO_2	750 ppm CO_2	375 ppm CO_2	750 ppm CO_2	375 ppm CO_2	750 ppm CO_2
Winter wheat										
1. Fredrick (North America I)	1.13	1.25	7.69	10.71***	22.32	23.64*	0.09	0.09	42.5	58.7
2. Karl (North America II)	1.08	1.13	5.05	5.67	18.53	18.98	0.06	0.06	81.2	92.5
3. Alba (Northern Europe)	1.00	1.04	5.25	6.53*	16.07	18.86**	0.05	0.06*	20.0	36.3
4. Mv 15 (Central Europe)	1.00	1.08	5.20	5.97	18.88	19.27	0.05	0.06*	81.2	91.2
5. Libellula (Southern Europe)	1.21	1.21	7.80	8.06	19.94	19.00	0.07	0.07	3.7	2.5
6. Thesee (Western Europe)	1.00	1.00	8.02	7.22	19.06	18.18	0.06	0.06	1.2	11.2
7. Bezostaya 1 (Eastern Europe)	1.08	1.08	7.11	7.22	19.34	19.22	0.06	0.07*	80.0	80.0
8. Gerek (South Eastern Europe)	1.08	1.00	6.57	6.29	18.49	18.16	0.05	0.05	3.7	31.2**
Winter triticale										
9. Presto	1.04	1.00	6.01	7.60*	16.56	18.29**	0.05	0.07*	38.7	63.7**
Winter rye										
10. Motto	1.13	1.25	9.95	9.74	15.24	18.94**	0.06	0.07*	93.7	94.1
Mean	1.07	1.10*	6.86	7.50*	18.44	19.25*	0.06	0.066*	44.6	56.1*
$LSD_{5\%}$ = between any two combinations	0.135		1.129		1.235		0.01		17.9	
$LSD_{5\%}$ = between the mean values	0.02		0.16		0.18		0.003		6.7	

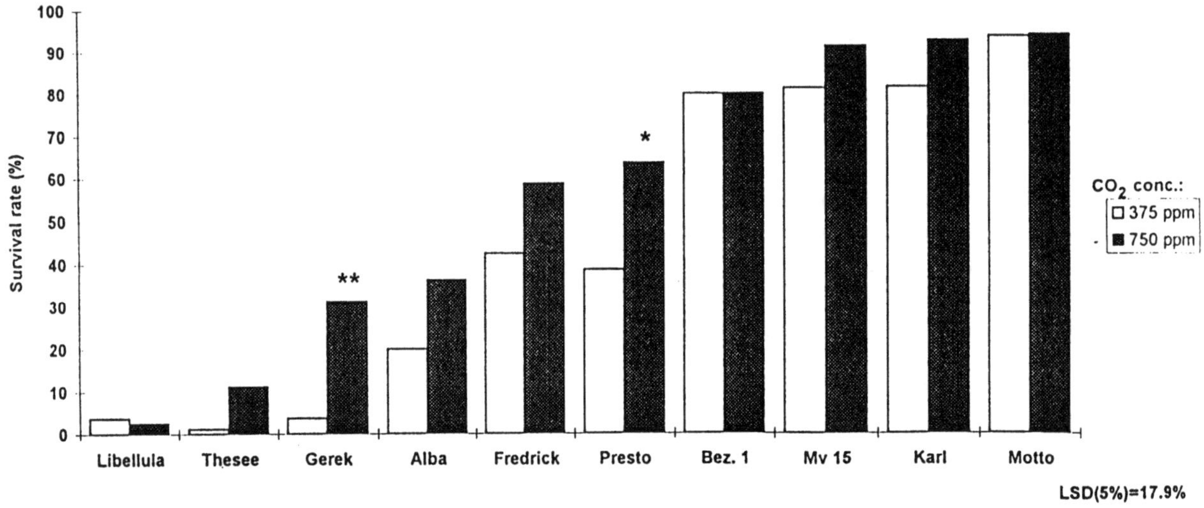

Figure 1. Plants surviving freezing (%).

tity of winter precipitation, the temperature during the winter period is predicted to be higher and a rise in the atmospheric CO_2 concentration is also to be expected (Czelnai et al., 1994).

The experimental results presented above confirm suggestions that all the changes will not necessarily be unfavourable for the living world (McKee & Woodward, 1994; Nicolas et al., 1993; Kimball, 1983). At the same time, it should not be forgotten that the combined effects of the changes will determine the final result, i.e. in the case of agriculture, the quantity and quality of crop production. The lack of precipitation may stunt plant development; in mild winters pathogens and pests also have a better chance to survive, thus causing damage leading to yield losses. This may mean that despite a rise in the number of plants overwintering, there may be a corresponding reduction in yield for the above reasons. A large number of further experiments will be required before this question can be adequately answered.

Acknowledgements

This work was supported by grants from the Hungarian National Scientific Research Foundation (OTKA 150).

References

Andrews, J.E., 1958. Controlled low temperature test of sprouted seeds as a measure of cold hardiness of winter wheat varieties. Can J Plant Sci 38: 1–17.

Andrews, C.J., M.K. Pomeroy & I.A. De La Roche, 1974. Changes in cold hardiness of overwintering winter wheat. Can J Plant Sci 54: 9–15.

Chen, C. & L.V. Gusta, 1978. The role of water in cold hardiness of winter cereals. In: Plant Cold Hardiness and Freezing Stress, pp. 165–174. Academic Press.

Czelnai, R., Zs. Harnos & G. Szász, 1994. Éghajlativáltozás és aszály. MTA Aszálybizottság jelentése (in press).

Gusta, L.V. & D.B. Fowler, 1976. Effects of temperature on dehardening and rehardening of winter cereals. Can J Plant Sci 56: 673–678.

Houghton, R.A. & G.M. Woodwell, 1989. Global climate change. Sci Amer 260: 18–26.

Kimball, B.A., 1983. Carbon dioxide and agricultural yield: an assemblage and analysis of 770 prior observations. WCL Report 14. Water Conservation Laboratory, Agricultural Research Service USDA, Phoenix, Arizona.

Mitchell, R.C., V.J. Mitchell, S.P. Driscoll, J. Franklin & D.W. Lawlor, 1993. Effects of increased CO_2 concentration and temperature on growth and yield of winter wheat at two levels of nitrogen application. Plant, Cell & Environment 16: 5, 521–529.

McKee, I.F. & F.I. Woodward, 1994. The effect of growth at elevated CO_2 concentrations on photosynthesis in wheat. Plant, Cell & Environment 17: 7, 853–859.

Nass, H.G., 1983. Relationship between plant water content and winter hardiness of winter wheat and fall rye grown in Atlantic Canada. Can J Plant Sci 63: 67–71.

Nicolas, M.E., R. Munns, A.B. Samarakoon & R.M. Gifford, 1993. Elevated CO_2 improves the growth of wheat under salinity. Australian J Plant Physiol 20: 3, 349–360.

Olien, C.R., 1967. Freezing stresses and survival. Ann Rev Plant Physiol 18: 387–408.

Rajki, E., 1980. Winter hardiness. Frost resistance. Acta Agronomica Academiae Scientiarum Hungaricae 29: 451–468.

Roberts, D.W.A. & M.N. Grant, 1968. Changes in cold hardiness accompanying development in winter wheat. Can J Plant Sci 48: 369–376.

Siminovitch, D., F. Gfeller & B. Rheaume, 1967. The multiple character of the biochemical mechanism of freezing resistance of plant cells. Cellular injury and resistance in freezing ogranisms. Proc Int Conf Low Temp Sci, Hokkaidó, Univ (Sapporo) 2: 93–117.

Smart, D.R., N.J. Chatterton & B. Bugbee, 1994. The influence of elevated CO_2 on non-structural carbohydrate distribution and fructan accumulation in wheat canopies. Plant, Cell & Environment 17: 4, 435–442.

Tischner, T., 1981. The simulation of environmental conditions in the Martonvásár Phytotron. In: K. Unger & G. Stöcker (Eds). Biophysikalische Ökologie und Ökosystemforschung, pp. 185–190. Akademie-Verlag, Berlin.

Thompson, G.B. & F.I. Woodward, 1994. Some influences of CO_2 enrichment, nitrogen nutrition and competition on grain yield and quality in spring wheat and barley. J Exp Bot 45: 276, 937–942.

Tyler, N.J., L.V. Gusta & D.B. Fowler, 1981. The effect of a water stress on the cold hardiness of winter wheat. Can J Bot 59: 1717–1721.

Veisz, O. & J. Sutka, 1991. Frost resistance studies with wheat in natural and artificial conditions In: Proceedings of International Symposium, Wheat Breeding Prospects and Approaches, pp. 12–17. Albena, Bulgária.

Veisz, O. & T. Tischner, 1989. Frost resistance research with winter wheat in the Martonvásár Phytotron. In: Abstract of International Conf Environmental Res with Plants in Closed Chambers. p. 82. München.

White, R.R., 1992. The road to Rio or the global environmental crisis and the emergence of different agendas for rich and poor countries. The International Journal of Environmental Studies 41: 187–201.

P.M.A. Tigerstedt (Ed.), *Adaptation in Plant Breeding*, 277–288, 1997.

Genetic resources in breeding for adaptation

Geoffrey Hawtin, M. Iwanaga & T. Hodgkin
International Plant Genetic Resources Institute, Rome, Italy

Key words: adaptation, conservation, landraces, plant genetic resources, wild relatives

Summary

A crop's ability to productively exploit its environment depends on many adaptive features which are controlled by multiple genes, interacting among themselves and with the environment in complex ways.

To promote widespread adoption, breeders frequently seek to develop broad adaptation in their varieties, often through the use of genes having a large effect on a single adaptive feature. Such genes may occur within the crop, its wild relatives, or unrelated taxa. Genes for many adaptive features (e.g. temperature tolerance) may be found in extreme environments. Others (e.g. photoperiod insensitivity) may have evolved away from primary centres of origin.

Well characterized and documented *ex situ* germplasm collections aim to serve plant breeders' need for genes. Molecular marker and geographic information system (GIS) techniques are proving useful for locating and characterizing genetic diversity. New techniques (e.g. core collections and electronic information systems) are adding to the value of collections. *Ex situ* and *in situ* conservation of wild crop relatives are receiving increased attention. With all organisms becoming a potential source of genes for breeding, *in situ* ecosystem conservation is assuming added significance.

Farmers, particularly in diverse, marginal environments in developing countries, continue to breed landraces adapted to their specific circumstances. In areas of high ecological diversity, a multitude of adaptive gene complexes have been selected within small geographic areas. Conventional breeding frequently neglects such farmers, and participatory methods based on locally adapted diversity, coupled with appropriate technical and policy support, may prove more effective in meeting their needs. Such dynamic, on-farm conservation and management systems would also enable genetic diversity to continue to evolve as a resource for conventional breeding.

For genetic resources to remain a foundation for future sustainable agricultural development, complementary conservation and breeding strategies are needed.

Introduction

The adaptation of a crop, its ability to survive in a particular environment and to exploit its various features productively, is under extremely complex genetic control. A plant must be able to withstand extremes of cold and heat, excess or lack of water, varying photoperiods and light intensity, and a range of soil physical and chemical conditions. It must be capable of exploiting both its physical and biological environments, and be able to remain productive under pest and disease pressures. It must thrive in a diversity of locations and under conditions that may vary widely, gradually or rapidly, throughout its life cycle.

The underlying genetic control of a crop's interaction with its environment is complex and often poorly understood. Response to a single factor of the environment, such as daylength or to the deficiency of a particular soil nutrient, might be under simple genetic control. Thus, four maturity genes control the response to daylength in sorghum (Quinby, 1974) and tolerance to copper toxicity in *Mimulus guttatus* is under major gene control (MacNair, 1983). The response to such single factors can mask responses to other factors and be overriding in determining a plant's adaptability. But the overall degree of fitness to the environment is much more than this. It also involves an interaction between and among complex gene systems and a multitude of environmental variables. Over the thousands of years of domestication and spread from their centres of origin, crops have become increasingly adapted to a wide range of environments. This broad adaptation has been both through the accumulation of genes responsible for specific adaptation to individual production environments and, more recently through the development of cultivars with wide adaptability to a great range of environments as in the case of modern wheat and rice cultivars (Evans, 1993).

The fact that adaptability is determined both by major and minor genes and that more or less complex 'co-adapted gene complexes' may develop which influence adaptability to specific environments (e.g. Allard, 1992) has important consequences for the conservation of genetic resources, the identification of suitable parental lines for hybridization programmes as well as for breeding and selection methods. While plant breeders in the formal sector have been able to exploit single genes for many characters (particularly those conferring pathogen resistance), and have been able to apply strong selection pressure for specific characters under complex genetic control such as yield related characters, traditional farmers have tended to use and manipulate more complex gene systems, with much lower levels of selection intensity, in the improvement of their crops. Both approaches have much to gain from a closer interaction between them.

Distribution of genetic diversity

Adaptation may result from a specific genetic response which confers a particular character such as tolerance of cold, frost, drought or a particular heavy metal. However, adaptability may also reflect the ability of a particular genotype to tolerate a wide range of different conditions, that is to produce well under a wide range of temperature or moisture conditions. This plasticity of crop response is the key to the large scale use of some varieties of major crops such as the products of CIMMYT's wheat cross 8156 or IRRI's IR 36 rice variety which have both been sown on more than 11 million ha in a single year (Evans, 1993). In breeding for adaptation, the starting point is the nature of the adaptive characteristic sought and the identification of suitable genes which will confer the desired character. Where specific adaptive characters are sought, the question arises as to how the desired genetic diversity is distributed among taxa, in space and in time. It is generally understood that genetic diversity is not uniformly distributed, and from an understanding of underlying patterns of distribution, the probability of locating genetically controlled variation in the desired characters, in existing collections or through targeted collecting, can be greatly increased.

Diversity among taxa. Harlan & de Wet (1971) provided a useful practical approach to classifying crop species and their wild relatives using the concepts of primary, secondary and tertiary genepools. The primary genepool comprises those species which cross freely with the crop, the secondary, those species which cross with difficulty, giving few fertile seeds, and the tertiary genepool includes those species which can only be crossed using artificial techniques such as embryo rescue. The increasing ability to make inter-specific crosses and to move genes among very different biological taxa using genetic engineering allows plants breeders to extend their search for adaptive characters to entirely unrelated species which may not even be part of the plant kingdom. In seeking specific adaptive characters, a functional approach is to consider four broad groups, each with its own adaptive features: non-related taxa, wild relatives, landraces and modern cultivars.

Non-related taxa can have genes of very considerable significance for adaptation. Single genes, or at best simple gene systems, can be transferred using modern genetic engineering techniques to the target crop. The transfer of genes for cold tolerance from the arctic char, or of genes for insect resistance from *Bacterium thuringienses* are examples of such transformations. These are likely to involve genes of large effect with relatively simple control systems. They may well prove useful where specific characteristics are required because of environmental pollution or to deal with soils containing toxic elements. The human metallothionein gene has been found to enhance cadmium tolerance in *Brassica* and *Nicotiana* (Misra & Gedamu, 1989). The extent to which these types of adaptive traits will be generally useful in plant breeding has been questioned but will ultimately be reflected in the number and success of varieties that are produced.

Wild relatives of crops which have survived under strong natural selection pressures can be particularly useful as a source of genes for specific adaptive traits. As Harlan & de Wet (1971) note the more closely related the species, the greater is the chance of being able to make crosses, and of the alien gene systems being effective within the crop genome. Wild relatives have proved particularly important as sources of novel pest and disease resistance genes and many examples of their use can be found in the literature (Stalker, 1980; Hawkes, 1983; Hodgkin & Debouck, 1992). They have also been used as sources of other adaptive characters. Salinity tolerance for tomato was found in *Lycopersicon cheesmanii* (Rick et al., 1987) and frost tolerance for potato was identified in *Solanum acaule* distributed around 3,500–4,000 m (Hawkes & Hjerting, 1989). Desirable adaptive traits have been reported in a wide range of other crop wild relatives, whenever specific surveys have been undertaken. Thus, for example, differences in water use efficiency between the domestic tomato and the more efficient *Lycopersicon penellii* were described by Martin & Thorstenson (1988). Tolerance to salinity and drought are also characters actively sought among the wild relatives of cereal crops.

The problems in the use of wild relatives appear to lie less in the identification of desirable characteristics as in their effective use and in the difficulty of introgressing the desired character into improved crop germplasm without introducing undesirable characteristics associated with the wild species. Molecular techniques may provide substantial opportunities for improving efficiencies in this regard. Using RFLP maps of the tomato Martin et al. (1989) were able to show that three loci could be used as predictors of water use efficiency in field grown tomato suggesting that the characteristic could be manipulated in inter-specific crossing programmes with fairly high efficiency.

Traditional agriculture is characterized by great spatial diversity of varieties and these are frequently well adapted to marginal environments and exhibit such features as the ability to be productive and stable under conditions of low soil fertility, strong competition from weeds, extremes of temperature and moisture and under pest and disease pressures. The underlying genetic control of such adaptive characteristics may be simple and the traits may be relatively easy to transfer through conventional breeding. Traditional varieties or landraces have played an important part in the introduction of improved adaptive characteristics. Most breeders would certainly prefer to use landraces than wild relatives and major investigations have often been undertaken to identify specific traits. Dorofeev (1975) tested 19,000 spring and autumn wheat accessions for drought resistance and identified 150 with the desired character. Interestingly, the desired accessions came from a wide range of countries including Uzbekistan, Syria, Afghanistan, India, Chile and Canada. In most cases the assumption that the stress resistance sought will be found in accessions exposed to the specific environmental stress seems to have been confirmed (e.g. Blum & Sullivan, 1986).

Some adaptive characteristics are extremely complex in the sense that a number of different mechanisms can provide the desired character. For example drought tolerance can be improved by improved water use efficiency, deep roots, thickened cuticles or by a variety of escape mechanisms such as earlier flowering (Boyer, 1992). Burton et al. (1957) demonstrated the value of deep rooted Bermuda grass rather than shallower rooted types and Hall & Grantz (1981) selected early flowering cowpeas (*Vigna unguiculata*) that escaped late season drought. Cold tolerance or frost resistance can be equally complex and the particular characteristics required by different crops may vary. In northern Britain, Brassica crops often need to cope with frequent thawing and freezing while, in a more continental environment, an ability to withstand prolonged freezing is required. It has generally been concluded that in cereals, which have been extensively studied, freezing stress resistance involves multiple genes with both additive and non-additive effects (Palta, 1992).

As in the case of wild relatives, transfer of adaptive characters from landraces to improved cultivars can present major problems. Once a broadly adapted genotype or one which meets a highly specialized market (e.g. malting barley) has been obtained, introducing new adaptive or other characters can present the breeder with severe difficulties. This is particularly so for characteristics under complex genetic control. One approach to this problem involves an improved definition of the nature and inheritance of the character which it is desired to manipulate. A good example of this has been the work described by Lawn et al. (1994) and by Summerfield et al. (this conference) to understand the underlying nature of variation in time to flower in soybean and other crops. Using the methods developed it is now possible to screen germplasm in different environments and obtain measures which can be used to predict time to flower in other environments. The breeder can thus select the accessions of most interest to his own production conditions. The development of molecular genetic methods, particularly those based on the use of increasingly informative RFLP maps, will certainly simplify the work of transferring desired characters to modern cultivars but, at present, it remains the case that landraces have probably been underused by breeders.

In some cases it has been considered necessary to embark on major programmes to incorporate new useful variation as in the case of sorghum in the U.S.A. Over 75% of accessions of sorghum are daylength sensitive and tall, and a major conversion programme has been undertaken to introduce new variation into daylength insensitive, dwarf, temperate genetic stocks to provide germplasm adapted to U.S.A. production (Stephens et al., 1967; Maunder, 1992). This is an interesting example of increasing the amount of variation available in adapted stocks.

Modern cultivars are usually the first choice for breeders seeking particular adaptive characters. They have normally already been selected for their high yielding ability under favourable agro-environmental conditions – good soils, adequate nutrients and water, and frequently with the chemical control of pests and diseases. The adaptive features commonly sought by breeders are frequently under relatively simple genetic control and have a large effect (e.g. the semi-dwarf genes in wheat from Norin-10) and can be manipulated with relative ease. Modern cultivars also tend to have a wide environmental adaptation through photoperiod insensitivity as in the case of sorghum, soybean and a number of other crops. They are able to capitalize on economies of scale and be competitive with other cultivars on the market over large geographic areas (as demonstrated in the case of the wheat and rice genotypes noted above).

The 'elite' genetic background of modern cultivars also makes them the first choice of many breeders as a source of genetic variability in their breeding programmes. This has sometimes led to an overdependance on modern cultivars which are not well adapted to the particular environment for which the breeding programme is designed. The use of adapted local varieties as the primary source of variation into which desired characters present in modern cultivars are introgressed (a kind of back conversion programme) may be an effective strategy for producing cultivars adapted to difficult production environments.

Modern cultivars often remain the first choice of breeders searching for specific adaptive traits. Once a desired trait has been introduced into a modern cultivar from either a crop relative or a traditional landrace the improved cultivar will become the primary source of the trait for other breeders. There remains substantial variation in the range of modern cultivars being grown in many crops (Allard, 1992) although in a number of crops the amount of variation may well be a limiting factor in their adaptation to specific stress environments.

Geographic distribution of diversity. Genetic diversity is not distributed uniformly throughout the range of environments in which a taxon is grown. Current evidence suggests that geographic distribution accounts for most of the observed variation in wild plant species (Hamrick & Godt, 1990). In crops, geographic distribution patterns reflect both the specific selection pressures prevailing in a particular environment as well as crop history.

The most widely studied distribution patterns are for disease resistance genes, and there are many examples in the literature. Resistance is most commonly found in regions where disease pressures are strongest and that coincide with centres of crop diversity, e.g. for Barley Yellow Dwarf Virus in the higher altitude regions of Ethiopia (Qualset, 1975), crown rust of oats in the coastal Mediterranean region (Dinoor, 1975), halo blight of beans from Mexico and El Salvador, and blast resistance in rice from the Philippines and Vietnam. However, while there is a clear tendency for distribution to be generally correlated with disease prevalence in areas of crop diversity, the association is not complete and there are many exceptions in the literature. Although there is little evidence (e.g. Allard, 1990), it is expected that the nature of the pathogen itself will also influence distribution patterns of resistance genes: resistance genes to air-borne foliar pathogens such as rusts are likely to be far more widely distributed than, say, resistance genes to soil borne pests such as nematodes.

In the case of genes for resistance to abiotic stresses, there is some support in the literature for the hypothesis that the occurrence of such traits can be correlated with the presence of the particular environmental factor concerned. In the case of salt tolerance, for example, Sayed (1985) screened 5,000 accessions of spring wheat from over 20 countries and found the highest proportion of tolerant lines from salt-affected areas of Egypt and the U.S.A. Rana (1986), from a study of bread wheat in India concluded '. . . the highest level of salt tolerance was found in indigenous cultivars adapted locally to specific areas of salt-affected soils.' However, in a separate study, Peeters et al. (1990) found that salinity tolerance in *Hordeum* was not confined to material collected from saline environments and that such accessions were often susceptible to high levels of salinity in their tests. For cold resistance, Elings (1992) in a study of Syrian durum wheat reported that tolerance to late frosts was highest in samples from areas where minimum winter temperatures were lowest.

A crop's origin and spread can also influence the distribution of adaptive traits. For example, *Solanum tuberosum* growing in Chiloe island is considered to be a derivative of *S. andigena*, which is adapted to the Andean highlands. The Chiloe potatoes have become adapted to the long-day conditions in the island and are an origin of present day modern potato cultivars in Europe and North America. Similarly, day neuturality has been reported in *macrosperma* lentil cultivars from Argentina that enable it to be grown under the short day conditions of Egypt, Ethiopia and the Indian subcontinent. Some bean cultivars in East Africa combine genes and cytoplasm from both Meso-American and Andean bean races, and such new recombinations may give rise to unique traits which are not seen in the primary centres of genetic diversity.

In the search for desirable traits, not only is the absolute geographic distribution important, but also the allelic frequency. Marshall & Brown (1975) recognized four classes of alleles:

a) common, widely distributed
b) common, locally distributed
c) rare, widely distributed
d) rare, locally distributed

They argued that collecting the first class represents no problem, and such alleles are likely to be included even within small samples collected from a few populations. If alleles are rare, whether or not they are widely distributed, the probability of their being included depends on the size of samples and number of populations that can be sampled. If locally distributed, a knowledge of the distribution would help, but sample sizes still have to be large. The most important class for targeted collecting are those that are common, but exist in localized areas.

Common locally distributed alleles are most likely to include those of adaptive significance which confer an advantage for the population which possesses them or are necessary for survival in a particular environment (see also Allard, 1992; for a discussion of the importance of conserving this class of allele). While some genes of this type may be easy to recognize as those conferring resistance to particular diseases or environmental stresses, the precise value of others may be less easy to determine. The evidence suggests that in may cases specific adaptation to particular environments is the result of the action of multiple genes and that individual effects of minor genes, and interactions among them, together determine environmental fitness (Allard, 1992). Thus Nevo et al. (1979, 1981, 1986) found that specific allozymes or combinations of allozymes were associated with combinations of rainfall, temperature, evaporation, soil type and exposure. The significance of coadapted gene complexes was shown by Allard and his collaborators (Allard, 1988; Zhang et al., 1990a,b; Garcia et al., 1991) who showed that changes occurred in the frequency of specific multilocus gametic types over the generations for which the composite cross population CCII of barley was grown. These changes reflected changes in the environment and transfer of the population to a different production environment resulted in further change in frequency of specific multilocus associations. Thus, coadapted gene complexes identified from isozyme studies clearly exist and have an adaptive significance. However, it is clear that in some cases it is possible to identify coadapted gene complexes in groups of allele whose distribution is associated with particular environments confirming the importance of developing conservation strategies which sample all environments in which the target species or crop is found.

Temporal distribution of diversity. There have been very few studies on the distribution of genetic diversity over time. The changes in gene frequency described in the various composite crosses in barley (Allard, 1988; Saghai-Maroof et al., 1990) provide almost the only example. However, as environments change, new natural selection pressures will inevitably result in shifts in the gene structure of populations. For example, the development of new virulences of a pathogen will, in turn, put new selection pressures on a crop in favour of increasing the frequency of alleles and allele complexes for resistance that may have arisen from mutation, from introgression or from new gene recombinations. Thus sampling at intervals of time, even from the same locations, might give rise to novel genotypes of use to breeders.

Evidence from studies on forest populations indicate that changes in gene frequencies are occurring as a result of increased amounts of aerial pollutants (Scholz et al., 1989). Overall, it appears that total genetic diversity of populations may be reduced in favour of specific genotypes most adapted to pollution. There is concern that changes of this type may be associated with climate change and that global warming will require development of crop varieties with quite different adaptive characteristics as cropping areas change to reflect alteration in the climate. However, many of our crops are grown over areas which have much larger ranges of temperature and moisture availability than that predicted under most climate change scenarios and the required characters probably already exist within modern cultivars within the exception of adaptation to longer days as cropping regions move north (Parry & Carter, 1990).

Perhaps the most obvious shift that occurs over time is the loss of diversity due to genetic erosion resulting from the destruction of habitats, overgrazing, changing farming systems, population movement, etc. The major cause of such erosion is the replacement of landraces by modern cultivars. The level at which genetic erosion occurs (single alleles; genotypes or allele complexes; landraces; whole populations or ecotypes; or even species) requires further definition and practical techniques are needed to estimate losses at the various levels and to further develop effective measures to counteract them. A general characteristic of this type of erosion would be the loss of genotypes specifically adapted for specific environments and their replacement with cultivars of wide general adaptability.

Techniques for locating diversity. Information on the distribution of useful diversity can be obtained from a wide variety of sources, for example from data on existing germplasm collections, herbaria studies, and from local indigenous knowledge. The probable existence of diversity for specific traits can also be postulated from specialized surveys, such as surveys on disease distribution and prevalence.

From an understanding of distribution patterns of genetic diversity, breeders can simplify their search for useful traits. Passport data providing information on collecting sites can provide a useful entry point to large germplasm collections. Knowledge of the spatial pattern of distribution of a crop or species, coupled with information on the geographic location of environments which have particular features of interest, can be used to identify those regions in which the desired adaptive trait is most likely to occur. Targeted collecting in such regions maximizes the probability of finding useful genetic diversity. Geographic Information Systems (GIS) provide a particularly valuable tool for combining diverse spatially-oriented data sets on plant distribution with environmental variables (Guarino, 1995). They are proving to be of increasing value to germplasm specialists and breeders as ever more data sets are assembled, and as the GIS techniques themselves become ever more sophisticated.

The need for information on the amount of genetic diversity and its distribution has sometimes been questioned. In relation to information on the amount of diversity in specific areas or populations Hawkes (1991) has asked '. . . although the diversity of isozymes may be intensive in certain areas, does that mean that useful genes for resistance to pests and diseases may also be found with greater intensity in these same areas? No-one has ever looked into this problem and indeed few have even asked the question.' The amount of diversity tends to differ more significantly between populations for self pollinated species than for cross pollinated species (Schoen & Brown, 1991) and Hawkes' comments are therefore primarily applicable to the former. There is certainly abundant evidence that useful adaptive traits have been found in centres of diversity and in highly variable populations. However for both conservation work and breeding work, the central concern is to ensure that the adaptive variation that occurs in specific environments is maintained and can be made available. On current evidence, both isozyme and molecular data seem to be generally correlated with variation at the agro-morphological level in crop plants and therefore provide good guidance on the distribution of useful variation as well as on the existence of coadapted gene complexes. Molecular genetic techniques (e.g. RFLPs, RAPDs, AFLPs, and microsatellite data) provide much more information about the distribution of diversity and can often be directly associated with variation in useful characters as in the case of water use efficiency in tomato as noted above.

Approaches to breeding and conservation

While there is a multiplicity of approaches to crop improvement around the world, they can be seen as falling into one or two broadly defined systems:

a) Formal systems, in which modern science is brought to bear on crop improvement by institutions such as government plant breeding stations, university departments and private breeding companies, with the aim of producing cultivars for wide, often commercial, distribution to farmers.
b) Informal systems, in which farmers and local communities, mainly in developing countries, breed and select cultivars primarily to meet their own needs and circumstances.

Both systems co-exist in many regions and each depends, to a greater or lesser extent, on the other.

In formal systems, breeders are generally concerned with adaptation in one of three ways:

a) to develop cultivars that are better adapted to the agro-environments in which the target crop is currently produced.
b) to develop cultivars that are adapted to new agro-environments (geographic areas or farming systems), or
c) to broaden adaptability in order to develop cultivars that are competitive over large areas and in the hope that they will exhibit greater stability across seasons.

The first of these is a more or less continuing part of most breeding programmes and includes breeding objectives such as the incorporation of improved frost resistance in northern areas or better drought tolerance in arid lands. The second objective has resulted in dramatic increases of area for a number of crops such as sorghum in the U.S.A. (Maunder, 1992), calabrese in N. Europe or canola in Canada and, more recently, China. The third has been a major objective of international breeding programmes in the 1960s and 1970s and resulted in such varieties as the IR36 rice with its 75% coverage of the S.E. Asian rice acreage. To achieve such objectives, use may be made of genes having a large effect on specific, identified adaptive features such as phenological characteristics, photoperiod response (e.g. daylength in sorghum), tolerance of extremes of temperature, soil moisture or soil chemical factors, or resistance to pests and diseases. However, a significant part is played by intensive directional selection for characters under largely additive genetic control such as flowering time, maturation period or by selection for broad adaptability as evidenced by low genotype × environment interaction as in the case of IR36 (Evans, 1993).

By contrast, farmer and community level crop improvement efforts are more concerned with adapting cultivars to local conditions and systems. In regions of high ecological diversity, a large number of micro-environments may have to be catered for, and farmer circumstances change over time. In continuing to adapt local cultivars to meet their needs, farmer selection pressure may have the effect of producing a series of small adaptive shifts, primarily within the context of complex gene systems. Farmer varieties or landraces may generally be characterized by their high levels of heterogeneity compared to modern commercial cultivars, comparative stability across seasons, location specificity and generalized, rather than highly specific, tolerances and resistances.

To serve these two broad crop improvement systems, different approaches are needed to the conservation and supply of the diversity for current and future genetic advance. As a generalization, *ex situ* methods were developed primarily to meet the needs of the formal crop improvement system. The importance of *in situ* conservation of crop wild relatives was recognized but, until recently, there were very few programmes which specifically targetted crop wild relatives. Within the informal system a more holistic approach has been advocated and activities aimed at supporting the conservation of traditional crop cultivars *in situ* have been initiated. As argued later, there is a great need to promote an integration of these approaches.

Ex situ conservation

Ex situ conservation comprises a wide range of techniques, from the storage of orthodox seeds in cold stores, the storage of tissues *in vitro* (either under slow growth conditions or cryopreserved in liquid nitrogen), to conservation *in vivo* in field genebanks. While for many species *ex situ* conservation techniques are well established, in each of these areas further research is needed, for example:

- to understand better the relationship between seed moisture and longevity, especially at very low seed moisture levels
- to develop low-cost seed drying methods
- to define more robust protocols for the slow growth of tissues *in vitro* and for cryopreserving tissues of a wider range of species and genotypes
- to develop better management techniques for field genebanks.

The screening of very large collections for specific traits can be very expensive and time-consuming. Attention has been focused in recent years on the development of core collections as a mechanism to facilitate their use (Hodgkin et al., 1995). Core collections are collections that aim to represent the full diversity spectrum of the parent collection within a manageable number of accessions thus improving access to the whole collection. In setting up a core collection, hierarchical approaches may be used, frequently with geographic origin as one of the primary levels of discrimination. Specific adaptive traits (such as maturity groups) can also be used to help stratify collections.

However conserved, to be of use to breeders germplasm collections must be well documented. Accurate passport data, including site descriptions, are useful as a basis for correlating origins with environmental parameters; characterization data (information on characters that are simply inherited and expressed in a wide range of environments such as major morphological traits) can assist in discriminating among accessions and can provide information on major adaptive features (e.g. phenological characteristics); evaluation data (usually involving characters important in crop production such as yield and its components, flowering time and plant height), especially if derived from several diverse locations, are perhaps the most useful overall in the search for special adaptive traits.

Local knowledge about the characteristics of particular landraces can provide very valuable leads as to possible sources of genes. Farmers typically have a very thorough knowledge about the attributes of their varieties – e.g. their phenology, reaction to prevalent pests and diseases and suitability for growing on the different soil types found in the vicinity. Local knowledge is only rarely sought during collecting, and greater efforts are needed to record such information (Guarino & Friis-Hansen, 1995). It has often been argued (IPGRI, 1993) that such information is under as great, or greater, threat as the germplasm itself and IPGRI has recently developed data collection sheets for plant germplasm collectors which include notes on ethnobotanic information that should be obtained during collecting missions (Eyzaguirre, 1995).

With recent advances in computer science, not only are germplasm documentation systems becoming more powerful and user-friendly, but data exchange and the sharing of information among different systems is becoming easier. One example of how the new technologies are being applied is the information system under development by the Centres of the Consultative Group on International Agricultural Research (CGIAR). Collectively, these Centres maintain about half a million germplasm accessions of the world's major food and forage crops. A project is underway to link their databases through an electronic information network known as SINGER (System-wide Information Network on Genetic Resources). It is planned to provide international access to the system via Internet.

In situ conservation

With the growing availability and use of techniques for wide crossing, and for transferring genes from non-related genera or even kingdoms, the search for useful genes has been broadened. This has resulted in an increase in activities devoted to the collection and maintenance of crop wild relatives (Ingram & Williams, 1987) and to a greater realization of the value of *in situ* techniques for ensuring the conservation of a large range of potentially useful genes for future use in breeding. Once considered primarily the domain of environmentalists and conservationists, *in situ* conservation is now becoming of increasing interest also to those concerned with crop improvement (Hodgkin, 1993).

While plant breeders can readily access germplasm maintained in *ex situ* collections, this is far more difficult in the case of material conserved *in situ*. Nevertheless, the amount of inter- and intra-specific diversity that can be conserved *ex situ* is a very small proportion of the total potentially useful variation. In addition, for technical reasons many wild species are very difficult to conserve *ex situ*. Thus conservation strategies must include both *ex situ* and *in situ* approaches.

Although there is a growing interest in the *in situ* conservation of genetic resources, most current *in situ* programmes target the preservation of ecosystems (often areas of outstanding natural beauty) or particular species (generally endangered animals or plants) rather than the intraspecific genetic diversity of plant species of potential interest for agriculture.

Options for *in situ* conservation range from nature reserves from which all human intervention is excluded, through national parks in which economic activities with a potential to disturb the natural ecosystems are carefully regulated, to the implementation of special management regimes in areas used primarily for agriculture and forestry. The identification of specific areas in which a deliberate attempt is made to increase and maintain intraspecific diversity of key species is another approach (Krugman, 1984) which is being tried in Turkey and in Mexico. The Man and Biosphere (MAB) programme of the United Nations Economic, Social and Cultural Organization (UNESCO) is perhaps the largest coordinated global attempt to establish *in situ* reserves, one of the objects of which is the conservation of natural areas and the genetic materials they contain. Under the MAB programme, more than 250 Biosphere Reserves have been established around the world.

On-farm management of genetic resources

Brush (1995) has convincingly shown that farmers in widely differing environments continue to grow and use traditional cultivars on at least part of their land, even when modern cultivars are available. Traditional cultivars may meet their specific requirements and may be better adapted to the environments in which they are grown. There is growing evidence (e.g. Riley, 1995) that throughout the world farmers, particularly in traditional agricultural systems, actively seek to improve their crops through exchange with other farmers and through selection within and among their landraces. There have been reports of farmers making crosses, and gene flow between crops and their wild relatives can provide additional sources of variation especially in centres of origin although the extent to which this occurs has been questioned (Wood & Lenne, 1993). Thus landraces, or farmer cultivars, are not static, but represent diverse and dynamic genepools that, under the pressure of both farmer and natural selection, evolve over time.

Small farmers, particularly those in marginal environments, have to cope with great environmental variability. They generally lack the economic and institutional resources to transform their environments to meet the requirements of their crops (Eyzaguirre & Iwanaga, 1995) and thus have to rely on within landrace variation or the inherent plasticity of their crops to ensure at least a minimum level of production over the seasons. For many, meeting immediate family needs and short or medium-term survival are primary objectives. Strategies for achieving them frequently depend on the maintenance and management of diversity. This occurs on a number of levels: within landraces, within crops, within cropping systems and within the total farming system. In applying selection pressures for particular traits, traditional farmers are generally aware of the need to maintain high levels of 'background' diversity.

Conservation *per se* is rarely a conscious objective of farmers; however, the ways in which traditional farmers manage their genetic diversity has the dual effect of conserving within the genepool many of the adaptive features of their crops either individually or as co-adapted gene complexes, and desired quality and agro-morphological characteristics, while at the same time evolving in other respects to meet new seeds.

Breeding for adaptation in the future

Formal systems of crop improvement normaly aim to produce elite, high-yielding cultivars that are broadly adapted across a wide range of agro-environments. Special attention is given to breeding cultivars with specific resistance to pests and diseases and tolerance to abiotic stresses. Quality characteristics are generally determined by the preferences of large consumer groups, often in importing countries, or by the demands of processors. New cultivars must meet legal requirements for distinctness, uniformity and stability in order to be officially registered. Plant breeding is an expensive process, and there is an ever-growing need to show commercial returns on investment in crop improvement from the public as well as private sectors. Under these circumstances it is perhaps not surprising that the majority of modern cultivars are of widely grown, or high value, crops and tend to have a relatively narrow genetic base.

In the formal sector, breeding for adaptation will continue to be concerned with improving adaptability in existing environments, extending the areas in which individual crops are grown or seeking improved stability across a range of environments. New variation will be sought and traditional cultivars and wild relatives will continue to provide the necessary variation so long as their conservation is secured. An improved knowledge of the distribution of diversity and of the effect of specific environmental variables on that distribution will improve our capacity to locate desired characters. An improved understanding of the significance and nature of coadapted gene complexes will enable breeders to use adapted germplasm with much greater efficiency.

By contrast, as Eyzaguirre & Iwanaga (1995) describe, farmers in traditional systems tend to favour adaptation to their specific micro-environments, stability across seasons, and quality and agro-morphological characteristics suited to their own local preferences and needs. They must balance, for example, straw versus grain yields, high yield versus the ability to yield under stress conditions and ease of hand harvesting versus optimum plant architecture. Under these circumstances, farmers' strategies for managing their genetic resources have resulted in an enormous diversity among and within their crops and cultivars.

Formal breeding, for economic reasons, has tended to neglect crops of only local importance and the needs of farmers in diverse marginal environments. However, for farmers in marginal areas such crops may be the basis of their survival. They have to depend on their own efforts and can rarely expect assistance from the formal system. It is highly unlikely that formal breeding wil ever be able to fully meet their diverse needs.

Plant breeding in the formal sector depends on four main sources of diversity: wild relatives (and increasingly non-related taxa), mutation (often artificially induced), landraces and modern cultivars. Of these four, it is argued that the variability in landraces is the most under-exploited. Isozyme and molecular data show that landraces and wild relatives remain the main reservoirs of genetic diversity in crop genepools (Miller & Tanksley 1990). Genetic erosion of these reservoirs continues, in some cases at alarming rates, and it is imperative that efforts to conserve them be strengthened. *Ex situ* genebanks alone are not sufficient, and it is in the best interests of both the formal system as well as the traditional farming communities themselves, that the dynamic 'cauldron' of genetic diversity be maintained on-farm, and be allowed to continue to evolve.

The need to produce more food for the ever-growing world population requires increased productivity from all sectors. Both formal and informal systems of crop improvement have a vital role to play, yet each tends to operate independently of the other. While there is a flow of materials from the informal to the formal system, its genetic diversity remains inadequately exploited by professional breeders. Materials also flow in the other direction; however there are significant opportunities to increase this to the immediate benefit of traditional farmers and, by eventual feedback, to the longer-term benefit of the formal sector itself.

Ways in which the formal system can assist the informal include:

- strengthening links between farmers and genebanks to ensure adequate long-term conservation of landraces, with systems to facilitate the restoration of landraces to communities that have lost them
- the provision by genebanks of adapted materials collected from one location to farmers in other, environmentally similar, locations for local testing and adaptation
- the provision to local communities of segregating populations and other breeding products derived from their own landraces
- the introduction into local landraces of specific genes of relevance to local circimstances
- the provision of a broad range of elite lines, within and among which farmers can select according to their needs
- training farmers in crop improvement techniques that are relevant to their own circumstances.

In several parts of the world participatory breeding approaches are being developed and tested (Ceccarelli, this Conference) in which formal sector breeders are working closely with farming communities. Through a greater understanding of the needs and aspirations of such communities, and through working in partnership, scientific expertise can best be brought to bear on the problems faced by those farmers that have been the most neglected up to now. Such approaches are expected not only to contribute to rural community development, but may also help to ensure that the large gene reservoir managed by the farmers continues to exist and to evolve, and remains a resource for all crop improvement efforts in the future.

References

Allard, R.W., 1988. Genetic changes associated with the evolution of adaptedess in cultivated plants and their wild progenitors. J Heredity 79: 225–238.

Allard, R.W., 1990. The genetics of host-pathogen coevolution: Implications for genetic resource conservation. J Heredity 79: 1–6.

Allard, R.W., 1992. Predictive methods for germplasm identification. In: H.T. Stalker & J.P.Murphy (Eds.), Plant Breeding in the 1990s, pp. 119–146. CAB International, Wallingford, Oxon, UK.

Blum, A. & C.Y. Cullivan, 1986. The comparative drought resistance of landraces of sorghum and millet from dry and humid regions. Ann Bot 57: 835–846.

Boyer, J.S., 1992. Mechanisms for obtaining water use efficiency and drought resistance. In: H.T. Stalker & J.P. Murphy (Eds.), Plant Breeding in the 1990s, pp. 1810–200. CAB International, Wallingford, Oxon, UK.

Brush, S., 1995. *In situ* conservation of landraces in centers of crop diversity. Crop Sci 35: 346–354.

Burton, G.W., G.M. Prine & J.E. Jackson, 1957. Studies of drought tolerance and water use of several southern grasses. Agronomy J 49: 498–503.

Dinoor, A., 1975. Evaluation of sources of disease resistance. In: O.H. Frankel & J.G. Hawkes (Eds.), Crop Genetic Resources for Today and Tomorrow, pp. 201–210. Cambridge University Press, Cambridge, UK.

Dorofeev, V.F., 1975. Evaluation of material for frost and drought resistance in wheat breeding. In: O.H. Frankel & J.G. Hawkes (Eds.), Crop Genetic Resources for Today and Tomorrow, pp. 211–222. Cambridge University Press, Cambridge, UK.

Elings, A., 1992. Evaluation methods for large germplasm collections; distribution, variation and evaluation of Syrian durum wheat landraces. Proefschrift Wageningen, Netherlands.

Evans, L.T., 1993. Crop evolution, adaptation and yield. Cambridge University Press, Cambridge, UK.

Eyzaguirre, P.B., 1995. IPGRI's revised collecting form: Ethnobotanical information in plant genetic resources collecting and documentation. Plant Gen Res Newsl 103: 45–46.

Eyzaguirre, P.B. & M.Iwanaga, 1995. Farmers' contribution to maintaining genetic diversity in crops, and its role within the total genetic resources system. In: P.B. Eyzaguirre & M. Iwanaga (Eds.), Participatory Plant Breeding. Proceedings of a Workshop on Participatory Plant Breeding, 26–29 July 1995, The Netherlands, pp. 9–18. IPGRI, Rome, Italy.

Garcia, P., M. Morris, L.E. Saenz-de-Miera, R.W. Allard, M. Perez de la Vega & G. Ladizinsky, 1991. Genetic diversity and adaptedness in tetraploid *Avena barbata* and its diploid ancestors, *A. hirtula* and *A. wiestii*. Proc Natl Acad Sci USA 88: 1207–1211.

Guarino, L., 1995. Geographic information systems and remote sensing for plant germplasm collectors. In: L. Guarino, V. Ramanatha Rao & R. Reid (Eds.), Collecting Plant Genetic Diversity Technical Guidelines, pp. 315–328. CAB International, Wallingford, Oxon, UK.

Guarino, L. & E. Friis-Hansen, 1995. Collecting plant genetic resources and documenting associated indigenous knowledge in the field: a participatory approach. In: L. Guarino, V. Ramanatha Rao & R. Reid (Eds.), Collecting Plant Genetic Diversity Technical Guidelines, pp. 345–366. CAB International, Wallingford, Oxon, UK.

Hall, A.E. & D.A. Grantz, 1981. Drought resistance of cowpea improved by selecting for early appearance of mature pods. Crop Sci 21: 461–464.

Hamrick, J.L. & M.J.W. Godt, 1990. Allozyme diversity in plant species. In: A.H.D. Brown, M.T. Clegg, A.L. Kahler & B.S. Weir (Eds.), Plant Population Genetics, Breeding and Genetic Resources, pp. 43–63. Sinauer Associates Inc., Sunderland, Massachusetts, USA.

Harlan, J.R. & J.M.J. de Wet, 1971. Toward a rational classification of cultivated plants. Taxon 20: 509–517.

Hawkes, J.G., 1983. The diversity of crop plants. Harvard University Press, Cambridge, Massachusetts, USA.

Hawkes, J.G. & J.P. Hjerting, 1989. The potatoes of Bolivia: their breeding value and evolutionary relationships. Clarendon Press, Oxford, UK.

Hawkes, J.G., 1991. Germplasm evaluation with special reference to the role of taxonomy in genebanks. In: J.M.M. Engels, J.G. Hawkes & M. Worede (Eds.), Plant Genetic Resources in Ethiopia, pp. 247–257. Cambridge University Press, UK.

Hodgkin, T. & D.G. Debouck, 1992. Some possible applications of molecular genetics in the conservation of wild species for crop improvement. In: R.P. Adams & J.E. Adams (Eds.), Conservation of Plant Genes, DNA Banking and *In Vitro* Biotechnology, pp. 153–182. Academic Press, Inc., San Diego, California.

Hodgkin, T., 1993. Wild relatives. Naturopa 73: 18.

Hodgkin, T., A.H.D. Brown, Th.J.L. van Hintum & E.A.V. Morales, 1995. Future directions. In: T. Hodgkin, A.H.D. Brown, Th.J.L. van Hintum & E.A.V. Morales (Eds.), Core Collections of Plant Genetic Resources, pp. 253–259. John Wiley and Sons, UK.

Ingram, G.B. & J.T. Williams, 1987. *In situ* conservation of wild relatives of crops. In: J.H.W. Holden & J.T. Williams (Eds.), Crop Genetic Resources: Conservation and Evaluation, pp. 163–179. George Allen & Unwin, London, UK.

IPGRI, 1993. Diversity for Development. The Strategy of the International Plant Genetic Resources Institute. IPGRI, Rome.

Krugman, S.L., 1984. Policies, strategies, and means for genetic conservation in forestry. In: C.W. Yeatman, D. Kafton & G. Wilkes (Eds.), Plant Genetic Resources a Conservation Imperative, pp. 71–78. Westview Press, Boulder, Colorado, USA.

Lawn, R.J., R.J. Summerfield, R.H. Ellis, A. Qi, E.H. Roberts, P.M. Chay, J.B. Brouwer, J.L. Rose & S.J. Yeates, 1994. Towards the reliable prediction of time to flowering in six annual crops. vi. applications in crop improvement. Expl Agric 1–40.

MacNair, M.R., 1983. The genetic control of copper tolerance in the yellow monkey flower, *Mimulus guttatus*. Heredity 50: 283–293.

Marshall, D.R. & A.H.D. Brown, 1975. Optimum sampling strategies in genetic conservation. In: O.H. Frankel & J.G. Hawkes (Eds.), Crop Genetic Resources for Today and Tomorrow, pp. 53–80. Cambridge University Press, UK.

Martin, B. & Y.R. Thorstenson, 1988. Stable carbon isotope composition (13C), water use efficiency, and biomass productivity of *Lycopersicon esculentum, Lycopersicon pennelli*, and the F1 hybrid. Plant Physiol 88: 213–217.

Martin, B., J. Nienhuis, G. King & A. Schaefer, 1989. Restriction fragment length polymorphisms associated with water use efficiency in tomato. Science (Washington, D.C.) 243: 1725–1728.

Maunder, A.B., 1992. Identification of useful germplasm for practical plant breeding programs. In: H.T. Stalker & J.P. Murphy (Eds.), Plant Breeding in the 1990s, pp. 147–169. CAB International, Wallingford, Oxon, UK.

Miller, J.C. & S.D. Tanksley, 1990. RFLP analysis of phylogenetic relationships and genetic variation in the genus *Lycopersicon*. Theor Appl Genet 80: 437–448.

Misra, S. & J. Gedamu, 1989. Heavy metal tolerant transgenic *Brassica napus* L. and *Nicotianan tabacum* L. plants. Theor Appl Genet 78: 161–168.

Nevo, E., D. Zohary, A.H.D. Brown & M. Haber, 1979. Genetic diversity and environmental associations of wild barley, *Hordeum spontaneum*, in Israel. Evolution 33: 825–833.

Nevo, E., A.H.D. Brown, D. Zohary, N. Storch & A. Beiles, 1981. Microgeographic edaphic differentiation in allozyme polymorphisms of wild barley (*Hordeum spontaneum, Poacheae*). Pl Syst Evol 138: 287–292.

Nevo, E., A. Beiles, D. Kaplan, E.M. Golenberg, L. Olsvig-Whittaker & Z. Naveh, 1986. Natural selection of allozyme polymorphisms: a microsite test revealing ecological genetic differentiation in wild barley. Evolution 40: 13–20.

Palta, J.P., 1992. Mechanisms for obtaining freezing stress resistance in herbaceous plants. In: H.T. Stalker & J.P. Murphy (Eds.), Plant Breeding in the 1990s, pp. 219–250. CAB International, Wallingford, Oxon, UK.

Parry, M.L. & T.R. Carter, 1990. An assessment of the effects of climatic change on agriculture. In: M.T. Jackson, B.V. Ford-Lloyd & M.L. Parry (Eds.), Climatic Change and Plant Genetic Resources, pp. 61–84. Belhaven Press, a division of Pinter Publishers, London and New York.

Peeters, J.P., H.G. Wilkes & N.W. Galwey, 1990. The use of ecogeographical data in the exploitation of variation from gene banks. Theor Appl Genet 80: 110–112.

Qualset, C.O., 1975. Sampling germplasm in a center of diversity: an example of disease resistance in Ethiopian barley. In: O.H. Frankel & J.G. Hawkes (Eds.), Crop Genetic Resources for Today and Tomorrow, pp. 81–96. Cambridge University Press, UK.

Quinby, J.R., 1974. Sorghum improvement and the genetics of growth. Texas A and M University Press, College Station, Texas, USA.

Rana, R.S., 1986. Evaluation and utilization of plant genetic resources for salt affected soils. Indian J Plant Genet Res 1: 76–88.

Rick, C.M., J.W. DeVerna, R.T. Chetelat & M.A. Stevens, 1987. Potential contributions of wide crosses to improvement of processing tomatoes. Acta Hort 200: 45–55.

Riley, K., 1995. Decentralized breeding and selection – a tool to link diversity and development. Presented at Working Seminar on Using Diversity: Enhancing and maintaining genetic resources on farm. New Delhi. June 19–21, 1995. Proceedings in press, IDRC, New Delhi.

Sayed, H., 1985. Diversity of salt tolerance in a germplasm collection of wheat (*Triticum* spp.). Theor Appl Genet 69: 651–657.

Saghai-Maroof, M.A., R.W. Allard & Q. Zhang, 1990. Genetic diversity and ecogeographical differentiation among ribosomal DNA alleles in wild and cultivated barley. Proc Natl Acad Sci USA 87: 8486–8490.

Schoen, D.J. & A.H.D. Brown, 1991. Intraspecific variation in population gene diversity and effective population size correlates with the mating system in plants. Proc Natl Acad Sci USA 88: 4494–4497.

Scholz, F., H.R. Gregorius & D. Rudin, 1989. Genetic effects of air pollutants in forest tree populations. Springer, Berlin, Germany.

Stalker, H.T., 1980. Utilization of wild species for crop improvement. Adv Agron 33: 111–147.

Stephens, J.C., F.R. Miller & D.T. Rosenow, 1967. Conversion of alien sorghums to early combine genotypes. Crop Sci 16: 553–556.

Storey, R., R.D. Graham & K.W. Shepherd, 1985. Modification of the salinity response of wheat by the genome of *Elytrigia elongatum*. Plant and Soil 83: 327–330.

Wood, D. & J. Lenne, 1993. Dynamic management of domesticated biodiversity by farming communities. In: UNEP/Norway Expert Conference on Biodiversity, Trondheim, Norway, pp. 1–26.

Zhang, Q., M.A. Saghai-Maroof & R.W. Allard, 1990a. Effects on adaptedness of variations in ribosomal DNA copy number in populations of wild barley (*Hordeum vulgare* spp. *spontaneum*). Proc Natl Acad Sci USA 87: 8741–8745.

Zhang, Q., M.A. Saghai-Maroof & R.W. Allard, 1990b. Worldwide patterns of multilocus structure in barley determined by discrete long-linear multivariate analysis. Theor Appl Genet 80: 121–128.

P.M.A. Tigerstedt (Ed.), *Adaptation in Plant Breeding*, 289–295, 1997.

Utilization of exotic germplasm in Nordic barley breeding and its consequences for adaptation

Merja Veteläinen[1,*], Eero Nissilä[2], P.M.A. Tigerstedt[3] & Roland von Bothmer[1]
[1] *The Swedish University of Agricultural Sciences, Department of Plant Breeding Research, S-268 31 Svalöv, Sweden; * present address: Nordic Gene Bank, P.O. Box 41, S-230 53 Alnarp, Sweden;* [2] *Boreal Plant Breeding, Myllytie 8, FIN-31 600 Jokioinen, Finland;* [3] *University of Helsinki, Department of Plant Biology, Plant and Tree Breeding, Viikki D, FIN-00014 University of Helsinki, Finland*

Key words: adaptation, agronomic performance, barley, breeding populations, *Hordeum vulgare*, exotic germplasm

Summary

Utilization of exotic germplasm is one way to broaden genetic variation in breeding populations. This approach has recently been adopted in Sweden and Finland, where experimental barley populations has been established for research and pre-breeding purposes. The aim of the project is threefold: (1) to increase overall genetic diversity of Nordic barley breeding material; (2) to develop breeding material which possesses a high level of resistance for various barley diseases; and (3) to study effects of exotic germplasm on adaptation and agronomic performance. Both the Finnish and the Swedish barley populations include the same exotic material i.e. unadapted landraces from different parts of Asia and wild barley (*Hordeum vulgare* ssp. *spontaneum*) accessions. Locally adapted high-yielding barley lines were included in the populations. The establishment of these populations involved six crossing generations in order to promote recombination and enhance the break-up of linkage blocks. The paper discusses the third aim of the project. Studies on agronomic performance and adaptation showed that (1) agronomically valuable genotypes can be constructed through recombination using exotic germplasm for Nordic conditions, (2) that incorporation of exotic material is most successful when made in a local genetic base and (3) that exotic germplasm has an effect on adaptation.

Introduction

Genetic variation serves as the basis for development of improved cultivars. Adequate genetic variation must be available in breeding stocks in order to make further improvements in crops. The most important sources of genetic variation are breeders' own breeding populations. Elite, adapted germplasm from parallel programmes within the same ecogeographical region can be intercrossed to facilitate incorporation of genotypic diversity valuable in the breeding programme. Only when sufficient variation is not available from these sources, do breeders turn to gene banks for exotic material (Baenziger & Peterson, 1991).

Exotic germplasm has been defined by Hallauer & Miranda (1981) as follows: 'All germplasm that does not have immediate usefulness without selection for adaptation to given area can be considered as exotic germplasm'. The definition itself reveals one of the main reasons for wariness to utilize exotic germplasm in breeding. It may affect adaptation, when introduced into a local genetic base. In some case, utilization of exotic germplasm is, however, important. This is the case when there is a need for extensive broadening of the local germplasm base. The aim is then to develop locally adapted populations with increased genetic variability, which will provide new desirable parents for future breeding programmes (Simmonds, 1993).

The effects of exotic germplasm to adaptation may be twofold. First, unimproved wild germplasm often possesses traits for effective dispersal and survival. In wild barley such characters are, for example, seed shattering and seed dormancy, which are negative traits for cultivation and they also delay progress in the breed-

Table 1. Crossing scheme for development of the populations

Generation	Design	No. of lines
0	–	40
I	Pairwise	20
II	Half diallel	190
III	Pairwise	95
IV	Pairwise	48
V	Pairwise	24
VI	Half diallel	276

Table 2. Phenotypic classes and score values for index in the glasshouse experiment

Trait	Phenotypic class	Score for index
Ear emergence,	30–55	4
days	56–80	3
	81–105	2
	106–120	1
Number of ears	1–5	1
per plant	6–10	3
	11–15	2
	15<	1
Straw length	60–80	4
cm	81–100	3
	101–120	2
	120–135	1
TKW, g	10–28	1
	29–38	2
	39–48	3
	49–65	4

ing procedure. Landraces are adapted to local, small-scale, low-input environments where the plant ideotype may differ considerably from that developed for modern agricultural systems. Thus exotic germplasm may have negative effects on agronomic performance, in other words, adaptation to modern farming systems in high-input agriculture. Secondly, exotic material may have effects on adaptation to extreme environmental conditions. In marginal areas of agriculture, as in the Nordic countries, adaptation to a short growing season is crucial for cultivation (Tigerstedt, 1994).

Nordic barley has its production area in marginal environments for the crop, with great variations even within the region. Specific adaptation for short growing season is needed almost for the whole area of the Nordic countries. In the northern parts early frosts in the autumn occur. In the southern part the growing season is 200–210 days, while in the northern parts it is usually less than 170 days (Mukula & Rantanen, 1987). Exotics brought from the more southern latitudes are often too late.

A population breeding project of barley in which exotic material has been used as a source of novel genetic variation has recently been initiated in two Nordic countries, Finland and Sweden. The project has three main aims: (1) to increase overall genetic diversity in Nordic barley breeding material, (2) to develop breeding material which posses a high level of resistance to various barley diseases and (3) to study effects of exotic germplasm on adaptation and agronomic performance (Veteläinen, 1994). The present paper will primarily deal with the third aim of the project.

Materials and methods

Development of the populations

Two barley populations ('dynamic genepools') were established, one for the northern (Finland) and one for the southern regions (Sweden). Local high-yielding breeding material was used as a base in each population, while identical exotic material from Pakistan, China, Syria and Jordan was incorporated in both populations. For full information on the exotic base material see Veteläinen (1995).

The establishment of the populations involved six crossing generations (Table 1) to promote recombination and enhance the break-up of linkage blocks. First, the forty parental lines selected for phenotypic diversity and resistance to various diseases, were intercrossed pairwise. The resulting twenty lines were crossed in a half-diallel design. Subsequently, four additional recombination cycles were done to produce new, unique gene combinations. No selection was carried out during the establishment of the populations.

Studies during the establishment phase

Two studies on the effects of exotic germplasm on agronomic performance and adaptation will be discussed in this paper:

Table 3. Crosses, their origin and parental lines (L = unadapted landrace, S = *Hordeum vulgare* ssp. *spontaneum* line) in the field experiment

Cross no.	Origin of the cross	Parent 1 (adapted)	Parent 2	Parent 3 (adapted)	Parent 4
Fi 1	Finland	Jo 1621	H 4241 (L)	Hja 80138	H 7614 (L)
Fi 2	Finland	Kalle	H 4248 (L)	Meltan	H 7405 (L)
Fi 3	Finland	Kinnan	H 4035 (L)	Pohto	H 3064 (S)
Fi 4	Finland	Hja 83054	Ob 264	Mette	H 3042 (S)
Fi 5	Finland	Arttu	H 4175 (L)	Jo 1545	H 4246 (L)
Sw 1	Sweden	Alfa	H 4241 (L)	Aisling	H 7614 (L)
Sw 2	Sweden	Korinna	H 4248 (L)	Sv 892052	H 7405 (L)
Sw 3	Sweden	Golf	H 4035 (L)	Jet	H 3064 (S)
SW 4	Sweden	Nancy	Sv 89412	Sv 898219	H 3042 (S)
Sw 5	Sweden	Ida	H 4175 (L)	Milka	H 4246 (L)

1. The base material and the Swedish population were tested after the third crossing generation for diversity and agronomic performance under controlled glasshouse conditions (experiment 1).
2. During the 1994 growing season, samples of crosses from the second crossing generation were tested against their adapted base lines in order to study their adaptation and agronomic performance. Both populations were tested at two locations in Finland (experiment 2).

Experiment 1

The base material comprised 40 lines of which 25 were spring barley varieties or lines adapted to Swedish conditions, and 15 exotic lines. The latter contained 10 cultivated landraces and 5 accessions of wild barley, *Hordeum vulgare* ssp. *spontaneum* (hereafter called *H. spontaneum*). The base material was intercrossed for three generations according the crossing scheme presented in Table 1. The resulting 95 hybrid lines used in the experiment contained from 25 to 50% exotic germplasm.

The experiment was conducted in a randomized block design in a glasshouse. Ten seeds from each of the base and hybrid line were sown in separate pots. The experiment was divided into ten blocks and each block was divided into two groups. Group A included all base lines and group B included the 8-way hybrids from the crossing generation 3. The experimental arrangement was designed to minimize interplant competition for light. A 18-hour photoperiod was used in the glasshouse with a day/night temperature 18/14° C to closely follow the Nordic growing season.

Four different agronomic traits were measured (Table 2). Agronomic performance was measured using an index (scale 4–15) composed of four traits:

$$\text{INDEX} = i_{\text{ear.em}} + i_{\text{no/ears}} + i_{\text{straw}} + i_{\text{TKW}}$$

Each trait was divided into four classes (Table 2). The class including the top lines scored 4, while class with the lowest values scored 1 for each index component. Thus, early heading plants with short straw and high TKW were considered most favorable. However, a moderate number of synchronous emerging ears per plant were considered best in this study. Therefore, the lowest and the highest class were treated similarly in the case of number of ears per plant, when calculating the index.

Experiment 2

Agronomic performance was studied more in detail in a field experiment during the summer 1994 in Finland. Five crosses from Finland and Sweden (Table 3) were tested against their adapted base lines. The Finnish base lines were extremely well-adapted to northern conditions with a short growing season, while the lines from Sweden presented a widely adapted southern Scandinavian base. Each of the crosses descended from four base lines, of which at least one was exotic i.e. unadapted landrace or wild barley (*H. sponteneum*). The crosses were selected so that a pair of Finnish and Swedish crosses had the same exotic base lines.

The experiment was conducted at two locations of which the first was sown on peat and the second on clay soil at Jokioinen in southern Finland. The seeds were space planted with two replications in each location.

Table 4. Phenotypic classes and score values for index computation from the field experiment

Trait	Phenotypic class	Score for index
Ear emergence,	1) ≤ 60	4
days	2) 61–70	3
	3) 71–80	2
	4) ≥ 81	1
Straw length,	1) 15–36	1
cm	2) 37–54	2
	3) 55–85	3
	4) 86–100	1
Number of ears/plant	1) ≤ 3	1
6-row	2) 4–8	3
	3) 9–13	2
	4) ≥ 14	1
Number of ears/plant	1) ≤ 2	2
2-row	2) 3–6	4
	3) 7–10	3
	4) ≥ 10	1
Heading synchrony	1) All the ears/plant emerge and ripen almost simultaneously	4
	2) Heading continues longer than in class 1, but almost all the ears ripen	3
	3.1) Heading continues for a long time and all the ears do not ripen	1
	3.2) Heading continues indefinitely	1
	4) Uniculm type	2
Yield,	1) 0–7	1
g/plant	2) 8–14	2
	3) 15–20	3
	4) 21–27	4
Seed shattering	1) Non-brittle spike	2
	2) Brittle spike	1
Persistence of awns	1) Non-persistent	3
in threshing	2) Partly persistent	2
	3) Persistent	1

Each plot consisted of six rows with 20 cm between rows and 10–15 cm between plants in a row. Two rows of a cross and one of each adapted base line of a cross were sown in a plot. Border rows were planted on the sides of the plot to eliminate side effects. Plots were randomized within a replication. Seven traits were measured (Table 4). In each location and replication, 50 random F_2 plants of each cross and 15 plants of each adapted base line were measured.

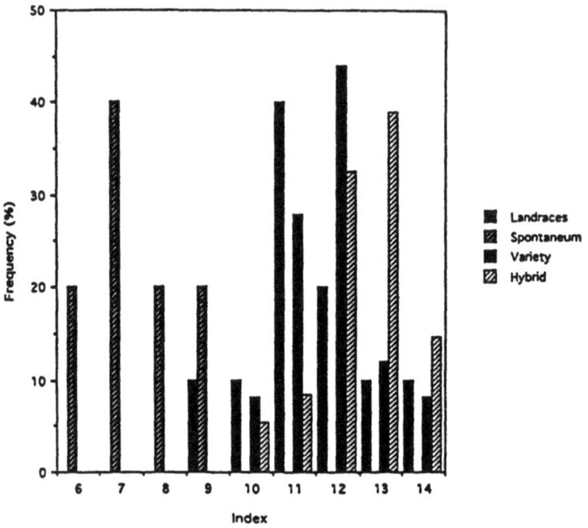

Figure 1. Frequency distribution of the 4-trait index measured in base material and hybrids in experiment 1.

A new index including seven traits was set up, separately for 2- and 6-row plants:

$$INDEX = i_{ear.em} + i_{straw} + i_{no/ears} + i_{synchrony} + i_{yield} + i_{shattering} + i_{persistence}$$

Earliness (ear emergence) was used as a measure of adaptation. The REG procedure of the Statistical Analyses System (SAS Inst., 1990) was used for regression analysis. χ^2-tests (Sokal & Rohlf, 1981) were used to test frequency distributions for indices.

Results and discussion

Agronomic performance was measured in both experiments using index scores. Frequency distribution for the 4-trait index in three types of base lines and hybrids in experiment 1 is shown in Figure 1. The distribution shows that Swedish adapted parents exceed exotic germplasm in overall performance. Apparently landraces may possess desirable trait combinations like earliness, short straw and high TKW. Twenty per cent of the landraces studied were in the classes (13, 14) with the highest indices. *Hordeum spontaneum* lines had a very poor combination of agronomic traits. Around 50% of the hybrids had indices 13 or 14 but only 20% of the adapted parents fell into the highest classes. It can be concluded that evaluation of agro-

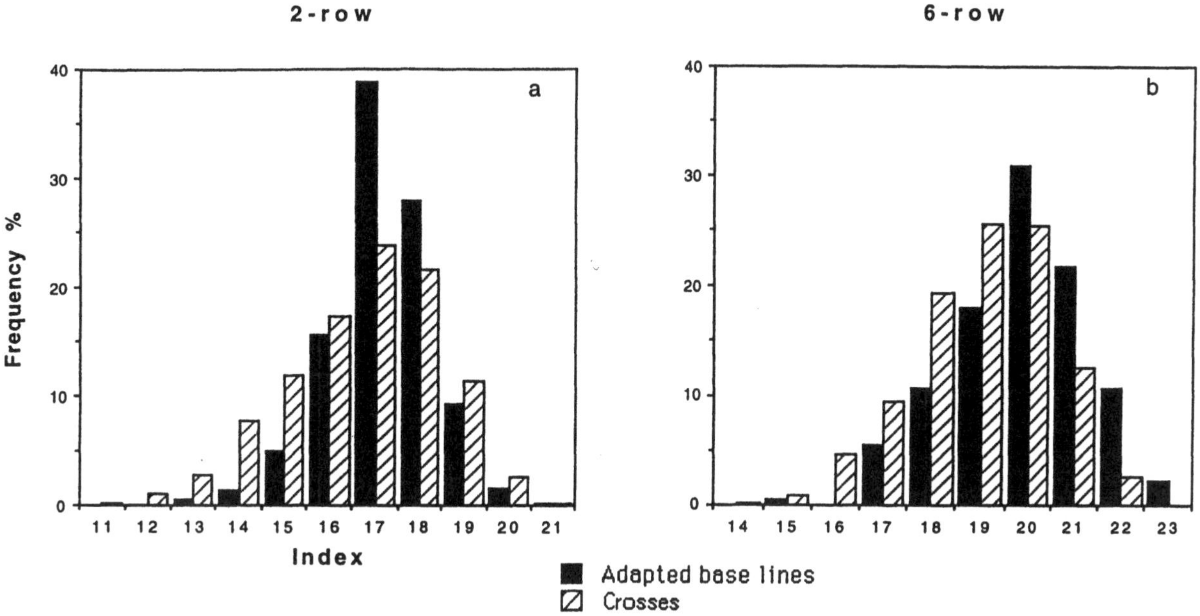

Figure 2. Frequency distributions for index in adapted base lines and crosses (a) 2-row and (b) 6-row progenies in experiment 2.

nomic performance of exotic parents did not give a clue on their progeny performance. The experiment showed also that exotic germplasm should not be discarded on the basis of its own poor or unknown performance, but evaluated when incorporated in the appropriate genetic background. Similar evidence has been found on studies on wild sorghum (Bramel-Cox & Cox, 1988) and oat (Frey et al., 1984). For wild barley, Rogers (1982) showed that their yields were quite low and that their correlations with the percentages of high-yielding segregates in backcross generations were low.

In experiment 2 agronomic value of the progenies were measured only against their adapted base lines. The frequency distributions for base lines and crosses are shown in Figure 2. Distributions were significantly different ($p = 0.000$) both in 2- and 6-rowed material. In 2-rowed material crosses included a higher frequency of plants with good agronomic quality than the adapted base lines, while in 6-rowed material the situation was reverse. In the 2-rowed case the distribution of crosses over the index spectrum was clearly inflated, indicating transgressive segregates on both flanks. It can be concluded from both of the experiments that acceptable agronomic performance can be achieved for Nordic conditions by using adapted × exotic crosses. This is in agreement with earlier studies (Vega & Frey, 1980; Rodgers, 1982; Frey et al., 1984), which show that wild barley may be used as a source when improving agronomic performance.

For marginal conditions some agronomic traits can limit adaptation (Ceccarelli, 1994). In the northern part of Scandinavia such a trait is earliness, specially studied in this project. In the first experiment, ear emergence was compared between different types of base material and hybrids (Figure 3). The earliest heading base material were landraces (with the mean value 51.5 days) followed by the adapted parents (56.3 days). Wild barley (91.8 days), the most exotic material was considerably later than the landraces and the adapted parents. Hybrids (48.1 days) were remarkably earlier than all the parental groups. Hence, it can be concluded that some exotics were good sources of earliness.

In experiment 2 earliness was studied by comparing a sample of crosses from the Swedish and Finnish experimental populations with their adapted base lines. Frequency distributions for ear emergence in Finnish (mean value 61.7 days) and Swedish (63.8 days) crosses are shown in Figure 4. Finnish progenies were more frequent in the early class (< 60 days to heading) than the Swedish progenies, which indicate the importance of a local genetic base, when utilizing exotic germplasm. However, the regression analysis (Figure 5) showed that only 39% of the variation in ear emergence in crosses could be explained by the means of the

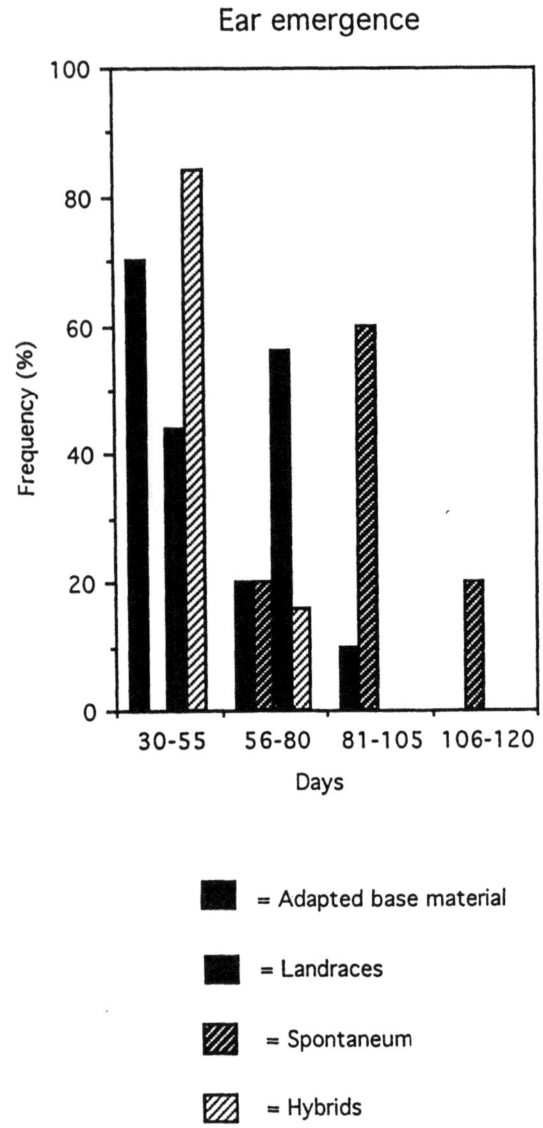

Figure 3. Frequency distribution for ear emergence in base material and hybrids in experiment 1.

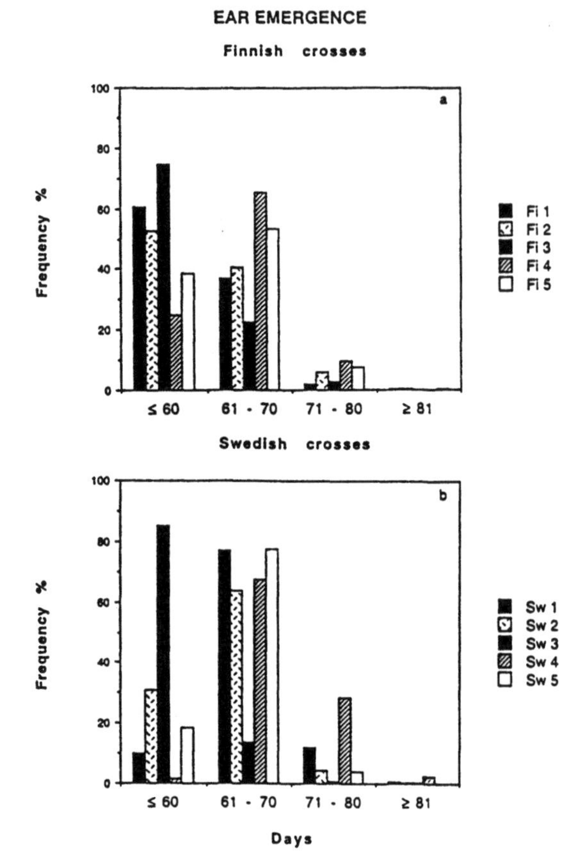

Figure 4. Frequency distribution for ear emergence in (a) Finnish and (b) Swedish progenies in experiment 2.

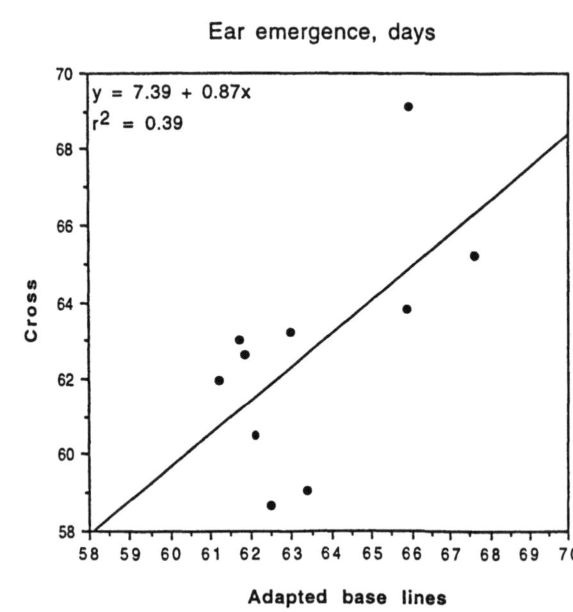

Figure 5. The relationship between parent and cross means in ear emergence in experiment 2.

adapted parents. Thus, the effect of exotic germplasm is also of considerable importance to adaptation.

Our studies indicate that agronomically valuable genotypes can be recovered through recombination of exotic and adapted germplasms. Incorporation of exotic material will be most successful when made in a local genetic base, especially when narrow adaptation is needed. Generally exotic germplasm has a negative effect on adaptation and rather cumbersome incorporation procedure, through repeated crossings, must be considered in relation to advantages for the breeder to work with a wider gene pool.

References

Baenziger, P.S. & C.J. Peterson, 1991. Genetic variation: Its origin and use for breeding self-pollinated species. In: H.T. Stalker & J.P. Murphy (Eds). Plant breeding in the 1990's. CAB International. pp. 69–92.

Bramel-Cox, P.J. & T.S. Cox, 1988. Use of wild germplasm in sorghum improvement. Proc. 43rd Annual Corn and Sorghum Res. Conf., pp. 13–25.

Ceccarelli, S., 1994. Specific adaptation and breeding for marginal conditions. In: O.A. Rongli, E. Solberg & I. Schjelderup (Eds). Developments in Plant Breeding: Breeding Fodder Crops for Marginal Conditions. Kluwer Academic Publishers. pp. 101–127.

Frey, K.J., T.S. Cox, D.M. Rodgers & P. Bramel-Cox, 1984. Increasing cereal yields with genes from wild and weedy species. In: V.L. Chopra, B.C. Joshi, R.P. Sharma & H.C. Bansal (Eds). Genetics: New Frontiers, Vol. IV, pp. 51–68.

Hallauer, A.R. & J.B. Miranda, 1981. Quantitative genetics in maize breeding. Iowa State University Press, Ames.

Mukula, J. & O. Rantanen, 1986. Climatic risks to yield and quality of crops in Finland. I Basic facts about Finnish field crop production. Ann. Agric. Fenn. 26: 1–8.

Rogers, D.M., 1982. Improvement of cultivated barley (*Hordeum vulgare*) with germplasm introgressed from *H. spontaneum*. Ph.D. dissertation, Iowa State.

SAS Institute, Inc., 1990. SAS User's Guide, Statistics, Version 5 edn. SAS Institute, Inc., Carey/NC.

Simmonds, N.W., 1993. Introgression and incorporation. Strategies for the use of crop genetic resources. Biol. Rev. 68: 539–562.

Sokal, R.R. & F.J. Rohlf, 1981. Biometry. 2nd ed. Freeman, San Francisco.

Tigerstedt, P.M.A., 1994. Adaptation, variation and selection in marginal areas. Euphytica 77: 171–174.

Vega, U. & K.J. Frey, 1980. Transgressive segregation in inter and intraspecific crosses of barley. Euphytica 29: 585–594.

Veteläinen, M. 1994. Widening of genetic variation in barley breeding programmes. Experiences from composite crosses to Nordic barley breeding populations. Acta Agric. Scand. Sect. B, Soil Plant Sci. 44: 129–135.

Veteläinen, M. 1995. Breeding Potential of Exotic Barley Germplasm. Proc. 2nd Int. Triticeae Symp. (in press).

P.M.A. Tigerstedt (Ed.), Adaptation in Plant Breeding, 297–302, 1997.

Exotic barley germplasms in breeding for resistance to soil-borne viruses

Frank Ordon, Jens Weyen, Michael Korell & Wolfgang Friedt
Institute of Crop Science and Plant Breeding, Justus-Liebig-University, Ludwigstr. 23, D-35390 Giessen, Germany

Key words: Hordeum vulgare s.l., barley, soil-borne yellow mosaic inducing viruses (BaMMV, BaYMV, BaYMV-2), resistance, germplasm, genetic analysis, molecular techniques (RFLP, RAPD)

Summary

Soil-borne mosaic inducing viruses, i.e., barley mild mosaic virus (BaMMV), barley yellow mosaic virus (BaYMV), and BaYMV-2, cause one of the most important diseases of winter barley in Western Europe. Since resistance of all commercial European barley cultivars is due to a single recessive gene (*ym4*) which is not effective against BaYMV-2, exotic barley germplasms (*Hordeum vulgare* L., *H. spontaneum* Koch) were screened for resistance to the different viruses and analyzed for genetic diversity concerning BaMMV resistance. In these studies it turned out that resistance to BaMMV is entirely inherited recessively and that a high degree of genetic diversity concerning resistance is present within the barley gene pool at least to BaMMV. Therefore, exotic barley germplasms are a very useful source for the incorporation of different resistance genes into barley breeding lines, thereby enabling the breeder to create cultivars adapted to cultivation in the growing area of fields infested by soil-borne viruses. Furthermore, in order to obtain more information on these germplasms they were evaluated for agronomic traits and isozyme, RFLP and RAPD analyses were carried out on these varieties to detect markers linked to the respective resistance genes and to obtain information on the genetic similarity between yellow mosaic resistant barley accessions derived from different parts of the world. Actual results of these studies are briefly reviewed.

Introduction

Barley yellow mosaic disease was first detected in Germany in 1978 (Huth & Lesemann, 1978). The disease is now recognized to be caused by a complex of at least three viruses, i.e., barley mild mosaic virus (BaMMV), barley yellow mosaic virus (BaYMV) and BaYMV-2 (Huth, 1990; Huth & Adams, 1990). In the course of the time it has become one of the most important diseases of winter barley in Western Europe. Due to transmission by the soil-borne fungus *Polymyxa graminis* (Toyama & Kusaba, 1970) chemical measures against the disease are neither efficient nor acceptable for ecological and economical reasons. Therefore, high yield losses in the increasing area of infested fields can only be prevented by growing resistant cultivars. As resistance of all commercial European barley cultivars is presumed to be due to a single recessive gene (*ym4*, Friedt et al., 1990) which is not effective against BaYMV-2, broadening of the genetic basis of resistance to these viruses is a permanent and important task for plant breeding in order to ensure winter barley cultivation in Western Europe. Screening of exotic germplasms for resistance and an analysis of genetic diversity concerning resistance to these viruses is a prerequisite for this purpose.

Resistance to the 'barley yellow mosaic virus complex'

In order to broaden the genetic base of resistance to the barley yellow mosaic virus complex intensive screening programmes were carried out using Turkish accessions of *H. spontaneum* and exotic barley germplasms (*H. vulgare*) mainly derived from East-Asia – where barley yellow mosaic disease is known since 1940 (Ikata & Kawai, 1940). By simultaneous mechanical inoculation in the greenhouse using BaMMV, which is mechanically transmissible efficiently (Friedt, 1983), and tests on a field infested with BaMMV/BaYMV and a second field infested with BaYMV-2 followed by DAS-ELISA different types of resistance to the soil-borne yellow mosaic inducing viruses were observed (Ordon et al., 1992, 1993; Götz & Friedt, 1993). Most of the germplasms tested were resistant to BaMMV as well as to BaYMV and BaYMV-2. Others were resistant to BaMMV only, while some germplasms turned out to be resistant to BaYMV and BaYMV-2 but susceptible to BaMMV. A fourth group reacted like resistant European cultivars, i.e., resistant to BaMMV and BaYMV but susceptible to BaYMV-2. These results lead to the conclusion that there is a very complex situation regarding varietal reaction to the yellow mosaic inducing viruses. However, it turned out that resistance to all of these viruses – even to the resistance breaking strain BaYMV-2 – is present in many exotic germplasms which are useful sources in breeding for resistance, therefore.

Genetics of resistance

Besides the resistance gene *ym4* which is located on the long arm of chromosome 3 (Kaiser & Friedt, 1989, 1992) the genes *Ym1* from 'Mokusekko 3' linked to the gene *K* (hooded lemma) on chromosome 4 and *Ym2* from 'Mihori Hadaka 3' linked to the gene *n* (naked karyopsis) on chromosome 1 have been described in Japan (Takahashi et al., 1973). Furthermore, another gene *ym3* was found in the gamma-ray induced mutant 'Ea 52' (Ukai, 1984) and an additional gene was described in 'Mokusekko 3' linked to Est1-Est2-Est4 on the long arm of chromosome 3 (Konishi et al., 1989). However, concerning BaMMV the resistance of 'Mokusekko 3' and 'Mihori Hadaka 3' seems to be allelic to *ym4*, and *ym3* is not effective against BaMMV (Götz & Friedt, 1993). Therefore, segregation analysis of crosses between resistant exotic germplasms (*H. vulgare* and *H. spontaneum*) and cultivars carrying *ym4* were initiated in order to broaden the genetic basis of resistance. Genetic analysis as early as in F_2 is still limited to BaMMV because this virus can be mechanically transmitted efficiently in contrast to BaYMV and BaYMV-2 (cf. Huth, 1990) and can be easily handled in the greenhouse under controlled conditions, therefore. Using mechanical inoculation of BaMMV followed by DAS-ELISA 18 exotic barley germplasms, i.e., 'Anson Barley', 'Bulgarian 347', 'Chikurin Ibaraki 1', 'Iwate Omugi 1', 'Krasnodar 1920', 'Maksimirski 452', 'Muju covered 2', 'Namji Milyang Native', 'NIR 55 MN-28', 'Ou 1', 'Russia 32', 'Russia 57', 'Rokkaku 1', 'Taihoku A', 'Turkey 235', '#9043', '#9048', '#10247' (Götz & Friedt, 1993) and 5 Turkish accessions of *H. spontaneum*, i.e., 'Candarli', 'Icemeler', 'Kupalan', 'Menemen' and 'Pinarbasi' (Erdogan et al., 1994) were shown to carry genes different from *ym4*. Some of them, e.g. 'Chikurin Ibaraki 1', 'Taihoku A', 'Ou 1' and others, are resistant to BaYMV and BaYMV-2 as well. These varieties are best suited to incorporate resistance to BaYMV-2 by simultaneously broadening the genetic base of BaMMV resistance. This may avoid the selection of new strains of BaMMV as already reported from Japan (Kashiwazaki et al., 1990). Further studies revealed that BaMMV resistance of these varieties is inherited entirely recessively (Table 1) and that a high degree of genetic diversity concerning BaMMV-resistance is present within those germplasms carrying genes different from *ym4* (Ordon & Friedt, 1993). Furthermore, first results on the genetics of BaYMV-2 resistance show that resistance to this virus, e.g. in 'Taihoku A' and 'Ou1', is also inherited recessively and that BaYMV-2 resistance genes different from each other are present within these germplasms. These exotic germplasms are a very useful source for broadening the genetic base of resistance to the soil-borne mosaic inducing viruses. Therefore, attempts to characterize these germplasms on the molecular level and to locate the respective resistance genes were carried out.

Characterization of exotic germplasms and localization of resistance genes

Since isozymes represent relatively simple and cheap markers provided that their chromosomal localization is known and polymorphism is present, first attempts to characterize these exotic germplasms were carried out using isozyme electrophoresis. However, out of 15 isozyme systems tested 12 were monomorphic (IDH, GPD, GPI, MDH, PGM, AAT, ADH, AMP, CAT, ENP, NDH, GDH) and only 3 were polymorphic (ACO, EST, PGD) on our set of exotic resistant germplasms and susceptible German varieties tested. Due to this very limited polymorphism no linkage between different resistance genes and isozymes was found, except the one between *ym4* and the esterase isozyme cluster Est1-Est2-Est4 on the long arm of chromosom 3 (Le Gouis et al., 1995). Therefore, in a next step RAPD and RFLP analysis were undertaken in cooperation with Dr. Andreas Graner (Federal

Table 1. Results of crosses between resistant exotic germplasms and susceptible winter barley cvs 'Alraune', 'Copia', 'Igri', 'Magie', 'Trixi' and 'Gerbel' (Ordon & Friedt, 1993; Erdogan et al., 1994)

Cross	F_1***	F_2 Segregation r : s	χ^2* 1r : 3s	χ^2* 7r : 9s	Infect.-rate (%) **
H. vulgare germplasms					
10247 × Trixi	s	28 : 86	0.461	–	97
9048 × Alraune	s	22 : 93	2.113	–	100
Anson Barley × Trixi	s	22 : 97	2.692	–	100
Bulgarian 347 × Alraune	s	41 : 135	0.928	–	98
Chikurin Ibaraki 1 × Magie	s	26 : 94	0.711	–	100
Iwate Omugi 1 × Magie	s	36 : 76	0.007	–	90
Krasnodar 1920 × Alraune	s	28 : 92	3.279	–	92
Maksimirski 452 × Trixi	s	34 : 79	0.001	–	93
NIR 55 MN-28 × Copia	s	32 : 84	0.414	–	100
Ou 1 × Alraune	s	37 : 80	2.738	–	100
Rokkaku 1 × Magie	s	35 : 83	1.367	–	100
Russia 32 × Trixi	s	24 : 96	1.600	–	100
Russia 57 × Alraune	s	30 : 120	2.000	–	100
Taihoku A × Magie	s	21 : 93	2.632	–	100
Turkey 235 × Alraune	s	32 : 85	0.345	–	100
9043 × Alraune	s	42 : 68	4.777	3.105	95
Namji Milyang Native × Igri	s	92 : 77	37.076	1.733	90
Muju covered 2 × Copia	s	99 : 131	17.824	1.988	93
Zairai Rokkaku × Igri	s	41 : 70	8.435	2.094	100
H. spontaneum germplasms					
Gerbel × Candarli	s	32 : 84	2.126	–	88
Gerbel × Menemen	s	58 : 61	–	0.718	98
Gerbel × Pinarbasi	s	50 : 63	–	0.011	100
Gerbel × Icemeler	s	58 : 58	–	1.231	98

r = resistant, s = susceptible; * = χ^2-values arithmetically corrected for the rate of infection according to J. Jensen (pers. comm.); ** = estimated by plants of the susceptible check cv 'Gerbel' and only concerning F_2-results, *** infection rate for all crosses 100%.

Centre for Breeding Research, Grünbach), as these techniques are much more powerful than isozyme electrophoresis and had proven their usefulness in detecting BaMMV-resistance genes already in the case of *ym4* (Graner & Bauer, 1993; Ordon et al., 1995). In these studies different resistance genes derived from exotic germplasms were localized on the respective chromosomes by RFLP-analysis, meanwhile (Graner et al., 1995). Promising results have also been obtained by RAPD analysis, already. In the end these markers are very useful tools for efficiently combining different resistance genes derived from exotic germplasms in one breeding line ('pyramiding' of resistance genes) and especially RAPD-marker will facilitate efficient and rapid marker based selection procedures in practical plant breeding needed for the effective incorporation of exotic resistance genes into adapted varieties.

In addition to the detection of markers for the respective resistance genes, the exotic germplasms were characterized by oligo and PCR fingerprinting in order to obtain information on the genetic similarity as the very few polymorphisms detected by isozyme electrophoresis between barley germplasms derived from different parts of the world were quite suprising. Using oligo fingerprinting out of the enzyme/probe combinations tested only $(GATA)_4$ and $(ACA)_6$ revealed clear polymorphism. However, although the bands were specific for each genotype, the patterns were very complex and extremely difficult to analyze. Therefore, no estimation of genetic similarity was carried out on this data set (Schiemann et al., 1995). However, by

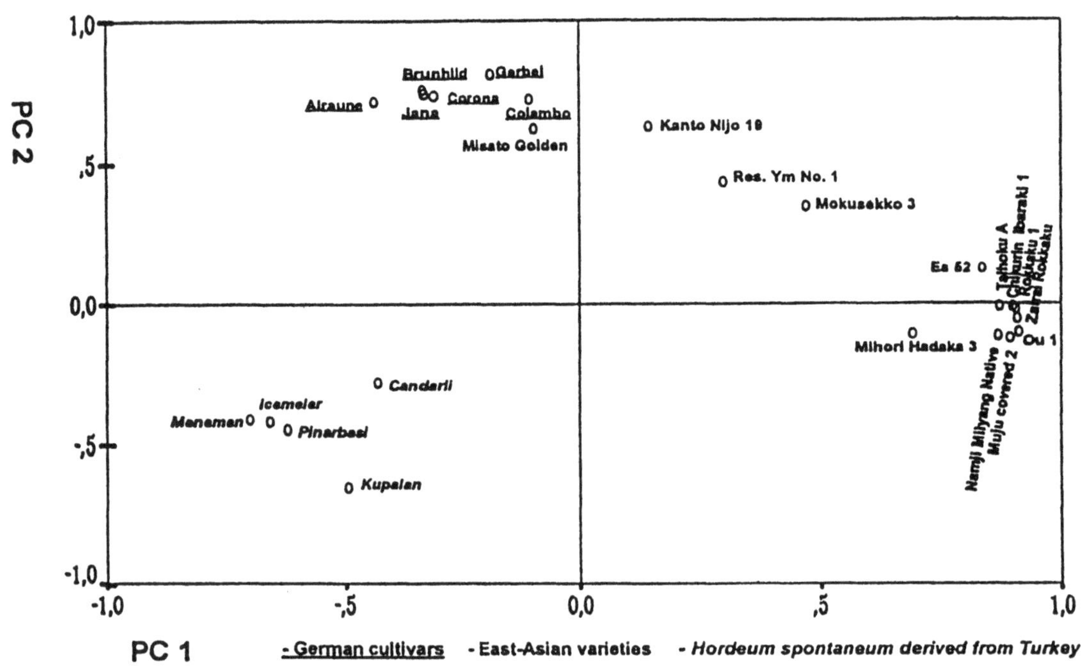

Figure 1. Association among resistant and susceptible German barley cultivars, resistant varieties derived from East-Asia and resistant accessions of *Hordeum spontaneum* Koch derived from Turkey revealed by principal coordinate analysis performed on genetic similarity (GS) estimates calculated from PCR data of 15 RAPD-primer (Schiemann et al., 1995).

using PCR-fingerprinting followed by principal coordinate analysis carried out on genetic similarity estimates according to Nei & Li (1979) preliminary results reveal a clear grouping of exotic germplasms according to their origin (Figure 1). While the first principal coordinate accounting for 39.3% of the total variation clearly separates Turkish varieties of *H. spontaneum* and commercial German cultivars from exotic germplasms derived from East-Asia, the second principal coordinate accounting for 23.6% of the total variation separates commercial German cultivars from *H. spontaneum* varieties. It is interesting to note, that germplasms like 'Res. Ym No. 1', 'Kanto Nijo 19' or 'Misato Golden' which derived their malting quality in a way from European spring barley cultivars and their virus resistance from 'Mokusekko 3' are more closely related to commercial German cultivars than other East Asian varieties. The results elucidate that despite the negative results of isozyme electrophoresis a large genetic diversity is present within germplasms resistant to soil-borne mosaic viruses. Genetic similarity estimates range from 0.67 to 0.89 (Schiemann et al., 1995).

Agronomic traits of exotic germplasms

Apart from genetics and marker based selection procedures resistance to soil-borne viruses is only one important goal in barley breeding. Therefore, resistant exotic germplasms were tested for agronomic traits in two years trials in order to obtain information on their usefulness in future breeding programmes (Ordon & Friedt, 1994). In these studies grain yield of those germplasms being resistant to BaMMV, BaYMV and BaYMV-2 and carrying genes different from *ym4* ranged from 40–54% only, in comparison to the adapted six-rowed cultivar Franka. Furthermore, it turned out that these germplasms are in general susceptible to lodging, winterkilling and scald (*Rhynchosporium secalis*). However, although all of the exotic germplasms are of inferior agronomic performance, these varieties are the only sources which combine resistance to BaYMV-2 with BaMMV-resistance genes different from *ym4*. Today the process of combining these resistance genes with superior agronomic performance may be abridged by the use of haploid techniques in combination with efficient marker based selection procedures (cf. Graner et al., 1995; Ordon et al., 1995a). Since most of the exotic germplasms were crossed to adapted cultivars and selected for yield and adaptiveness in the offspring, some of these different resistance genes are already available in breeding lines yielding about 90% in comparison to commercial cultivars. Consequently, these breeding strains are well suited to be used in practical breeding programmes as cross parents, therefore.

Conclusions

The results presented elucidate that exotic *Hordeum vulgare* germplasms are a very useful source for broadening the genetic base of resistance of barley to soil-borne mosaic inducing viruses, thereby enabling the breeder to create cultivars adapted to cultivation in the growing area of infested fields. Today, due to the results of genetic analyses, genetic fingerprinting and information on agronomic traits, different well characterized germplasms may be used for this purpose and different resistance genes may be incorporated efficiently using marker based selection procedures.

Acknowledgements

We would like to thank Mrs. Ines Müller, Mrs. Christiane Happel and Mr. Martin Seim for their excellent technical assistance. Thanks are due to Dr. Winfried Huth, Federal Biological Research Center, Braunschweig for providing ELISA-antisera and to Dr. Reinhard Hemker, Nickerson Pflanzenzucht GmbH, for conducting the BaYMV-2 tests at Schladen. Furthermore, we thank the Gemeinschaft zur Förderung der Deutschen Pflanzenzüchtung (GFP) and the Deutsche Forschungsgemeinschaft (DFG) for financial support.

References

Erdogan, M., F. Ordon & W. Friedt, 1994. Genetics of resistance of *Hordeum spontaneum* Koch from Turkey to the barley yellow mosaic virus complex. Barley Genet Newsletter 23: 41–43.

Friedt, W., 1983. Mechanical transmission of soil-borne barley yellow mosaic virus. Phytopath Z 106: 16–22.

Friedt, W., F. Ordon, R. Götz & R. Kaiser, 1990. Bodenbürtige Krankenheiten, eine fortdauernde Herausforderung für die Pflanzenzüchtung – beleuchtet am Beispiel der Gelbmosaikvirose der Gerste. Ber Arbeitstag Saatzuchtl Gumpenstein 40: 27–38.

Götz, R. & W. Friedt, 1993. Resistance to the barley yellow mosaic virus complex – Differential genotypic reactions and genetics of BaMMV resistance of barley (*Hordeum vulgare* L.). Plant Breeding 111: 125–131.

Graner, A. & E. Bauer, 1993. RFLP mapping of the *ym4* virus resistance gene in barley. Theor Appl Genet 86: 689–693.

Graner, A., E. Bauer, A. Kellermann, G. Proeseler, G. Wenzel & F. Ordon, 1995. RFLP analysis of resistance to the barley yellow mosaic virus complex. Agronomie 15: 475–479.

Huth, W., 1990. The yellow mosaic inducing viruses of barley in Germany. p. 113–115. In: R. König (Ed). Proc 1st Symp Int Working Group on Plant Virus with Fungal Vectors. Schriftenreihe Dt Phytomed Ges.

Huth, W. & D.E. Lesemann, 1978. Eine für die Bundesrepublik neue Virose an Wintergerste. Nachrichtenbl Deut Pflanzenschutzd 30: 184–185.

Huth, W. & M.J. Adams, 1990. Barley yellow mosaic virus (BaYMV) and BaYMV-M: two different viruses. Intervirology 31: 38–42.

Ikata, A. & I. Kawai, 1940. Studies on wheat yellow mosaic disease. Noji Kairyo Shiryo 154: 1–123.

Kaiser, R. & W. Friedt, 1989. Chromosomal location of resistance to barley yellow mosaic virus in German winter barley identified by trisomic analysis. Theor Appl Genet 77: 241–245.

Kaiser, R. & W. Friedt, 1992. Gene for resistance to barley mild mosaic virus in German winter barley located on chromosome 3L. Plant Breeding 108: 169–172.

Kashiwazaki, S., K. Nomura, K. Watanabe, I. Toshima, Y. Iida, T. Usugi, K. Ogawa, H. Hibino & T. Tsuchizaki, 1990. Barley yellow mosaic virus and barley mild mosaic virus: strains and host resistance. p. 105–108. In: R. König (Ed). Proc 1st Symp Int Working Group on Plant Virus with Fungal Vectors. Schriftenreihe Dt Phytomed Ges.

Konishi, T., N. Kawada, H. Yoshida & K. Soutome, 1989. Linkage relationships between two loci for the barley yellow mosaic resistance of Mokusekko 3 and esterase isozymes in barley (*Hordeum vulgare* L.). Japan J Breed 39: 423–430.

Le Gouis, J., M. Erdogan, W. Friedt & F. Ordon, 1995. Potential and limitations of isozymes for chromosomal localization of resistance genes against barley mild mosaic virus (BaMMV). Euphytica 82: 25–30.

Nei, M. & W.H. Li, 1979. Mathematical model for studying genetic variation in terms of restriction endonucleases. Proc Natl Acad Sci USA 76: 5269–5273.

Ordon, F. & W. Friedt, 1993. Mode of inheritance and genetic diversity of BaMMV resistance of exotic barley germplasms carrying genes different from '*ym4*'. Theor Appl Genet 86: 229–233.

Ordon, F. & W. Friedt, 1994. Agronomic traits of exotic barley germplasms resistant to soil-borne mosaic-inducing viruses. Genetic Resources and Crop Evolution 41: 43–46.

Ordon, F., M. Erdogan & W. Friedt, 1992. Genetics of resistance of barley to soil-borne viruses. Reproductive Biology and Plant Breeding, Book of Poster Abstracts, XIIIth EUCARPIA Congress, 6–11 July. Angers, France, pp 707–708.

Ordon, F., R. Götz & W. Friedt, 1993. Genetic stocks resistant to barley yellow mosaic virus (BaMMV, BaYMV, BaYMV-2). Barley Genet Newsletter 22: 46–50.

Ordon, F., E. Bauer, W. Friedt & A. Graner, 1995a. Marker-based selection for the *ym4* BaMMV-resistance gene in barley using RAPDs. Agronomie 15: 481–485.

Ordon, F., E. Bauer, K.J. Dehmer, A. Graner & W. Friedt, 1995. Identification of a RAPD-marker linked to the BaMMV/BaYMV resistance gene *ym4*. Barley Genet Newsletter 24: 123–126.

Schiemann, A., K.J. Dehmer, W. Friedt & F. Ordon, 1995. Characterization of barley germplasms resistant to soil-borne mosaic inducing viruses by oligo and PCR fingerprinting. Proc Int Symp

75 Years of Pytopathological and Resistance Research at Aschersleben, Züchtungsforschung 2: 152–155.
Takahashi, R., J. Hayashi, T. Inouye, I. Moriya & C. Hirao, 1973. Studies on resistance to yellow mosaic disease in barley. I. Tests for varietal reactions and genetic analysis of resistance to the disease. Ber Ohara Inst 16: 1–17.
Toyama, A. & T. Kusaba, 1970. Transmission of soil-borne barley yellow mosaic virus. 2. *Polymyxa graminis* Led. as vector. Ann Phytopath Soc Japan 36: 223–229.
Ukai, Y., 1984. Genetic analysis of a mutant resistant to barley yellow mosaic virus. Barley Genet Newsletter 14: 31–33.

P.M.A. Tigerstedt (Ed.), Adaptation in Plant Breeding, 303–308, 1997.

Phenological adaptation to cropping environment. From evaluation descriptors of times to flowering to the genetic characterisation of flowering responses to photoperiod and temperature

R.J. Summerfield, R.H. Ellis & P.Q. Craufurd
Plant Environment Laboratory, Department of Agriculture, The University of Reading, Cutbush Lane, Shinfield, Reading RG2 9AD, Berkshire, United Kingdom

Key words: adaptation, flowering, photoperiod, temperature, germplasm, characterisation

Summary

Descriptor lists for the major cereals and grain legumes include information on the duration from sowing to flowering. This is because the first step towards maximizing crop yield by agronomic management or plant breeding is to ensure that the phenology of the crop is well matched to the resources and constraints of the production environment. In this context, durations from sowing to flowering are of critical importance if crops sown on the appropriate date and at the appropriate density are to have the potential to yield well in a given environment. In most annual crops, the timing of phenological events is modulated primarily by responsiveness to photoperiod and temperature with large differences in sensitivity among genotypes. Irrespective of the crop, many advantages accrue from analyses of these photothermal responses not in terms of the evaluation descriptor 'time from sowing to flowering (*f*)' but in terms of the 'rate of progress towards flowering (1/*f*)'. A computer program *RoDMoD* has been developed to convert evaluation descriptors for times of flowering to characterisation descriptors of the flowering response of a genotype to photoperiod and temperature. The program and the associated phenological model were developed from research undertaken in controlled environments and then validated in the field over wide ranges of locations and seasons. These advances should contribute to the development of cultivars phenologically well adapted to their target environments.

Introduction

Organisations concerned with the conservation and utilisation of genetic resources commonly recognise four categories of information – passport descriptors, management descriptors, characterisation descriptors, and evaluation descriptors. Passport descriptors comprise information on systematics and provenence; management descriptors comprise information on where the material is stored, the quantity of material in store, and its condition (e.g. seed viability); and characterisation descriptors comprise information on strongly heritable characters which are independent of the environment, e.g. taxonomic characters. However, it is evaluation descriptors which are often the traits of principal concern in plant breeding programmes and which can be thought of as inherited agronomic qualities. Evaluation descriptors involve those genes which are among the most important to plant breeders, agronomists, and farmers. Unfortunately, these descriptors are also the most problematic because expression depends upon genotype, environment, and (in the context of the entire germplasm collection) the interaction between genotype and environment.

Descriptor lists for many crops, especially the major cereals and grain legumes, include information on the duration from sowing to flowering. This is because the first step towards maximizing crop yield by agronomic management or plant breeding is to ensure that the phenology of the crop is well matched to the resources and constraints of the production environment (Buddenhagen & Richards, 1988; Richards, 1989; Shorter et al., 1991). In this context, durations from sowing to flowering are of critical importance if crops sown on the appropriate date and at the appropriate density are to have the potential to yield well in a given environment (Bunting, 1975). In most annual crops, the timing of phenological events is modulated primarily by responsiveness to photoperiod and temperature with strong differences in responsiveness among genotypes (Roberts & Summerfield, 1987). The implications of different genotypic responses to the photothermal environment for adaptation with respect to latitude, altitude and sowing date are substantial. Nevertheless, it has not been possible until recently to develop quantitative models sufficiently reliable to predict when genotypes will flower in any field environment.

The general photothermal model

The fundamental concept essential to the general photothermal model of development is to consider rates of progress from sowing towards flowering, rather than times taken to flower. This, we believe, not only relates more closely to the underlying biological events which lead to flowering but the approach has several other advantages. Amongst these are that in most species it is found that when rates are considered the responses to both photoperiod and temperature are linear. Furthermore, while there are powerful interactions between photoperiod and temperature when effects on time taken to flower are examined, over wide and specified ranges of conditions these interactions disappear when rates are analysed. This means that the effects of those genes affecting photoperiod-sensitivity can be separated from those which affect temperature-sensitivity, thus allowing rational genetic analysis (Summerfield et al., 1991, 1993).

The model can be applied in contrasting genotypes of both long-day plants (LDP) and short-day plants (SDP) (Roberts & Summerfield, 1987). It is convenient to consider first the case where photoperiod-sensitivity genes are either absent (i.e. in day-neutral plants) or do not interfere, i.e. in circumstances when for SDP the photoperiod is shorter than the critical photoperiod ($P < P_c$), or for LDP when the photoperiod is longer than the critical photoperiod ($P > P_c$). The rate of progress towards flowering is then determined within wide temperature limits (i.e. between the base, T_b, and optimum, T_o, temperatures) according to the Equation:

$$1/f = a + bT \qquad (1)$$

where f is the number of days from sowing to the appearance of the first flower, T is the mean pre-flowering temperature (° C) and a and b are genotype-specific constants.

When a plant contains genes which confer photoperiod sensitivity, they are expressed in SDP only when $P > P_c$ (the critical photoperiod) and in LDP only when $P < P_c$. Between P_c and P_{ce} (the ceiling photoperiod) the delay in rate of progress towards flowering is described by the following photothermal regression plane:

$$1/f = a' + b'T + c'P \qquad (2)$$

in which P is the mean pre-flowering photoperiod (hd^{-1}) and a', b' and c' are genotype-specific constants. The critical photoperiod occurs as a consequence of the thermal and photothermal planes: that is, it becomes evident under photothermal conditions when Equation 2 predicts a greater delay in flowering than Equation 1, and is therefore at the intersect of these two planes. In LDP the values of c' (sensitivity of rate of progress to flowering to photoperiod) are positive, while in SDP they are negative.

The other photoperiod limit to Equation 2 occurs at the ceiling photoperiod. Here, a plane of maximum delay to flowering is exposed; it is a zone within which, in certain SDPs at least (e.g., soyabean), $1/f$ is insensitive to both P and T (Major, 1980; Hadley et al., 1984; Beech et al., 1988) and so when $P > P_{ce}$, then:

$$1/f = d' \qquad (3)$$

Hence, photothermal flowering responses can be quantified by three intersecting planes which, within wide photothermal limits (defined in detail by Watkinson et al. 1994), relate $1/f$ to P and T. These three planes can be characterised for any genotype by six constants (a, b, a', b', c' and d') from which f can be predicted.

Converting evaluation descriptors to characterisation descriptors

Although it was essential to use controlled environments to develop the robust and biologically meaningful model provided by Equations 1, 2 and 3, once it had been developed it was possible to use field data to determine the values of the coefficients which quantify the responses of any genotype and so predict the time it will take to flower in any environment. Figure 1 shows the reponses of two cultivars of soy-

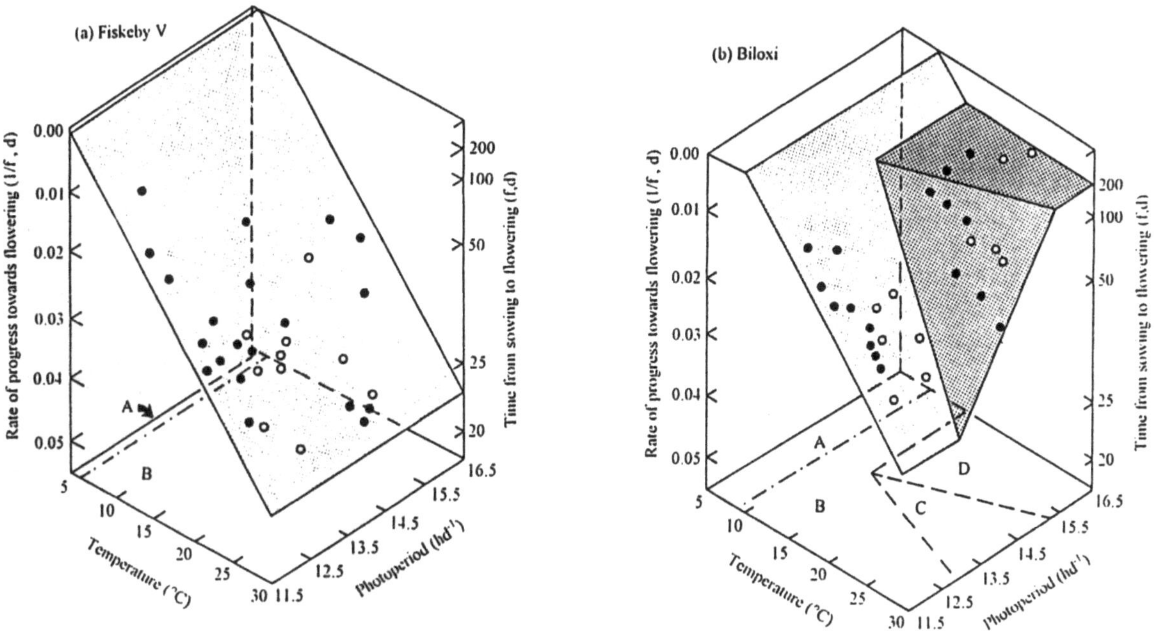

Figure 1. Photothermal flowering responses of two genotypes of the SDP soyabean determined from observations from crops sown on various dates at six sites in Australia in 1986–1988 (•) and at one site in Australia and two in Taiwan in 1989–1990 (○). The vertical scale ($1/f$) on the left-hand ordinate is converted to f as a non-linear scale on the right of each Figure. The base of each graph has been divided into up to four Sectors by vertically projecting: (A) the area where flowering cannot occur ($1/f = 0$), because the termperature is cooler than the base temperature; (B) the area projected by the thermal response plane described by Equation 1; (C) the area projected by the photothermal response plane described by Equation 2; and (D) the area projected by the plane of maximum delay described by Equation 3. The line between Sectors A and B represents the base temperature; that between B and C represents the critical photoperiod; while that between C and D represents the ceiling photoperiod. From Summerfield et al. (1993).

abean (*Glycine max*) which were generated from multi-location trials in Australia and Taiwan. Fiskeby V is a photoperiod-insensitive cultivar and so the response surface is affected by temperature but not daylength. In contrast Biloxi is a photoperiod-sensitive cultivar which Garner & Allard (1920) used in their classical studies on photoperiodism. This cultivar still shows the underlying plane of temperature sensitivity similar to Fiskeby V, but its photoperiod-sensitivity genes cause a *delay* in flowering when daylengths are longer than the critical photoperiod; ultimately there comes a point of maximum delay, the ceiling photoperiod, beyond which further increases in daylength have no additional effect. While each of the symbols shown in Figure 1 represents a combination of evaluation descriptors (time to flowering) and site descriptors (mean pre-flowering values of temperature and photoperiod), the parameters defining these fitted planes represent characterisation descriptors for particular genotypes. This is because the parameter values are not influenced by environment; instead they quantify the response to environment. Indeed, the value of these parameters for genetic characterisation has been shown in studies of heritability in mungbean (*Vigna radiata*) genotypes and their hybrids (Imrie & Lawn, 1990). Then again, the characterisation of accessions by determining the values of the parameters of the model for close to 300 representatives of the world lentil (*Lens culinaris*) germplasm has identified clear trends in these values with crop domestication and dissemination (Erskine et al., 1990, 1994).

In studies of the effects of individual so-called 'maturity genes' on progress to flowering in an isogenic background, Upadhyay et al. (1994) have shown that the triple intersecting-plane photothermal model fitted the results well (R^2 varied between 0.89 to 0.93 amongst the isolines). Furthermore, although temperature increased the rate of progress towards flowering under all photoperiods shorter than the ceiling value there were no significant differences in response to temperature amongst the isolines. Under conditions where photoperiod modulates flowering, i.e. when

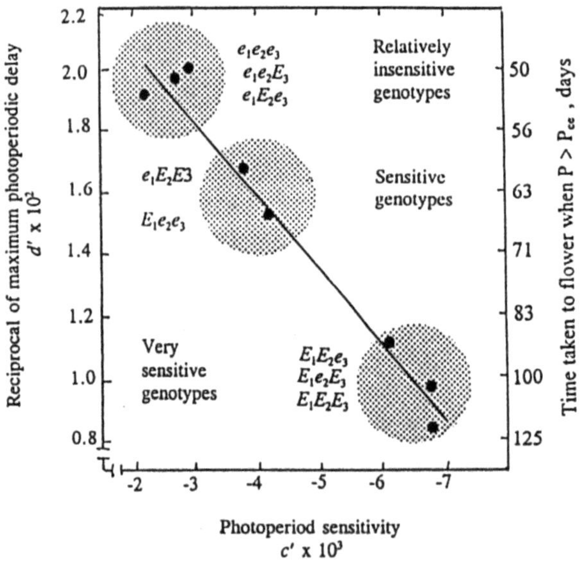

Figure 2. The relation between photoperiod sensitivity (c') and the reciprocal of the maximum delay in flowering caused by photoperiod (d') among eight isolines of soyabean cv. Clark with the gene combinations indicated. The genotypes fall into three main groups of photoperiod sensitivity (stippled) as described in the text. From Upadhyay et al. (1994).

daylength is longer than the critical photoperiod but shorter than the ceiling photoperiod, there was no interaction in any genotype between temperature and photoperiod. Consequently it is concluded that flowering responses to photoperiod and temperature are under separate genetic control, and that the genes E_1, E_2 and E_3 affect photoperiod sensitivity but not temperature sensitivity. Accordingly, providing appropriate analytical precautions are taken to discount temperature effects, it is possible to determine the individual and epistatic effects of these genes and to quantify the phenotypic responses to photoperiod which they express. The maturity genes affect the values of both c' and d' consistently as illustrated in Figure 2; with increasing photoperiod sensitivity there is also increased maximum delay in flowering when the photoperiod exceeds the ceiling value. Further analysis shows that the isolines fall into three groups (stippled in Figure 2) in which photoperiod sensitivity is significantly different among but not within the groups. It is therefore concluded that neither E_2 nor E_3 on its own has any effect since the phenological responses in isolines $e_1E_2e_3$ and $e_1e_2E_3$ were similar to that of $e_1e_2e_3$. But E_2 and E_3 interact positively and their combined effect is similar to E_1 alone. The combined effects of E_2 or E_3 with E_1 are also important since they interact positively with this allele. But the presence of all three dominant alleles, i.e. $E_1E_2E_3$, is not significantly different from either E_1 with E_2 or E_1 with E_3.

RoDMoD – Rate of development model

Careful examination of Figure 1b illustrates a difficulty in fitting the three-plane model. In photoperiod-sensitive genotypes it is not necessarily obvious whether a particular observation is within the thermal plane defined by Equation 1, the photothermal plane defined by Equation 2, or the plane of maximum delay defined by Equation 3. Similarly, the observations available for a particular genotype may not encompass all three planes; instead, they may encompass only one or two planes – because, for example, the range of environments over which observations had been obtained is limited, or the range of photoperiods between P_c and P_{ce} is considerable. In these cases it may not be clear as to how many and which planes the observations span.

In order to solve this 'bottleneck' to the utilisation of the three-plane photothermal flowering model, Watkinson et al. (1994) have developed the computer program *RoDMoD* (Rate of Development Model). *RoDMoD* is a computer software package that provides an objective procedure for fitting the three planes, defined by Equations 1–3, that describe the flowering responses of a genotype to photoperiod and temperature over very wide ranges of photothermal regimes. The *RoDMoD* package includes a menu-driven routine and other associated routines along with extensive documentation and tutorials to facilitate its use. The system requirements are an IBM or IBM-compatible Personal Computer, with an 8088, 8086, 80286, or 80486 main processor, 512 K RAM, DOS version 2.11 or later, and either a hard disk with 1.2 MB free space or two floppy disc drives at least one of which is a high density drive (Watkinson et al., 1994).

The number of observations required for genetic characterisation is an important aspect of the economy and thus the feasibility of such research. For example, it is not possible to fit all three planes if fewer than seven observations are available. General advice on this topic is provided in the *RoDMoD* documentation, but we can be rather more specific for two LDPs of importance in the highlands of West Asia and North Africa.

Research with both chickpea (*Cicer arietinum*) and lentil has shown that the photothermal plane quantified by Equation 2 applies across a very wide range of photoperiods, such that it is difficult to determine the values of the boundary photoperiods P_c and P_{ce} if observations are obtained solely in agriculturally-realistic environments (Summerfield et al., 1985; Ellis et al., 1994b). Hence, in these two crops at least and for most practical purposes, it is necessary only to estimate the values of the parameters a', b', and c' of Equation 2 which quantify the photothermal plane. This requires only four or five observations for each accession in carefully selected photothermal environments.

In lentil, this approach involving a comparatively limited number of observations for each of a large number of accessions selected from the world germplasm has been very effective in characterising a core collection (Erskine et al., 1990, 1994). Moreover, the resultant parameter values from Erskine et al. (1990) – despite being obtained from results in just four, admittedly carefully selected environments – have predicted flowering responses in the field in the highlands of West Asia and North Africa with pleasing precision (Keatinge et al., 1996).

In addition to soyabean, mung bean, chickpea, and lentil (all already cited), genetic characterisation of photothermal flowering responses has so far been shown to be possible and successful in cowpea (*Vigna unguiculata*), common bean (*Phaseolus vulgaris*), bambara groundnut (*Vigna subterranea*), faba bean (*Vicia faba*), field pea (*Pisum sativum*), groundnut (*Arachis hypogaea*), pigeonpea (*Cajanus cajan*), lupin (*Lupinus angustifolius*), clover (*Trifolium subterraneum*), rice (*Oryza sativa*), maize (*Zea mays*), wheat (*Triticum aestivum*), barley (*Hordeum vulgare*), sorghum (*Sorghum bicolor*), sesbania (*Sesbania* spp.) and meadow foam (*Limnanthus alba*) – and the list continues to grow.

Acknowledgements

We thank many colleagues in the International Agricultural Research Centres of the Consultative Group on International Agricultural Research, the CSIRO Division of Tropical Crops and Pastures, Australia, and the University of Reading for their scientific contributions. Our research has been funded by the Overseas Development Administration of the UK Foreign and Commonwealth Office and by the International Plant Genetic Resources Institute.

References

Beech, D.F., A.L. Garside & I.M. Wood, 1988. Response of soyabeans to sowing date during the wet season in the Ord Irrigation Area, Western Australia. Aust J Expl Agric 28: 357–365.

Buddenhagen, I.W. & R.A. Richards, 1988. Breeding cool season food legumes for improved performance in stress environments. p. 81–95. In: R.J. Summerfield (Ed). World Crops: Cool Season Food Legumes. Kluwer Academic Publishers, Dordrecht, The Netherlands.

Bunting, A.H., 1975. Time, phenology and the yields of crops. Weather 30: 312–325.

Ellis, R.H., R.J. Lawn, R.J. Summerfield, A. Qi, E.H. Roberts, P.M. Chay, J.B. Brouwer, J.L. Rose, S.J. Yeates & S. Sandover, 1994. Towards the reliable prediction of time to flowering in six annual crops. V. Chickpea (*Cicer arietinum*). Expl Agric 30: 271–282.

Erskine, W., R.H. Ellis, R.J. Summerfield, E.H. Roberts & A. Hussain, 1990. Characterisation of responses to temperature and photoperiod for time to flowering in a world lentil collection. Theor Appl Genet 80: 193–199.

Erskine, W., A. Hussain, M. Tahir, A. Bahksh, R.H. Ellis, R.J. Summerfield & E.H. Roberts, 1994. Field evaluation of a model of photothermal flowering responses in a world lentil collection. Theor Appl Genet 88: 423–428.

Garner, W.W. & H.A. Allard, 1920. Effect of relative length of day and night and other factors of the environment on growth and reproduction of plants. J Agric Res 18: 553–606.

Hadley, P., E.H. Roberts, R.J. Summerfield & F.R. Minchin, 1984. Effects of temperature and photoperiod on flowering in soyabean (*Glycine max* (L.) Merrill). Ann Bot 53: 669–681.

Imrie, B.C. & R.J. Lawn, 1990. Time to flowering of mungbean (*Vigna radiata*) genotypes and their hybrids in response to photoperiod and temperature. Expl Agric 26: 307–318.

Keatinge, J.D.H., A. Qi, I. Kusmenoglu, R.H. Ellis, R.J. Summerfield, W. Erskine & S.P.S. Beniwal, 1996. Using genotypic variation in flowering responses to temperature and photoperiod to select lentil for the West Asian Highlands. Agric & Forest Meteorol 78: 53-65.

Major, D.J., 1980. Photoperiod response characteristics controlling flowering of nine crop species. Can J Plant Sci 60: 777–784.

Richards, R.A., 1989. Breeding for drought resistance – physiological approaches. p. 65–79. In: F.W.G. Baker (Ed). Drought Resistance in Cereals. CAB International, Wallingford.

Roberts, E.H. & R.J Summerfield, 1987. Measurement and prediction of flowering in annual crops. p. 17–50. In: J.G. Atherton (Ed). Manipulation of Flowering. Butterworths, London.

Shorter, R., R.J. Lawn & G.L. Hammer, 1991. Improving genetic adaptation in crops – a role for breeders, physiologists and modellers. Exp. Agric 27: 155–176.

Summerfield, R.J., E.H. Roberts, W. Erskine & R.H. Ellis, 1985. Effects of temperature and photoperiod on flowering in lentils (*Lens culinaris*). Ann Bot 56: 659–671.

Summerfield, R.J., E.H. Roberts, R.H. Ellis & R.J. Lawn, 1991. Towards the reliable prediction of time to flowering in six annual crops. I. The development of simple models for fluctuating field environments. Expl Agric 27: 11–31.

Summerfield, R.J., R.J. Lawn, A. Qi, R.H. Ellis, E.H. Roberts, P.M. Chay, J.B. Brouwer, J.L. Rose, S. Shanmugasundaram, S.J. Yeates & S. Sandover, 1993. Towards the reliable prediction of time to flowering in six annual crops. II. Soyabean (*Glycine max*). Expl Agric 29: 253–289.

Upadhyay, A.P., R.H. Ellis, R.J. Summerfield, E.H. Roberts & A. Qi, 1994. Characterisation of photothermal flowering responses in maturity isolines of soyabean (*Glycine max* (L.) Merrill) cv. Clark. Ann Bot 74: 87–96.

Watkinson, A.R., R.J. Lawn, R.H. Ellis, A. Qi & R.J. Summerfield, 1994. RoDMoD: A computer program for characterising genotypic variation in flowering responses to photoperiod and temperature. CSIRO, Australia, 49pp. + diskette. ISBN O 643 005715 3.

MIX
Papier aus verantwortungsvollen Quellen
Paper from responsible sources
FSC® C105338

If you have any concerns about our products,
you can contact us on
ProductSafety@springernature.com

In case Publisher is established outside the EU,
the EU authorized representative is:
Springer Nature Customer Service Center GmbH
Europaplatz 3, 69115 Heidelberg, Germany

Printed by Libri Plureos GmbH
in Hamburg, Germany